Toxicogenomics in Predictive Carcinogenicity

Issues in Toxicology

Series Editors:
Professor Diana Anderson, *University of Bradford, UK*
Dr Michael D. Waters, *Michael Waters Consulting, N. Carolina, USA*
Dr Timothy C. Marrs, *Edentox Associates, Kent, UK*

Advisor to the Board:
Dr Alok Dhawan, *Ahmedabad University, India*

Titles in the Series:
1: Hair in Toxicology: An Important Bio-Monitor
2: Male-mediated Developmental Toxicity
3: Cytochrome P450: Role in the Metabolism and Toxicity of Drugs and other Xenobiotics
4: Bile Acids: Toxicology and Bioactivity
5: The Comet Assay in Toxicology
6: Silver in Healthcare
7: *In Silico* Toxicology: Principles and Applications
8: Environmental Cardiology
9: Biomarkers and Human Biomonitoring, Volume 1: Ongoing Programs and Exposures
10: Biomarkers and Human Biomonitoring, Volume 2: Selected Biomarkers of Current Interest
11: Hormone-Disruptive Chemical Contaminants in Food
12: Mammalian Toxicology of Insecticides
13: The Cellular Response to the Genotoxic Insult: The Question of Threshold for Genotoxic Carcinogens
14: Toxicological Effects of Veterinary Medicinal Products in Humans: Volume 1
15: Toxicological Effects of Veterinary Medicinal Products in Humans: Volume 2
16: Aging and Vulnerability to Environmental Chemicals: Age-related Disorders and their Origins in Environmental Exposures
17: Chemical Toxicity Prediction: Category Formation and Read-Across
18: The Carcinogenicity of Metals: Human risk through occupational and environmental exposure
19: Reducing, Refining and Replacing the Use of Animals in Toxicity Testing
20: Advances in Dermatological Sciences
21: Metabolic Profiling: Disease and Xenobiotics
22: Manganese in Health and Disease
23: Toxicology, Survival and Health Hazards of Combustion Products
24: Masked Mycotoxins in Food: Formation, Occurence and Toxicological Relevance
25: Aerobiology: The Toxicology of Airborne Pathogens and Toxins

26: Chemical Warfare Toxicology, Volume 1: Fundamental Aspects
27: Chemical Warfare Toxicology, Volume 2: Management of Poisoning
28: Toxicogenomics in Predictive Carcinogenicity

How to obtain future titles on publication:
A standing order plan is available for this series. A standing order will bring delivery of each new volume immediately on publication.

For further information please contact:
Book Sales Department, Royal Society of Chemistry, Thomas Graham House, Science Park, Milton Road, Cambridge, CB4 0WF, UK
Telephone: +44 (0)1223 420066, Fax: +44 (0)1223 420247
Email: booksales@rsc.org
Visit our website at www.rsc.org/books

Toxicogenomics in Predictive Carcinogenicity

Edited by

Michael D. Waters
Michael Waters Consulting, Hillborough, NC, USA
E-mail: mdwaters@centurylink.net

Russell S. Thomas
US Environmental Protection Agency, Research Triangle Park, NC, USA
E-mail: thomas.russell@epa.gov

ROYAL SOCIETY OF CHEMISTRY

THE QUEEN'S AWARDS FOR ENTERPRISE: INTERNATIONAL TRADE 2013

Issues in Toxicology No. 28

Print ISBN: 978-1-78262-162-1
PDF eISBN: 978-1-78262-405-9
EPUB eISBN: 978-1-78262-819-4
ISSN: 1757-7179

A catalogue record for this book is available from the British Library

Published by The Royal Society of Chemistry,
Thomas Graham House, Science Park, Milton Road,
Cambridge CB4 0WF, UK

Registered Charity Number 207890

For further information see our web site at www.rsc.org

Printed in the United Kingdom by CPI Group (UK) Ltd, Croydon, CR0 4YY, UK

Preface

The concept for this book grew out of an Applied Symposium presented in two parts by many of the chapter authors on September 19, 2014, at the 45th Anniversary Annual Meeting of the Environmental Mutagenesis and Genomics Society in Orlando (FL, USA): "Toxicogenomics in the Assessment of Mutagenicity and Carcinogenicity: Implications for Risk Assessment (Parts 1 and 2)"; chairs: Michael Waters, Russell Thomas and Francina Webster.

In attempting to address the topic of the book, predictive carcinogenicity, we have followed essentially the same format as the symposium. The earlier chapters document the status of toxicogenomics methods as they have been applied in the fields of predictive carcinogenicity testing from the year 1999 to the present. These chapters describe the development and application of toxicogenomics *in vitro* and *in vivo* in ways that recapitulate the 45 year history of genetic toxicology bioassays in the screening and identification of genotoxic and nongenotoxic substances. With the recognition of the full power of genomic and bioinformatics technologies, toxicogenomics applications have rapidly expanded to include both safety and health risk assessment. Thus, subsequent chapters address toxicogenomics dose–response analysis as well as mode of action—all of which are subjects of major regulatory significance for drugs and environmental chemicals. To further represent the breadth and depth of coverage of the applications of toxicogenomics, there are chapters on the effects of xenobiotics on the genome and epigenome of stem cells, and the use of toxicogenomics data by the International Agency for Research on Cancer in the evaluation of carcinogenic hazards to humans. In three additional chapters, we have included extensive details of research to evaluate chemicals to which humans are widely exposed (the conazoles, benzene, and furan), so that the scope and extent of knowledge that has been and can be gained from

Issues in Toxicology No. 28
Toxicogenomics in Predictive Carcinogenicity
Edited by Michael D. Waters and Russell S. Thomas

Published by the Royal Society of Chemistry, www.rsc.org

toxicogenomics research may be fully appreciated. Finally, there are two concluding chapters, one representing the parallelogram approach applied to acetaminophen (paracetamol), an extrapolation concept from the past that continues to be useful in this toxicogenomics-based instance, and a broad and heavily illustrated survey of the use bioinformatics of genomics in the assessment of cancer.

Throughout the book we have tried to emphasize best-of-class investigations that point the way to future scientific developments. With all of this, we hope to bring the reader up to date on progress in the field of predictive carcinogenicity that has resulted from the application of toxicogenomics since its inception just over 15 years ago. We would like to acknowledge the pioneering work of the National Center for Toxicogenomics at the National Institute of Environmental Health Sciences in Research Triangle Park (NC, USA), where the word "toxicogenomics" was coined, many important early studies were performed, and the first public toxicogenomics knowledgebase (Chemical Effects in Biological Systems) was built. What has resulted from these beginnings and other pioneering research efforts worldwide is truly remarkable.

Michael D. Waters,
Michael Waters Consulting, North Carolina, USA

Russell S. Thomas,
National Center for Computational Toxicology, US EPA,
North Carolina, USA

Contents

Chapter 1 Introduction to Predictive Toxicogenomics for Carcinogenicity 1
Michael D. Waters

1.1 Background on -omics Technologies Applied in Toxicology 1
1.1.1 Conventional Toxicity Testing 5
1.1.2 Genomic and Postgenomic Technologies 7
1.2 The Relative Roles of Toxicogenomics, Conventional Toxicity Testing, and High-throughput Screening 12
1.3 Predictive Toxicology 15
1.4 Systems Toxicology 16
1.4.1 Dosimetry 16
1.4.2 Adverse *vs.* Homeostatic Responses 18
1.4.3 Phenotypic Anchoring 18
1.4.4 Genetic Variation 19
1.4.5 Validation 19
1.4.6 Classes of Chemicals and Prototypic Compounds Studied to Date 20
1.4.7 Target Organs Studied 21
1.5 Predictive Carcinogenicity 21
Acknowledgements 26
References 26

Issues in Toxicology No. 28
Toxicogenomics in Predictive Carcinogenicity
Edited by Michael D. Waters and Russell S. Thomas

Published by the Royal Society of Chemistry, www.rsc.org

Chapter 2 Genomic Biomarkers in Cell-based Drug Screening 39
Heng-Hong Li

2.1 Genotoxicity and the Traditional Testing Battery 39
2.2 Mechanisms of Action for Genotoxicity and the Genotoxic Stress Responses 40
2.2.1 Categorization by Mechanisms of Action 40
2.2.2 Signaling and Transcriptional Responses upon Genotoxic Stress 42
2.3 Expression Profiling and Toxicogenomics 43
2.3.1 Genotoxicity Assays Based on Transcriptional Responses 43
2.3.2 Toxicogenomics and Genomic Biomarkers 45
2.3.3 Considerations in Biomarker Identification 47
2.4 The Genotoxicity Biomarker TGx-28.65: Identification and Application 48
2.4.1 Choice of the Cell Line and Toxicants 48
2.4.2 Dose and Treatment Time Parameters 49
2.4.3 Global Transcriptional Analysis 51
2.4.4 Delineation of Gene Subclusters Using a Biclustering Approach 53
2.4.5 Development of a Genomic Biomarker for Genotoxicity 56
2.4.6 Validation and a Case Study on Utility of the TGx-28.65 Biomarker in Human Health Risk Assessment 57
2.5 Summary and Perspectives 58
References 67

Chapter 3 Toxicogenomics *In vitro*: Gene Expression Signatures for Differentiating Genotoxic Mechanisms 76
Julie K. Buick and Carole L. Yauk

3.1 Introduction 76
3.2 Predictive Toxicogenomics 78
3.2.1 Overview of Predictive Toxicogenomics 78
3.2.2 Rationale for the Need for Toxicogenomic Predictors of Genotoxicity 79
3.3 Toxicogenomic Predictors of Genotoxicity 80
3.3.1 Development and Validation of *In vitro* Toxicogenomic Predictors of Genotoxic MoAs 80

3.3.2 Integration of Metabolic Activation 93
3.3.3 Additional Considerations for Experimental Design of Predictive Toxicogenomics Studies 95
3.4 Summary and Conclusions 97
Acknowledgements 99
References 99

Chapter 4 ***In vivo* Signatures of Genotoxic and Non-genotoxic Chemicals** **113**
Scott S. Auerbach

4.1 Introduction 113
4.2 General Signature of Genotoxicity 116
4.3 Liver 129
4.3.1 Rat 130
4.3.2 Mouse 133
4.3.3 Human 135
4.4 Kidney 135
4.4.1 Rat 135
4.4.2 Mouse 137
4.4.3 Human 137
4.5 Heart 137
4.5.1 Rat 138
4.5.2 Mouse 138
4.5.3 Human 139
4.6 Skeletal Muscle 140
4.6.1 Rat 140
4.6.2 Mouse 140
4.6.3 Human 140
4.7 Bone Marrow and Blood 140
4.7.1 Rat 141
4.7.2 Mouse 141
4.7.3 Human 141
4.8 Spleen 143
4.8.1 Rat 143
4.8.2 Mouse 143
4.8.3 Human 143
4.9 Other Tissues 144
4.9.1 Rat 144
4.9.2 Mouse 144
4.9.3 Human 144

4.10 Study Design 145
4.10.1 Dose 145
4.10.2 Duration 146
4.10.3 Tissue Selection 146
4.11 Conclusion 147
Acknowledgements 147
References 147

Chapter 5 Transcriptomic Dose–Response Analysis for Mode of Action and Risk Assessment **154**
Russell S. Thomas and Michael D. Waters

5.1 Introduction 154
5.2 Traditional Statistical Methods for Analyzing Transcriptomic Dose–Response Data 156
5.3 A Benchmark Dose Method for Analyzing Transcriptomic Data 156
5.4 Application of Transcriptomic Dose–Response Analysis to MoA Assessment 159
5.4.1 Case Study: Formaldehyde Exposure in Rat Nasal Epithelium 160
5.4.2 Case Study: β-chloroprene in Mouse and Rat Lung 162
5.4.3 Case Study: Naphthalene in the Rat Nasal and Olfactory Epithelium 163
5.4.4 Case Study: Fenofibrate and Methapyrilene for Receptor-mediated Nongenotoxic Carcinogenesis 165
5.5 Applications of Transcriptomic Dose–Response Analysis to Assess Cross-species Extrapolation 166
5.6 On the use of Toxicity Pathways in Toxicity Testing, Transcriptional BMD Analysis, and Risk Assessment 168
5.7 Comparison of Traditional Risk Assessment Approaches with Those Applying Transcriptomics 173
5.8 Significance of Integrating Transcriptomic Data into Risk Assessment 175
References 176

Chapter 6 Using Transcriptomics to Evaluate Thresholds in Genotoxicity Dose–Response 185
Patrick D. McMullen, Salil Pendse, Yeyejide Adeleye, Paul L. Carmichael, Melvin E. Andersen and Rebecca A. Clewell

6.1 Introduction 185
6.1.1 Challenges in Assessing the Safety of Potential Genotoxicants 185
6.1.2 Understanding the Nature of the Threshold-shaped Dose–Response Curve 187
6.2 Methods for Evaluating Low-dose Genotoxicity 189
6.3 Micronucleus Dose Response 190
6.4 Comparing Gene Signatures for Chemicals Causing Different Types of DNA Damage 191
6.5 Integrating Dose–Response Trends for Transcriptomic and Micronucleus Endpoints 196
6.6 Homeostasis, Transcriptional Regulation, and Post-translational Activity 199
6.7 Suitability of Gene Expression Changes for Genotoxic Adversity 201
References 203

Chapter 7 Dissecting Modes of Action of Non-genotoxic Carcinogens 209
Mirjam M. Schaap, Jan van Benthem, Miriam N. Jacobs, Annamaria Colacci, Anne S. Kienhuis, Harry van Steeg and Mirjam Luijten

7.1 Introduction 209
7.1.1 Current Regulatory Requirements and Difficulties with Respect to Non-genotoxic Carcinogens 209
7.1.2 What are Non-genotoxic Carcinogens? 210
7.1.3 Possible Methods for the Identification of Non-genotoxic Carcinogens 215
7.1.4 Comparison Approach 215
7.2 Improving the Comparison Approach: A Case Study 217
7.2.1 Cyclosporin A and Tacrolimus as Model Compounds 217
7.2.2 Concentration Selection 218
7.2.3 Microarray Analyses 218

7.2.4 Comparison Approach on a Concentration Range 219
7.2.5 Biological Response at the Pathway Level 220
7.2.6 Biological Relevance of Comparison Approach 225
7.3 Discussion and Future Prospects 228
References 230

Chapter 8 Human Embryonic Stem Cells as Biological Models to Examine the Impact of Xenobiotics on the Genome and Epigenome 236
Leslie Recio

8.1 Introduction 236
8.2 hES Cells as Biological Models in Toxicology 238
8.3 Developing hES Cells into a Toxicology Testing Platform 239
8.4 Stem Cells as a Biological Platform to Examine the Impact of Xenobiotics on the Genome and Epigenome 241
8.5 The miRNA and lncRNA Epigenome 241
8.6 Epigenomic Biomarkers and Toxicity Testing in the 21st Century 243
Acknowledgements 243
References 244

Chapter 9 Novel Data Streams in the Assessment of Mutagenicity and Carcinogenicity: Implications for Cancer Hazard Assessment 247
Kathryn Z. Guyton and Michael D. Waters

9.1 Integrating Across Multiple Data Streams to Reach Hazard Conclusions: Mechanistic Data can be Critical When Human Evidence is Less than Sufficient 247
9.1.1 Introduction to the IARC Monographs Hazard Classification Process 248
9.1.2 Recent Examples 250
9.2 Predicting Chemical Carcinogenicity Using Mechanistic Data 257
9.2.1 Toxicogenomics Studies to Predict Carcinogenicity 258
9.2.2 *In vitro* Testing for Carcinogenicity—Moving Beyond Batteries 258

9.2.3 Using Biomarkers of Cancer Risk in Molecular Epidemiology 259
9.2.4 Filling Data Gaps: Predicting Chemical Carcinogenicity Using Modeling can Aid in Prioritization Exercises 260
9.2.5 Addressing Human Variability 261
9.3 Future Directions in Mechanistic Data Evaluations to Support IARC Monographs 264
9.3.1 Systematic Approaches to Identifying and Presenting Mechanistic Evidence in Human Health Assessments 264
9.3.2 High-throughput Data Analysis 265
9.4 Conclusions and Future Directions 267
Appendix 9.1: Classification Categories for the Overall Evaluation for the IARC Monographs (IARC 2006) 268
References 269

Chapter 10 Conazoles and Cancer: A Review 280
Stephen Nesnow

10.1 Introduction and Background 280
10.2 Inhibition of CYP51 by Conazoles 281
10.2.1 Introduction to CYP51 281
10.2.2 Mechanism of the Inhibitory Action of CYP51 284
10.2.3 Inhibition of Fungal CYP51 285
10.2.4 Comparison of the Inhibition of Fungal and Human CYP51 287
10.2.5 Inhibition of Other Mammalian CYPs 288
10.3 Induction of CYPs by Conazoles 293
10.3.1 Introduction to Nuclear Receptors 293
10.3.2 Induction of CYPs 294
10.4 Hepatic Effects Induced by Conazoles 294
10.4.1 Increased Liver Weight and Hypertrophy 294
10.4.2 Cell Proliferation 298
10.5 Effects on Serum Cholesterol and Triglyceride Levels 300
10.6 Genotoxicity 302
10.7 Tumorigenic Effects of Conazoles 302
10.8 Toxicogenomic Studies in Mice 305
10.8.1 Genomic Studies Using Liver Samples from Mice Treated with Conazoles 306

10.8.2 Proteomic Studies Using Liver Samples from Mice Treated with Conazoles 309
10.8.3 Metabolomic Studies Using Liver Samples from Mice Treated with Conazoles 310
10.9 Toxicogenomic Studies in Rats 312
10.9.1 Toxicological Studies in Rats 312
10.9.2 Genomic Studies Using Liver Tissues from Rats Treated with Conazoles 313
10.9.3 Genomic Studies Using Thyroid Tissues from Rats Treated with Conazoles 313
10.10 Mode of Action of Propiconazole: Introduction 314
10.10.1 Mode of Carcinogenic Action of Propiconazole, a Series of Key Events Leading to Cancer 314
10.10.2 Key Event: Activate Nuclear Receptors 316
10.10.3 Key Event: CYP Induction 316
10.10.4 Key Event: Induction of ROS and Oxidative Stress 317
10.10.5 Key Event: Increase in Endogenous DNA Adduct Levels and Mutations 319
10.10.6 Key Event: Increased atRA Metabolism 320
10.10.7 Key Event: Decreased Hepatic Levels of atRA 320
10.10.8 Key Event: Inhibit CYP51 Activity 320
10.10.9 Key Event: Dysregulation of Cholesterol Biosynthesis and Metabolism 321
10.10.10 Key Event: Increase in Cell Proliferation by Decreased Levels of atRA and by a Mevalonic Acid/Cholesterol Biosynthesis Feedback Mechanism 321
10.10.11 Key Event: Hepatocellular Tumors 324
10.11 Mode of Carcinogenic Action of Propiconazole: Discussion 325
10.12 Mode of Carcinogenic Action: Human Relevance 327
10.12.1 Species Comparison Across Key Events 327
10.12.2 Comparison of Toxicologic and Genomic Studies Comparing Mouse Liver Responses to Conazoles and Phenobarbital 328
10.13 Conclusions 334
Acknowledgements 335
References 335

Chapter 11 Application of Transcriptomics in Exposed Human Populations: Benzene as an Example **352**
Cliona M. McHale, Martyn T. Smith and Luoping Zhang

11.1 Application of Toxicogenomics in Occupational Benzene Exposure 352
11.1.1 Mechanisms and Biomarkers of Benzene Toxicity 354
11.1.2 Discerning Low-dose Effects is a Challenge in Risk Assessment 354
11.1.3 Toxicogenomic Studies 355
11.2 Transcriptomic Studies of Occupational Benzene Exposure 356
11.2.1 The Complex Human Transcriptome and Its Analysis 356
11.2.2 Microarrays 357
11.2.3 RNA-Seq 361
11.2.4 NanoString 365
11.2.5 L1000 and S1500 Platforms 365
11.2.6 Transcriptomic Platform Choice and Study Design Considerations 366
11.3 Future Directions and Translation 368
11.4 Conclusion 369
Acknowledgements 370
References 370

Chapter 12 Toxicogenomics Case Study: Furan **390**
A. Francina Webster, Iain B. Lambert and Carole L. Yauk

12.1 Introduction 390
12.1.1 Chemical Testing and Toxicogenomics 391
12.1.2 Reducing Animal use in Toxicity Testing 393
12.2 Toxicogenomics Case Study: Furan 396
12.2.1 Liver Physiology and Hepatocarcinogenesis 396
12.2.2 Test Article: Furan 398
12.2.3 Quantitative Toxicogenomics 402
12.2.4 Predictive Toxicogenomics 403
12.2.5 Mechanistic Toxicogenomics 405
12.2.6 FFPE Toxicogenomics 408

12.3 Role for Toxicogenomics in Chemical Risk Assessment 409

12.3.1 Guidelines for Using Toxicogenomics Data in Formal Risk Assessment 410

12.3.2 FFPE Genomics in Risk Assessment 411

12.3.3 Adverse Outcome Pathways 412

12.3.4 Toxicogenomics in Tiered Testing Strategies 413

12.4 Concluding Remarks 414

References 415

Chapter 13 The Parallelogram Approach to Assess Human Relevance of Toxicogenomics-derived Toxicity Pathways in Human Health Risk Assessment 423

Anne S. Kienhuis, Wim C. Mennes, Marja Driessen, Leo T. M. van der Ven, Jan van Benthem and Mirjam Luijten

13.1 Human Health Risk Assessment 423

13.2 Toxicogenomics in Risk Assessment 425

13.2.1 Toxicogenomics in Hazard Identification 426

13.2.2 Toxicogenomics in Dose–Response Modeling 426

13.3 Assessing Biological Significance and Human Relevance of Toxicogenomics Data 427

13.3.1 The Parallelogram Approach 427

13.3.2 The Concordance Model 431

13.4 Discussion 435

13.4.1 Pathway Analysis 436

13.4.2 Toxicity Pathway-derived BMDs 436

13.4.3 Apical Endpoints 437

13.4.4 Adversity 437

13.5 Conclusion 438

References 438

Chapter 14 Bioinformatics of Genomics in the Assessment of Cancer 442

Pierre R. Bushel

14.1 Introduction 442

14.1.1 Classification and Prediction 445

14.1.2 Over-represented Pathways, Enriched Gene Sets and Gene Regulatory Networks 453

14.1.3 Bioinformatics and Computational Biology for Integrative Genomics 468
14.2 The Future can be Now 473
Acknowledgements 475
References 475

Subject Index **484**

CHAPTER 1

Introduction to Predictive Toxicogenomics for Carcinogenicity

MICHAEL D. WATERS

Integrated Laboratory Systems (ILS), Inc., PO Box 13507, Research Triangle Park, NC 27709, USA
Email: mdwaters@centurylink.net

1.1 Background on -omics Technologies Applied in Toxicology

The term "toxicogenomics" (TGx) has been applied to the combined technologies of transcriptomics, proteomics, metabolomics, and epigenomics, as they are used in the field of toxicology to study the expression of genes, proteins, metabolites, and epigenetic modifications, respectively.[1] The goal of this chapter is to place the TGx of predictive carcinogenicity in perspective, with both the relevant conventional toxicological methodologies that have come before genomics and those that coexist at present and quite likely will be used together in the future to study relevant biological processes, including pathways of toxicity and chemical modes of action.

Transcriptomics (commonly referred to as "gene-expression profiling" or "transcript profiling") involves measuring the relative abundance of potentially thousands of RNA transcripts present in a sample from one tissue extract at one point in time and at a single dose of a chemical. Transcriptomics is typically performed using microarray technology. Microarrays

Issues in Toxicology No. 28
Toxicogenomics in Predictive Carcinogenicity
Edited by Michael D. Waters and Russell S. Thomas

Published by the Royal Society of Chemistry, www.rsc.org

have advantages in speed and ease of sample preparation, low per-sample cost, and well-established protocols and methods for data analysis and data normalization. Microarrays are available for transcriptomics (*e.g.*, messenger (m) ribonucleic acids (RNAs), micro (mi)RNAs and long non-coding (lnc)RNAs) as well as for epigenomics (*e.g.*, DNA methylation microarrays).

Whole-transcriptome next-generation sequencing (RNA-seq) offers an alternative method for estimating transcript abundance in gene expression studies. RNA-seq has the potential to overcome many of the limitations associated with microarrays as it does not rely on predetermined probe sequences for expression measurements, and it is based on simple counting of reads that can be reliably aligned to a reference sequence.[2] The use of molecular expression analytical technologies, including microarray or RNA-seq, in TGx is based on the assumption that chemicals exhibiting similar types and degrees of toxicity will induce similar profiles of gene, protein, or metabolite expression. When used in conjunction with conventional toxicity testing methods and/or histopathology, to provide context, these snapshots or images of molecular expression can be assembled sequentially into a series of high-resolution profiles of adaptive and adverse effects, including causal events that occur as a function of dose and time and may reflect particular chemical modes of action.

Another set of analytical technologies, high-throughput screening (HTS) methods, were developed in the mid-1980s in the pharmaceutical industry to rapidly screen libraries of candidate small molecules or drugs for specific types of biological activity or disease processes. This set of analytical technologies has been applied in multi-endpoint toxicity screening for the past 10 years.[3] A consortium of US government agencies, referred to as Tox21,[4] consisting initially of the US Environmental Protection Agency (EPA) ToxCast program,[4] the National Toxicology Program (NTP), the National Institutes of Health Chemical Genomics Center (NCGC), and later the US Food and Drug Administration (FDA), has used *in vitro* assays in robotic HTS to study chemical perturbations of biological pathways which can become pathways of toxicity. Chemical–protein interactions have been measured and bioinformatics has been used to delineate the toxic response by making use of known links between genes, proteins, and diseases.[5] Although some predictive patterns or "signatures" of toxicity have been identified, a comprehensive chemical toxicity predictive capability has been elusive and HTS remains at a developmental stage for many toxicological outcomes.

Toxicogenomics, which originated in name in 1999,[6] also has been slow to achieve its anticipated predictive and mechanistic analytical capabilities. While microarray and RNA-seq based TGx methods clearly provide useful information about how biological systems respond to chemical compounds, currently they are too expensive to serve in a screening role. Combining HTS and TGx technologies can provide the opportunity both to screen for perturbations in biological, toxicological, and disease pathways and to carefully investigate multiple potential modes of chemical action. Each technology therefore can help to define the dose–response behavior of drugs and chemicals, and, when used together, they ultimately should be able to accurately predict toxicological and disease outcomes and likely modes of action.

The fundamental question to be addressed in this introductory chapter is how best to leverage data and information coming from these technologies to develop integrated strategies for the evaluation of chemicals for potential human health risks. This question may be answered by better understanding the goals and challenges of TGx and HTS.

The stated goal of EPA's ToxCast program[3] is to screen large numbers of environmental chemicals using *in vitro* HTS methods and to prioritize them for further testing based on predictive scores for disease-related signatures and corresponding estimates of human exposure. In the longer term, however, the goal of ToxCast is to improve the efficiency and reliability of current toxicity test methods. This could be accomplished by using both HTS and TGx technologies in *in vitro* assays to delineate putative modes or mechanisms of action relevant to *in vivo* chemical toxicity. There are several challenges to be overcome before such a goal can be realized, and its current status is reflected in discussions throughout this book.

Chapter 2 by Li provides details on the evolution and current status of *in vitro* screening technologies with an emphasis on biomarkers or signatures for that can be further developed for HTS applications in drug safety assessment. There have been several transcriptomic signatures identified for assessing genotoxicity, but the one (TGx-28.65) developed by Li (chapter 2), and also discussed in chapter 3 by Buick and Yauk, shows convincing inter- and intra-laboratory reproducibility and performs robustly and consistently on different transcriptomic platforms.

A common criticism of the *in vitro* to *in vivo* prediction approach is that a chemical's toxicity may depend on unique properties of intact tissues and organisms that are not exhibited by genes or cells *in vitro*. In addition, biotransformation of parent compounds into metabolites that are more or less active than the parent must be considered in each assay or model system. Development of tests in which realistic levels and kinds of biotransformation occur *in vitro,* with fidelity to what occurs *in vivo,* is a challenge. Furthermore, understanding the correlation between *in vitro* lowest-effective-dose values and corresponding chemical concentrations in blood or tissues is crucial in extending either HTS or *in vitro* TGx studies to inform quantitative assessments of health risk.

In principle, challenges such as those described may be addressed by performing parallel rodent TGx *in vitro* and *in vivo* studies with human *in vitro* assays and by confirming the findings with human clinical and epidemiological studies (*i.e.*, using a parallelogram method).[7] This approach proposed originally by Sobels[8,9] has the advantage of simultaneously addressing the question of rodent-to-human extrapolation and the issue of internal-tissue dosimetry. Several published TGx investigations appear to validate this thinking, as discussed in chapter 13 by Kienhuis *et al.* Information garnered from parallelogram-based TGx investigations can benefit both HTS and TGx in the development of quantitative data for use in risk assessment. Furthermore, comparative genomics approaches have improved and extended the parallelogram approach to a variety of species (see chapter 7 by Schaap *et al.* and chapter 13 by Kienhuis *et al.*).

Another challenge is how to determine what short-timescale (hours to days) *in vitro* assays can reveal about long-timescale (months to years) processes that lead to *in vivo* chronic toxicity outcomes such as cancer. This challenge has been addressed by applying TGx together with (within) standard toxicological and chronic bioassays such that details of changes in molecular expression occurring early in the time course of chronic disease can be defined and used to predict of chronic disease outcomes. Alternatively, technology now exists to retrospectively examine formalin-fixed paraffin-embedded archival tissues to discover the profiles of gene expression that correspond to histopathological phenotypes associated with various disease processes and outcomes. The NTP archives contain tissue blocks from cancer bioassays on hundreds of compounds that could be studied in this way. This approach was successfully applied in TGx studies on furan,[10] as discussed by Webster, *et al.* in chapter 12.

One further issue relates to the relative priority being given to HTS screening and TGx analysis of the various chemicals of regulatory concern. Although HTS aims to predict the potential for chemicals to affect human health based on results of *in vitro* tests, essentially all of the current *in vivo* data being used to develop prediction models are from high-dose animal testing using conventional methods (as discussed in the following section). For future HTS efforts, priority should be given to chemicals to which humans are exposed where there is evidence of bioaccumulation (*i.e.*, body burdens). Such exposure estimates are routinely developed by the EPA, and blood and tissue analyses are performed by the US Agency for Toxic Substances and Disease Registry. If HTS shows that chemicals activate toxicity pathways at concentrations similar to those found in human biomonitoring studies, they should be given higher priority for more detailed testing using TGx together with conventional test methods.

If challenges such as those previously described can be met, the HTS approach in toxicity testing can become a practical solution in evaluating the backlog of thousands of untested environmental chemicals. Furthermore, the identification of molecular signatures and toxicity pathways with TGx can be used to develop predictive toxicological methods and essential information to substantiate putative modes of action in risk assessment. While it is generally acknowledged that TGx approaches cannot yet replace conventional toxicity tests, data derived through the use of TGx studies within the framework of conventional testing have enabled a more predictive toxicology, provided a better understanding of the mode of action of drugs and chemicals,[11,12] and helped to improve hazard and risk assessment, as discussed by Auerbach, Thomas and Waters, and Guyton and Waters (in chapters 4, 5, and 9, respectively). In subsequent chapters, that which has been accomplished in this regard for several well-studied model rodent carcinogens, including benzene, the conazoles, furan, and acetaminophen (paracetamol), is reviewed by Nesnow, McHale *et al.*, Webster *et al.*, and Kienhuis *et al.* (chapters 10–13, respectively). Our plan is to work through these prototypic compounds to understand how gene response reflects causal events on pathways of toxicity and disease. This approach is consistent with current regulatory thinking and can facilitate the eventual integration of HTS and TGx data for predictive purposes. When causal events are linked to toxicity and disease pathways, as a function of dose–response, they provide regulatory agencies with essential data with which to determine under what conditions environmental exposures pose quantitative risks to human health.

The US National Research Council report *Toxicity Testing in the 21st Century: A Vision and a Strategy*[13] overviewed by Andersen and Krewski[14] proposed that new genomic technologies could both dramatically increase the number of chemicals comprehensively evaluated for toxicity and broaden and improve the human relevance of toxicity endpoints assessed. The use of genomic technologies in toxicity testing holds great promise for informing and significantly improving health risk assessments for a broad range of environmental chemicals. The sections below trace developments in the field of toxicity testing that have led to where we are today. In addition, greater definition of the subjects that will be covered in subsequent chapters of the book is provided.

1.1.1 Conventional Toxicity Testing

Since the 1950s conventional toxicity testing involving chemical, physical, or biological agents has relied on high-dose studies in laboratory animals to assess potential human toxicity. The results of such studies are extrapolated to environmental or occupational exposure levels to predict human health outcomes. When this approach was developed, little was known about the modes of action by which chemicals caused toxicological responses in animals. Since that time, *in vivo* toxicity tests have been standardized and *in vitro* tests have been developed to characterize toxic effects at the cellular level (*e.g.*, cytotoxicity). As our understanding of the genetic basis of toxicity (in particular, carcinogenicity) has grown, a battery of *in vivo* and *in vitro* tests to detect mutations and DNA damage has also been developed and standardized.[15–19]

Conventional toxicity testing currently involves using *in vivo* and *in vitro* tests to screen chemicals for defined toxic effects (*e.g.*, neurotoxicity, developmental toxicity, and carcinogenicity) or modes of action relevant to a specific outcome (*e.g.*, mutagenicity, cytotoxicity, and regenerative cell proliferation relevant to genotoxic carcinogenicity). Intact-animal models are the only surrogates for human clinical studies that offer reasonable biochemical, metabolic and physiological fidelity to the human response. *In vitro* test systems can represent investigator-selected biological processes with various levels of fidelity under carefully controlled experimental conditions.

To profile the toxicity of a chemical in standard *in vivo* assays, the chemical typically is tested in rodents for acute toxicity (following a single or short-term exposure); subchronic toxicity (following exposure for 14–90 days); chronic toxicity (including carcinogenicity) following long-term exposure (up to 2 years); eye and skin irritation; reproductive and developmental toxicity (in breeding studies); hypersensitivity (immune response); and, sometimes, phototoxicity (toxicity in response to light). Toxicokinetic studies also are conducted to characterize absorption, distribution, metabolism, and excretion of the chemical following *in vivo* exposure.

Toxicology studies generally use a series of dose levels, ranging from doses where no effects are expected to doses where frank clinical or histopathologic changes are expected. Typically, the highest dose at which no overt toxicity occurs in a 90-day subchronic toxicity study (the "maximum tolerated dose") is used to establish dose levels for long-term bioassays, which measure endpoints such as clinical signs of toxicity, body- and organ-weight changes, clinical chemistry, and histopathological responses, and which provide insight into potential latent effects, including cancer, reproductive or developmental toxicity, or immunotoxicity.

The relationship between exposure to toxicants and adverse health effects is also examined through epidemiological studies. These studies attempt to assess the relationship between exposure levels and the likelihood of adverse effects ("exposure–response relationships") by comparing estimated exposure levels and disease distributions in human populations. However, exposure can be difficult to estimate, and such studies are complicated by confounding factors such as co-exposure to other potential toxicants such as cigarette smoke or pesticides.

Most of our understanding of chemical toxicity has come from data obtained through conventional *in vivo* testing. However, *in vivo* toxicity testing is costly and time consuming, requiring large numbers of animals. A 2-year rodent study requires the use of more than 800 mice and rats and the histopathological examination of more than 40 tissues. Full characterization of the potentially hazardous properties of the many thousands of environmental chemicals through conventional toxicity testing is thus severely constrained by financial and other resource limitations. In practice, a tier-based testing framework generally has been used for evaluation of environmental chemicals, whereby the choice of specific tests for a given chemical is guided by observations of toxicity in early-stage tests and information on potential human exposure (routes and levels).

Another limitation of conventional toxicity testing is that even extensive animal testing does not fully explain the mechanisms of toxicity in humans. Detailed knowledge concerning risk to humans is still inadequate for many chemicals. The need for more mechanistic data and a "theoretical framework for rational decision making" was identified in the early 1980s.[20]

1.1.2 Genomic and Postgenomic Technologies

The discovery in 1944 that genetic information was transmitted by deoxyribonucleic acid (DNA), the discovery in 1953 of the structure of DNA, and the cracking of the genetic code in 1966—showing how the sequence of chemical bases in DNA codes for the synthesis of proteins—opened the door to an explosion of knowledge in molecular biology. In the nucleus of the cell, a gene's DNA serves as a template for synthesis of mRNA; mRNA then travels from the nucleus to the cytoplasm and serves, in turn, as a template for the synthesis of a protein molecule. Proteins are involved in virtually every biological function, including cell growth, differentiation, and death, regulation and coordination of physiological processes, metabolism, immune function, and disease processes. Gene expression or transcription refers to the construction of mRNA molecules (transcripts) from the template provided by a specific gene. Knowledge of how gene expression is controlled and coordinated, as well as the roles of specific proteins in cellular functions and in disease processes, has informed our understanding of the modes and mechanisms of action by which chemicals produce toxic effects in cells and whole organisms.

The genome is the entire complement of genetic information found in each cell of an organism, and genomics refers to the analysis of the entire genome of an organism (or cell or tissue). The introduction of high-throughput dideoxynucleotide DNA sequencing in 1977[21] made it feasible to determine the sequences of entire genomes, ushering in the era of genomics. This technology was used in the Human Genome Project, which was proposed in the late 1980s and begun in 1990. The reference human genome sequence—consisting of about 23 000 genes that code for proteins, made up of 3 billion chemical base pairs—was made available in 2000.[22] Since then, high-throughput genome sequencing has become faster and less expensive.[23] By 2010 a whole human genome could be sequenced on the Illumina platform for less than $14 500[24] and today the basic cost approaches one-tenth that amount.

To capitalize on the enormous potential of genome-wide DNA sequence information, new molecular technologies and bioinformatics tools (to be discussed later) have been developed that make it possible to generate and analyze biological datasets of unprecedented magnitude and complexity.[13] Beyond DNA-sequence analysis, postgenomic technologies enable analysis of gene expression or the production of mRNA transcripts (transcriptomics); the production of the proteins themselves (proteomics); the endogenous production of metabolites (metabolomics); and the transmissible epigenetic modifications of chromatin and DNA (epigenomics).

1.1.2.1 Toxicogenomics (TGx)

Simply stated, TGx is the application of genomic technologies to the study of adverse effects of toxicants. It aims to study the response of the entire genome to toxicants or environmental stressors. In 2002, scientists at the National Institute of Environmental Health Sciences (NIEHS) National Center for Toxicogenomics (NCT) proposed that patterns of induced gene expression changes are characteristic of specific classes of toxic compounds and that their distinctive molecular expression fingerprints can help in classifying agents with different mechanisms of action.[25,26] Since then, TGx technologies have made it possible to use genotypes and toxicant-induced gene expression, protein, and metabolite profiles to screen chemicals in hazard identification, to monitor exposure, to measure dose–response curves at the cellular level, to elucidate mechanisms of action, and to predict individual variability in sensitivity.

Whereas genomes are relatively static, transcriptomes, proteomes, metabolomes, and epigenomes are dynamic—displaying moment-to-moment changes in response to diet, stress, disease processes, and exposure to toxicants and stressors—and their analysis must be linked to the state and condition of the biological system under investigation. Toxicity coincides with changes in specific mRNAs, proteins, metabolites, and epigenetic modifications. These changes observed under defined conditions of cellular location, dose level, time, and biological context can provide meaningful information about biological responses to toxic insult. Typically, adverse effects can be detected at subcellular and molecular levels at time points before they are manifested at the level of tissues, organs, or the whole organism. Toxicant-specific alterations in gene expression, protein synthesis, and metabolite production correspond with observable or phenotypic responses of cells, tissues, and organisms. The process of relating molecular expression data to toxicity and pathology observed in conventional toxicology tests and using histopathological evaluation is referred to as "phenotypic anchoring".[27,28] This subject is covered in greater detail later.

1.1.2.1.1 Transcriptomics. Transcriptomics, the measurement of the relative abundance of mRNA transcripts in TGx investigations, has typically been measured using microarrays—glass slides or silicon chips on which arrays of thousands of DNA (copy (c)DNA) probes are attached, which react (hybridize) with RNA transcripts in the biological sample. Hybridization is detected by labeling of the RNA with fluorescent dye, and the intensity of fluorescence indicates the level of gene expression. The results of microarray assays have been shown to correlate well with measurements of the expression of single genes by methods such as quantitative real-time (qRT) polymerase chain reaction (PCR, a technology that "amplifies" DNA segments, creating large numbers of copies to facilitate analysis).

Modern microarray technology uses oligonucleotide probes (shorter DNA sequences), allowing over a million probes per chip.[29–31] Microarrays corresponding to essentially all known human genes and representing the genomes of animal models used in conventional toxicology are available from commercial vendors (such as Affymetrix, Agilent, and Applied

Biosystems). However, only those mRNAs, miRNAs or lncRNAs for which there is a complementary sequence on the microarray can be hybridized and measured in a given experiment. As discussed earlier, microarrays have advantages in speed and ease of sample preparation, low per-sample cost, and well-established protocols and methods for data analysis and data normalization. However, microarrays also have limitations, including a finite dynamic range due to the limitations of fluorescence and saturation of binding sites within the probe sets printed on the microarray. At low levels of expression, microarrays often have poor resolution of transcript binding due to high background effects and nonspecific binding.

RNA-seq offers an alternative method for estimating transcript abundance and has the potential to overcome many of the limitations associated with microarrays: it does not rely on predetermined probe sequences for expression measurements and is based on simple counting of reads that can be reliably aligned to a reference sequence. In providing count data, RNA-Seq has effectively no limit to the dynamic range of signal detection, and, in theory, can provide a higher degree of accuracy and precision in estimating relative expression levels.[2] However, RNA-seq data also has potential challenges which remain less well explored. Reliable quantification of expression levels appears highly dependent on read depth, and low transcript abundances are characterized by high variance. Thus, the effective or useful lower limit of the dynamic range of RNA-seq is not clear. Additionally, methods for normalization and statistical analysis of RNA-seq data are less mature, and no established best practices exist for RNA-seq data analyses.[2]

1.1.2.1.2 Proteomics. Proteomics is the study of collections of proteins in living systems (*i.e.*, the complete set of proteins found in a particular cell type at a given time). Because a given protein may exist in several different forms, proteomes are more complex than the genomes and transcriptomes that code for them. Exposure to toxicants can result in two types of proteome changes: (1) changes in protein levels due to changes in gene expression, mRNA stability, protein stability, or some combination of these; and (2) changes in the relative levels of two or more modified forms of a protein, which may be more critical to function than the absolute protein levels.[32]

Proteomics relies on a combination of technologies to separate and identify proteins and measure their levels. Because there is no technology analogous to PCR that can amplify proteins, they must be analyzed at their native concentrations, which span more than six orders of magnitude. As a result, it is difficult to detect less-abundant proteins in complex mixtures. Another challenge to proteome analysis is that proteins may be damaged by reactive chemical intermediates produced from toxic chemicals or as a result of oxidative stress. Most analytical proteomic methods are not truly parallel, but involve elaborate serial analyses. Two major approaches are gel-based proteomics and "shotgun" proteomics.

In gel-based proteomics, proteins are separated by electrophoresis (*e.g.*, two-dimensional (2D)) sodium dodecyl sulfate polyacrylamide gel electrophoresis (SDS-PAGE), and selected proteins are identified by digestion to peptides, mass spectrometry (MS) analysis, and database searching. Comparative 2D SDS-PAGE with differential fluorescent labeling (for example, differential gel electrophoresis) allows effective quantitative comparisons of proteomes.[33,34] Modified and unmodified forms of proteins often can be resolved for separate characterization and quantitative analysis. Although 2D gel electrophoresis has been used mostly for global analyses of complex proteomes, the method is also useful for comparative analyses of smaller subproteomes.

In shotgun proteomic analysis, protein mixtures are digested to form complex mixtures of peptides, which are analyzed by liquid-chromatography-coupled MS.[35] Databases are then searched to match the resulting peptide tandem mass spectrometry spectra with corresponding peptide sequences, and software is used to reassemble the collection of peptide sequences into proteins. Shotgun proteomics is the most effective technology for automated analysis of complex peptide mixtures.[36,37]

A key issue in proteomics is the standardization of data analysis methods formats for data representation and reporting. Another unresolved issue is the fact that different database search algorithms can yield different identifications of proteins and peptides.

1.1.2.1.3 Metabolomics. Metabolomics is the study of collections of molecules (intermediates and products) generated through metabolic processes in an organism or cell at a given point in time. (Some authors distinguish between "metabolomics" and "metabonomics"; however, the distinction in the literature is not consistent, and this chapter uses the term "metabolomics".) Metabolites reflect the actions of proteins in biochemical pathways; metabolomes thus represent biologic states (phenotypes) analogous to proteomes.

Metabolomic analysis relies mainly on nuclear magnetic resonance (NMR) spectroscopy and MS (both gas and liquid chromatography MS). NMR-based technology has been shown to be robust and reproducible in laboratories that follow similar analytical protocols,[38] and consensus standards for analytical standardization and data representation in metabonomic analyses have been agreed.[39] NMR is valuable for identifying patterns of spectra reflecting global metabolic changes, while MS-based analyses offer the advantage of greater sensitivity. Both technologies can detect differences in metabolic profiles that correspond to various modes of toxicity, but integration of the technologies will allow a more comprehensive approach.

1.1.2.1.4 Biomarkers or Signatures. In conventional toxicology, the term "biomarker" generally refers to biological indicators of exposure to specific toxicants. However, in TGx, the term "biomarker" has been used to refer to a gene-expression, protein, or metabolic profile that serves as an indicator of toxicity. Such a "signature" or "fingerprint" represents the results of complex interactions within the organism. If shown to be reproducible and correlated with a toxicity endpoint, the molecular signature potentially can serve as a predictive biomarker. Several review articles explore issues related to biomarker assay development and provide examples of the biomarker development process.[40–42] TGx biomarkers or gene signatures *in vitro* and *in vivo* are discussed extensively in chapters 2, 3, and 4 of this volume.

1.1.2.2 Bioinformatics

The magnitude and complexity of the data generated by genomic technologies, as well as the need to integrate genomic data with data from other disciplines, requires the use of advanced computational techniques. "Bioinformatics" is the branch of computational biology focused on the collection, management, analysis, and integration of numerical biological data and it is essential to genomic analyses. Bioinformatics encompasses the integration of data across genomic technologies, the integration of genomic data with data from other observations and measurements (including HTS data), and the integration of all these data in databases and related information resources. At a basic level, bioinformatics is represented by information resources such as GenBank, *i.e.*, repositories of gene sequence data and associated information, structured for easy retrieval. At an intermediate level are tools, such as BLAST or SAGEmap, that perform insightful sequence alignment and function and structure analysis. Finally, sophisticated information systems (*e.g.*, expert systems) integrate data from numerous sources to solve multifaceted problems. An excellent illustrated introduction to bioinformatics and various bioinformatics methods applicable to TGx and predictive carcinogenicity is provided by Bushel in chapter 14.

1.1.2.3 High-Throughput Screening (HTS)

As previously discussed, microarray- or RNA-seq-based TGx methods provide useful information about how thousands of genes in biological systems respond to chemical compounds, but they are too expensive to serve as high-volume screening tests for the effects of thousands of chemicals in numerous cell types at a range of dose levels and time points. In contrast, individual HTS methods look at only one or a few genes at a time, but allow thousands of chemicals to be screened in one day over a wide range of concentrations. Compared with more extensive conventional toxicology and TGx tests, *in vitro* screening tests as used in the triage of molecular libraries for particular molecular or biological activities typically employ higher and fewer doses of the chemical, fewer test subjects, shorter observation periods, and less extensive evaluation of the toxic outcomes. HTS thus provides a practical method to investigate more than 100 000 compounds per day in miniaturized *in vitro* assays in order to identify those with the potential to cause adverse effects.[43] For safety evaluation and toxicity testing "hits" in the screening assays correspond to biological pathways that are known to lead to adverse outcomes. With sufficient accumulated data, it may be possible to use structure–activity analysis to predict HTS hits, so that potential targets can be predicted prior to screening. The application of robotic HTS as a useful complement to conventional toxicology has been expanding.[3,44–46]

HTS assays fall into two broad categories: biochemical assays and cell-based assays. Biochemical (cell-free) assays generally measure direct effects on specific molecular targets of interest. These assays have been used to measure enzymatic activity,[47–49] binding of substances to receptors,[50] ion-channel activity,[51] nuclear receptor activity,[52] and protein–protein interactions.[53] Because they involve homogenous reactions, biochemical assays are readily miniaturized. However, not all targets can be prepared satisfactorily for biochemical testing. Furthermore, a chemical's activity measured in cell-free assays does not necessarily correspond to its activity in the intact cell, which may be affected by the presence of intracellular cofactors, issues of membrane permeability, cytotoxicity, and other influences on the target molecule. In contrast, cell-based assays measure the effects of chemicals on pathways of interest in the physiological environment of a cell, without the need to specify a molecular target. Examples include functional assays,[54,55] reporter-gene assays (which use "marker" genes to signal activation of target genes),[56,57] and phenotypic assays for processes such as cell migration[58] or division.[59] Because cell-based assays measure effects on entire pathways, perturbations can be assessed at more than one step in a pathway. Cell-based HTS in 1536- or even 3456-well plate formats is not uncommon.[60–62] Unfortunately, space does not permit more extensive coverage of HTS methodology in this volume.

1.2 The Relative Roles of Toxicogenomics, Conventional Toxicity Testing, and High-throughput Screening

Conventional toxicity testing has served regulatory risk assessment reasonably well since its inception in the early 1970s. However, while conventional tests have become more refined and more definitive, they have not been able to meet the demand for higher throughput at lower cost. Recent innovations in HTS technologies promise higher throughput and lower cost; however, a key question is whether they can significantly improve risk assessment in support of regulatory decision-making. TGx technologies potentially can improve predictive toxicology (and thus risk assessment) by improving our understanding of dose–response relationships, cross-species extrapolation, exposure quantification, mechanisms of action, and variations in individual susceptibility. Although current risk assessment processes use mechanistic information in hazard identification and in understanding dose–response relationships, the information on mechanisms of toxicity often is incomplete or inconclusive. TGx is well suited to the classification of compounds by mode of action and to prediction of toxicological outcomes based on perturbation of known pathways of toxicity and disease. Applied in parallel with standard toxicological assays, TGx is quite useful in defining the details of mechanisms of toxicity and in providing early predictions of chronic disease outcomes (*e.g.*, cancer). TGx technologies also hold promise for improving exposure assessment. Because exposure to environmental chemicals is rarely measured directly, exposure is estimated typically by mathematical models based on factors that would affect exposure, such as typical water consumption, respiratory rates, and activity patterns. Following are specific roles that have been proposed for TGx studies and HTS tests in risk assessment:

- Exposure assessment: the high level of information inherent in gene-expression analysis may enable various types of exposure to be distinguished. In addition, proteomic analysis of biofluids may provide a means for noninvasive identification of biomarkers of exposure.
- Hazard screening: HTS tests are ideally suited for rapid testing of large numbers of compounds and may also be useful for preliminary hazard classification based on the perturbation of well-understood pathways of toxicity.
- Hazard classification and mode-of-action triage: identification of TGx signatures and corresponding toxicity pathways can allow various toxic and nontoxic endpoints to be distinguished. Mathematical modeling can be used to identify signatures for known toxicity outcomes and then to predict the effects of unknown toxicants in medium-throughput assays. Signature genes can be used, for example, to identify potential genotoxic and nongenotoxic carcinogens, aneugens, and cytotoxic agents.[63,64]
- Mechanistic information: TGx studies offer the opportunity to screen, or to evaluate in detail, molecular mechanisms of toxic action. When applied to the study of large classes of chemicals, TGx information can be used to globally define modes or mechanisms of toxic action. Significant changes in biologic mechanisms may occur with increasing dose, so that the TGx profiles are altered qualitatively as well as quantitatively.[65] Although some mechanistic information can be garnered from results of HTS *in vitro,* this is more readily accomplished at present by TGx methods, because of the density and highly parallel nature of TGx datasets. Furthermore, the mechanistic insights provided by TGx data can help refine the models used to predict target-organ doses, thus increasing the accuracy of dose–response assessment.
- Cross-species extrapolation: TGx methods can be used to evaluate the degree to which the biologic response pathways responsible for toxicity are conserved across species. Comparative TGx studies in two different species can identify orthologous genes (*i.e.*, genes in different species

that have evolved from a common ancestral gene by speciation and generally retain a similar function).

- Dose–response relationships: TGx studies are usually performed at several different doses and in a series of time points, enabling construction of dose–response curves over dose ranges that may be relevant to human exposures. There have been significant advances in this area of application of TGx methods.[66–68] In addition, HTS offers a significant potential advantage by enabling studies over an extensive range of doses in each individual assay.
- Developmental exposures: TGx techniques can be used in conjunction with conventional developmental toxicology methods to examine the highly time- and stage-sensitive effects of exposure on developmental processes *in vivo*.
- Variability in susceptibility: TGx can examine variability in gene expression due to mutations in genes or in the regions of chromosomes that regulate gene expression, and other factors that modify expression, such as epigenetic effects (reversible heritable changes in gene expression caused by mechanisms other than changes in DNA sequence).
- Mixtures: environmental exposure of humans typically is to complex mixtures, rather than to single agents. TGx and HTS technologies can be used to evaluate typical environmental mixtures, albeit with special requirements for sample preparation.
- Gene–environment interactions: TGx can provide insight into the responses of biological pathways that reflect the interactions between genetic makeup and environmental conditions and exposures.

The various potential roles of TGx has been discussed in numerous scientific reviews and commentaries over the past decade and a half,[10–12,66–92] which attests to the evolving reality that TGx is enhancing the ability of scientists to study and estimate the risks that chemicals pose to human health and the environment. Furthermore, it is becoming clear that HTS approaches using *in vitro* and computational methods can reduce the use of animals in toxicity testing, perhaps eventually replacing some uses. An example of a computational systems model currently being developed is the EPA's Virtual Liver project.[93] The aim is to develop models for predicting liver injury resulting from chronic chemical exposure by simulating the key chemical-induced events that result in changes in liver cells and tissue. Many scientists question whether this approach can completely replace the use of animals in regulatory testing. These scientists see the pathway-based screening approach as useful mainly to prioritize chemicals for further testing and perhaps to verify the relevance of animal toxicity testing methods to humans. However, the limited gain in knowledge from conventional animal studies and significant uncertainties regarding hazard identification hamper appropriate risk evaluation without efforts in pathway-based screening. Programs such as the ToxCast project and other alternative testing strategies were initiated to obtain more meaningful data that are beyond the limited information gathered from EPA or Organisation for Economic Co-operation and Development (OECD) guideline studies.[94] By 2010, nearly 500 assays that focus particularly on pathway perturbations had been developed under ToxCast.[95]

A very difficult challenge for *in vitro* testing is that it ideally must encompass the full range of toxicokinetic and toxicocodynamic phenomena that contribute to whole-animal toxicity.[96] However, combining the new technologies with refined approaches to whole-animal toxicity testing (*e.g.*, eliminating the need for 2-year rodent bioassays) could be one of the major decision-making opportunities in implementing the National Research Council's vision for the 21st century.[97] In any event, the goal in incorporating new technologies into risk assessment must be better informed decisions concerning potential adverse human health outcomes than is possible through conventional toxicity testing and risk assessment practices.

1.3 Predictive Toxicology

Predictive toxicology is the study of how toxic effects observed in model systems (or humans) can be used to predict pathogenesis, assess risk, and prevent human disease. Improving risk assessment is an essential aim of predictive toxicology. Information gaps and inconsistencies[98] include:

- the need for more toxicity screening data, data on effects in humans and on human exposure levels;
- information on the relevance of animal data to humans;
- exposure–response data (especially at low, environmental exposure levels);
- data on different routes of exposure;
- data on the effects of co-exposure to more than one chemical;
- data on human variability in susceptibility to toxicants; and
- information to resolve or explain inconsistencies in data from various animal models.

TGx methods are capable of addressing many, if not all, of these information gaps. However, fully integrating TGx technologies into predictive toxicology will require a coordinated effort to identify key datasets and to extract the essential data and information that can be used for predictive purposes. In this regard, there is a continuing need for better bioinformatics, statistical, and computational approaches and software to analyze TGx datasets as described in subsequent chapters. The resulting datasets should be deposited into curated public databases[99–101] with appropriate documentary information so that they can be readily analyzed and compared. There is also a need for public databases to facilitate sharing and use of TGx data and for tools to mine these databases.[1] The concerted efforts of government and the private sector are necessary to address these needs and propel the field forward.[102]

1.4 Systems Toxicology

A long-term goal of TGx is to achieve a systems-level understanding of the responses of organisms to toxicants.[1] This "systems toxicology" approach attempts to synthesize many different types of data in order to more completely understand of the biological response of a cell, organ, or organism to a particular toxicant, leading to the creation of predictive biomathematical models. Figure 1.1 illustrates the sequence of events between initial exposure to a toxicant and final disease outcome. After exposure, the absorption, distribution, metabolism, and excretion systems of the body control local concentrations of a chemical stressor in various body compartments. The internal dose in a given body compartment is affected by genetics through the involvement of specific forms of genes that code for various transporters and metabolizing enzymes. Mathematical models such as exposure models, physiologically based pharmacokinetic (PBPK) models, and biologically based dose–response (BBDR) models can be used to approximate these processes. PBPK models are a set of differential equations structured to provide a time course of a chemical's mass-balance disposition (wherein all inputs, outputs, and changes in total mass of the chemical are accounted for) in preselected anatomical compartments. BBDR models are dose–response models that are based on underlying biological processes. Once the target tissue is exposed to a local stressor, the cells respond and either adapt or undergo a toxic response; this process can be modeled using systems toxicology approaches. Finally, the disease outcome itself can be mimicked by genetic or chemically-induced models of particular diseases.[1]

1.4.1 Dosimetry

An advantage of HTS technologies is their ability to rapidly and efficiently test chemicals over a wide range of doses and to describe the dose–response curve at concentrations that better reflect real-world human exposure to environmental chemicals. Low-dose effects observed in HTS systems will need to be interpreted in the context of no-effect and effect-level doses reported in whole-animal toxicity tests. Preferably, low-dose effects would be addressed through the use of TGx methods in concert with conventional animal testing.[84,103]

Conventional toxicity studies of environmental chemicals generally express dose simply as external exposure levels, whereas the biologically relevant doses are those delivered to the target tissues. Recent advances in analytical technologies provide an opportunity to capture internal dosimetric information in conventional toxicology studies.[104] Integration of animal internal dosimetry findings with human biomonitoring data and with dose–response information from HTS and TGx technologies will provide information of great value in health risk assessment. Such opportunities are illustrated by recent efforts to develop "biomonitoring equivalents" (benchmarks for relating safety to concentrations of chemicals in biological

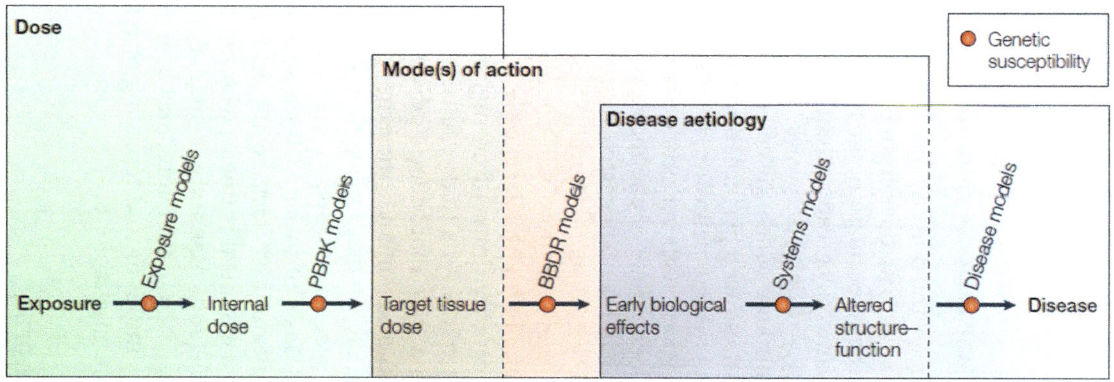

Figure 1.1 The sequence of events between initial exposure to a toxicant and final disease outcome; note that genetic susceptibility influences every level of toxicological analysis. The colored boxes show the type of toxicologically relevant information that can be obtained from each set of models. PBPK: physiologically based pharmacokinetic; BBDR: biologically based dose–response. Reproduced from Waters and Fostel, 2004.[1]

specimens, rather than to estimated daily intakes). Efforts to relate human exposure to doses in toxicity test systems and regulatory exposure standards will aid in development of approaches for efficient selection of environmental chemicals for detailed toxicological assessment.[105,106] Biomonitoring equivalents can inform selection of concentrations used in TGx and HTS, to ensure that they correspond to meaningful human risk assessment benchmarks, such as reference doses.[107] These concepts are discussed in detail by Thomas and Waters in chapter 5.

1.4.2 Adverse *vs.* Homeostatic Responses

The adage that "the dose makes the poison", credited to Paracelsus, implies distinguishing a nontoxic or subtoxic dose from a toxic dose of a chemical. The increasing focus on toxicity evaluations at low, environmentally relevant, doses must be accompanied by greater attention to what constitutes an "adverse" response. Initially, "toxicity pathways" are normal physiological and biochemical pathways through which the cell or organism is attempting to compensate for chemically induced perturbations. If chemicals tested are simply described as affecting toxicity pathways, it may be inferred that such effects must be adverse. However, such perturbations may represent normal homeostatic responses to interactions of chemicals with cellular systems at low doses.[106,108] Indeed, it has been suggested that the "exposure-adverse effect continuum" is in reality a "discontinuum",[108] because responses that remain within homeostatic tolerance limits will not result in adverse effects, even at sensitive life stages and in highly susceptible individuals. Research is needed on the range and tolerance limits of normal homeostatic responses and on the dose-dependent transitions that result in shifts from a normal state to an adaptive state to a state that reflects adverse effects.

An important aspect of risk assessment is the extrapolation of results from high-dose animal toxicity studies to human health risks, taking into account differences among animal species, strains, or sexes. In the absence of information on the biological basis of toxicity, risk assessors must rely upon default assumptions or "uncertainty" factors; for example, it may be assumed that the response to a chemical is linear, with no threshold. Debates surrounding the use of default assumption may delay regulatory decisions.[109] An important role for TGx, therefore, may be in testing the validity of currently used default assumptions and replacing or refining them,[110] as discussed by McMullen *et al.* in chapter 6.

1.4.3 Phenotypic Anchoring

Phenotypic anchoring[27,28] relates specific changes in gene-expression profiles to specific typically adverse effects observed in conventional toxicity testing. This process generally is based on histopathology (microscopic anatomical changes in tissues) or clinical chemistry (chemical analysis of fluids or tissues). Tissue histopathology is regarded as the most precise conventional indicator of toxicity. Ideally, tissues are examined by a board-certified pathologist, whose interpretation is recorded through the use of a controlled vocabulary that matches an image, as may be available in an online atlas of histopathological images. Alternatively, clinical chemistry indicators are appropriate for other parameters of toxicity, such as oxidative stress. Phenotypic anchoring is necessary in order to clearly distinguish between gene-expression changes that are associated with adverse effects and changes that are incidental or are associated with normal homeostatic responses (*i.e.*, physiological adjustments to maintain the equilibrium of internal systems).

1.4.4 Genetic Variation

The same level of exposure to a chemical may produce different biologic effects in different individuals. TGx offers opportunities to more fully characterize genetic variation in human susceptibility to toxic effects of chemicals, through studies on how individual variations in gene sequence or epigenetic modifications influence the response to chemicals. Variability in gene expression reflects individual variability as a result of genetic polymorphisms (the existence of alternative forms of a particular gene, which produce different phenotypes) or mutations in genes or in DNA that regulate gene expression, as well as other factors, such as epigenetic alterations. The bases for human variability in responses to toxicants can be explored by studying differences in levels of expression of a given gene, as well as differences in which genes are expressed.

1.4.5 Validation

For any screening test to be useful, it must be sensitive (able to detect the state being tested where it truly exists) and specific (responding only to the specific state being tested). A challenge for the design and validation of TGx screening tests is identification of a "gold standard"—an indicator of the true state of toxicity against which the screening test's sensitivity and specificity can be measured. Setting a relatively low threshold for a positive result provides greater sensitivity and lower specificity (that is, fewer false-negative and more false-positive results), whereas setting a high threshold provides lower sensitivity and higher specificity (that is, more false-negative and fewer false-positive results). In the case of environmental chemicals, false-positive results can lead to inappropriate regulatory restrictions on their use, and may divert resources from other public health issues, while false-negative results may delay or prevent responses needed to protect public health.

Therefore, caution is required in any transition from *in vivo* to predominantly *in vitro* testing. The focus must be on identifying true human health risks with a higher degree of confidence than is possible with conventional test systems. Confidence in HTS and computational profiling methods will depend on evaluation of the tests' relevance, reliability, sensitivity, and specificity. The National Research Council[110] has suggested approaches for validation of TGx technologies, and the OECD has provided principles and guidance for the validation of quantitative structure–activity relationships[111] and evidence-based toxicology.[112,113] The OECD *Guidance Document on the Design and Conduct of Chronic Toxicity and Carcinogenicity Studies* was released in September 2014[114] and current OECD activities related to molecular screening and toxicogenomics are described on the OECD website (www.oecd.org/chemicalsafety/testing/toxicogenomics.htm, 2015).

In 2012, the OECD launched a new program on the development of adverse outcome pathways.[115] An adverse outcome pathway is defined as "an analytical construct that describes a sequential chain of causally linked events at different levels of biological organization that lead to an adverse health or ecotoxicological effect". Adverse outcome pathways are the central components of a toxicological knowledge framework being constructed to support chemical risk assessment based on mechanistic reasoning, and are equally applicable to TGx or HTS investigations.[116]

1.4.6 Classes of Chemicals and Prototypic Compounds Studied to Date

The selection of classes of chemicals and prototype compounds for study by industry and government in TGx and HTS studies has been driven largely by the investigative focus. Thus the NIEHS NCT targeted NTP-tested hepatotoxicants and nephrotoxicants, including environmental chemicals and generic drugs for which there are substantial in-life, clinical, and histopathology data, as well as human exposure information.[117] The decision to do so was taken because of the requirement for phenotypic anchoring of the TGx data and the anticipated direct applicability of the resulting datasets to public health.

In candidate drug screening and subsequent TGx investigations, pharmaceutical companies have addressed mainly libraries of candidate drugs from their development pipelines, and have compared them to prototypic industrial chemicals and failed drugs for purposes of drug safety assessment.

For HTS, the EPA National Center for Computational Toxicology and ToxCast selected food-use pesticides and other chemicals for which there exist animal toxicology and human exposure information. The corresponding NTP compound collection for HTS consists of solvents, fire retardants, dyes, preservatives, plasticizers, therapeutic agents, inorganic and organic pollutants, drinking water disinfection byproducts, pesticides and natural products, partly based on the availability of toxicological data from standard tests of carcinogenicity, genotoxicity, immunotoxicity and/or reproductive and developmental toxicity. The NIH NCGC Pharmaceutical Collection (Ruili and colleagues, unpublished data) contains small molecules or compounds that have been evaluated in clinical trials and approved by the FDA in the United States or elsewhere.

1.4.7 Target Organs Studied

Both TGx and HTS efforts to date have emphasized studies on hepatotoxicity and hepatocarcinogenicity. Prototypic compounds have included direct- and indirect-acting genotoxins, nongenotoxic carcinogens, and noncarcinogens that target the liver.[117] The NTP has identified and studied numerous hepatoxicants and hepatocarcinogens in rats and mice. The rationale for the selection of the liver is that it is the principal metabolic organ of the body and that tissue and primary cells are readily available from mice, rats, and humans. The availability of several metabolically active rodent and human liver cell lines make *in vitro* TGx and HTS assays feasible. In chapter 3 Buick and Yauk address specifically the issue of metabolic competency of mammalian and human cell systems. Subsequent chapters in this book discuss predictive TGx in substantial detail with the liver as a model target organ.

1.5 Predictive Carcinogenicity

Predictive carcinogenicity focuses on cancer as the disease entity and conventional animal testing has dominated the field for more than four decades. It may be surprising to learn that the highly resource intensive 2-year rodent cancer bioassay was originally established by the National Cancer Institute (NCI) as a screening test to identify potential carcinogens for further analysis in human epidemiological studies.[118] The 2-year rodent NTP bioassay has evolved as the primary means and gold standard for determining the carcinogenic potential of a chemical by providing the dose–response information that is required for risk assessment. However, it is so costly and time consuming that as of 2007 only 1547 chemicals had been tested, including about 560 tested in the NTP carcinogen bioassay program.[119,120] As of early 2015 there are 582 technical reports available from the NTP carcinogen bioassay program.[121]

Initially, chemicals selected for testing in the NCI/NTP rodent carcinogenicity bioassay program were those suspected to be carcinogenic based on expert knowledge. Later, the selection process changed to include chemicals based on widespread human exposure and high production volume. To determine the dependence of positive results in the carcinogenicity bioassay on these chemical selection criteria, Fung *et al.*[122] analyzed the results from bioassays of 400 chemicals tested since 1995. Of 267 (67%) chemicals selected on the basis of suspect carcinogenicity, 187 (70%) were carcinogenic. Of 133 chemicals selected based only on exposure or production volume, the majority (80%) were not carcinogenic in animals, even when tested at the maximum tolerated dose. Of the total 400 chemicals, 210 (53%) caused tumors in at least one organ of one sex of one species (among male and female mice and rats). Only 92 chemicals (23%) were carcinogenic in both species, the criterion used by the International Agency for Research on Cancer (IARC) to determine whether a chemical is likely to be carcinogenic in humans. Based on this analysis, the authors concluded that fewer than 5–10% of the 75 000 chemicals in commercial use might be reasonably anticipated to be carcinogenic to humans. In fact, to date IARC has classified 116 agents as group 1 "known human carcinogens" (http://monographs.iarc.fr/, 2015).

The likelihood that a relatively small percentage of the hundreds of thousands of chemicals in industrial use will cause cancer favored the development and application of highly predictive short-term genotoxicity screening tests beginning in the late 1960s and early 1970s. Publications in the 1980s on short-term tests resulting from EPA's GENE-TOX program[15–19] defined the basic protocols and performance characteristics of these tests and helped to begin the gradual process of reducing their numbers from more than 200 to the handful of guideline tests used today.

Among the early short-term bioassays, the Ames test for gene mutation (point mutation) in bacteria held the greatest promise as a predictor of carcinogenicity.[123,124] This test emerged as the screening test of choice for potential genotoxic carcinogens. However, the Ames test could not be used blindly, because the bacterial standard tester strains did not respond positively to certain classes of carcinogens (*e.g.*, inorganic metal and halogenated organic compounds) and displayed poor specificity for others (*e.g.*, nitrogen- and sulfur-containing organic compounds).[125] Despite these limitations, the test was so successful that in the early 1990s, some scientists tended to categorize as nongenotoxic ("Ames negative") chemicals that gave positive results in other assays for various types of genetic damage.[126]

Butterworth[127] emphasized that the "primary biological activity of the [genotoxic] chemical or a metabolite is alteration of the information encoded in the DNA ... point mutations, insertions, deletions or changes in chromosome structure or number ... Nongenotoxic chemicals ... may yield genotoxic events as a secondary result of other induced toxicity, such as forced cellular growth, but their primary action does not involve reactivity with the DNA". From a regulatory perspective, the gradual recognition of nongenotoxic mechanisms of carcinogenesis[128] complicated the established relationship between genotoxicity and carcinogenicity and also challenged the conventional interpretation of rodent carcinogenicity results in terms of relevance to human cancer.[63] Because of the default assumption in regulatory decision-making regarding the presumed linearity of the dose–response curve for genotoxic carcinogens, the classification of carcinogens as genotoxic or nongenotoxic became an essential but highly debatable component of cancer risk assessment. As science advances with new technologies, it is important to recognize that the makeup of the reference set of chemicals and databases can dramatically influence chemical classification. Because of the purely statistical approaches frequently used in interpreting the results of highly parallel and high-content assays, these technologies can be very susceptible to misinterpretation due to inappropriate composition of the training sets of chemicals.[129]

We have learned over the years that of all compounds tested in rodents, about half are carcinogenic,[130] and roughly half of these are putatively nongenotoxic.[131] Rat liver appears to be particularly sensitive to nongenotoxic carcinogens, and has been studied extensively to understand the potential mechanisms involved. However, it is clear that both genotoxic and nongenotoxic chemicals induce cancer in a variety of target sites in rodents.

The eight most frequent target sites in both rats and mice are liver, lung, mammary gland, stomach, vascular system, kidney, hematopoietic system, and urinary bladder. There are some species differences in results for particular target organs, including the liver, Zymbal's gland, and kidney.[132] However, there is no support for any systematic inter-species differences in tissue distribution and pharmacokinetics between genotoxic and nongenotoxic agents, nor for the idea that these two categories of agents induce tumors in different target organs.[132] The central dogma in the etiology of carcinogenesis is that DNA damage resulting in mutation and other events that enhance cell proliferation are both required. Many carcinogens are cytotoxic and induce compensatory cell proliferation. Thus in predicting carcinogenicity it is appropriate to further study this phenomenon *in vitro* and comparatively *in vivo* using the appropriate TGx methods to identify relevant and predictive pathways.

The investigations *in vivo*, reviewed in Waters *et al.*, 2010[129] and updated in chapter 4 by Auerbach, have identified cancer-relevant gene signatures or biomarkers that discriminate between direct and indirect genotoxic carcinogens, nongenotoxic carcinogens, and noncarcinogens. In an early series of studies by Ellinger-Ziegelbauer and colleagues,[63,133,134] a strong DNA-damage response at the gene-expression level suggested direct DNA modification, whereas increased expression of genes involved in cell-cycle progression appeared characteristic of indirect-acting agents. Metabolism genes were prominently represented among gene-expression signatures that discriminated nongenotoxic modes of action (*e.g.*, cytotoxicity and regenerative proliferation, xenobiotic receptor agonists, peroxisome-proliferator-activated receptors, or hormone-mediated processes[77,135]). The preponderance of accumulated evidence suggested that gene-expression profiles reflect underlying modes or mechanisms of action and are therefore useful in predicting chemical carcinogenicity in rodents, especially in conjunction with conventional short-term tests for gene mutation and other forms of DNA damage.[136]

A more difficult task is distinguishing nongenotoxic carcinogens from noncarcinogens. Evidence to date suggests that some modes of action of nongenotoxic carcinogenicity such as the induction of oxidative stress exhibit definitive signatures as early as 24 h following single dosing in animals.[77,135,137] Furthermore, it seems clear that false-positive TGx signatures may be resolved by repeat dosing up to 28 days. Because of multiple modes of action, the resolution of classification of nongenotoxic carcinogens *vs.* noncarcinogens cannot be accomplished without efforts to clarify what combinations of marker gene sets related to specific pathophysiological processes of carcinogenesis. This will take time and highly directed efforts.

The majority of *in vivo* studies reviewed in Waters *et al.*, 2010[129] were performed in the liver, with the notable exception of studies by Thomas and colleagues[79,138] in mouse lung. It is important to extend these investigations to other target organs and to identify within these organs the target cell populations from which tumors develop. For TGx studies to be broadly predictive, *in vivo* studies should be performed simultaneously in several relevant metabolically active target organs. In such studies, it is important to distinguish between a tissue carcinogen and a tissue toxin, since not even all hepatotoxicants cause liver cancer. Five tissue sites (liver, lung, mammary gland, kidney, and the hematopoietic system) account for the positive responses for about half of the chemicals identified by the NTP as carcinogens and positive results have been observed at 24 tissue sites for five or more chemicals in at least one species and sex.[138] Developing gene-expression biomarkers for each of the top five tumor sites in mice and rats should provide an efficient means to prioritize chemicals for further testing. As suggested by Thomas *et al.*,[138] it may be useful in the longer term to develop biomarkers for each of the 24 main target tissues, which may facilitate replacement of the rodent cancer bioassay. It may also be of value to use TGx methods to better understand the mechanistic basis for species differences between rats and mice in particular target organs, including the liver, Zymbal's gland, and kidney.[132]

Thomas *et al.*[138] advocated applying the TGx approach in the preclinical phase of drug and chemical development to discriminate compounds likely to be human carcinogens. Such testing could provide an assessment of product safety earlier in the development pipeline, leading to substantial monetary savings and reduced time to market. The prevalence of potential nongenotoxic carcinogens in the drug-development pipeline has been one of the primary motivators for the pharmaceutical industry to develop TGx approaches for predictive carcinogenicity. The work of Fielden *et al.*[135] and Nie *et al.*[77] suggests that transcription profiling in appropriate target organs *in vivo* after short-term treatment (up to 14 days) has the potential to predict putative non-DNA-reactive mechanisms. Indeed, it may be possible to use TGx methods to exclude from further consideration DNA-reactive mechanisms for compounds for which positive results are observed only at high concentrations in *in vitro* gene mutation or chromosome damage assays. However, the results of the Fielden and Nie studies also raise the question of why the signatures of nongenotoxic carcinogenicity were so different between the studies and yet apparently predictive in both. Clearly when such predictive approaches are combined with standardized test procedures in prospective interlaboratory validation studies, their accuracy and potential utility in carcinogenicity evaluation can be better ascertained.[139,140]

Many types of commercial and industrial chemicals are not required to be tested for carcinogenicity unless evidence for adverse health effects is obtained. For those that do require further testing, TGx approaches would seem particularly valuable when used together with range-finding toxicity (14-day and 90-day) studies, as currently performed in conjunction with the rodent carcinogenicity bioassay. A report by Auerbach *et al.*[141] focused primarily on defining classification models with the best cross-validation based on results from 2-, 14-, and 90-day studies. Most of the TGx studies reviewed by Waters *et al.*[129] identified liver carcinogenicity signatures from animals exposed to chemicals for 28 days or less, a duration that is referred to in traditional toxicology as "subacute". Auerbach *et al.*[141] hypothesized that exposures up to 90 days would accentuate the expression of genes related to carcinogenic activity and therefore allow the models to achieve a higher degree of certainty when making predictions. Auerbach *et al.*[141] suggested that longer exposure durations would limit the influence of "mode of action" genes and allow for better identification of predictive genes with biology related to processes involved in the formation of neoplasms that are typically seen subsequent to the primary toxicity. Auerbach *et al.*[141] also noted that the idea of a shared precancerous biology (that is independent of a specific chemical challenge) is not unreasonable, since the process of cancer manifestation is a continuum, and most types of cancer share a degree of universal biology that is manifested in their gene expression.[142] Auerbach *et al.*[141] concluded that the duration of exposure is the primary factor affecting the utility of the models, and that a 90-day exposure provides superior data.

Indeed, carcinogenesis is a complex and protracted multi-step process. It could easily be argued that a focus on transcription profiling with exposures of up to 14 days or even 90 days cannot hope to predict the outcome of a 2-year rodent carcinogenicity bioassay. To better understand what transpires in the complex process of carcinogenesis, it would seem most appropriate to "profile to the phenotype"; *i.e.*, to perform transcription profiling in conjunction with the pathological evaluation of target tissues (even if this were to be done using archival material, such as is available in the extensive NTP archives). Admittedly, this is a diagnostic (or retrospective) approach, as opposed to prognostic or predictive carcinogenicity methodology. However, the information potentially to be gained could lead to the ability to predict much earlier based on "shared precancerous biology" whether neoplastic lesions will progress to cancer over the lifetime of the animal. If data are encouraging, similar information developed based on studies of archived human tissues could readily be performed.

A multi-laboratory project coordinated by the Health and Environmental Sciences Institute (HESI) Committee on the Application of Genomics in Mechanism-based Risk Assessment evaluated gene-expression profiles of TK6 cells treated with model genotoxic agents using a targeted high-density RT-PCR. This study adds to the increasing body of evidence indicating that TGx analysis of cellular stress responses provides insight into mechanisms of action of genotoxicants.[63] The reproducibility of data across collaborating laboratories indicates that expression analysis of a relevant gene set is capable of distinguishing compounds that cause DNA adducts or double strand breaks from those that interfere with mitotic spindle function or that cause chromosome damage as a consequence of cytotoxicity.

TGx studies of chemicals that are both rodent and human carcinogens could identify biomarkers with more direct relevance to human health.[138] Compounds that do not produce positive test results in the conventional genotoxicity assays and that do not exhibit biomarkers of genotoxicity in TGx methods are very unlikely to pose a genotoxic carcinogenic risk to humans. The same cannot be said for putative nongenotoxic carcinogens that are identified through the use of TGx methods. However, it should be possible in such cases to use TGx methods to characterize their likely modes of action by comparison with previously well-studied chemicals, as demonstrated by Fielden and colleagues[90,135] and Uehara and colleagues,[87,137] and, with more experience, to predict relevance to humans.

The potential of -omics technologies to explore transcriptional regulation (including epigenetics and miRNA) as well as downstream events (proteomics and metabolomics) in evaluating mechanisms of genotoxicity and carcinogenicity must also be investigated.[75,143] No single organization has the resources to accomplish all of this independently. Therefore, collaborative efforts that include scientists from academia, industry, and regulatory agencies, such as the HESI Genomics Committee, the Critical Path Initiative in the United States, and the Innovative Medicines Initiative in Europe, are essential for developing standardized testing protocols and critically needed reference data.[63] If the TGx approach proves to be more broadly applicable through such efforts, it has the potential to become an efficient and economical alternative to the rodent cancer bioassay, potentially reducing the use of experimental animals while increasing the efficiency of predictive carcinogenicity. Clearly, the investigations discussed here and to be discussed in subsequent chapters have opened the door to a paradigm shift in chemical safety testing and health risk assessment.

Acknowledgements

The author acknowledges funding from Health Canada Existing Substances Risk Assessment Bureau, to Integrated Laboratory Systems (ILS), Inc., and significant contributions of colleagues at ILS, including Marcus Jackson, Isabel Lea, H. Frank Stack, and Susan Dakin, in the preparation of background information (Interim Report, *Approach to Development and Integration of Toxicogenomics and High-Throughput Screening in Support of Human Health Risk Assessment*, submitted January 24, 2011, (Health Canada Contract Reference Number 4500247514)) included in part in this chapter by permission.

References

1. M. D. Waters and J. M. Fostel, Toxicogenomics and systems toxicology: aims and prospects, *Nat. Rev. Genet.*, 2004, **5**, 936–948.

2. M. B. Black, B. B. Parks, L. Pluta, T. M. Chu, B. C. Allen, R. D. Wolfinger and R. S. Thomas, Comparison of microarrays and RNA-seq for gene expression analyses of dose-response experiments, *Toxicol. Sci.*, 2014, **137**, 385–403.
3. D. J. Dix, K. A. Houck, M. T. Martin, A. M. Richard, R. W. Setzer and R. J. Kavlock, The ToxCast program for prioritizing toxicity testing of environmental chemicals, *Toxicol. Sci.*, 2007, **95**, 5–12.
4. C. P. Austin, R. J. Kavlock and R. Tice, Tox21: Putting a Lens on the Vision of Toxicity Testing in the 21st Century, Office of Research and Development, National Center for Computational Toxicology, 2008.
5. H. H. Li, D. R. Hyduke, R. Chen, P. Heard, C. L. Yauk, J. Aubrecht and A. J. Fornace Jr., Development of a toxicogenomics signature for genotoxicity using a dose-optimization and informatics strategy in human cells, *Environ. Mol. Mutagen.*, 2015, **56**, 505–519.
6. E. F. Nuwaysir, M. Bittner, J. Trent, J. C. Barrett and C. A. Afshari, Microarrays and toxicology: the advent of toxicogenomics, *Mol. Carcinog.*, 1999, **24**, 153–159.
7. D. Anderson, M. Sorsa and M. D. Waters, The parallelogram approach in studies of genotoxic effects, *Mutat. Res.*, 1994, **313**, 101–115.
8. F. H. Sobels, Some problems associated with the testing for environmental mutagens and a perspective for studies in "comparative mutagenesis", *Mutat. Res.*, 1977, **46**, 245–260.
9. F. H. Sobels, Models and assumptions underlying genetic risk assessment, *Mutat. Res.*, 1989, **212**, 77–89.
10. A. F. Jackson, A. Williams, L. Recio, M. D. Waters, I. B. Lambert and C. L. Yauk, Case study on the utility of hepatic global gene expression profiling in the risk assessment of the carcinogen furan, *Toxicol. Appl. Pharmacol.*, 2014, **274**, 63–77.
11. N. L. Chepelev, I. D. Moffat, S. Labib, J. Bourdon-Lacombe, B. Kuo, J. K. Buick, F. Lemieux, A. I. Malik, S. Halappanavar, A. Williams and C. L. Yauk, Integrating toxicogenomics into human health risk assessment: Lessons learned from the benzo[a]pyrene case study, *Crit. Rev. Toxicol.*, 2015, **45**, 44–52.
12. I. Moffat, N. L. Chepelev, S. Labib, J. Bourdon-Lacombe, B. Kuo, J. K. Buick, F. Lemieux, A. Williams, S. Halappanavar, A. I. Malik, M. Luijten, J. Aubrecht, D. R. Hyduke, A. J. Fornace Jr., C. D. Swartz, L. Recio and C. L. Yauk, Comparison of toxicogenomics and traditional approaches to inform mode of action and points of departure in human health risk assessment of benzo[a]pyrene in drinking water, *Crit. Rev. Toxicol.*, 2015, **45**, 1–43.
13. NRC, *Toxicity Testing in the 21st Century: a Vision and a Strategy*, The National Academies Press, Washington, D.C., 2007.
14. M. E. Andersen and D. Krewski, Toxicity testing in the 21st century: bringing the vision to life, *Toxicol. Sci.*, 2009, **107**, 324–330.
15. A. E. Auletta, L. D. Kier and A. D. Mitchell, Current status of the Gene-Tox Program, *Prog. Clin. Biol. Res.*, 1990, **340D**, 273–281.

16. D. Brusick and A. Auletta, Developmental status of bioassays in genetic toxicology. A report of Phase II of the U.S. Environmental Protection Agency Gene-Tox Program, *Mutat. Res.*, 1985, **153**, 1–10.
17. S. Nesnow and H. Bergman, An analysis of the Gene-Tox Carcinogen Data Base, *Mutat. Res.*, 1988, **205**, 237–253.
18. V. A. Ray, L. D. Kier, K. L. Kannan, R. T. Haas, A. E. Auletta, J. S. Wassom, S. Nesnow and M. D. Waters, An approach to identifying specialized batteries of bioassays for specific classes of chemicals: class analysis using mutagenicity and carcinogenicity relationships and phylogenetic concordance and discordance patterns. 1. Composition and analysis of the overall data base. A report of phase II of the U.S. Environmental Protection Agency Gene-Tox Program, *Mutat. Res.*, 1987, **185**, 197–241.
19. M. D. Waters and A. Auletta, The GENE-TOX program: genetic activity evaluation, *J. Chem. Inf. Comput. Sci.*, 1981, **21**, 35–38.
20. A. N. Rowan, *Alternatives: interaction between science and animal welfare*, Mary Ann Leibert, Inc., 1983.
21. F. Sanger, S. Nicklen and A. R. Coulson, DNA sequencing with chain-terminating inhibitors, *Proc. Natl. Acad. Sci. U. S. A.*, 1977, **74**, 5463–5467.
22. J. C. Venter, M. D. Adams, E. W. Myers, P. W. Li, R. J. Mural, G. G. Sutton, H. O. Smith, M. Yandell, C. A. Evans, R. A. Holt, J. D. Gocayne, P. Amanatides, R. M. Ballew, D. H. Huson, J. R. Wortman, Q. Zhang, C. D. Kodira, X. H. Zheng, L. Chen, M. Skupski, G. Subramanian, P. D. Thomas, J. Zhang, G. L. Gabor Miklos, C. Nelson, S. Broder, A. G. Clark, J. Nadeau, V. A. McKusick, N. Zinder, A. J. Levine, R. J. Roberts, M. Simon, C. Slayman, M. Hunkapiller, R. Bolanos, A. Delcher, I. Dew, D. Fasulo, M. Flanigan, L. Florea, A. Halpern, S. Hannenhalli, S. Kravitz, S. Levy, C. Mobarry, K. Reinert, K. Remington, J. Abu-Threideh, E. Beasley, K. Biddick, V. Bonazzi, R. Brandon, M. Cargill, I. Chandramouliswaran, R. Charlab, K. Chaturvedi, Z. Deng, V. Di Francesco, P. Dunn, K. Eilbeck, C. Evangelista, A. E. Gabrielian, W. Gan, W. Ge, F. Gong, Z. Gu, P. Guan, T. J. Heiman, M. E. Higgins, R. R. Ji, Z. Ke, K. A. Ketchum, Z. Lai, Y. Lei, Z. Li, J. Li, Y. Liang, X. Lin, F. Lu, G. V. Merkulov, N. Milshina, H. M. Moore, A. K. Naik, V. A. Narayan, B. Neelam, D. Nusskern, D. B. Rusch, S. Salzberg, W. Shao, B. Shue, J. Sun, Z. Wang, A. Wang, X. Wang, J. Wang, M. Wei, R. Wides, C. Xiao, C. Yan, A. Yao, J. Ye, M. Zhan, W. Zhang, H. Zhang, Q. Zhao, L. Zheng, F. Zhong, W. Zhong, S. Zhu, S. Zhao, D. Gilbert, S. Baumhueter, G. Spier, C. Carter, A. Cravchik, T. Woodage, F. Ali, H. An, A. Awe, D. Baldwin, H. Baden, M. Barnstead, I. Barrow, K. Beeson, D. Busam, A. Carver, A. Center, M. L. Cheng, L. Curry, S. Danaher, L. Davenport, R. Desilets, S. Dietz, K. Dodson, L. Doup, S. Ferriera, N. Garg, A. Gluecksmann, B. Hart, J. Haynes, C. Haynes, C. Heiner, S. Hladun, D. Hostin, J. Houck, T. Howland, C. Ibegwam, J. Johnson, F. Kalush, L. Kline, S. Koduru, A. Love, F. Mann, D. May, S. McCawley, T. McIntosh, I. McMullen,

M. Moy, L. Moy, B. Murphy, K. Nelson, C. Pfannkoch, E. Pratts, V. Puri, H. Qureshi, M. Reardon, R. Rodriguez, Y. H. Rogers, D. Romblad, B. Ruhfel, R. Scott, C. Sitter, M. Smallwood, E. Stewart, R. Strong, E. Suh, R. Thomas, N. N. Tint, S. Tse, C. Vech, G. Wang, J. Wetter, S. Williams, M. Williams, S. Windsor, E. Winn-Deen, K. Wolfe, J. Zaveri, K. Zaveri, J. F. Abril, R. Guigo, M. J. Campbell, K. V. Sjolander, B. Karlak, A. Kejariwal, H. Mi, B. Lazareva, T. Hatton, A. Narechania, K. Diemer, A. Muruganujan, N. Guo, S. Sato, V. Bafna, S. Istrail, R. Lippert, R. Schwartz, B. Walenz, S. Yooseph, D. Allen, A. Basu, J. Baxendale, L. Blick, M. Caminha, J. Carnes-Stine, P. Caulk, Y. H. Chiang, M. Coyne, C. Dahlke, A. Mays, M. Dombroski, M. Donnelly, D. Ely, S. Esparham, C. Fosler, H. Gire, S. Glanowski, K. Glasser, A. Glodek, M. Gorokhov, K. Graham, B. Gropman, M. Harris, J. Heil, S. Henderson, J. Hoover, D. Jennings, C. Jordan, J. Jordan, J. Kasha, L. Kagan, C. Kraft, A. Levitsky, M. Lewis, X. Liu, J. Lopez, D. Ma, W. Majoros, J. McDaniel, S. Murphy, M. Newman, T. Nguyen, N. Nguyen, M. Nodell, S. Pan, J. Peck, M. Peterson, W. Rowe, R. Sanders, J. Scott, M. Simpson, T. Smith, A. Sprague, T. Stockwell, R. Turner, E. Venter, M. Wang, M. Wen, D. Wu, M. Wu, A. Xia, A. Zandieh and X. Zhu, The sequence of the human genome, *Science*, 2001, **291**, 1304–1351.

23. J. Shendure, G. J. Porreca, N. B. Reppas, X. Lin, J. P. McCutcheon, A. M. Rosenbaum, M. D. Wang, K. Zhang, R. D. Mitra and G. M. Church, Accurate multiplex polony sequencing of an evolved bacterial genome, *Science*, 2005, **309**, 1728–1732.
24. *In Sequence*, Pacific Biosciences, 8 June 2010.
25. H. K. Hamadeh, P. R. Bushel, S. Jayadev, O. DiSorbo, L. Bennett, L. Li, R. Tennant, R. Stoll, J. C. Barrett, R. S. Paules, K. Blanchard and C. A. Afshari, Prediction of compound signature using high density gene expression profiling, *Toxicol. Sci.*, 2002, **67**, 232–240.
26. H. K. Hamadeh, P. R. Bushel, S. Jayadev, K. Martin, O. DiSorbo, S. Sieber, L. Bennett, R. Tennant, R. Stoll, J. C. Barrett, K. Blanchard, R. S. Paules and C. A. Afshari, Gene expression analysis reveals chemical-specific profiles, *Toxicol. Sci.*, 2002, **67**, 219–231.
27. R. Paules, Phenotypic anchoring: linking cause and effect, *Environ. Health Perspect.*, 2003, **111**, A338–339.
28. R. W. Tennant, The National Center for Toxicogenomics: using new technologies to inform mechanistic toxicology, *Environ. Health Perspect.*, 2002, **110**, A8–10.
29. S. P. Fodor, R. P. Rava, X. C. Huang, A. C. Pease, C. P. Holmes and C. L. Adams, Multiplexed biochemical assays with biological chips, *Nature*, 1993, **364**, 555–556.
30. R. J. Lipshutz, S. P. Fodor, T. R. Gingeras and D. J. Lockhart, High density synthetic oligonucleotide arrays, *Nat. Genet.*, 1999, **21**, 20–24.
31. A. C. Pease, D. Solas, E. J. Sullivan, M. T. Cronin, C. P. Holmes and S. P. Fodor, Light-generated oligonucleotide arrays for rapid DNA sequence analysis, *Proc. Natl. Acad. Sci. U. S. A.*, 1994, **91**, 5022–5026.

32. M. Mann and O. N. Jensen, Proteomic analysis of post-translational modifications, *Nat. Biotechnol.*, 2003, **21**, 255–261.
33. R. Tonge, J. Shaw, B. Middleton, R. Rowlinson, S. Rayner, J. Young, F. Pognan, E. Hawkins, I. Currie and M. Davison, Validation and development of fluorescence two-dimensional differential gel electrophoresis proteomics technology, *Proteomics*, 2001, **1**, 377–396.
34. F. Von Eggeling, A. Gawriljuk, W. Fiedler, G. Ernst, U. Claussen, J. Klose and I. Romer, Fluorescent dual colour 2D-protein gel electrophoresis for rapid detection of differences in protein pattern with standard image analysis software, *Int. J. Mol. Med.*, 2001, **8**, 373–377.
35. J. R. Yates 3rd, S. F. Morgan, C. L. Gatlin, P. R. Griffin and J. K. Eng, Method to compare collision-induced dissociation spectra of peptides: potential for library searching and subtractive analysis, *Anal. Chem.*, 1998, **70**, 3557–3565.
36. M. P. Washburn, R. Ulaszek, C. Deciu, D. M. Schieltz and J. R. Yates 3rd, Analysis of quantitative proteomic data generated *via* multidimensional protein identification technology, *Anal. Chem.*, 2002, **74**, 1650–1657.
37. D. A. Wolters, M. P. Washburn and J. R. Yates 3rd, An automated multidimensional protein identification technology for shotgun proteomics, *Anal. Chem.*, 2001, **73**, 5683–5690.
38. J. C. Lindon, J. K. Nicholson, E. Holmes, H. Antti, M. E. Bollard, H. Keun, O. Beckonert, T. M. Ebbels, M. D. Reily, D. Robertson, G. J. Stevens, P. Luke, A. P. Breau, G. H. Cantor, R. H. Bible, U. Niederhauser, H. Senn, G. Schlotterbeck, U. G. Sidelmann, S. M. Laursen, A. Tymiak, B. D. Car, L. Lehman-McKeeman, J. M. Colet, A. Loukaci and C. Thomas, Contemporary issues in toxicology the role of metabonomics in toxicology and its evaluation by the COMET project, *Toxicol. Appl. Pharmacol.*, 2003, **187**, 137–146.
39. J. C. Lindon, J. K. Nicholson, E. Holmes, H. C. Keun, A. Craig, J. T. Pearce, S. J. Bruce, N. Hardy, S. A. Sansone, H. Antti, P. Jonsson, C. Daykin, M. Navarange, R. D. Beger, E. R. Verheij, A. Amberg, D. Baunsgaard, G. H. Cantor, L. Lehman-McKeeman, M. Earll, S. Wold, E. Johansson, J. N. Haselden, K. Kramer, C. Thomas, J. Lindberg, I. Schuppe-Koistinen, I. D. Wilson, M. D. Reily, D. G. Robertson, H. Senn, A. Krotzky, S. Kochhar, J. Powell, F. van der Ouderaa, R. Plumb, H. Schaefer and M. Spraul, Summary recommendations for standardization and reporting of metabolic analyses, *Nat. Biotechnol.*, 2005, **23**, 833–838.
40. W. A. Colburn, Biomarkers in drug discovery and development: from target identification through drug marketing, *J. Clin. Pharmacol.*, 2003, **43**, 329–341.
41. R. Frank and R. Hargreaves, Clinical biomarkers in drug discovery and development, *Nat. Rev. Drug Discovery*, 2003, **2**, 566–580.
42. J. A. Wagner, Overview of biomarkers and surrogate endpoints in drug development, *Dis. Markers*, 2002, **18**, 41–46.

43. D. A. Pereira and J. A. Williams, Origin and evolution of high throughput screening, *Br. J. Pharmacol.*, 2007, **152**, 53–61.
44. N. Bhogal, C. Grindon, R. Combes and M. Balls, Toxicity testing: creating a revolution based on new technologies, *Trends Biotechnol.*, 2005, **23**, 299–307.
45. A. F. Fliri, W. T. Loging, P. F. Thadeio and R. A. Volkmann, Analysis of drug-induced effect patterns to link structure and side effects of medicines, *Nat. Chem. Biol.*, 2005, **1**, 389–397.
46. R. Kikkawa, M. Fujikawa, T. Yamamoto, Y. Hamada, H. Yamada and I. Horii, *In vivo* hepatotoxicity study of rats in comparison with *in vitro* hepatotoxicity screening system, *J. Toxicol. Sci.*, 2006, **31**, 23–34.
47. S. Burns, J. Travers, I. Collins, M. G. Rowlands, Y. Newbatt, N. Thompson, M. D. Garrett, P. Workman and W. Aherne, Identification of small-molecule inhibitors of protein kinase B (PKB/AKT) in an AlphaScreenTM high-throughput screen, *J. Biomol. Screening*, 2006, **11**, 822–827.
48. K. Sudo, K. Yamaji, K. Kawamura, T. Nishijima, N. Kojima, K. Aibe, K. Shimotohno and Y. Shimizu, High-throughput screening of low molecular weight NS3-NS4A protease inhibitors using a fluorescence resonance energy transfer substrate, *Antiviral Chem. Chemother.*, 2005, **16**, 385–392.
49. S. Swaney, M. McCroskey, D. Shinabarger, Z. Wang, B. A. Turner and C. N. Parker, Characterization of a high-throughput screening assay for inhibitors of elongation factor p and ribosomal peptidyl transferase activity, *J. Biomol. Screening*, 2006, **11**, 736–742.
50. M. Allen, J. Reeves and G. Mellor, High throughput fluorescence polarization: a homogeneous alternative to radioligand binding for cell surface receptors, *J. Biomol. Screening*, 2000, **5**, 63–69.
51. J. Xu, X. Wang, B. Ensign, M. Li, L. Wu and A. Guia, Ion-channel assay technologies: quo vadis?, *Drug Discovery Today*, 2001, **6**, 1278–1287.
52. G. J. Parker, T. L. Law, F. J. Lenoch and R. E. Bolger, Development of high throughput screening assays using fluorescence polarization: nuclear receptor-ligand-binding and kinase/phosphatase assays, *J. Biomol. Screening*, 2000, **5**, 77–88.
53. C. H. Kenny, W. Ding, K. Kelleher, S. Benard, E. G. Dushin, A. G. Sutherland, L. Mosyak, R. Kriz and G. Ellestad, Development of a fluorescence polarization assay to screen for inhibitors of the FtsZ/ZipA interaction, *Anal. Biochem.*, 2003, **323**, 224–233.
54. C. Chambers, F. Smith, C. Williams, S. Marcos, Z. H. Liu, P. Hayter, G. Ciaramella, W. Keighley, P. Gribbon and A. Sewing, Measuring intracellular calcium fluxes in high throughput mode, *Comb. Chem. High Throughput Screening*, 2003, **6**, 355–362.
55. I. I. Kariv, M. E. Stevens, D. L. Behrens and K. R. Oldenburg, High Throughput Quantitation of cAMP Production Mediated by Activation of Seven Transmembrane Domain Receptors, *J. Biomol. Screening*, 1999, **4**, 27–32.

56. V. Beck, A. Pfitscher and A. Jungbauer, GFP-reporter for a high throughput assay to monitor estrogenic compounds, *J. Biochem. Biophys. Methods*, 2005, **64**, 19–37.
57. X. Li, F. Shen, Y. Zhang, J. Zhu, L. Huang and Q. Shi, Functional characterization of cell lines for high-throughput screening of human neuromedin U receptor subtype 2 specific agonists using a luciferase reporter gene assay, *Eur. J. Pharm. Biopharm.*, 2007, **67**, 284–292.
58. J. C. Yarrow, G. Totsukawa, G. T. Charras and T. J. Mitchison, Screening for cell migration inhibitors via automated microscopy reveals a Rho-kinase inhibitor, *Chem. Biol.*, 2005, **12**, 385–395.
59. U. S. Eggert, A. A. Kiger, C. Richter, Z. E. Perlman, N. Perrimon, T. J. Mitchison and C. M. Field, Parallel chemical genetic and genome-wide RNAi screens identify cytokinesis inhibitors and targets, *PLoS Biol*, 2004, **2**, e379.
60. J. Bradley, J. Gill, F. Bertelli, S. Letafat, R. Corbau, P. Hayter, P. Harrison, A. Tee, W. Keighley, M. Perros, G. Ciaramella, A. Sewing and C. Williams, Development and automation of a 384-well cell fusion assay to identify inhibitors of CCR5/CD4-mediated HIV virus entry, *J. Biomol. Screening*, 2004, **9**, 516–524.
61. P. E. Brandish, C. S. Chiu, J. Schneeweis, N. J. Brandon, C. L. Leech, O. Kornienko, E. M. Scolnick, B. Strulovici and W. Zheng, A cell-based ultra-high-throughput screening assay for identifying inhibitors of D-amino acid oxidase, *J. Biomol. Screening*, 2006, **11**, 481–487.
62. F. Wunder, J. P. Stasch, J. Hutter, C. Alonso-Alija, J. Huser and E. Lohrmann, A cell-based cGMP assay useful for ultra-high-throughput screening and identification of modulators of the nitric oxide/cGMP pathway, *Anal. Biochem.*, 2005, **339**, 104–112.
63. H. Ellinger-Ziegelbauer, J. Aubrecht, J. C. Kleinjans and H. J. Ahr, Application of toxicogenomics to study mechanisms of genotoxicity and carcinogenicity, *Toxicol. Lett.*, 2009, **186**, 36–44.
64. H. Ellinger-Ziegelbauer, J. M. Fostel, C. Aruga, D. Bauer, E. Boitier, S. Deng, D. Dickinson, A. C. Le Fevre, A. J. Fornace Jr., O. Grenet, Y. Gu, J. C. Hoflack, M. Shiiyama, R. Smith, R. D. Snyder, C. Spire, G. Tanaka and J. Aubrecht, Characterization and interlaboratory comparison of a gene expression signature for differentiating genotoxic mechanisms, *Toxicol. Sci.*, 2009, **110**, 341–352.
65. H. C. Keun, T. M. Ebbels, M. E. Bollard, O. Beckonert, H. Antti, E. Holmes, J. C. Lindon and J. K. Nicholson, Geometric trajectory analysis of metabolic responses to toxicity can define treatment specific profiles, *Chem. Res. Toxicol.*, 2004, **17**, 579–587.
66. R. S. Thomas, H. J. Clewell 3rd, B. C. Allen, S. C. Wesselkamper, N. C. Wang, J. C. Lambert, J. K. Hess-Wilson, Q. J. Zhao and M. E. Andersen, Application of transcriptional benchmark dose values in quantitative cancer and noncancer risk assessment, *Toxicol. Sci.*, 2011, **120**, 194–205.

67. R. S. Thomas, M. A. Philbert, S. S. Auerbach, B. A. Wetmore, M. J. Devito, I. Cote, J. C. Rowlands, M. P. Whelan, S. M. Hays, M. E. Andersen, M. E. Meek, L. W. Reiter, J. C. Lambert, H. J. Clewell 3rd, M. L. Stephens, Q. J. Zhao, S. C. Wesselkamper, L. Flowers, E. W. Carney, T. P. Pastoor, D. D. Petersen, C. L. Yauk and A. Nong, Incorporating new technologies into toxicity testing and risk assessment: moving from 21st century vision to a data-driven framework, *Toxicol. Sci.*, 2013, **136**, 4–18.
68. R. S. Thomas, S. C. Wesselkamper, N. C. Wang, Q. J. Zhao, D. D. Petersen, J. C. Lambert, I. Cote, L. Yang, E. Healy, M. B. Black, H. J. Clewell 3rd, B. C. Allen and M. E. Andersen, Temporal concordance between apical and transcriptional points of departure for chemical risk assessment, *Toxicol. Sci.*, 2013, **134**, 180–194.
69. M. J. Aardema and J. T. MacGregor, Toxicology and genetic toxicology in the new era of "toxicogenomics": impact of "-omics" technologies, *Mutat. Res.*, 2002, **499**, 13–25.
70. R. P. Amin, H. K. Hamadeh, P. R. Bushel, L. Bennett, C. A. Afshari and R. S. Paules, Genomic interrogation of mechanism(s) underlying cellular responses to toxicants, *Toxicology*, 2002, **181–182**, 555–563.
71. W. J. Bailey and R. Ulrich, Molecular profiling approaches for identifying novel biomarkers, *Expert Opin. Drug Saf.*, 2004, **3**, 137–151.
72. S. W. Burchiel, C. M. Knall, J. W. Davis 2nd, R. S. Paules, S. E. Boggs and C. A. Afshari, Analysis of genetic and epigenetic mechanisms of toxicity: potential roles of toxicogenomics and proteomics in toxicology, *Toxicol. Sci.*, 2001, **59**, 193–195.
73. M. R. Fielden and T. R. Zacharewski, Challenges and limitations of gene expression profiling in mechanistic and predictive toxicology, *Toxicol. Sci.*, 2001, **60**, 6–10.
74. T. W. Gant, Novel and future applications of microarrays in toxicological research, *Expert Opin. Drug Metab. Toxicol.*, 2007, **3**, 599–608.
75. W. H. Heijne, A. S. Kienhuis, B. van Ommen, R. H. Stierum and J. P. Groten, Systems toxicology: applications of toxicogenomics, transcriptomics, proteomics and metabolomics in toxicology, *Expert Rev. Proteomics*, 2005, **2**, 767–780.
76. D. L. Mendrick, Genomic and genetic biomarkers of toxicity, *Toxicology*, 2008, **245**, 175–181.
77. A. Y. Nie, M. McMillian, J. B. Parker, A. Leone, S. Bryant, L. Yieh, A. Bittner, J. Nelson, A. Carmen, J. Wan and P. G. Lord, Predictive toxicogenomics approaches reveal underlying molecular mechanisms of nongenotoxic carcinogenicity, *Mol. Carcinog.*, 2006, **45**, 914–933.
78. W. Pennie, S. D. Pettit and P. G. Lord, Toxicogenomics in risk assessment: an overview of an HESI collaborative research program, *Environ. Health Perspect.*, 2004, **112**, 417–419.
79. R. S. Thomas, W. Bao, T. M. Chu, M. Bessarabova, T. Nikolskaya, Y. Nikolsky, M. E. Andersen and R. D. Wolfinger, Use of short-term transcriptional profiles to assess the long-term cancer-related

safety of environmental and industrial chemicals, *Toxicol. Sci.*, 2009, **112**, 311–321.
80. V. Thybaud, M. Aardema, J. Clements, K. Dearfield, S. Galloway, M. Hayashi, D. Jacobson-Kram, D. Kirkland, J. T. MacGregor, D. Marzin, W. Ohyama, M. Schuler, H. Suzuki and E. Zeiger, Strategy for genotoxicity testing: hazard identification and risk assessment in relation to *in vitro* testing, *Mutat. Res.*, 2007, **627**, 41–58.
81. D. Gusenleitner, S. S. Auerbach, T. Melia, H. F. Gomez, D. H. Sherr and S. Monti, Genomic models of short-term exposure accurately predict long-term chemical carcinogenicity and identify putative mechanisms of action, *PloS One*, 2014, **9**, e102579.
82. M. Chen, M. Zhang, J. Borlak and W. Tong, A decade of toxicogenomic research and its contribution to toxicological science, *Toxicol. Sci.*, 2012, **130**, 217–228.
83. J. Eichner, C. Wrzodek, M. Romer, H. Ellinger-Ziegelbauer and A. Zell, Evaluation of toxicogenomics approaches for assessing the risk of nongenotoxic carcinogenicity in rat liver, *PloS One*, 2014, **9**, e97678.
84. H. J. Clewell, A. Efremenko, J. L. Campbell, D. E. Dodd and R. S. Thomas, Transcriptional responses in the rat nasal epithelium following subchronic inhalation of naphthalene vapor, *Toxicol. Appl. Pharmacol.*, 2014, **280**, 78–85.
85. F. Yamada, K. Sumida, T. Uehara, Y. Morikawa, H. Yamada, T. Urushidani and Y. Ohno, Toxicogenomics discrimination of potential hepatocarcinogenicity of non-genotoxic compounds in rat liver, *J. Appl. Toxicol.*, 2013, **33**, 1284–1293.
86. M. Zhang, M. Chen and W. Tong, Is toxicogenomics a more reliable and sensitive biomarker than conventional indicators from rats to predict drug-induced liver injury in humans?, *Chem. Res. Toxicol.*, 2012, **25**, 122–129.
87. T. Uehara, Y. Minowa, Y. Morikawa, C. Kondo, T. Maruyama, I. Kato, N. Nakatsu, Y. Igarashi, A. Ono, H. Hayashi, K. Mitsumori, H. Yamada, Y. Ohno and T. Urushidani, Prediction model of potential hepatocarcinogenicity of rat hepatocarcinogens using a large-scale toxicogenomics database, *Toxicol. Appl. Pharmacol.*, 2011, **255**, 297–306.
88. H. Matsumoto, Y. Yakabe, F. Saito, K. Saito, K. Sumida, M. Sekijima, K. Nakayama, H. Miyaura, M. Otsuka and T. Shirai, New short term prediction method for chemical carcinogenicity by hepatic transcript profiling following 28-day toxicity tests in rats, *Cancer Inf.*, 2011, **10**, 259–271.
89. Y. Low, T. Uehara, Y. Minowa, H. Yamada, Y. Ohno, T. Urushidani, A. Sedykh, E. Muratov, V. Kuz'min, D. Fourches, H. Zhu, I. Rusyn and A. Tropsha, Predicting drug-induced hepatotoxicity using QSAR and toxicogenomics approaches, *Chem. Res. Toxicol.*, 2011, **24**, 1251–1262.
90. M. R. Fielden, A. Adai, R. T. Dunn 2nd, A. Olaharski, G. Searfoss, J. Sina, J. Aubrecht, E. Boitier, P. Nioi, S. Auerbach, D. Jacobson-Kram, N. Raghavan, Y. Yang, A. Kincaid, J. Sherlock, S. J. Chen, B. Car and

Consortium Working Group, Development and evaluation of a genomic signature for the prediction and mechanistic assessment of non genotoxic hepatocarcinogenesis in the rat, *Toxicol. Sci.*, 2011, **124**, 54–74.

91. R. S. Thomas, H. J. Clewell 3rd, B. C. Allen, L. Yang, E. Healy and M. E. Andersen, Integrating pathway-based transcriptomic data into quantitative chemical risk assessment: a five chemical case study, *Mutat. Res.*, 2012, **746**, 135–143.
92. J. P. Bercu, R. A. Jolly, K. M. Flagella, T. K. Baker, P. Romero and J. L. Stevens, Toxicogenomics and cancer risk assessment: a framework for key event analysis and dose-response assessment for nongenotoxic carcinogens, *Regul. Toxicol. Pharmacol.*, 2010, **58**, 369–381.
93. EPA, *science in ACTION*, Accessed 4 July 2015.
94. A. Rohrbeck, G. Salinas, K. Maaser, J. Linge, S. Salovaara, R. Corvi and J. Borlak, Toxicogenomics applied to *in vitro* carcinogenicity testing with Balb/c 3T3 cells revealed a gene signature predictive of chemical carcinogens, *Toxicol. Sci.*, 2010, **118**, 31–41.
95. R. S. Judson, K. A. Houck, R. J. Kavlock, T. B. Knudsen, M. T. Martin, H. M. Mortensen, D. M. Reif, D. M. Rotroff, I. Shah, A. M. Richard and D. J. Dix, In vitro screening of environmental chemicals for targeted testing prioritization: the ToxCast project, *Environ. Health Perspect.*, 2010, **118**, 485–492.
96. J. S. MacDonald and R. T. Robertson, Toxicity testing in the 21st century: a view from the pharmaceutical industry, *Toxicol. Sci.*, 2009, **110**, 40–46.
97. NRC, *Toxicity Testing in the 21st Century: A Vision and a Strategy*, Washington, DC, 2007.
98. NRC, *Toxicogenomic Technologies and Risk Assessment of Environmental Carcinogens, A Workshop Summary*, National Academies Press, Washington, DC, 2005.
99. A. Brazma, H. Parkinson, U. Sarkans, M. Shojatalab, J. Vilo, N. Abeygunawardena, E. Holloway, M. Kapushesky, P. Kemmeren, G. G. Lara, A. Oezcimen, P. Rocca-Serra and S. A. Sansone, ArrayExpress–a public repository for microarray gene expression data at the EBI, *Nucleic Acids Res.*, 2003, **31**, 68–71.
100. R. Edgar, M. Domrachev and A. E. Lash, Gene Expression Omnibus: NCBI gene expression and hybridization array data repository, *Nucleic Acids Res.*, 2002, **30**, 207–210.
101. M. Waters, S. Stasiewicz, B. A. Merrick, K. Tomer, P. Bushel, R. Paules, N. Stegman, G. Nehls, K. J. Yost, C. H. Johnson, S. F. Gustafson, S. Xirasagar, N. Xiao, C. C. Huang, P. Boyer, D. D. Chan, Q. Pan, H. Gong, J. Taylor, D. Choi, A. Rashid, A. Ahmed, R. Howle, J. Selkirk, R. Tennant and J. Fostel, CEBS–Chemical Effects in Biological Systems: a public data repository integrating study design and toxicity data with microarray and proteomics data, *Nucleic Acids Res.*, 2008, **36**, D892–900.

102. M. Waters and C. Yauk, Consensus recommendations to promote and advance predictive systems toxicology and toxicogenomics, *Environ. Mol. Mutagen.*, 2007, **48**, 400–403.
103. H. J. Clewell 3rd, M. E. Andersen and H. A. Barton, A consistent approach for the application of pharmacokinetic modeling in cancer and noncancer risk assessment, *Environ. Health Perspect.*, 2002, **110**, 85–93.
104. S. A. Saghir, A. L. Mendrala, M. J. Bartels, S. J. Day, S. C. Hansen, J. M. Sushynski and J. S. Bus, Strategies to assess systemic exposure of chemicals in subchronic/chronic diet and drinking water studies, *Toxicol. Appl. Pharmacol.*, 2006, **211**, 245–260.
105. S. M. Hays, R. A. Becker, H. W. Leung, L. L. Aylward and D. W. Pyatt, Biomonitoring equivalents: a screening approach for interpreting biomonitoring results from a public health risk perspective, *Regul. Toxicol. Pharmacol.*, 2007, **47**, 96–109.
106. B. Meek and J. Doull, Pragmatic challenges for the vision of toxicity testing in the 21st century in a regulatory context: another Ames test? … or a new edition of "the Red Book"?, *Toxicol. Sci.*, 2009, **108**, 19–21.
107. L. L. Aylward and S. M. Hays, ToxCast Data Anlysis Summit, 2009.
108. J. S. Bus and R. A. Becker, Toxicity testing in the 21st century: a view from the chemical industry, *Toxicol. Sci.*, 2009, **112**, 297–302.
109. NRC, *Committee on Improving Risk Analysis Approaches used by the U.S. EPA Board of Environmental Studies and Toxicology*, National Academies Press, Washington DC, 2008.
110. NRC, *Applications of Toxicogenomic Technologies to Predictive Toxicology and Risk Assessment*, http://www.nap.edu/catalog/12037.html.
111. OECD, *Guidance Document on the Validation of (Quantitative) Structure-Activity Relationships [(Q)Sar] Models*.
112. P. S. Guzelian, M. S. Victoroff, N. C. Halmes, R. C. James and C. P. Guzelian, Evidence-based toxicology: a comprehensive framework for causation, *Hum. Exp. Toxicol.*, 2005, **24**, 161–201.
113. S. Hoffmann and T. Hartung, Toward an evidence-based toxicology, *Hum. Exp. Toxicol.*, 2006, **25**, 497–513.
114. OECD, *Guidance Document 116 on the Conduct and Design of Chronic Toxicity and Carcinogenicity Studies, Supporting Test Guidelines 451, 452 and 453, ENV/JM/MONO(2011)47, Second edition,13 April 2012*, 2014.
115. OECD, Guidance Document on Developing and Assessing Adverse Outcome Pathways, *OECD Environment, Health and Safety Publications. Series on Testing and Assessment, No.184, ENV/JM/MONO(2013)6, Paris: OECD.184*, *ENV/JM/MONO(2013)6, Paris:OECD*., 2013.
116. C. L. Yauk, J. Bishop, K. L. Dearfield, G. R. Douglas, B. F. Hales, M. Luijten, J. M. O'Brien, B. Robaire, R. Sram, J. van Benthem, M. G. Wade, P. A. White and F. Marchetti, The development of adverse outcome pathways for mutagenic effects for the organization for economic co-operation and development, *Environ. Mol. Mutagen.*, 2013, **54**, 79–81.

117. M. D. Waters, K. Olden and R. W. Tennant, Toxicogenomic approach for assessing toxicant-related disease, *Mutat. Res.*, 2003, **544**, 415–424.
118. J. R. Bucher and C. Portier, Human carcinogenic risk evaluation, Part V: The national toxicology program vision for assessing the human carcinogenic hazard of chemicals, *Toxicol. Sci.*, 2004, **82**, 363–366.
119. L. S. Gold, N. B. Manley, T. H. Slone, L. Rohrbach and G. B. Garfinkel, Supplement to the Carcinogenic Potency Database (CPDB): results of animal bioassays published in the general literature through 1997 and by the National Toxicology Program in 1997–1998, *Toxicol. Sci.*, 2005, **85**, 747–808.
120. L. S. Gold, *Summary by Chemical of Carcinogenicity Results in the Carcinogen Potency Data Base (CPDB) from Technical Reports of the NCI/NTP*, http://toxnet.nlm.nih.gov/cpdb/.
121. M. Ring and B. M. Eskofier, Data Mining in the U.S. National Toxicology Program (NTP) Database Reveals a Potential Bias Regarding Liver Tumors in Rodents Irrespective of the Test Agent, *PloS One*, 2015, **10**, e0116488.
122. V. A. Fung, J. C. Barrett and J. Huff, The carcinogenesis bioassay in perspective: application in identifying human cancer hazards, *Environ. Health Perspect.*, 1995, **103**, 680–683.
123. J. McCann, E. Choi, E. Yamasaki and B. N. Ames, Detection of carcinogens as mutagens in the *Salmonella*/microsome test: assay of 300 chemicals, *Proc. Natl. Acad. Sci. U. S. A.*, 1975, **72**, 5135–5139.
124. J. McCann and B. N. Ames, Discussion paper: the detection of mutagenic metabolites of carcinogens in urine with the *Salmonella*/microsome test, *Ann. N. Y. Acad. Sci.*, 1975, **269**, 21–25.
125. L. D. Claxton, A. G. Stead and D. Walsh, An analysis by chemical class of *Salmonella* mutagenicity tests as predictors of animal carcinogenicity, *Mutat. Res.*, 1988, **205**, 197–225.
126. M. A. Jackson, H. F. Stack and M. D. Waters, The genetic toxicology of putative nongenotoxic carcinogens, *Mutat. Res.*, 1993, **296**, 241–277.
127. B. E. Butterworth, Consideration of both genotoxic and nongenotoxic mechanisms in predicting carcinogenic potential, *Mutat. Res.*, 1990, **239**, 117–132.
128. M. D. Waters, H. F. Stack and M. A. Jackson, Genetic toxicology data in the evaluation of potential human environmental carcinogens, *Mutat. Res.*, 1999, **437**, 21–49.
129. M. D. Waters, M. Jackson and I. Lea, Characterizing and predicting carcinogenicity and mode of action using conventional and toxicogenomics methods, *Mutat. Res.*, 2010, **705**, 184–200.
130. M. Kinoshita and M. Miyata, Underexpression of mRNA in human hepatocellular carcinoma focusing on eight loci, *Hepatology*, 2002, **36**, 433–438.
131. R. D. Snyder and J. W. Green, A review of the genotoxicity of marketed pharmaceuticals, *Mutat. Res.*, 2001, **488**, 151–169.

132. L. S. Gold, T. H. Slone, B. R. Stern and L. Bernstein, Comparison of target organs of carcinogenicity for mutagenic and non-mutagenic chemicals, *Mutat. Res.*, 1993, **286**, 75–100.
133. H. Ellinger-Ziegelbauer, B. Stuart, B. Wahle, W. Bomann and H. J. Ahr, Comparison of the expression profiles induced by genotoxic and non-genotoxic carcinogens in rat liver, *Mutat. Res.*, 2005, **575**, 61–84.
134. H. Ellinger-Ziegelbauer, H. Gmuender, A. Bandenburg and H. J. Ahr, Prediction of a carcinogenic potential of rat hepatocarcinogens using toxicogenomics analysis of short-term *in vivo* studies, *Mutat. Res.*, 2008, **637**, 23–39.
135. M. R. Fielden, R. Brennan and J. Gollub, A gene expression biomarker provides early prediction and mechanistic assessment of hepatic tumor induction by nongenotoxic chemicals, *Toxicol. Sci.*, 2007, **99**, 90–100.
136. D. A. Eastmond, A. Hartwig, D. Anderson, W. A. Anwar, M. C. Cimino, I. Dobrev, G. R. Douglas, T. Nohmi, D. H. Phillips and C. Vickers, Mutagenicity testing for chemical risk assessment: update of the WHO/IPCS Harmonized Scheme, *Mutagenesis*, 2009, **24**, 341–349.
137. T. Uehara, M. Hirode, A. Ono, N. Kiyosawa, K. Omura, T. Shimizu, Y. Mizukawa, T. Miyagishima, T. Nagao and T. Urushidani, A toxicogenomics approach for early assessment of potential non-genotoxic hepatocarcinogenicity of chemicals in rats, *Toxicology*, 2008, **250**, 15–26.
138. R. S. Thomas, L. Pluta, L. Yang and T. A. Halsey, Application of genomic biomarkers to predict increased lung tumor incidence in 2-year rodent cancer bioassays, *Toxicol. Sci.*, 2007, **97**, 55–64.
139. C. Magkoufopoulou, S. M. Claessen, D. G. Jennen, J. C. Kleinjans and J. H. van Delft, Comparison of phenotypic and transcriptomic effects of false-positive genotoxins, true genotoxins and non-genotoxins using HepG2 cells, *Mutagenesis*, 2011, **26**, 593–604.
140. C. Magkoufopoulou, S. M. Claessen, M. Tsamou, D. G. Jennen, J. C. Kleinjans and J. H. van Delft, A transcriptomics-based *in vitro* assay for predicting chemical genotoxicity *in vivo*, *Carcinogenesis*, 2012, **33**, 1421–1429.
141. S. S. Auerbach, R. R. Shah, D. Mav, C. S. Smith, N. J. Walker, M. K. Vallant, G. A. Boorman and R. D. Irwin, Predicting the hepatocarcinogenic potential of alkenylbenzene flavoring agents using toxicogenomics and machine learning, *Toxicol. Appl. Pharmacol.*, 2010, **243**, 300–314.
142. M. L. Whitfield, L. K. George, G. D. Grant and C. M. Perou, Common markers of proliferation, *Nat. Rev. Cancer*, 2006, **6**, 99–106.
143. M. Romer, J. Eichner, U. Metzger, M. F. Templin, S. Plummer, H. Ellinger-Ziegelbauer and A. Zell, Cross-platform toxicogenomics for the prediction of non-genotoxic hepatocarcinogenesis in rat, *PloS One*, 2014, **9**, e97640.

CHAPTER 2

Genomic Biomarkers in Cell-based Drug Screening

HENG-HONG LI

Georgetown University, Department of Biochemistry, Molecular and Cellular Biology, 3970 Reservoir Rd, Washington, DC 20007, USA
Email: HL234@georgetown.edu

2.1 Genotoxicity and the Traditional Testing Battery

According to Chemical Abstracts Service (CAS; www.cas.org/about-cas/cas-fact-sheets/registry-fact-sheet), more than 15 000 new substances are being added to the CAS registry each day. Regulatory agencies require thorough assessment of the health effects of chemicals present in the environment and marketplace. Genotoxicity is one of the major concerns, since chemicals with DNA-damaging potential lead to genetic alterations that may cause adverse health effects. Chemicals that interact with DNA can cause direct damage by covalent modifications, such as adducts or strand breaks, or can perturb DNA and chromatin function by non-covalent binding. Both general types of interactions lead to genetic changes such as stalled replication, cell-cycle delays, cytotoxicity, mutations, and consequent genomic instability that can contribute to cancer development or other health problems. Thus, genotoxicity testing has become a crucial component of safety evaluation for new drugs and chemicals. Compared to 2 year animal carcinogenicity trials, the standard *in vitro* genotoxicity testing battery[1,2] provides sensitive, relatively simple, fast, and economical tools for detection of genetic damage. Since its conception in 1970s, the genotoxicity battery has effectively assured genetic safety of consumer chemicals and/or drugs. However, the current

Issues in Toxicology No. 28
Toxicogenomics in Predictive Carcinogenicity
Edited by Michael D. Waters and Russell S. Thomas

Published by the Royal Society of Chemistry, www.rsc.org

testing paradigm features relatively low specificity with many false positives.[3] This is especially noticeable in case of *in vitro* chromosome aberration assays that yield a false-positive rate of up to 30%.[4] The risk management of compounds with positive *in vitro* findings is a major challenge for both industry and regulatory agencies, and requires complicated, time consuming, and costly follow-up strategies, including animal testing.

The current carcinogenicity risk assessment is typically based on linear extrapolation approaches. This might be appropriate for agents that directly interact with DNA (*e.g.* adduct-forming), since a single mutation theoretically may initiate cancer formation. Nevertheless, there is ample evidence suggesting that some agents may exhibit a threshold dose response.[5] In these instances, chromosome damage as a consequence of cytotoxicity *via* non-DNA-reactive mechanisms in the *in vitro* test system might be considered biologically irrelevant.[6,7] To clearly differentiate a true threshold dose response from the effect of simply reaching an assay detection limit, a proper understanding of the underlying genotoxic mechanisms is essential. However, investigation of genotoxic mechanisms is often laborious and time-consuming, with an uncertain outcome, and can lead to significant delays in drug development. Thus, the advancement of alternative experimental approaches capable of evaluating a whole range of genotoxic mechanisms is extremely important. In this regard, "omic" approaches that utilize the measurement of global cellular responses at RNA, protein, or metabolite levels have been evaluated. As will be discussed, transcriptional stress responses when integrated with pathway analysis may provide valuable insights into the genotoxic mechanism.

2.2 Mechanisms of Action for Genotoxicity and the Genotoxic Stress Responses

2.2.1 Categorization by Mechanisms of Action

Many compounds that cause DNA damage and related perturbations share common features in their structure and biochemical effects, and these parameters can be used to help categorize agents by mechanisms of action. From the view of bond force and structure, interactions between a compound and DNA can be either covalent or non-covalent. A covalent interaction is irreversible, although there are some exceptions: *e.g.* O6-alkylguanine can be directly reversed with the "suicide" repair protein O6-alkylguanine-DNA alkyltransferase.[8] Unrepaired DNA base damage can lead to potentially mutagenic and lethal consequences, such as by mismatch or production of apurinic/apyrimidinic sites. In addition, covalent bulky adducts can cause DNA backbone distortion, which in turn can affect both transcription and replication, such as by disrupting protein complex recruitment. Interactions between DNA and non-covalent agents are through van der Waals forces, hydrogen bonding, hydrophobic, and/or charge transfer forces, and as such are reversible. Non-covalent agents can be

Table 2.1 Examples of DNA-interacting agents.[a]

Non-covalent DNA interacting agents		Covalent DNA-adducts and direct DNA damage[b]
Groove-binding agents	DNA intercalator	
Berenil	Aminoacridines	Busulfan
Bisbenzimadoles	Arylaminoalcohols	Camptothecin
Bleomycin[c]	Coumarins	Chlorambucil
Chloroquine	Cystodytin J	*cis*-Platinum
Chromomycin A3	Diplamine	Clomesone
DAPI[d]Distamycin A	Doxorubicin[c]	Cyclodisone
Guanyl bisfuramidine	Daunomycin[c]	Ionizing radiation
Hoechst 33258	Echinomycin	Nitrogen mustard
Mithramycin	Ethidium bromide	Nitrosoureas:
Netamycin	Indoles	BCNU[e]
Netropsin	M-AMSA	CCNU[f]
Pentamidine	Mitoxantrone	PCNU[g]
Pilcamycin	Naphthalimides	Oxidative-stress agents
SN6999	Phenanthridines	UV radiation
SN7167	Proflavine	
	Quinolines and Quinoxalines	
	YO-1 and YOYO-1	
	Chlorpheniramine	
	Methapyrilene	
	Tamoxifen	

[a]Atypical DNA intercalating agents are shown in italic type.
[b]A limited number of examples of well-known agents that damage DNA by covalent modifications such as adducts and other types of DNA damage, *e.g.* strand breaks.
[c]Many non-covalent binding agents have other properties that result in DNA damage; *e.g.* while bleomycin is a groove-binding agent, it also generates oxidative damage which directly damages DNA.
[d]Diamidine-2-phenylindole.
[e]1,3-bis(2-Chloroethyl)-1-nitrosourea.
[f]1-(2-Chloroethyl)-3-cyclohexyl-nitrosourea.
[g]1-(2-Chloroethyl)3-(2,6-dioxo-3-piperidyl)-1-nitrosourea.

classified into groove-binding agents and DNA intercalators.[9] Some representative examples for both types are shown in Table 2.1 along with examples of agents producing covalent damage. Groove binders are usually crescent shaped and generally not particularly toxic. They can block DNA protein interactions: *e.g.* groove-binding agents, such as diamidine-2-phenylindole (DAPI) and distamycin,[10] can act as catalytic topoisomerase inhibitors by blocking the docking site on DNA for topoisomerase 2. Intercalators are planar aromatic molecules that insert between bases; they can cause DNA backbone distortion and frameshift mutations during DNA replication. By their winding/unwinding effects on supercoiled DNA, intercalators can change the normal torsional strain in cellular DNA. Some can act as topoisomerase poisons, with resultant strand breaks, and this is probably a major cause for their genotoxicity in mammalian cells.[9] Some molecular features, including a *N*-dialkyl group on the side chain, are associated with higher affinity binding to DNA. Bis-intercalating compounds also show higher DNA binding activity, thereby contributing to

Table 2.2 Prediction of mechanism of action using biological parameters.[a]

Alkylating agents
Topoiosomerase 1 inhibitors
Topoisomerase 2 inhibitors
RNA/DNA antimetabolites
DNA antimetabolites
Antimitotic agents

[a]Employing a high-throughput screen, cytotoxicity, as measured by growth inhibition, was compared in 60 human tumor cell lines of the National Cancer Institute's antineoplastic drug screen (NCI-ADS) panel using 134 genotoxic agents with known mechanisms of action. The 134 agents thought to be in one of the categories noted above were chosen from 141 compounds. The mechanisms of action for these agents were divided into the six general categories above. This is an early example where a biological parameter had predictive value in delineating general categories of mechanism of action On the basis of later analyses of approximately 10^5 damaging agents in the NCI-ADS screen, the Unidrug drug set of 1247 agents was defined, having differing signatures, and thus represents this number of potential differing mechanisms of action.

higher toxicity. An important complication in the characterization of non-covalent DNA-interacting agents, which include DNA groove-binding chemicals and DNA intercalators, is the inclusion of other damaging properties. For example, psoralens are generally considered non-toxic even though they can intercalate between DNA bases, while activation of their covalent crosslinking capability has profound effects on cell viability, DNA replication, and gene expression; bleomycin, which binds to the major groove of DNA, also generates free radicals that directly target DNA and chromatin.[9,11,12]

It is generally accepted that biological responses to damaging agents can be used to categorize these agents by their biochemical mechanism(s) of action. In a large panel of yeast mutants, toxicity modulation has been used to assign protein networks involved in damage by a simple alkylating agent, an agent producing bulky DNA-base adducts, UV radiation, and an oxidizing agent.[13] In Table 2.2, an early cytotoxicity approach in mammalian cells using a short-term cytotoxicity assay is summarized. Here 60 human tumor lines were assayed for cytotoxicity in a short-term assay and their responses were compared using 141 genotoxic agents with known mechanisms of action. The patterns or "fingerprints" were then delineated for six general mechanisms of action. A neural-network approach was used to assign these agents to the correct category and successfully grouped 129 agents.[14] This approach has been used to assign unknown agents to these, and subsequently other, categories.[15]

2.2.2 Signaling and Transcriptional Responses upon Genotoxic Stress

Genotoxicity causes mutation, DNA stand breaks, changes in DNA conformation, and/or DNA torsional tension, which may interrupt protein–DNA interaction and potentially lead to indirect DNA strand breaks. All of these events can have substantial effects on gene expression. An elaborate network of sensors and response pathways has evolved to respond to genotoxic and other stresses. DNA damage, such as strand breaks, is detected by sensor molecules including the MRN complex, Rad17, 9-1-1 complex, MDC1, 53BP1, TopBP1, Claspin, and BRCA1.[16–18] The recruited DNA-damage sensor proteins and complexes trigger kinases such as ATM, ATR, DNA-PK, Chk1, and Chk2, which then leads to regulation of various transcriptional factors. Among these transcription factors, p53 is known to be a major mediator and contributor to many genotoxic stress responses.[19] DNA-intercalating acridine derivatives including 9-aminoacridine, amsacrine, quinacrine, and acridine orange, have been shown to induce p53 transcriptional activity.[20] A broad spectrum of p53 downstream genes, such as Cdkn1a (p21/WAF1/CIP1), Gadd45a, BTG2, cyclin G1, Bax, PIG3, Puma, and Wig1, play important roles in orchestrating cell cycle checkpoint activation and programmed cell death responses to stress. Cdkn1a has important roles in cell-cycle control and can be regulated by both p53-dependent and p53-independent mechanisms: *e.g.* Cdkn1a-promoter activity is repressed by c-Myc in proliferating cells, and down-regulation of c-Myc by a Top2 inhibitor releases this suppression, with consequent induction of Cdkn1a.[21,22] p53 can have wide-ranging effects on many cellular parameters and there are probably several hundred mammalian genes containing p53-binding sites;[23–25] the number of p53-responsive genes is probably even greater.[26,27]

2.3 Expression Profiling and Toxicogenomics

2.3.1 Genotoxicity Assays Based on Transcriptional Responses

Cellular stress induced transcriptional activation has been utilized in the development of a variety of molecular toxicology assays. For instance, transcriptional activation of the SOS pathway in bacteria and RAD54 pathway in yeast has been used to monitor genotoxicity using single-gene promoter reporter assay systems.[28–31] In mammalian cells, discovery of the *Gadd45a* and other stress genes,[32,33] and an appreciation of the central role of p53 in many DNA damage response pathways and cancer development[34,35] provided the foundation for development of pre-genomic approaches for the detection of genotoxicity in mammalian cells. The resulting mammalian-cell based assays detected genotoxicity as activation of stress-gene promoters, *e.g.* human *GADD45A*, *via* a reporter molecule such as chloramphenicol acetyltransferase or green fluorescent protein.[36,37] Furthermore, the monitoring of cellular p53 protein level has also been evaluated as a tool for detection of genotoxicants and carcinogens.[38,39] Since the p53 pathway can be also activated by a variety of other stresses including hypoxia, nutrient starvation, alteration of the ubiquitin pathway, or ribonucleotide depletion,[40–42] investigation of genotoxic mechanisms requires monitoring of more than a single gene or pathway. To this end, several promoter-reporter biosensor-carrying cell lines were combined in the Cat-tox (L) assay and treated in parallel with tested agents.[37] In this assay, the DNA damage was detected as activation of promoters containing a p53 response element, *Gadd45a*, and/or cFos; protein damage was measured by activation of the GRP78 and HSP70 promoters; heavy metal exposure was measured by methallothionine IIA promoter, and oxidative stress was measured by glutathione S transferase Ya subunit promoters. This approach has been shown to provide useful mechanistic information applicable for drug development[43] and detection of pollutants.[44] In contrast to a limited number of gene promoters/pathways in these assays, the more recent availability of transcriptomics technologies such as microarray and RNAseq to measure stress transcriptional responses at the whole-genome level has made these previous biosensor-based approaches less attractive.

Various stresses can trigger distinct cellular responses. In the case of genotoxic stimuli, even very early studies showed differences in the spectrum of genes induced. For example, UV radiation-induced gene alterations showed quantitative and qualitative differences from the ones caused by the alkylating agent, methylmethane sulfonate.[32] Early studies in eukaryotes were modeled on the SOS response in bacteria where many DNA repair genes exhibited induction by genotoxic agents such as UV radiation.[45] The situation in eukaryotes is more complex. In yeast, many genes required for protection against classic stressors, such as ionizing radiation (IR), UV radiation, hydrogen peroxide, or cisplatin, have been reported to not be inducible.[46] Even early studies showed that the situation in yeast is complex with many genes showing genotoxic stress responsiveness as well as differences in responses to different types of DNA-damaging agents.[47] For mammalian cells, genotoxic stress responsiveness is even more complex, with at least hundreds of genes showing inducibility,[19] and some reports put the total number near a thousand.[26] While some DNA repair genes have been shown to be DNA-damage-inducible,[48] many inducible genes mediate regulatory processes involved in cell cycle control, apoptosis, senescence, and inflammation.[19,49] The cellular response to stresses, such as radiation, involves activation of several pathways including p53, mitogen-activated protein kinase cascades, nuclear factor-κB and the activating protein-1 transcription factors.[34,50,51] Key regulatory factors, such as p53, show important quantitative and qualitative differences in their activation by different types of genotoxic stress. For example, the relative levels of post-translational modifications of p53 at various sites in its regulatory regions have been shown to vary for IR, UV radiation, or adriamycin,[52] and there is a marked difference in the spectrum of sites modified after oncogenic stress (activated Ras) compared to UV radiation.[53] Considering that genotoxic stress responses can be regulated at the level of transcription, post-translational changes, and/or *via* specific molecular interactions, the investigation of complex networks of molecular pathways is likely to provide insight into mechanisms of genotoxicity.

2.3.2 Toxicogenomics and Genomic Biomarkers

Toxicogenomics is the application of genomics in toxicology, which reveals the changes of the global transcriptome in response to xenobiotics and offers genomic signatures to distinguish different types of damaging agents. Global expression profiling has revolutionized our ability to monitor for transcriptional responses. The field has exploded since the first report.[54] Hundreds of yeast genes have been identified after chemical exposures[55–57] and have been grouped by gene ontology. Similar studies have been performed in a wide variety of other eukaryotes and there has been substantial focus on responses in a toxicological context (see reviews[35,58,59]). Transcriptome data have been used to classify pharmacologic agents with different mechanisms of action, such as in Lamb *et al.*[103] where a collection of drug-response transcriptome profiles was termed the Connectivity Map (CMAP). *In vivo* toxicogenomics efforts are underway in a variety of laboratories: *e.g.* grouping of hepatotoxins has been performed in rat liver and other tissues with signatures for different types of damaging agents.[60–63] The results from *in vivo* systems are typically more complicated than for tissue culture where a single cell type is analyzed. Most genotoxin-responsive genes in cultured cells are frequently found with *in vivo* studies.

In the case of genotoxic agents, toxicogenomics is an emerging approach potentially useful for investigating genotoxic mechanisms.[58] In mammalian cell lines including HepG2 cells,[64] TK6,[65,66] and L5178Y,[67,68] signatures for genotoxic stress responses have been reported. In a study with 14 test agents, genotoxic agents could be distinguished from non-genotoxic agents by expression profiling in TK6 cells and its p53-null derivative NH32.[66] Surprisingly for these agents with known mechanisms of action, classification by type of stress could be distinguished even in the p53-null derivative, which indicated that remaining pathways still support sufficient responsiveness to genotoxicity. In multiple cases, responses were more robust in the p53 wild-type line and known p53-regulated genes could be distinguished. With the larger data set used by Amundson *et al.*,[66] stress responses could be further categorized using a gene-selection approach combining maximum pairwise *t*-statistics, distance-based method, class correlation and Wilcoxon or Kruskal–Wallis statistics algorithms, and this result is reproduced in Figure 2.1. With a multidimensional scaling approach, the similarity of the overall responses can be estimated by the proximity of single data points, which each represent results for microarrays performed with a specific agent in either TK6 or NH32 cells. As seen in Figure 2.1, the stresses could be further divided into four subgroups based on general type of damaging agent and mechanisms of action. Gene-based biomarker panels have been explored for classifying compounds according to whether they will elicit a specific toxic response. Typically, the biomarker panels consisting of small gene sets are identified *via* statistical algorithms based on the ability of the panel to discriminate between samples of different experimental groups. This approach rarely incorporates biological

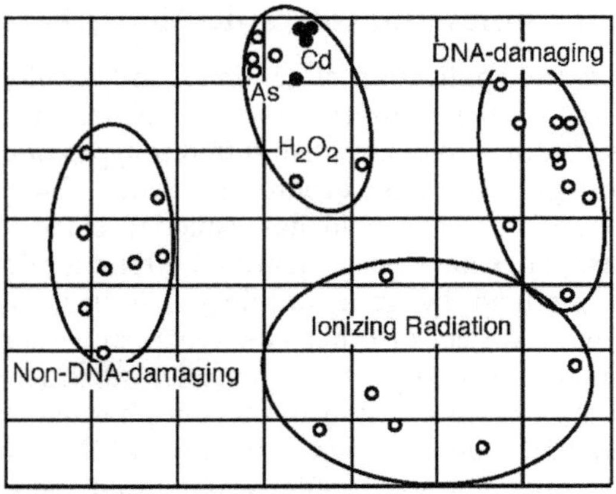

Figure 2.1 Example of similarities in overall gene expression signatures for agents with similar mechanisms of action. Two-dimensional projections of multidimensional scaling (MDS) analysis of gene expression in p53 wild-type cells (TK6) and its p53-null derivative (NH32) are shown for treatments for 4 h with diverse stress agents. Stress agents were grouped into four broad categories of action: non-DNA-damaging stresses (heat shock, osmotic shock, and TPA), ionizing radiations (neutron and γ-ray exposures), other DNA-damaging agents (UV radiation, methyl methanesulfonate (MMS), adriamycin, camptothecin, and cisplatin), and agents causing mainly oxidative stress (arsenite and H_2O_2). Gene selection identified a set of 346 genes that enhanced the separation of the four agent groups; cells were subsequently treated with cadmium chloride, and the data (filled circles) were added to the MDS analysis using the same 346-gene signature.
(This example is reproduced from figure 2(c) in the previous study.[66] Reprinted by permission from Macmillan Publishers Ltd: Oncogene. Ref. 66, copyright 2005.)

knowledge during the gene selection process. Nevertheless, toxic agents often display pleiotropic (Figure 2.2) effects, inducing multiple stress-response pathways.[69,70] For example, methylmethane sulfonate (MMS) damages DNA and results in the induction of a DNA damage response, in parallel with inducing the unfolded protein response and other endoplasmic reticulum (ER)-associated signaling events.[71] In addition, gene panels that provide mechanistic insights (*i.e.* mechanism-based genomic biomarkers) should be better suited to deciphering the complex array of induced stress-response pathways, and help delineate those associated with genotoxicity.

Pathway analysis tools are used in conjunction with biomarker panels to develop hypotheses for the underlying regulatory events and response pathways perturbed by a specific chemical. Unfortunately, these methods often generate multiple hypotheses that are not easily testable given the promiscuity of genes. Integration of multiple omics types, including delineation of coordinately regulated transcriptomic responses, may aid in classifying the particular stress-response pathways that are invoked

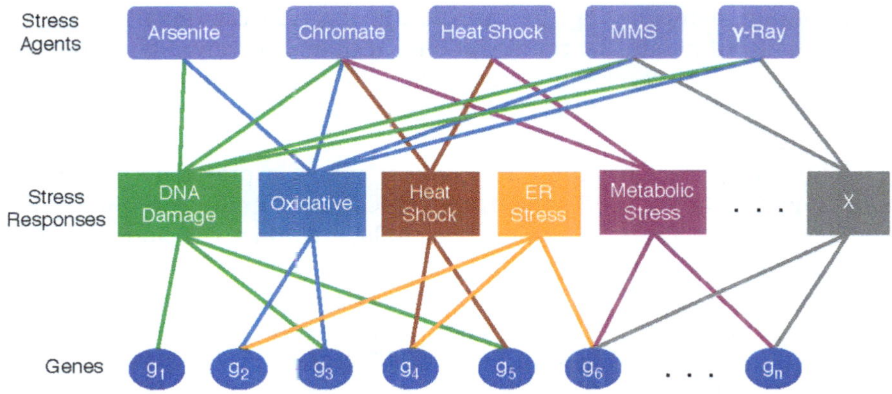

Figure 2.2 Stress agents often induce multiple stress responses, and often participate in multiple stress-response pathways. For example, chromate can activate stress response pathways associated with genotoxicity, oxidative stress, heat shock, and metabolism.

following chemical exposure, providing insight into underlying mechanisms of action and pathways of toxicity.[69,72–74]

2.3.3 Considerations in Biomarker Identification

The key considerations in setting up a protocol for genomic biomarker identification include the choice of cells, toxicant selection, duration of response time, and dose optimization. Cells from a variety of tissue origins have been used in toxicogenomics studies.[75–77] In the effort to identify biomarkers for specific toxicity, the cells need to have the intact signaling pathways and effective transcriptional regulation that are responsive to the targeted toxicity. Thus, cell lines with high genomic stability and that are easy to maintain are preferable, since many tumor lines have lost key response elements, such as p53.

Complex cellular stress responses following treatment with chemicals are time- and dose-dependent. However, toxicogenomic studies evaluating both time course and dose responses across a large set of compounds are not economically feasible. Therefore, the experimental design necessitated selecting a single dose and time point post-exposure for sample collection. In the majority of toxicogenomics studies, the dosing of cells with test chemicals was either selected in an *ad hoc* fashion or was based on the results of cytotoxicity tests, and no systematic effort was made to optimize dose selection. While most stress-response genes show responses that are roughly proportional to dose, there are exceptions and the argument has been made that there may be unique responses at low doses in some cases.[78] Without dose optimization, there is a risk that the transcriptomic response will be either negligible due to under-dosing or obscured by apoptotic and other general responses due to overdosing. In addition, alkylating agents and other protein-damaging agents can disrupt the transcriptional machinery and actually attenuate the transcriptional response at very high doses.[70,79] As the genomic biomarker is a transcriptional endpoint, it is critical to determine and use the dose that induces robust transcriptional responses for each tested toxicant.

Cellular responses to stress detected *via* mRNA expression changes are also time dependent. The response includes many immediate-early genes and other genes whose transcripts accumulate within a few hours after exposure to genotoxic[35,66,71,79,80] and non-genotoxic agents (*e.g.* heat shock).[81] The early induced stress response can lead to a cascade of events that cause later cytotoxicity and secondary pathway changes[66] that show a similar profile across variety of toxic mechanisms.[80] In fact, the lack of the appropriate dose-setting metrics and time of exposure cause major problems with interpretation of toxicogenomic studies. Thus, it is critical to establish a standardized dose-setting paradigm in order to ensure that comparisons can be made across chemicals and studies.

2.4 The Genotoxicity Biomarker TGx-28.65: Identification and Application

Our laboratory and collaborators developed a genotoxicity biomarker comprising 65 genes responsive to DNA damage that was named TGx-28.65 using our standardized protocol. Our studies[82,83] have shown that TGx-28.65 is useful for risk assessment of genotoxic effects; in particular the biomarker is capable of differentiating compounds with irrelevant false positive findings in chromosome aberration assays from true DNA-damaging agents.

2.4.1 Choice of the Cell Line and Toxicants

Our approach measured transcriptomic perturbations in the human lymphoblastoid-derived TK6 cell line because it is p53 proficient, well characterized, extensively used in toxicologic studies, and robustly responsive in previous stress-signaling studies.[65,66,84] We developed our model using a diverse set of 28 model agents representing genotoxic and non-genotoxic modes of action. Genotoxic agents were considered to be those causing direct DNA damage or those leading to DNA perturbations indirectly. Direct-acting genotoxic agents in this group include alkylating agents and other DNA strand breaking agents. The indirect-acting agents are topoisomerase I and II inhibitors, and antimetabolites interfering with DNA synthesis; some of the latter agents also affect RNA synthesis. These indirect-acting genotoxicants typically induce strand breaks indirectly by blocking resealing of transient topoisomerase strand breaks or by accumulation of gaps and single-strand DNA during DNA synthesis. Non-genotoxic agents included ER stress agents, energy metabolism inhibitors, Histone deacetylase (HDAC) inhibitors, heat shock, and ethanol. The heavy metal group has diverse mechanisms of action, including DNA damaging activity (*e.g.* DNA–protein crosslinks) by chromate and arsenite, and non-genotoxic damage by cadmium. Finally, microtubule inhibitors were considered non-genotoxic because they cause aneugenicity through spindle interference, rather than DNA damage and mutations. The mechanisms of action for these agents are well established and were derived from the literature, as listed in Table 2.3.

2.4.2 Dose and Treatment Time Parameters

Since early gene expression changes have been shown to be indicative of initial damage and not to be influenced by subsequent molecular processes such as apoptosis,[80] a 4 h post-exposure time point was selected for the development of the biomarker.[66,71,85]

To enable comparison of transcriptome profiles across the whole set of agents at a single dose per chemical, and to establish a strategy for setting doses for new test compounds, we developed a quantitative real-time (qRT)-polymerase chain reaction (PCR) molecular phenotyping protocol based on expression changes in three known stress-response genes in the dose-setting experiments; these genes included *ATF3*, *GADD45A* and *CDKN1A* (also referred to as p21, Cip1, or Waf1). Previous studies have demonstrated that these genes exhibit strong responses to a wide variety of agents.[66,71,85] Both *GADD45A* and *CDKN1A* are known to be p53-regulated, and show robust mRNA induction by genotoxic agents, while we have found that *ATF3* is responsive to a wide variety of stress agents in TK6 cells.[66] Despite the fact that qRT-PCR can measure significant increases that are less than two-fold, we arbitrarily used a threshold of two-fold to indicate robust induction. As a representative example of our results, the qRT-PCR results for bleomycin showed an expected dose-dependent increase in *ATF3*, *GADD45A*, and *CDKN1A* transcript levels (Figure 2.3A). The full molecular phenotyping results for the agents are summarized in Figure 2.3B. In general, treatment with genotoxic agents led to the induction of all three genes, whereas the non-genotoxic treatments primarily induced *ATF3* and *GADD45A* (Figure 2.3B). The doses that showed the highest induction of one or more of these genes for each agent were selected for full transcriptomic analysis, after considering measures of cytotoxicity. The doses selected showed no appreciable cytotoxicity at 4 h (Figure 2.4) and only moderate effects on viability at 24 h (Figure 2.4). When more than one dose met the criteria and behaved similarly, the lower concentration was selected. For example, all doses of bleomycin showed increases in relative mRNA levels for all three genes (several-fold increases or more), but the magnitude of *GADD45A* induction decreased slightly at doses greater than 10 $\mu g\, mL^{-1}$. Since all doses showed strong responses for the indicator genes, the 10 $\mu g\, mL^{-1}$ dose was chosen for microarray analysis. All model agents, except for paclitaxel, showed at least a two-fold increase in one or more of the three-gene battery. In the case of paclitaxel, induction was less than 1.5-fold so a biologically effective dose from previous *in vitro* studies using cultured cells was selected.[86]

Table 2.3 Standard agents used in the TGx-28.65 genotoxicity assessment study.[82]

Categories	Compound names	Solvent	Dose range	Concentration for array
Alkylating agents	Cisplatin	0.9% NaCl	10–80 μM	80 μM
	Methyl methane sulfonate (MMS)	H_2O	20–200 $\mu g\,mL^{-1}$	100 $\mu g\,mL^{-1}$
Topoisomerase I inhibitors	Camptothecin	DMSO	62.5–500 nM	125 nM
Topoisomerase II inhibitors	Etoposide	DMSO	50–400 nM	200 nM
RNA/DNA antimetabolites	5-Fluorouracil (5-FU)	DMSO	6.25–50 $\mu g\,mL^{-1}$	25 $\mu g\,mL^{-1}$
	Methotrexate	DMSO	0.05–1 mM	100 μM
DNA antimetabolites	Arabinofuranosyl cytidine (AraC)	H_2O	12.5–50 μM	50 μM
	Hydroxyurea	H_2O	0.25–1 mM	0.5 mM
Causing DNA strand break by other mechanisms	γ-Rays	N/A[a]	4 Gy	4 Gy
	Bleomycin	H_2O	5–40 $\mu g\,mL^{-1}$	10 $\mu g\,mL^{-1}$
	Hydrogen peroxide	N/A[a]	20–80 μM	80 μM
Antimitotic agents	Colchicine	Ethanol	62.5–1000	250 $ng\,mL^{-1}$
	Docetaxel	DMSO	25–100 nM	50 nM
	Paclitaxel	DMSO	12.5–200 nM	50 nM
	Vinblastin	DMSO	50–800 $ng\,mL^{-1}$	200 $ng\,mL^{-1}$
Histone modification inhibitors	Trichostatin A (TSA)	DMSO	5–80 $ng\,mL^{-1}$	20 $ng\,mL^{-1}$
	Apicidin	DMSO	0.25–4 $\mu g\,mL^{-1}$	1 $\mu g\,mL^{-1}$
	HC toxin	Methanol	5–80 $ng\,mL^{-1}$	20 $ng\,mL^{-1}$
	Oxamflatin	DMSO	0.25–4 μM	1 μM
Endoplasmic reticulum modulator	Tunicamysin	Methanol	1.25–10 $\mu g\,mL^{-1}$	2.5 $\mu g\,mL^{-1}$
	Thapsigargin	Ethanol	62.5–500 nM	250 nM
Glycolysis inhibitor	2-deoxy-D-glucose (2-DG)	H_2O	0.16–20 μM	20 μM
Energy metabolism inhibitor (uncoupling agent)	Antimycin A	Ethanol	25–200 μM	100 μM
Heavy metals	Cadmium chloride	H_2O	50–800 μM	50 μM
	Potassium chromate(VI)	H_2O	25–400 μM	100 μM
	Sodium arsenite	H_2O	10–90 μM	30 μM
Other stresses	Heat shock	N/A[a]	47 °C	47 °C
	Ethanol	N/A[a]	2%, 4%	2%, 4%

[a]Not applicable.

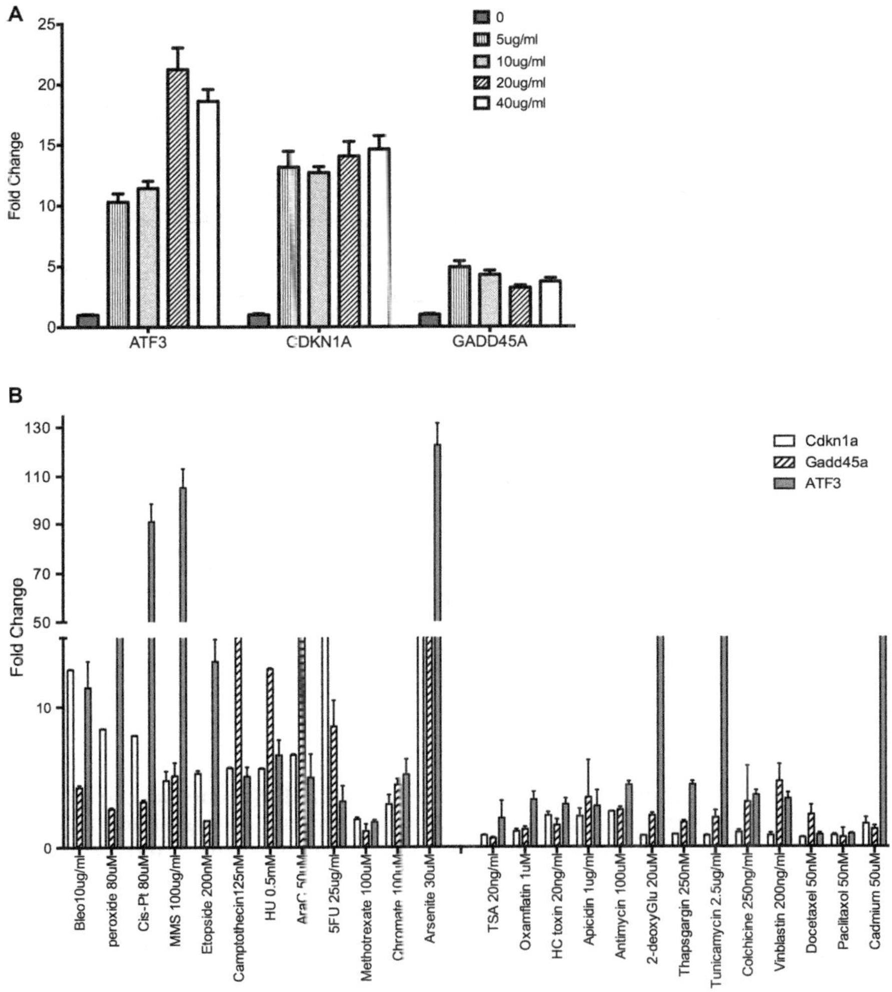

Figure 2.3 Quantitative real-time (qRT)-polymerase chain reaction (PCR) measurements for a panel of known stress response genes (*ATF3*, *CDKN1A*, and *GADD45A*) were used to guide stress agent dose selection. Doses were selected by measuring perturbations of this panel as a function of agent dose; fold change is relative to vehicle control cells assayed at the same time. The assay was performed in triplicate cultures within one experiment. (A) The dose–response profile for bleomycin illustrates increased expression of the stress response genes for 5, 10, 20, and 40 μg mL^{-1} with an observable reduction in *GADD45A* up-regulation at the two highest doses. (B) Stress gene panel expression profiles for the doses of chemical agents eliciting a robust response.

2.4.3 Global Transcriptional Analysis

After selecting doses that elicited robust expression of genes in our qRT-PCR battery, we measured global transcriptomic response using Agilent (Santa

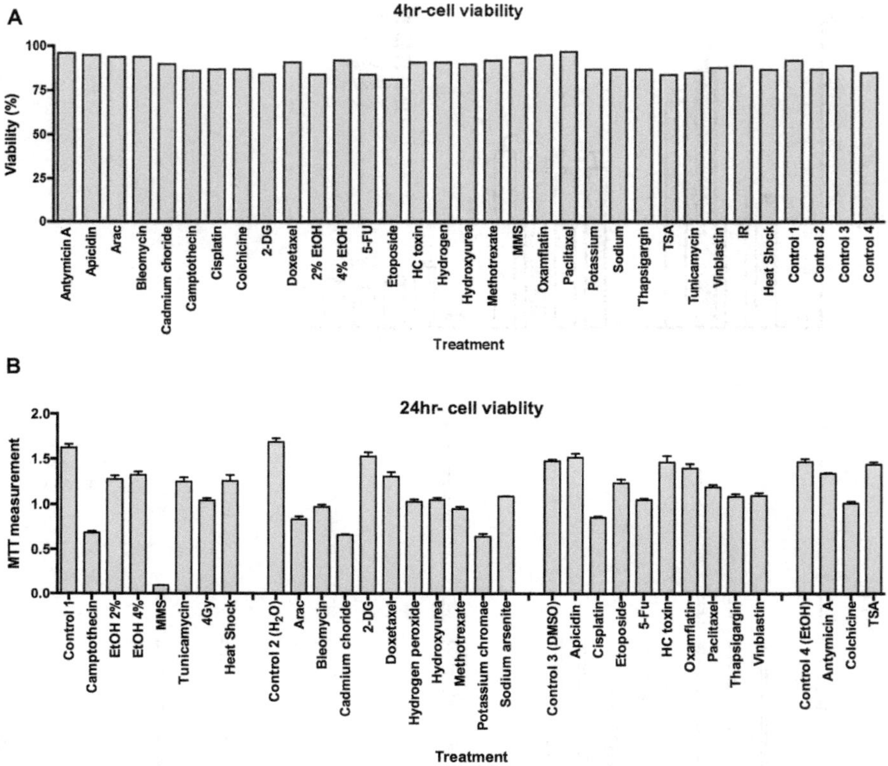

Figure 2.4 Cell viability results of treatments with the agents at the dose for microarray. (A) Trypan blue exclusion test at 4 h of treatment. (B) MTT assay at 24 h of treatment. The *y*-axis represents arbitrary scale of readout.

Clara, CA, USA) human whole-genome oligonucleotide arrays. To reduce the effects associated with different labeling efficiencies, we used a two-channel dye-swapping configuration[87] where the vehicle control was used as a reference sample. Subsequent studies have shown that the use of modern single-channel/single-dye approaches yield comparable results (unpublished). Visualization of the expression profiles highlighted the pleiotropic nature of these agents (*i.e.* gene expression profiles consist of more than one defined gene cluster (Figure 2.5)). In the case of genotoxic agents, hierarchical clustering grouped all 13 toxicants together on the right-hand side of this heatmap. Distinct subsets of genes showed coordinate expression, such as those in the region designated by numerous red sidebars. The non-genotoxic agents represent diverse toxicants but those with similar mechanisms of action showed other subsets of responsive genes. For example, the HDAC inhibitors (apicidin, oxamflatin, HC toxin, and trichostatin A) are known to have a similar mechanism of action and elicited highly similar transcriptome perturbations (error-weighted Pearson correlation coefficient range 0.86–0.92) with prominent groupings of induced or repressed genes on the left side of the heatmap. The transcriptome profiles of the more general category of non-genotoxic agents were not as strongly correlated (error-weighted Pearson correlation coefficient range 0.01–0.85), but other clusters can be appreciated by visual inspection. For example, the ER stress agents thapsigargin and tunicamycin grouped together by hierarchical clustering. Interestingly, a subset of induced genes (designated by the arrow in Figure 2.5) for these ER stress agents were also strongly induced by a few other agents including MMS, which directly damages proteins. Many toxicants can trigger multiple stress response pathways (Figure 2.2), and one of the most striking examples is the case of chromate, which is known to damage DNA upon reduction to chromic anion within the cell.[88–90] As expected, chromate induced a set of genes in common with the other genotoxic agents (Figure 2.5, red bars). However, chromate also strongly induced expression changes in a set of genes that was only strongly induced by cadmium and heat shock (which did not elicit genotoxic stresses), while these genes were strongly repressed by some of the other genotoxic stress agents (bleomycin, cisplatin, MMS, and arsenite) (Figure 2.5, blue bars).

2.4.4 Delineation of Gene Subclusters Using a Biclustering Approach

Although two-dimensional clustering can provide biological insight by organizing large data sets for visual inspection, it does not indicate the extent to which genes participate in multiple pathways (Figure 2.2). To account for the fact that the response to toxicity can involve overlapping subsets of genes, a variety of biclustering methods have been developed. One of the most promising methods is coupled two-way clustering (CTWC) using superparamagnetic clustering (SPC).[91] Not only does this method account for gene products mediating multiple stress-response pathways, it also provides a metric for assessing the statistical significance of each cluster of genes.

When we applied the CTWC/SPC biclustering algorithm to our transcriptomic data, we identified 67 gene groups with stability scores ≥40 (Figure 2.6); these groups will subsequently be referred to as subclusters to avoid confusion with results from hierarchical clustering. These gene subclusters were each comprised of 20–193 genes, with over 50% of the genes participating in multiple clusters and a single gene participating in as many as eight clusters. This method revealed a subcluster of genes induced by chromate and the genotoxic agents that was further analyzed using a hierarchical cluster approach (Figure 2.7A), as well as a set of genes induced by chromate, cadmium, and heat shock (Figure 2.7B). Ingenuity pathway analysis indicated that the chromate/genotoxic agent subcluster was enriched for DNA-damage responsive genes, some of which are known to be regulated by p53, such as *AEN*, *C12orf5*, *EI24*, and *PHLDA3* (data not shown), whereas

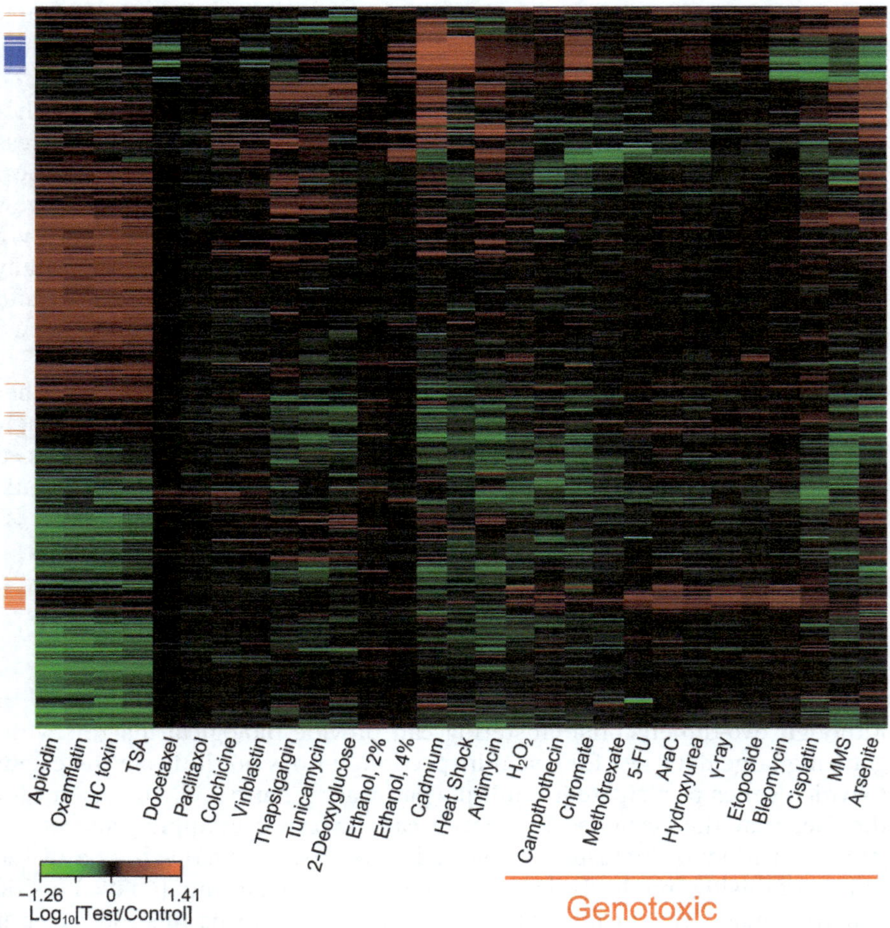

Figure 2.5 Heatmap of transcriptome profiling data illustrating co-expressed sets of genes associated with various toxicants, such as genotoxic agents (red side bars). Some agents also exhibit obvious pleiotropy: *e.g.* chromate induced expression of genes associated with genotoxic agents (red side bars) and genes associated with cadmium and heat-shock (blue side bars). This heatmap contains 1628 genes that were significantly ($p < 0.01$, *t*-test) perturbed at least 1.7-fold, relative to the control, by at least one stress agent. The genes in the heatmap were organized by hierarchical clustering with complete linkage based on their error-weighted Pearson distances. The genotoxic (red side bars) and cadmium and heat-shock (blue side bars) agents were identified using coupled two-way clustering. The arrow highlights a cluster of genes responding to endoplasmic reticulum stress agents thapsigargin and tunicamycin, as well as certain other agents.

the chromate/cadmium/heat shock cluster included genes involved in amino acid metabolism, amino acid transportation, and the unfolded protein response (data not shown).

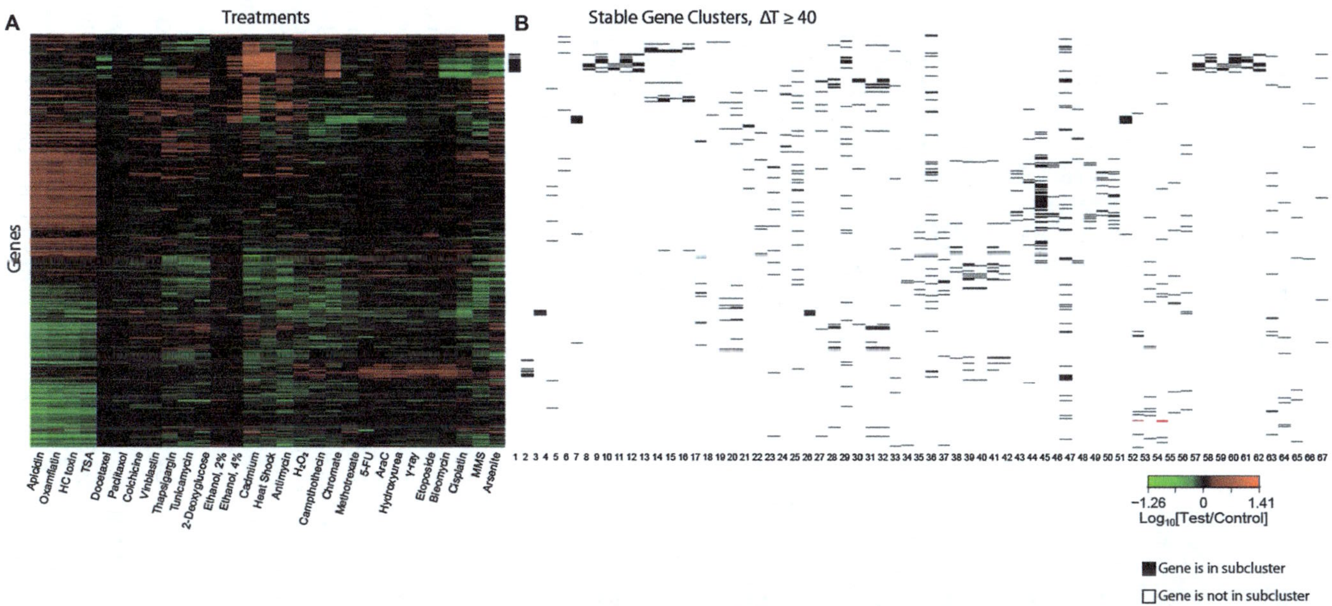

Figure 2.6 Stable gene subclusters are not readily apparent from a heatmap of genes organized by hierarchical clustering. (A) The heatmap contains 1682 genes that were significantly ($p<0.01$, t-test) perturbed at least 1.7-fold, relative to the control, by one or more stress agents. The genes in the heatmap were organized by hierarchically clustering with complete linkage based on their error-weighted Pearson distances. (B) Stable gene subclusters ($\Delta T \geq 40$) were identified from the 1682 genes using coupled two-way clustering (CTWC) with superparamagnetic clustering. The columns represent gene subclusters and the rows represent genes that are ordered to match (A). Minimum gene subcluster size was set to 15 and CTWC was run with a gene depth of 5 and sample depth of 1.

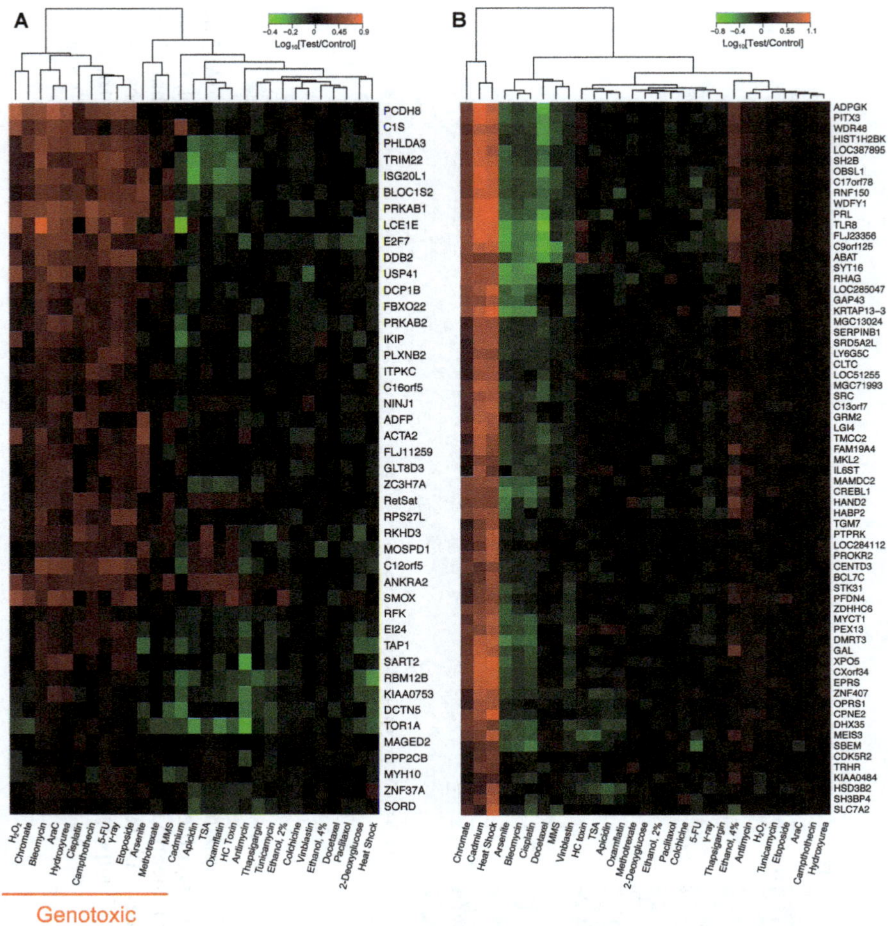

Figure 2.7 Chromate induces pleiotropic stress responses that involve genotoxic and non-genotoxic modes. (A) Cluster of chromate-responsive genes that tend to be up-regulated in response to genotoxic agents. (B) Genes that are up-regulated in response to chromate, cadmium, and heat-shock, but are either repressed or not perturbed by other genotoxic agents. The genes and toxicants in the heatmaps were organized by hierarchically clustering with complete linkage based on their error-weighted Pearson distances after selection using the coupled two-way clustering/super-paramagentic clustering biclustering algorithm.

2.4.5 Development of a Genomic Biomarker for Genotoxicity

To identify an mRNA signature indicative of genotoxicity, we applied the nearest shrunken centroids method[92] to our transcriptome data. Briefly, the standardized centroid computed for each class within a training set, where the standardized centroid is the mean expression level for each gene in a class divided by its within-class standard deviation. The standard centroid for each class is shrunken toward the overall centroid to produce the nearest shrunken centroid. The method employs a shrinkage parameter that is used to control the number of features used to construct the classifier. As a result, we identified a panel of 65 transcripts whose expression resulted in 100% accuracy as assessed by 10-fold cross-validation.[93] This gene panel, designated TGx-28.65, displayed a clear-cut separation between genotoxic and non-genotoxic agents (Figure 2.8A). A number of these genes, such as *CDKN1A*, *GADD45A*, and *TRIM22*, are regulated by p53.[24] For a complete list of the 65 genes in the TGx-28.65 biomarker, refer to Table 2.4.

2.4.6 Validation and a Case Study on Utility of the TGx-28.65 Biomarker in Human Health Risk Assessment

To assess the practical utility of this genomic biomarker-based approach in the assessment of genotoxic hazards, we evaluated the gene expression profiles of three well-characterized agents: 3-nitropropionic acid (3-NP), isopropyl methanesulfonate (iPMS), and tri-methylxanthine (caffeine). 3-NP is an irreversible inhibitor of succinate dehydrogenase (also known as complex II of the mitochondrial respiratory chain) that participates in the tricarboxylic acid cycle and electron transport chain. It forms a covalent adduct with Arg297 in the active site of succinate dehydrogenase resulting in severe energy (ATP) impairment; therefore, 3-NP is considered to be a non-genotoxic agent based on our definition of genotoxicity. iPMS belongs to a class of DNA-alkylating chemicals that are positive in standard *in vitro* gene mutation and chromosome aberration assays and thus should be readily detectable with our genomic biomarker approach as genotoxic. In contrast, the genotoxicity profile of caffeine is typical of compounds that present challenges for assessment of potential genotoxic hazard. Specifically, caffeine gives negative results in bacterial mutation assays, positive findings in *in vitro* chromosome aberration assays, and is not genotoxic *in vivo* (discussed by Goodsaid *et al.*[94]).

In order to find the appropriate doses for transcriptomic profiling of these agents, we first assessed mRNA levels of *ATF3*, *CDKN1A*, and *GADD45A* in cells treated in increasing concentrations of the agents. The concentrations that showed the most robust induction of one or more of these genes were selected and it was ensured that these were not overtly cytotoxic (as described above). As shown in Figure 2.9, the optimal dose for transcriptomic profiling was 3 mM for 3-NP, 250 $\mu g\,mL^{-1}$ for iPMS and 2 mM for caffeine. Comparison of the gene expression profiles of these test agents with the TGx-28.65 gene panel classified 3-NP as non-genotoxic and iPMS as genotoxic. Analysis of the gene expression profile of caffeine resulted in it being classified as a non-genotoxic agent using the TGx-28.65 gene classifier (Figure 2.8B).

2.5 Summary and Perspectives

TGx-28.65 is a robust transcriptomic signature for the assessment of genotoxicity using a refined *in vitro* approach. Importantly, a protocol has been established for optimal dose selection that can be used to ensure that cells are not over- or under-dosed and enables more effective comparison across toxicants. This biomarker, derived from both DNA reactive and DNA

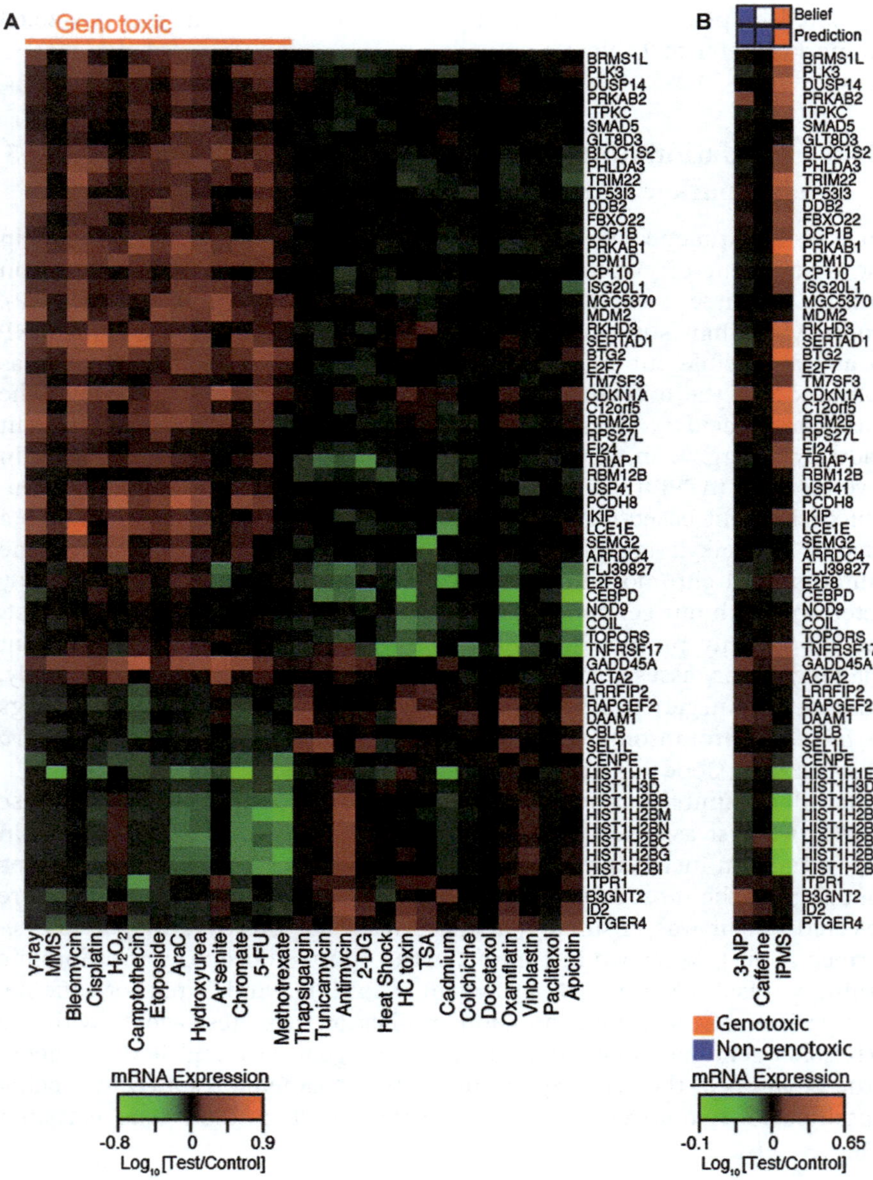

non-reactive genotoxic agents (Table 2.3), clearly distinguished the 13 genotoxic agents from the 15 non-genotoxic agents with widely varying mechanisms of action. It resulted in 100% accuracy in classifying genotoxic and non-genotoxic agents as assessed by 10-fold cross-validation.

To evaluate the practical utility of this 65-gene panel, three additional agents with known mechanisms of action were tested, including caffeine, which is known to be positive in the *in vitro* chromosome aberration assay,[95] but is negative in *in vivo* genotoxicity tests.[94] iPMS was chosen as a prototypical representative of DNA alkylating agents that is positive for all assays in the standard genotoxicity testing battery *in vitro* and *in vivo.*[96] While 3-NP is considered non-genotoxic,[97] a contaminant in previous studies may have led to a positive bacterial mutation data reported in the earlier literature.[98,99] Therefore, we carried out a *Salmonella* mutation assay to test the genotoxicity of this agent and the results were negative (data not shown). As we expected, both agents with well-characterized mechanisms were correctly classified, supporting that TGx-28.65 can be used to distinguish genotoxic and non-genotoxic agents (Figure 2.8B). Caffeine is well known to show positive findings in the *in vitro* chromosome aberration assay;[95] however, there was no genotoxic response in cells treated with caffeine, suggesting that the clastogenicity observed in the *in vitro* mammalian chromosome damage assay is not likely to be caused by DNA reactive mechanisms and is probably the consequence of general toxicity or other molecular processes. Finally, intra- and inter-laboratory reproducibility, and the reproducibility on different technical platforms have been carefully validated by us and by a second laboratory.[83] We also note that the TGx-28.65 biomarker classifies genotoxicity correctly for chemicals requiring metabolic activation (benzo[*a*]pyrene, aflatoxin B1, dexamethasone, and phenobarbital) in TK6 cells co-exposed to the aforementioned chemicals and rat liver S9.[83]

One important consideration is that our genotoxic classifier does not include some prominent aneugens in the antimitotic agent group. While the

Figure 2.8 Discrimination of genotoxic and non-genotoxic agents based on gene expression. (A) Heatmap of a 65-gene panel that was able to discriminate genotoxic from non-genotoxic agents with 100% accuracy after 10-fold cross-validation. The genes were identified applying the nearest shrunken centroids (NSC) algorithm to the 28 treatments and 1628 genes that were significantly ($p<0.01$, t-test) perturbed at least 1.7-fold, relative to the control, by at least one stress agent. The NSC threshold was set at 2.2 to ensure 100% accuracy. The genes in the heatmap were organized by hierarchically clustering with complete linkage based on their Pearson distances. (B) Classification of three test agents using the TGx-28.65 gene panel. The agents 3-nitroproprionic acid (3-NP) and isopropyl methanesulfonate (iPMS) showed correct classification compared to their known mechanisms of action. Caffeine was classified as a non-genotoxic agent. Bars above the heatmap: belief indicates the genotoxicity of agents based on their known mechanisms; prediction designates the probability of genotoxicity or non-genotoxicity using Tgx28.65. Red: genotoxic; blue: non-genotoxic; white: unclassified.

Table 2.4 Genes in TGx-28.65 classifier.

Entrez ID	Gene symbol	Response[a]	p53-regulated
59	ACTA2	↑	yes
64782	AEN	↑	yes
7832	BTG2	↑	yes
57103	C12orf5	↑	yes
1026	CDKN1A	↑	yes
1643	DDB2	↑	yes
11072	DUSP14	↑	yes
144455	E2F7	↑	yes
9538	EI24	V	yes
26263	FBXO22	↑	yes
1647	GADD45A	↑	yes
121457	IKBIP	↑	yes
4193	MDM2	↑	yes
23612	PHLDA3	↑	yes
8493	PPM1D	↑	yes
51065	RPS27L	↑	yes
50484	RRM2B	↑	yes
9540	TP53I3	↑	yes
51499	TRIAP1	↑	yes
10346	TRIM22	↑	yes
91947	ARRDC4	↑	
10678	B3GNT2	↓	
282991	BLOC1S2	↑	
84312	BRMS1L	↑	
868	CBLB	V	
9738	CCP110	↑	
1052	CEBPD	V	
1062	CENPE	↓	
8161	COIL	V	
23002	DAAM1	V	
196513	DCP1B	↑	
79733	E2F8	↑	
139285	FAM123B	V	
283464	GXYLT1	V	
3008	HIST1H1E	↓	
3018	HIST1H2BB	↓	
8347	HIST1H2BC	↓	
8339	HIST1H2BG	↓	
8346	HIST1H2BI	↓	
8342	HIST1H2BM	↓	
8341	HIST1H2BN	↓	
8351	HIST1H3D	↓	
3398	ID2	V	
80271	ITPKC	↑	
3708	ITPR1	V	
353135	LCE1E	↑	
9209	LRRFIP2	V	
84206	MEX3B	↑	
79671	NLRX1	V	
5100	PCDH8	↑	
1263	PLK3	↑	
5564	PRKAB1	↑	

Table 2.4 (*Continued*)

Entrez ID	Gene symbol	Response[a]	p53-regulated
5565	PRKAB2	↑	
5734	PTGER4	V	
9693	RAPGEF2	↓	
389677	RBM12B	V	
6400	SEL1L	V	
6407	SEMG2	↑	
29950	SERTAD1	↑	
4090	SMAD5	↑	
51768	TM7SF3	↑	
608	TNFRSF17	↑	
10210	TOPORS	V	
373856	USP41	↑	

[a]Observed transcriptomics responses to genotoxic agents by members of the TGx-28.65 dataset. ˆ represents those showing more than 1.5-fold induction for more than one third of tested genotoxic agents; ↓ designates those showing more than 1.5-fold repression in more than one third of tested genotoxic agents; remaining genes showed more variable responses and are designated with "V". Genes known to be p53-regulated are indicated in the last column.

taxols and vinblastine are effective cancer chemotherapy agents, their properties are distinct from typical cytotoxic chemotherapy agents that damage DNA. For example, the cellular responses in the NCI60 cell line panel substantially differed as exemplified by lack of correlation with p53 status with cytotoxicity for antimitotic agents.[100,101]

The strategy described here builds on our laboratory's earlier report in TK6 cells and its p53-null derivative NH32 cells where seven genotoxic agents were compared to three non-genotoxic agents and three oxidizing agents. In this earlier study using a smaller custom cDNA array with 7.7 k features,[66] some separation between agent classes was observed, but the contribution by p53 was minimized by inclusion of the p53-null cell line, and a robust classifier for genotoxicity was not further developed. When a 16-gene signature was defined to discriminate p53 status, only four genes overlapped with our current 65-gene classifier. This earlier limited study did not employ dose optimization or the current bioinformatics approach, but did assess early responses to stress at 4 h. As seen previously, there was also no appreciable cytotoxicity detected by trypan blue staining at 4 h (Figure 2.4) and assessment of viability at 24 h showed only moderate effects on viability for most agents (Figure 2.4) in the present study. A collaborative effort[83] demonstrated the utility of our TGx-28.65 biomarker by accurately classifying benzo[*a*]pyrene and aflatoxin B1 as genotoxic and dexamethasone and phenobarbital as non-genotoxic in the presence of rat liver S9 at 4 h and 8 h in TK6 cells. These results support the use of early time points (up to 8 h) in the presence of S9, but suggest that later time points (24 h) may reveal false classification for non-genotoxic agents under conditions of high cytotoxicity. However, the study also reveals that TGx-28.65 is 100% accurate in classifying genotoxicity in human liver HepaRG cells at much later time points.[83]

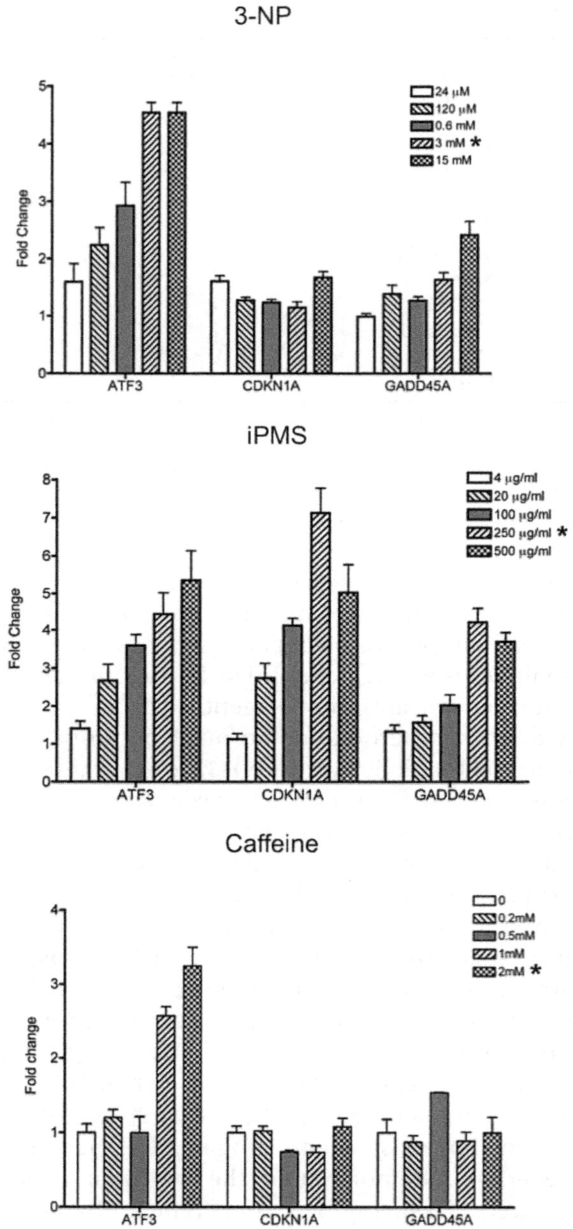

Figure 2.9 Quantitative real-time (qRT)-polymerase chain reaction (PCR) measurements for a panel of known stress-response genes (*ATF3*, *CDKN1*, and *GADD45A*) for dose–response profiles of testing agents. The doses that showed the most robust induction of one or more genes out of these three stress-response genes were selected for subsequent microarray analyses. The selected doses are indicated with asterisks (3 mM for 3-nitroproprionic acid (3-NP), 250 μg mL^{-1} for isopropyl methanesulfonate (iPMS), and 2 mM for caffeine).

Compared to previous studies (discussed in Natsoulis *et al.*[62] Li *et al.*,[70] and Godderis *et al.*[102]), our approach benefited by combining dose optimization, an informatics approach that takes into account overlapping signaling events (Figure 2.2), duration of response time (4 h), and choice of cell line. As discussed,[66,70] many toxicants, including DNA-damaging agents, trigger transcriptional and post-transcriptional responses resulting in high and frequently maximal mRNA accumulation within several hours. These toxicants usually show a dose-dependence up to very high concentrations, at which point transcriptional machinery or cell integrity is compromised. The aim here is to generate clear responses and accurate stress-response signatures rather than assess cellular parameters such as lethality and cell cycle redistribution *per se*. Many studies have relied on toxicological parameters, such as micronucleus formation, mutagenesis, or long-term (>24 h) cytotoxicity for dose determination, so subtle changes in gene expression can impact these parameters but give an unreliable signal for transcriptomics. Thus, a few marginal or sub-threshold responses can compromise classifier development, and even with very large surveys,[103] lead to some false negatives. The qRT-PCR dose-optimization approach was effective for both genotoxic and non-genotoxic agents using only three well-characterized stress genes, and, as shown in Figure 2.3B, the pattern of responses tended to differ for these two classes particularly with more robust *CDKN1A* induction for genotoxic agents. While *GADD45A* is a p53-regulated gene,[104] it can be induced by many stresses and was initially isolated in p53-deficient rodent cells.[32] *GADD45A* mRNA has been found to increase following exposure to many non-genotoxic agents, but elements of this gene's promoter may show more specificity.[105] Taken together, the dose optimization procedure provides a standard condition for every agent's effect at the selected dose, and decreases the likelihood of false negatives.

Several mammalian cell lines are commonly used in toxicology screening assays, but TK6 cells have characteristics that make them attractive for transcriptomic genotoxicity testing. For example, the commonly used mouse lymphoma and HepG2 lines are tumor-derived with compromised tumor suppressor pathways, while TK6 is a spontaneously transformed lymphoid line (discussed by Amundson *et al.*[66] and Li *et al.*[70]). Moreover, the mouse lymphoma line[106] is p53-deficient and HepG2 cells exhibit less robust responses of stress genes (in particular those where p53 contributes to their regulation; data not shown). While TK6 cells have limited capacity for metabolic activation, the approach by Buick *et al.*[83] clearly shows its broad utility for genotoxicity assessment in the presence of rat liver S9. Our approach can be applied in future to screen for additional effective lines from other tissue types.

In order to associate cellular function related to particular gene clusters, sets of genes from some of the major clusters in Figure 2.5 were subjected to ingenuity pathway analysis (Table 2.5). The pathways that were affected in the various agent clusters differed significantly, and suggest that pathway analysis can provide mechanistic insight into the action of the agents based

Table 2.5 Ingenuity pathway analysis for the specific classes. Ingenuity Pathway Analysis is shown for subset of genes from Figure 2.5, whose response was common for a particular class of stress agents. Pathways showing significant overrepresentation, $p \leq 0.05$, for each class, are shown.

Category	Overrepresented pathways	– log (*p*-value)
Genotoxicity	p53 signaling	6.79
	Cell cycle: G2/M DNA damage checkpoint regulation	3.10
	VDR/RXR activation	2.25
	Eicosanoid signaling	1.50
Endoplasmic reticulum	Endoplasmic reticulum stress pathway	6.14
	NRF2-mediated oxidative stress response	3.10
	Glycine, serine and threonine metabolism	2.56
	One carbon pool by folate	2.40
	Alanine and aspartate metabolism	1.89
	Neurotrophin/TRK signaling	1.56
Microtubule inhibitor	Cell cycle: G2/M DNA damage checkpoint regulation	1.98
	Lysine biosynthesis	1.45
Histone deacetylase inhibitors	Death receptor signaling	2.52
	Xenobiotic metabolism signaling	2.34
	Endoplasmic reticulum stress pathway	2.25
	p38 MAPK signaling	2.10
	Huntington's disease signaling	2.06
	Synaptic long term potentiation	1.83
	Sphingolipid metabolism	1.79
	Fc epsilon RI signaling	1.69

Table 2.6 Ingenuity pathway analysis for TGx-28.65: the top five overrepresented ingenuity pathways are shown for the 65 genes in TGx-28.65.

Overrepresented pathways	– log (*p*-value)
p53 Signaling	6.44
Cell cycle: G2/M DNA damage checkpoint regulation	3.50
ATM signaling	3.10
GADD45 signaling	2.78
Hereditary breast cancer signaling	2.28

on the cellular responses. The pathways of "p53 signaling" and "cell cycle: G2/M DNA damage checkpoint regulation" are highly overrepresented in our genotoxic signatures, with *p*-values of 1.6×10^{-7} and 7.9×10^{-4}, respectively. The same pathways were also the most enriched in our TGx-28.65 classifier gene set (Table 2.6). It is possible that two or more categories affect the same function, but from a different aspect. For example, functional analysis revealed that both genotoxic agents and microtubule inhibitors affect genes involved in DNA replication, recombination, and repair. However in the genotoxic group, the processes involved in this function are DNA damage recognition and processing, while in the microtubule group the processes are involved in chromosomal structure and organization.

Transcriptome data have been used to classify pharmacological agents with different mechanisms of action, such as in Lamb *et al.*,[103] where a collection of drug–response transcriptome profiles was termed the Connectivity Map (CMAP). While no dose optimization was carried out, it was possible to use CMAP to identify compounds that may have similar effects on a set of genes selected using the gene set enrichment algorithm (GSEA).[107] GSEA identifies genes that are strongly expressed, or repressed, across a set of transcriptome profiles. While GSEA will indicate whether a set of genes is strongly perturbed in a transcriptome database, it is not designed to account for genes participating in multiple gene sets. In this study we used a deconvolution method to address the challenge encountered with multiple mechanisms activated by one compound, by using a CTWC/SPC algorithm to define the stable subclusters. Biclustering approaches, such as CTWC/SPC,[91] allow genes to participate in multiple subclusters in line with pleotropic activities. Pathway analysis of gene subclusters may serve to more effectively associate them with existing biological knowledge.

Even though non-genotoxic agents represent a wide diversity of stresses and probably a myriad of pathways of toxicity, our approach should have utility in the development of robust classifiers for distinct stress responses. For example, since HDAC inhibitors are known to trigger broad transcriptional responses,[103] they showed a pronounced signature in Figure 2.5, as expected. Other signatures or subclusters can also be discerned visually in Figure 2.5 or by CTWC/SPC, such as for ER stress agents or heat shock. Unanticipated subclusters were also delineated including a prominent one triggered by Cd, Cr, and a rather severe heat shock treatment,[81] and probably represent a response to some perturbation of protein structure and/or function. While TK6 cells lack many of metabolic pathways present in hepatic cells, alcohol did elicit an appreciable response and a clear non-genotoxic signature. An ER stress response signature is unsurprising considering the relatively high concentrations of alcohol. In summary, our group has developed a toxicogenomics approach for differentiation of genotoxic agents from other toxicants based on the analysis of cellular stress responses. Our approach demonstrates the importance of the qRT-PCR molecular dose-selection protocol that established the appropriate dose by monitoring the expression of several known stress response genes to exposure with tested agents in a dose setting experiment. The approach yielded a panel of 65 biologically relevant genes that differentiate whether a particular agent is genotoxic or non-genotoxic. Our method was further independently confirmed by Buick *et al.*[83] in the presence of metabolic activation. While limited in scope, our approach should have the ability to classify such toxicants considering the diversity of non-genotoxic agents assessed in this study, especially as additional agents are assessed in the future.

We propose that our TGx-28.65 genomic biomarker is useful for risk assessment of genotoxic effects, in particular to provide mechanistic insight into positive findings in the *in vitro* chromosome aberration assays for

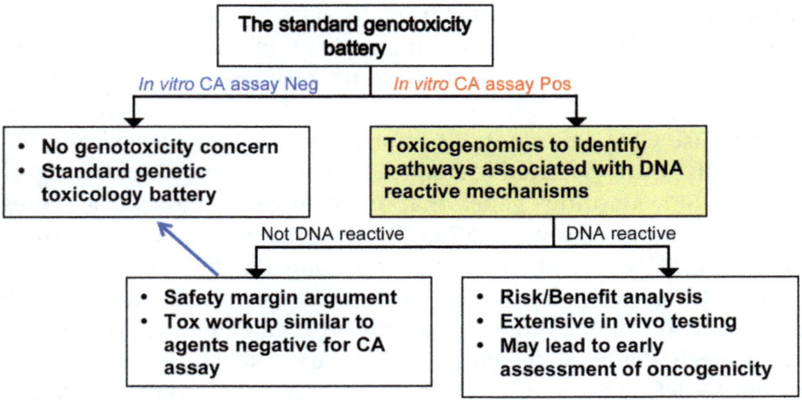

Figure 2.10 Schematic diagram for incorporation of transcriptomic biomarker component into genotoxicity risk assessment. With the toxicogenomics approach, compounds that are not DNA-reactive would be treated similarly to those that are negative in the chromosome aberration assay (designated by blue arrow).

compounds with otherwise clean genotoxicity profiles. Since the majority of such compounds are also negative in carcinogenicity tests, the risk assessment of these positive *in vitro* findings is a challenge to industry and regulatory agencies.[94] Considering that many drug candidates with otherwise clean genotoxicity profiles in the *Salmonella* assay and in the *in vivo* micronucleus assay show positive test results in the *in vitro* chromosome aberration assays, the application of genomic biomarker-derived information should add significant value to the current genotoxicity testing. In our case study, caffeine as a prototypical agent with positive findings in the chromosomal aberration assay and absence of documented human cancer risk, was evaluated by applying our toxicogenomics approach as shown schematically in Figure 2.10. In this approach the absence of the genotoxic signature is considered as a lack of genotoxicity relevant to human health.[94] Although additional studies are necessary, the genomic biomarker approach as demonstrated has the potential to complement and/or replace a number of assays currently used for developing a "weight of evidence" assessment by offering essential mechanistic information that is difficult or impossible to obtain using current methods.

There have been several transcriptomic signatures identified for assessing genotoxicity, but TGx-28.65 is the first one that shows convincing inter- and intra-laboratory reproducibility, and performs robustly and consistently on different assay platforms. Since this biomarker is comprised of only 65 genes, it is feasible to develop TGx-28.65 into the high-throughput screening application. This approach is simple, inexpensive and rapid and can be easily integrated into the evaluation of lead compounds and chemical series in early research stages of drug discovery. The broad incorporation of a genomic biomarker-based genotoxic risk assessment into the new product development process would significantly reduce animal testing. Furthermore, the strategies and protocols that were used in identifying TGx-28.65 can serve as a prototype for developing genomic biomarkers for other toxicity.

References

1. L. Muller, Y. Kikuchi, G. Probst, L. Schechtman, H. Shimada, T. Sofuni and D. Tweats, ICH-harmonised guidances on genotoxicity testing of pharmaceuticals: evolution, reasoning and impact, *Mutat. Res.*, 1999, **436**(3), 195–225.
2. H. Ellinger-Ziegelbauer, J. Aubrecht, J. C. Kleinjans and H. J. Ahr, Application of toxicogenomics to study mechanisms of genotoxicity and carcinogenicity, *Toxicol. Lett.*, 2009, **186**(1), 36–44.
3. D. Kirkland, M. Aardema, L. Henderson and L. Muller, Evaluation of the ability of a battery of three in vitro genotoxicity tests to discriminate rodent carcinogens and non-carcinogens I. Sensitivity, specificity and relative predictivity, *Mutat. Res.*, 2005, **584**(1–2), 1–256.
4. R. D. Snyder and J. W. Green, A review of the genotoxicity of marketed pharmaceuticals, *Mutat. Res.*, 2001, **488**(2), 151–169.
5. L. Henderson, S. Albertini and M. Aardema, Thresholds in genotoxicity responses, *Mutat. Res.*, 2000, **464**(1), 123–128.
6. D. J. Kirkland and L. Muller, Interpretation of the biological relevance of genotoxicity test results: the importance of thresholds, *Mutat. Res.*, 2000, **464**(1), 137–147.
7. M. Kirsch-Volders, A. Vanhauwaert, U. Eichenlaub-Ritter and I. Decordier, Indirect mechanisms of genotoxicity, *Toxicol. Lett.*, 2003, **140–141**, 63–74.
8. A. E. Pegg and T. L. Byers, Repair of DNA containing O6-alkylguanine, *FASEB J.*, 1992, **6**(6), 2302–2310.
9. R. D. Snyder and L. B. Hendry, Toward a greater appreciation of non-covalent chemical/DNA interactions: application of biological and computational approaches, *Environ. Mol. Mutagen.*, 2005, **45**(2–3), 100–105.
10. J. M. Woynarowski, M. McHugh, R. D. Sigmund and T. A. Beerman, Modulation of topoisomerase II catalytic activity by DNA minor groove binding agents distamycin, Hoechst 33258, and 4′,6-diamidine-2-phenylindole, *Mol. Pharmacol.*, 1989, **35**(2), 177–182.
11. H. Kappus, Oxidative stress in chemical toxicity, *Arch. Toxicol.*, 1987, **60**(1–3), 144–149.
12. P. C. Dedon and I. H. Goldberg, Free-radical mechanisms involved in the formation of sequence-dependent bistranded DNA lesions by the antitumor antibiotics bleomycin, neocarzinostatin, and calicheamicin, *Chem. Res. Toxicol.*, 1992, **5**(3), 311–332.
13. M. R. Said, T. J. Begley, A. V. Oppenheim, D. A. Lauffenburger and L. D. Samson, Global network analysis of phenotypic effects: protein networks and toxicity modulation in Saccharomyces cerevisiae, *Proc. Natl. Acad. Sci. U. S. A.*, 2004, **101**(52), 18006–18011.

14. J. N. Weinstein, K. W. Kohn, M. R. Grever, V. N. Viswanadhan, L. V. Rubinstein, A. P. Monks, D. A. Scudiero, L. Welch, A. D. Koutsoukos, A. J. Chiausa and K. D. Paull, Neural computing in cancer drug development: predicting mechanism of action, *Science*, 1992, **258**(5081), 447–451.
15. J. N. Weinstein and J. K. Buolamwini, Molecular targets in cancer drug discovery: cell-based profiling, *Curr. Pharm. Des.*, 2000, **6**(4), 473–483.
16. C. J. Bakkenist and M. B. Kastan, Initiating cellular stress responses, *Cell*, 2004, **118**(1), 9–17.
17. M. O'Driscoll and P. A. Jeggo, The role of double-strand break repair – insights from human genetics, *Nat. Rev. Genet.*, 2006, **7**(1), 45–54.
18. M. Maugeri-Sacca, M. Bartucci and R. De Maria, DNA damage repair pathways in cancer stem cells, *Mol. Cancer Ther.*, 2012, **11**(8), 1627–1636.
19. S. A. Amundson and A. J. Fornace, Jr, in *3* ed. M. Karin, *et al.*, Academic Press, San Diego, CA, 2003, pp. 179–184.
20. W. Wang, W. C. Ho, D. T. Dicker, C. MacKinnon, J. D. Winkler, R. Marmorstein and W. S. El-Deiry, Acridine derivatives activate p53 and induce tumor cell death through Bax, *Cancer Biol. Ther.*, 2005, **4**(8), 893–898.
21. H. R. Lu, L. H. Meng, M. Huang, H. Zhu, Z. H. Miao and J. Ding, DNA damage, c-myc suppression and apoptosis induced by the novel topoisomerase II inhibitor, salvicine, in human breast cancer MCF-7 cells, *Cancer Chemother. Pharmacol.*, 2005, **55**(3), 286–294.
22. J. Horiguchi-Yamada, S. Fukumi, S. Saito, R. Nakayama, S. Iwase and H. Yamada, DNA topoisomerase II inhibitor, etoposide, induces p21WAF1/CIP1 through down-regulation of c-Myc in K562 cells, *Anticancer Res.*, 2002, **22**(6C), 3827–3832.
23. T. Tokino, S. Thiagalingam, W. S. el-Deiry, T. Waldman, K. W. Kinzler and B. Vogelstein, p53 tagged sites from human genomic DNA, *Hum. Mol. Genet.*, 1994, **3**(9), 1537–1542.
24. T. Riley, E. Sontag, P. Chen and A. Levine, Transcriptional control of human p53-regulated genes, *Nat. Rev. Mol. Cell Biol.*, 2008, **9**(5), 402–412.
25. B. Wang, Z. Xiao and E. C. Ren, Redefining the p53 response element, *Proc. Natl. Acad. Sci. U. S. A.*, 2009, **106**(34), 14373–14378.
26. T. F. Burns and W. S. El-Deiry, Microarray analysis of p53 target gene expression patterns in the spleen and thymus in response to ionizing radiation, *Cancer Biol. Ther.*, 2003, **2**(4), 431–443.
27. S. T. Younger, D. Kenzelmann-Broz, H. Jung, L. D. Attardi and J. L. Rinn, Integrative genomic analysis reveals widespread enhancer regulation by p53 in response to DNA damage, *Nucleic Acids Res.*, 2015, **43**(9), 4447–4462.
28. P. A. Cahill, A. W. Knight, N. Billinton, M. G. Barker, L. Walsh, P. O. Keenan, C. V. Williams, D. J. Tweats and R. M. Walmsley, The GreenScreen genotoxicity assay: a screening validation programme, *Mutagenesis*, 2004, **19**(2), 105–119.

29. Y. Oda, S. Nakamura, I. Oki, T. Kato and H. Shinagawa, Evaluation of the new system (umu-test) for the detection of environmental mutagens and carcinogens, *Mutat. Res.*, 1985, **147**(5), 219–229.
30. P. Quillardet and M. Hofnung, The SOS chromotest: a review, *Mutat. Res.*, 1993, **297**(3), 235–279.
31. D. van der Lelie, L. Regniers, B. Borremans, A. Provoost and L. Verschaeve, The VITOTOX test, an SOS bioluminescence Salmonella typhimurium test to measure genotoxicity kinetics, *Mutat. Res.*, 1997, **389**(2–3), 279–290.
32. A. J. Fornace, Jr, I. J. Alamo and M. C. Hollander, DNA damage-inducible transcripts in mammalian cells, *Proc. Natl. Acad. Sci. U. S. A.*, 1988, **85**, 8800–8804.
33. A. J. Fornace, Jr, J. Jackman, M. C. Hollander, B. Hoffman-Liebermann and D. A. Liebermann, Genotoxic-stress-response genes and growth-arrest genes; the gadd, MyD, and other genes induced by treatments eliciting growth arrest, *Ann. N. Y. Acad. Sci.*, 1992, **663**, 139–153.
34. Z. G. Liu, R. Baskaran, E. T. Lea-Chou, L. D. Wood, Y. Chen, M. Karin and J. Y. Wang, Three distinct signalling responses by murine fibroblasts to genotoxic stress, *Nature*, 1996, **384**, 273–276.
35. P. Jennings, A. Limonciel, L. Felice and M. O. Leonard, An overview of transcriptional regulation in response to toxicological insult, *Arch. Toxicol.*, 2013, **87**(1), 49–72.
36. P. W. Hastwell, L. L. Chai, K. J. Roberts, T. W. Webster, J. S. Harvey, R. W. Rees and R. M. Walmsley, High-specificity and high-sensitivity genotoxicity assessment in a human cell line: validation of the Green-Screen HC GADD45a-GFP genotoxicity assay, *Mutat. Res.*, 2006, **607**(2), 160–175.
37. M. D. Todd, M. J. Lee, J. L. Williams, J. M. Nalezny, P. Gee, M. B. Benjamin and S. B. Farr, The CAT-Tox (L) assay: a sensitive and specific measure of stress-induced transcription in transformed human liver cells, *Fundam. Appl. Toxicol.*, 1995, **28**(1), 118–128.
38. J. Yang and P. Duerksen-Hughes, A new approach to identifying genotoxic carcinogens: p53 induction as an indicator of genotoxic damage, *Carcinogenesis*, 1998, **19**(6), 1117–1125.
39. P. J. Duerksen-Hughes, J. Yang and O. Ozcan, p53 induction as a genotoxic test for twenty-five chemicals undergoing in vivo carcinogenicity testing, *Environ. Health Perspect.*, 1999, **107**(10), 805–812.
40. S. A. Amundson, T. G. Myers and A. J. Fornace, Jr, Roles for p53 in growth arrest and apoptosis: Putting on the brakes after genotoxic stress, *Oncogene*, 1998, **17**, 3287–3300.
41. L. J. Ko and C. Prives, p53: puzzle and paradigm, *Genes Dev.*, 1996, **10**(9), 1054–1072.
42. U. G. Lopes, P. Erhardt, R. Yao and G. M. Cooper, p53-dependent induction of apoptosis by proteasome inhibitors, *J. Biol. Chem.*, 1997, **272**(20), 12893–12896.

43. J. Aubrecht, R. K. Narla, P. Ghosh, J. Stanek and F. M. Uckun, Molecular genotoxicity profiles of apoptosis-inducing vanadocene complexes, *Toxicol. Appl. Pharmacol.*, 1999, **154**(3), 228–235.
44. R. Vincent, P. Goegan, G. Johnson, J. R. Brook, P. Kumarathasan, L. Bouthillier and R. T. Burnett, Regulation of promoter-CAT stress genes in HepG2 cells by suspensions of particles from ambient air, *Fundam. Appl. Toxicol.*, 1997, **39**(1), 18–32.
45. M. D. Sutton, B. T. Smith, V. G. Godoy and G. C. Walker, The SOS response: recent insights into umuDC-dependent mutagenesis and DNA damage tolerance, *Annu. Rev. Genet.*, 2000, **34**, 479–497.
46. G. W. Birrell, J. A. Brown, H. I. Wu, G. Giaever, A. M. Chu, R. W. Davis and J. M. Brown, Transcriptional response of Saccharomyces cerevisiae to DNA-damaging agents does not identify the genes that protect against these agents, *Proc. Natl. Acad. Sci. U. S. A.*, 2002, **99**(13), 8778–8783.
47. S. W. Ruby and J. W. Szostak, Specific Saccharomyces cerevisiae genes are expressed in response to DNA-damaging agents, *Mol. Cell Biol.*, 1985, **5**(1), 75–84.
48. S. A. Amundson, A. Patterson, K. T. Do and A. J. Fornace, Jr, A nucleotide excision repair master-switch: p53 regulated coordinate induction of global genomic repair genes, *Cancer Biol. Ther.*, 2002, **1**(2), 145–149.
49. M. B. Kastan and J. Bartek, Cell-cycle checkpoints and cancer, *Nature*, 2004, **432**(7015), 316–323.
50. A. R. Snyder and W. F. Morgan, Gene expression profiling after irradiation: clues to understanding acute and persistent responses?, *Cancer Metastasis Rev.*, 2004, **23**(3–4), 259–268.
51. S. A. Amundson, M. Bittner and A. J. Fornace, Jr, Functional genomics as a window on radiation stress signaling, *Oncogene*, 2003, **22**(37), 5828–5833.
52. S. Saito, H. Yamaguchi, Y. Higashimoto, C. Chao, Y. Xu, A. J. Fornace, Jr, E. Appella and C. W. Anderson, Phosphorylation site interdependence of human p53 post-translational modifications in response to stress, *J. Biol. Chem.*, 2003, **278**(39), 37536–37544.
53. D. V. Bulavin, O. N. Demidov, S. Saito, P. Kauraniemi, C. Phillips, S. A. Amundson, C. Ambrosino, G. Sauter, A. R. Nebreda, C. W. Anderson, A. Kallioniemi, A. J. Fornace, Jr and E. Appella, Amplification of PPM1D in human tumors abrogates p53 tumor-suppressor activity, *Nat. Genet.*, 2002, **31**, 210–215.
54. M. Schena, D. Shalon, R. W. Davis and P. O. Brown, Quantitative monitoring of gene expression patterns with a complementary DNA microarray, *Science*, 1995, **270**(5235), 467–470.
55. T. R. Hughes, M. J. Marton, A. R. Jones, C. J. Roberts, R. Stoughton, C. D. Armour, H. A. Bennett, E. Coffey, H. Dai, Y. D. He, M. J. Kidd, A. M. King, M. R. Meyer, D. Slade, P. Y. Lum, S. B. Stepaniants, D. D. Shoemaker, D. Gachotte, K. Chakraburtty, J. Simon, M. Bard and S. H. Friend, Functional discovery via a compendium of expression profiles, *Cell*, 2000, **102**(1), 109–126.

56. S. A. Jelinsky, P. Estep, G. M. Church and L. D. Samson, Regulatory networks revealed by transcriptional profiling of damaged Saccharomyces cerevisiae cells: Rpn4 links base excision repair with proteasomes, *Mol. Cell Biol.*, 2000, **20**, 8157–8167.
57. E. Caba, D. A. Dickinson, G. R. Warnes and J. Aubrecht, Differentiating mechanisms of toxicity using global gene expression analysis in Saccharomyces cerevisiae, *Mutat. Res.*, 2005, **575**(1–2), 34–46.
58. J. Aubrecht and E. Caba, Gene expression profile analysis: an emerging approach to investigate mechanisms of genotoxicity, *Pharmacogenomics*, 2005, **6**(4), 419–428.
59. R. D. Irwin, G. A. Boorman, M. L. Cunningham, A. N. Heinloth, D. E. Malarkey and R. S. Paules, Application of toxicogenomics to toxicology: basic concepts in the analysis of microarray data, *Toxicol. Pathol.*, 2004, **32**(Suppl 1), 72–83.
60. L. D. Kier, R. Neft, L. Tang, R. Suizu, T. Cook, K. Onsurez, K. Tiegler, Y. Sakai, M. Ortiz, T. Nolan, U. Sankar and A. P. Li, Applications of microarrays with toxicologically relevant genes (tox genes) for the evaluation of chemical toxicants in Sprague Dawley rats in vivo and human hepatocytes in vitro, *Mutat. Res.*, 2004, **549**(1–2), 101–113.
61. H. K. Hamadeh, P. R. Bushel, S. Jayadev, O. DiSorbo, L. Bennett, L. Li, R. Tennant, R. Stoll, J. C. Barrett, R. S. Paules, K. Blanchard and C. A. Afshari, Prediction of compound signature using high density gene expression profiling, *Toxicol. Sci.*, 2002, **67**(2), 232–240.
62. G. Natsoulis, L. El Ghaoui, G. R. Lanckriet, A. M. Tolley, F. Leroy, S. Dunlea, B. P. Eynon, C. I. Pearson, S. Tugendreich and K. Jarnagin, Classification of a large microarray data set: algorithm comparison and analysis of drug signatures, *Genome Res.*, 2005, **15**(5), 724–736.
63. Y. Cui and R. S. Paules, Use of transcriptomics in understanding mechanisms of drug-induced toxicity, *Pharmacogenomics*, 2010, **11**(4), 573–585.
64. J. H. van Delft, E. van Agen, S. G. van Breda, M. H. Herwijnen, Y. C. Staal and J. C. Kleinjans, Discrimination of genotoxic from non-genotoxic carcinogens by gene expression profiling, *Carcinogenesis*, 2004, **25**(7), 1265–1276.
65. M. Islaih, B. Li, I. A. Kadura, J. L. Reid-Hubbard, J. T. Deahl, J. L. Altizer, D. E. Watson and R. K. Newton, Comparison of gene expression changes induced in mouse and human cells treated with direct-acting mutagens, *Environ. Mol. Mutagen.*, 2004, **44**(5), 401–419.
66. S. A. Amundson, K. T. Do, L. Vinikoor, C. A. Koch-Paiz, M. L. Bittner, J. M. Trent, P. Meltzer and A. J. J. Fornace, Stress-specific signatures: expression profiling of p53 wild-type and -null human cells, *Oncogene*, 2005, **24**(28), 4572–4579.
67. D. A. Dickinson, G. R. Warnes, G. Quievryn, J. Messer, A. Zhitkovich, E. Rubitski and J. Aubrecht, Differentiation of DNA reactive and non-reactive genotoxic mechanisms using gene expression profile analysis, *Mutat. Res.*, 2004, **549**(1–2), 29–41.

68. T. Hu, D. P. Gibson, G. J. Carr, S. M. Torontali, J. P. Tiesman, J. G. Chaney and M. J. Aardema, Identification of a gene expression profile that discriminates indirect-acting genotoxins from direct-acting genotoxins, *Mutat. Res.*, 2004, **549**(1–2), 5–27.
69. V. S. Wilson, N. Keshava, S. Hester, D. Segal, W. Chiu, C. M. Thompson and S. Y. Euling, Utilizing toxicogenomic data to understand chemical mechanism of action in risk assessment, *Toxicol. Appl. Pharmacol.*, 2013, **271**(3), 299–308.
70. H. H. Li, J. Aubrecht and A. J. J. Fornace, Toxicogenomics: overview and potential applications for the study of non-covalent DNA interacting chemicals, *Mutat. Res.*, 2007, **623**(1–2), 98–108.
71. D. R. Hyduke, S. A. Amundson and A. J. J. Fornace, *Regulation of Organelle and Cell Compartment Signaling*, ed. R. A. Bradshaw and E. A. Dennis, Academic Press, San Diego, Complexity of stress signaling and responses, pp. 187–206.
72. C. Magkoufopoulou, S. M. Claessen, D. G. Jennen, J. C. Kleinjans and J. H. van Delft, Comparison of phenotypic and transcriptomic effects of false-positive genotoxins, true genotoxins and non-genotoxins using HepG2 cells, *Mutagenesis*, 2011, **26**(5), 593–604.
73. D. Jennen, A. Ruiz-Aracama, C. Magkoufopoulou, A. Peijnenburg, A. Lommen, J. van Delft and J. Kleinjans, Integrating transcriptomics and metabonomics to unravel modes-of-action of 2,3,7,8-tetrachlorodibenzo-p-dioxin (TCDD) in HepG2 cells, *BMC Syst. Biol.*, 2011, **5**, 139.
74. T. Hartung and M. McBride, Food for Thought. on mapping the human toxome, *ALTEX*, 2011, **28**(2), 83–93.
75. Q. Zhang, H. Xu, S. Zheng, M. Su and J. Wang, Genotoxicity of mesoporous silica nanoparticles in human embryonic kidney 293 cells, *Drug Test. Anal.*, 2015.
76. K. Mathijs, K. J. Brauers, D. G. Jennen, D. Lizarraga, J. C. Kleinjans and J. H. van Delft, Gene expression profiling in primary mouse hepatocytes discriminates true from false-positive genotoxic compounds, *Mutagenesis*, 2010, **25**(6), 561–568.
77. K. J. Vinoth, J. Manikandan, S. Sethu, L. Balakrishnan, A. Heng, K. Lu, M. P. Hande and T. Cao, Evaluation of human embryonic stem cells and their differentiated fibroblastic progenies as cellular models for in vitro genotoxicity screening, *J. Biotechnol.*, 2014, **184**, 154–168.
78. M. A. Coleman, E. Yin, L. E. Peterson, D. Nelson, K. Sorensen, J. D. Tucker and A. J. Wyrobek, Low-dose irradiation alters the transcript profiles of human lymphoblastoid cells including genes associated with cytogenetic radioadaptive response, *Radiat. Res.*, 2005, **164**(4 Pt 1), 369–382.
79. A. J. J. Fornace, D. W. Nebert, M. C. Hollander, J. D. Luethy, M. Papathanasiou, J. Fargnoli and N. J. Holbrook, Mammalian genes coordinately regulated by growth arrest signals and DNA-damaging agents, *Mol. Cell. Biol.*, 1989, **9**(10), 4196–4203.

80. H. Ellinger-Ziegelbauer, J. M. Fostel, C. Aruga, D. Bauer, E. Boitier, S. Deng, D. Dickinson, A. C. Le Fevre, A. J. J. Fornace, O. Grenet, Y. Gu, J. C. Hoflack, M. Shiiyama, R. Smith, R. D. Snyder, C. Spire, G. Tanaka and J. Aubrecht, Characterization and interlaboratory comparison of a gene expression signature for differentiating genotoxic mechanisms, *Toxicol. Sci.*, 2009, **110**(2), 341–352.
81. A. J. Fornace, Jr, I. J. Alamo, M. C. Hollander and E. Lamoreaux, Induction of heat shock protein transcripts and B2 transcripts by various stresses in Chinese hamster cells, *Exp. Cell Res.*, 1989, **182**(1), 61–74.
82. H. H. Li, D. R. Hyduke, R. Chen, P. Heard, C. L. Yauk, J. Aubrecht and A. J. J. Fornace, Development of a toxicogenomics signature for genotoxicity using a dose-optimization and informatics strategy in human cells, *Environ. Mol. Mutagen.*, 2015, **56**(6), 505–519.
83. J. K. Buick, I. Moffat, A. Williams, C. D. Swartz, L. Recio, D. R. Hyduke, H. H. Li, A. J. J. Fornace, J. Aubrecht and C. L. Yauk, Integration of metabolic activation with a predictive toxicogenomics signature to classify genotoxic versus nongenotoxic chemicals in human TK6 cells, *Environ. Mol. Mutagen.*, 2015, **56**(6), 520–534.
84. G. S. Akerman, B. A. Rosenzweig, O. E. Domon, L. J. McGarrity, L. R. Blankenship, C. A. Tsai, S. J. Culp, J. T. MacGregor, F. D. Sistare, J. J. Chen and S. M. Morris, Gene expression profiles and genetic damage in benzo(a)pyrene diol epoxide-exposed TK6 cells, *Mutat. Res.*, 2004, **549**(1–2), 43–64.
85. F. Fan, S. Jin, S. A. Amundson, T. Tong, W. Fan, H. Zhao, X. Zhu, L. Mazzacurati, X. Li, K. L. Petrik, A. J. Fornace, Jr, B. Rajasekaran and Q. Zhan, ATF3 induction following DNA damage is regulated by distinct signaling pathways and over-expression of ATF3 protein suppresses cells growth, *Oncogene*, 2002, **21**(49), 7488–7496.
86. F. Fallo, C. Pilon, L. Barzon, M. Pistorello, U. Pagotto, G. Altavilla, M. Boscaro and N. Sonino, Effects of taxol on the human NCI-H295 adrenocortical carcinoma cell line, *Endocr. Res.*, 1996, **22**(4), 709–715.
87. T. A. Patterson, E. K. Lobenhofer, S. B. Fulmer-Smentek, P. J. Collins, T. M. Chu, W. Bao, H. Fang, E. S. Kawasaki, J. Hager, I. R. Tikhonova, S. J. Walker, L. Zhang, P. Hurban, F. de Longueville, J. C. Fuscoe, W. Tong, L. Shi and R. D. Wolfinger, Performance comparison of one-color and two-color platforms within the MicroArray Quality Control (MAQC) project, *Nat. Biotechnol.*, 2006, **24**(9), 1140–1150.
88. K. Salnikow and A. Zhitkovich, Genetic and epigenetic mechanisms in metal carcinogenesis and cocarcinogenesis: nickel, arsenic, and chromium, *Chem. Res. Toxicol.*, 2008, **21**(1), 28–44.
89. R. F. Whiting, H. F. Stich and D. J. Koropatnick, DNA damage and DNA repair in cultured human cells exposed to chromate, *Chem. Biol. Interact.*, 1979, **26**(3), 267–280.
90. A. J. J. Fornace, D. S. Seres, J. F. Lechner and C. C. Harris, DNA-protein cross-linking by chromium salts, *Chem Biol Interact*, 1981, **36**(3), 345–354.

91. G. Getz, E. Levine and E. Domany, Coupled two-way clustering analysis of gene microarray data, *Proc. Natl. Acad. Sci. U. S. A.*, 2000, **97**(22), 12079–12084.
92. R. Tibshirani, T. Hastie, B. Narasimhan and G. Chu, Diagnosis of multiple cancer types by shrunken centroids of gene expression, *Proc. Natl. Acad. Sci. U. S. A.*, 2002, **99**(10), 6567–6572.
93. T. Hastie, R. Tibshirani and J. H. Friedman, *The Elements of Statistical Learning: Data Mining, Inference, and Prediction: With 200 Full-Color Illustrations*, Springer, New York, 2001, vol. xvi, p. 533.
94. F. M. Goodsaid, S. Amur, J. Aubrecht, M. E. Burczynski, K. Carl, J. Catalano, R. Charlab, S. Close, C. Cornu-Artis, L. Essioux, A. J. J. Fornace, L. Hinman, H. Hong, I. Hunt, D. Jacobson-Kram, A. Jawaid, D. Laurie, L. Lesko, H. H. Li, K. Lindpaintner, J. Mayne, P. Morrow, M. Papaluca-Amati, T. W. Robison, J. Roth, I. Schuppe-Koistinen, L. Shi, O. Spleiss, W. Tong, S. L. Truter, J. Vonderscher, A. Westelinck, L. Zhang and I. Zineh, Voluntary exploratory data submissions to the US FDA and the EMA: experience and impact, *Nat. Rev. Drug Discovery*, 2010, **9**(6), 435–445.
95. D. Weinstein, I. Mauer, M. L. Katz and S. Kazmer, The effect of methylxanthines on chromosomes of human lyphocytes in culture, *Mutat. Res.*, 1975, **31**(1), 57–61.
96. A. Segal, I. Seidman, S. Melchionne, R. E. Albert and A. C. Upton, Rapid induction of thymic lymphomas by isopropyl methanesulfonate: a preliminary report, *Proc. Soc. Exp. Biol. Med.*, 1986, **183**(1), 132–135.
97. Y. Oshiro, C. E. Piper, P. S. Balwierz and S. G. Soelter, Chinese hamster ovary cell assays for mutation and chromosome damage: data from non-carcinogens, *J. Appl. Toxicol.*, 1991, **11**(3), 167–177.
98. E. Zeiger, B. Anderson, S. Haworth, T. Lawlor and K. Mortelmans, Salmonella mutagenicity tests: IV. Results from the testing of 300 chemicals, *Environ. Mol. Mutagen.*, 1988, **11**(Suppl 12), 1–157.
99. T. J. Hansen, Ames mutagenicity tests on purified 3-nitropropionic acid, *Food Chem. Toxicol.*, 1984, **22**(5), 399–401.
100. J. N. Weinstein, T. G. Myers, P. M. O'Connor, S. H. Friend, A. J. Fornace, Jr, K. W. Kohn, T. Fojo, S. E. Bates, L. V. Rubinstein, N. L. Anderson, J. K. Buolamwini, W. W. Osdol, A. P. Monks, D. A. Scudiero, E. A. Sausville, D. W. Zaharevitz, B. Bunow, G. S. Johnson, R. E. Wittes and K. D. Paull, An Information Intensive Approach to the Molecular Pharmacology of Cancer, *Science*, 1997, **275**, 343–349.
101. P. M. O'Connor, J. Jackman, I. Bae, T. G. Myers, S. Fan, M. Mutoh, D. A. Scudiero, A. Monks, E. A. Sausville, J. N. Weinstein, S. Friend, A. J. J. Fornace and K. W. Kohn, Characterization of the p53 tumor suppressor pathway in cell lines of the National Cancer Institute anticancer drug screen and correlations with the growth-inhibitory potency of 123 anticancer agents, *Cancer Res.*, 1997, **57**, 4285–4300.
102. L. Godderis, R. Thomas, A. E. Hubbard, A. M. Tabish, P. Hoet, L. Zhang, M. T. Smith, H. Veulemans and C. M. McHale, Effect of chemical

mutagens and carcinogens on gene expression profiles in human TK6 cells, *PLoS One*, 2012, 7(6), e39205.

103. J. Lamb, E. D. Crawford, D. Peck, J. W. Modell, I. C. Blat, M. J. Wrobel, J. Lerner, J. P. Brunet, A. Subramanian, K. N. Ross, M. Reich, H. Hieronymus, G. Wei, S. A. Armstrong, S. J. Haggarty, P. A. Clemons, R. Wei, S. A. Carr, E. S. Lander and T. R. Golub, The Connectivity Map: using gene-expression signatures to connect small molecules, genes, and disease, *Science*, 2006, **313**(5795), 1929–1935.

104. M. B. Kastan, Q. Zhan, W. S. el-Deiry, F. Carrier, T. Jacks, W. V. Walsh, B. S. Plunkett, B. Vogelstein and A. J. J. Fornace, A mammalian cell cycle checkpoint pathway utilizing p53 and GADD45 is defective in ataxia-telangiectasia, *Cell*, 1992, **71**(4), 587–597.

105. A. M. Lynch, J. C. Sasaki, R. Elespuru, D. Jacobson-Kram, V. Thybaud, M. De Boeck, M. J. Aardema, J. Aubrecht, R. D. Benz, S. D. Dertinger, G. R. Douglas, P. A. White, P. A. Escobar, A. J. Fornace, M. Honma, R. T. Naven, J. F. Rusling, R. H. Schiestl, R. M. Walmsley, E. Yamamura, J. van Benthem and J. H. Kim, New and emerging technologies for genetic toxicity testing, *Environ. Mol. Mutagen.*, 2011, **52**(3), 205–223.

106. R. D. Storer, A. R. Kraynak, T. W. McKelvey, M. C. Elia, T. L. Goodrow and J. G. DeLuca, The mouse lymphoma L5178Y Tk+/− cell line is heterozygous for a codon 170 mutation in the p53 tumor suppressor gene, *Mutat. Res.*, 1997, **373**(2), 157–165.

107. A. Subramanian, P. Tamayo, V. K. Mootha, S. Mukherjee, B. L. Ebert, M. A. Gillette, A. Paulovich, S. L. Pomeroy, T. R. Golub, E. S. Lander and J. P. Mesirov, Gene set enrichment analysis: a knowledge-based approach for interpreting genome-wide expression profiles, *Proc. Natl. Acad. Sci. U. S. A.*, 2005, **102**(43), 15545–15550.

CHAPTER 3

Toxicogenomics In vitro*: Gene Expression Signatures for Differentiating Genotoxic Mechanisms*

JULIE K. BUICK AND CAROLE L. YAUK*

Mechanistic Studies Division, Environmental Health Science and Research Bureau, Environmental and Radiation Health Sciences Directorate, Healthy Environments and Consumer Safety Branch, Health Canada, 50 Colombine Driveway, PL0803A, Ottawa, Ontario K1A 0K9, Canada
*Email: carole.yauk@canada.ca; carole.yauk@hc-sc.gc.ca

3.1 Introduction

Toxicological risk assessment of chemicals in food, pharmaceuticals, consumer products, occupational settings, and the environment is critical to public health. Current approaches are both time- and cost-intensive and they rely heavily on high-dose animal studies, measurement of apical effects, and extrapolation methods for human exposure and relevance. Advanced scientific methods involving rapidly evolving molecular biology, genomics, biotechnology, and bioinformatics fields are required to derive high-quality scientific data that are relevant to human health in a timely and cost-effective manner.

Over the past decade government agencies worldwide have begun to invest significant efforts into the development of new approaches to chemical

Issues in Toxicology No. 28
Toxicogenomics in Predictive Carcinogenicity
Edited by Michael D. Waters and Russell S. Thomas

Published by the Royal Society of Chemistry, www.rsc.org

evaluations, as the number of substances currently requiring assessment greatly exceeds the capabilities of the existing system.[1,2] Major reports have been published that call for an evolution towards higher throughput, targeted *in vitro* pathway- and mechanism-based approaches that leverage the significant progress made in the fields of genomics and bioinformatics.[3–5] Indeed, high-throughput techniques, computational toxicology, toxicogenomics, and systems biology have significantly advanced these efforts in the last decade,[6] enabling a movement towards targeted *in vitro* assays in relevant cell lines to detect pathway perturbations in a more timely and cost-effective manner.[7] However, a lack of standardized test methods, and challenges with biological interpretation and data analysis have hindered the integration of toxicogenomics data into routine regulatory decision making.[8,9]

Toxicogenomics is the application of 'omic'-based technologies (genomics, proteomics, and metabolomics) and advanced statistical and bioinformatics practices to the field of toxicological research.[10] More than 15 years ago, it was proposed that toxicogenomics would be extremely valuable for the purposes of mechanistic and predictive toxicology.[11–13] Indeed, toxicogenomics changes should precede an adverse event and thus be useful as an early surrogate marker of toxicological effects.

Today, characterization of global gene expression profiles is primarily used to investigate mode-of-action (MoA). In combination with traditional toxicity endpoints, analysis of genomic perturbations can provide rich mechanistic information to inform dose–response relationships following chemical insult to complement current testing approaches. Toxicogenomics experiments can be used to streamline further chemical testing and thus may offer considerable cost- and time-savings relative to applying the full suite of toxicological tests. Moreover, higher throughput gene expression profiling techniques are being developed, further reducing time and cost for MoA discovery.[14–18] Indeed, it is widely believed that toxicogenomics analysis has the potential to improve the efficiency of chemical screening through the application of signatures that are predictive of specific toxicological outcomes.[19–21] However, the technology has yet to be fully implemented and the maximum potential realized.[22]

In this chapter, we summarize progress made in the development of *in vitro* toxicogenomics signatures to predict genotoxicity and explore methods to address the major shortcoming of work with cells in culture: the lack of or reduced endogenous metabolic capabilities of the models. We begin by defining and describing progress in predictive toxicogenomics in general, and provide a rationale for focusing on genetic toxicology in this chapter. We briefly review existing toxicogenomics signatures for genotoxicity and describe work addressing metabolic issues for toxicogenomics *in vitro* models. We devote some attention to describing our own research focus, which has been on identifying and developing human cell-culture models that integrate metabolic activation (human TK6 cells + S9), or models that express both phase I and phase II metabolic enzymes (human HepaRG cells), with toxicogenomics signatures of genotoxicity. We finish by considering additional factors that must be taken into account as we move forward in implementing predictive *in vitro* signatures, and present our opinions on the state of the science for application of existing signatures in chemical screening.

3.2 Predictive Toxicogenomics

3.2.1 Overview of Predictive Toxicogenomics

As described in Chapter 1, predictive toxicogenomics refers to the development and validation of gene expression 'signatures' of specific MoAs or adverse effects. These signatures are derived from controlled experiments on well-characterized chemicals demonstrated to operate through specific MoAs (or inducing a particular effect), in parallel with analysis of agents that are known to not induce that specific MoA/effect. This group of positive and negative control chemicals is collectively called the 'training set' of agents. Typically, cells in culture or animals are exposed to the training set of chemicals. RNA is extracted from the exposed cells or tissues and is subject to gene expression analysis (usually by DNA microarrays, or real-time reverse transcriptase quantitative polymerase chain reaction). Advanced statistical modeling and bioinformatics techniques are used to identify gene expression changes that can be used to classify the chemicals into their correct categories (*e.g.* the correct MoA or toxicity). External testing, using agents outside the training set, is applied to determine the predictive accuracy of the expression signatures.

In order to transition the field of predictive toxicogenomics from proof-of-concept studies to routine incorporation in chemical testing, it is critical to develop standardized test procedures (addressing, for example, choice of cell model, selection of appropriate test concentrations, selection of time points, RNA isolation procedures, gene expression platform, and bioinformatics approach) and to ensure full disclosure (*i.e.* public access) of experimental data and analytical methods so that they can be widely utilized. Method refinement and definitive validation studies are also required to rigorously evaluate potential gene expression signatures (also referred to as genomic biomarkers) to ensure that classification accuracy, precision, specificity, and sensitivity are maintained regardless of minor variations in test conditions across laboratories. While many published gene expression signatures have been generated to predict diverse toxicological endpoints, including hepatotoxicity,[23,24] nephrotoxicity,[25,26] developmental toxicity,[27] specific exposures,[28,29] and carcinogenicity,[30–33] none have undergone rigorous and thorough validation. In addition, many of these signatures have been developed in rodent cells and tissues, so they do not address the need for predictive *in vitro* signatures in human cells. Clearly, validated/accepted genomic signatures that capture diverse toxicological outcomes and MoAs in human cells in culture are urgently needed by regulatory agencies for the purposes of chemical screening for human health risk assessment.

3.2.2 Rationale for the Need for Toxicogenomic Predictors of Genotoxicity

Genotoxicity testing is a critical component of chemical evaluation, as gene mutations and chromosomal aberrations can lead to adverse human health outcomes including cancer and inherited genetic disease.[34–36] Presently, the standard genotoxicity testing battery requires: (i) a bacterial mutagenesis assay; (ii) an *in vitro* cytogenetics assay to assess chromosome damage, or an *in vitro* mammalian mutation assay; and (iii) an *in vivo* rodent genotoxicity assay to measure chromosomal damage or gene mutation.[37–40] This battery is used by international regulatory agencies as it is efficient, economical, and sensitive in the detection of genotoxic compounds. However, this approach is hampered by low specificity leading to a high rate of false positive results, especially in mammalian cell assays.[39–43] Mechanistic insight into positive genotoxicity findings would improve scientific understanding of these results and could be used in a weight-of-evidence approach to better inform human health risk assessment.[36] Indeed, parallel analyses of gene expression changes and genetic toxicology endpoints are increasingly done to evaluate the relevant mechanisms that underlie genotoxic hazards.[44–47]

Genotoxic stress initiates a complex transcriptional response at the cellular level that alters the expression of genes associated with many biological pathways including DNA damage, DNA repair, cell cycle, and apoptosis.[10,48–51] If chemicals with similar mechanisms of genotoxic action induce comparable gene expression changes, then it is possible to predict the genotoxicity of unknown compounds by comparing gene expression signatures (genomic biomarkers) to those with known mechanisms of action.[52]

As genotoxicity testing is a critical component of chemical screening, human toxicogenomics signatures that predict genotoxic MoAs would be of tremendous value to advance current regulatory testing strategies. For example, the Health and Environmental Sciences Institute's (HESI) Committee on the Relevance and Follow-Up of Positive Results in *in Vitro* Genetic Toxicity Testing (IVGT) is working towards improving the scientific understanding of positive results derived from standard genetic toxicology tests by investigating different follow-up strategies (*i.e.* tests that may be applied to provide insight into the mechanisms operating to produce a positive genotoxic result) to determine relevance to human health. New technologies proposed by the committee to provide increased insight into positive *in vitro* genetic toxicology findings include high content imaging, quantitative high-throughput screening using the Tox21 strategy, and the use of genomic signatures to classify chemicals as genotoxic/non-genotoxic and to derive MoA and information regarding biological perturbation.[53] The need to add weight of evidence and provide MoA insight for improved genetic toxicity testing for human health was also highlighted by this committee; this may be achievable by the addition of gene expression analysis to the standard test battery.[52]

3.3 Toxicogenomic Predictors of Genotoxicity

While toxicogenomics has proven to be a very powerful tool that has revolutionized the field of toxicology, it has taken longer than expected to translate these results into human health risk assessment paradigm changes.[20] The potential of global gene expression profiling and predictive toxicogenomics to identify mechanisms underlying chemically induced genotoxicity and carcinogenicity has been favorably reviewed.[10,20,50] Moreover, analysis of the current state of the science in this area supports that toxicogenomics classifiers in general are robust and can be applied across technologies and models.[20,54–57] Thus, there is great promise in this field. Although numerous efforts have focused on developing predictive signatures for categorizing genotoxic and non-genotoxic compounds in rat liver,[58–63] the focus of this chapter is on *in vitro* models for chemical screening; below, we specifically review and describe method development and application of *in vitro* signatures for predictive toxicogenomics for genotoxicity. This is an important gap that our research is attempting to address.

3.3.1 Development and Validation of *In vitro* Toxicogenomic Predictors of Genotoxic MoAs

Early studies demonstrated the ability of predictive signatures to distinguish genotoxic (or DNA-reactive), and non-genotoxic (or non-DNA-reactive) chemicals in different *in vitro* cell systems including yeast,[64] LY5178 cells,[65–70] HepG2 cells,[71–73] and TK6 cells.[48,67,74,75] The majority of these studies anchored the gene expression data to molecular phenotypes through the measurement of frank genotoxicity, such as the induction of micronuclei and DNA adducts, and/or measures of cytotoxicity. These early studies produced gene sets ranging from three to 346 genes, and demonstrated a varying degree of predictive accuracy. Importantly, these pioneering studies suggested promise in the approach and led to further efforts in developing predictive biomarkers in various *in vitro* cell systems for this purpose.

More recent efforts have focused on the need to ensure an intact and robust p53 DNA damage response within the cell culture system used for genotoxicity prediction. A wide variety of cell culture systems are used to test genotoxicity endpoints *in vitro*, including cultured mammalian primary cells, peripheral blood lymphocytes, different rodent cell lines [Chinese hamster ovary (CHO), Chinese hamster lung (CHL), V79, L5178Y], and human cell lines such as TK6 cells.[76,77] The Organisation for Economic Co-operation and Development (OECD) test guidelines for *in vitro* genotoxicity testing state that the choice of cell type must be justified. There is consensus that it is important to choose a cell line that is p53-competent, as this is an important pathway involved in the DNA damage response. Fowler *et al.* found that p53-deficient rodent cells (V79, CHO and CHL) were more likely to result in misleading positive test results, as they are more susceptible to cytotoxic insult and induction of micronuclei.[41] This is consistent with requirements for model selection for assessment of toxicogenomics response to genotoxins. Although many important signaling molecules are stimulated by DNA-damaging agents, p53 has been shown to be 'a universal sensor of genotoxic stress'[78] that is important in tumor suppression, cell cycle arrest, and cell death. Predictive toxicogenomics studies for genotoxicity to date have been undertaken in a wide variety of human cell lines and primary cells that are p53 proficient, including TK6 cells,[79–83] HepG2 cells,[84–88] HepaRG cells,[85,89,90] human embryonic stem cells,[89,91] and primary cells derived from mouse,[92–94] rat,[89,95,96] and human.[89,97] Table 3.1 provides a summary of studies that developed *in vitro* predictive gene expression signatures for differentiating genotoxic mechanism in various mammalian cell culture systems.

We identified 32 studies in which the primary focus was developing *in vitro* signatures to predict whether a compound is genotoxic or not (see Table 3.1). Twelve of the studies were performed on rodent cells in culture, including primary hepatocytes, and most frequently mouse L5178Y cells. Twenty-one studies developed or tested signatures of genotoxicity in human cells in culture. The most frequently used human cells were HepG2, which were applied in nine studies. TK6 cells were included in eight studies, and rat liver S9 was used in combination with these cells for metabolic activation in one of the studies. Thus, a significant amount of research has been done on signature development in both rodent and human cell-culture models. However, as described above, a primary determinant of accuracy and reproducibility of a predictive signature is the training set of chemicals on which development was based. Indeed, a diverse spectrum of genotoxic and non-genotoxic MoAs should be included within the training and test sets to achieve the most robust results. Unfortunately, the vast majority of the studies we reviewed used fewer than five chemicals per test category (*e.g.* fewer than five genotoxic agents and/or fewer than five non-genotoxic agents) in the development of their training set. Indeed, only 10 of these studies included five or more chemicals per category, and only three had 10 or more chemicals per category. Predictive accuracy is best determined by external validation using compounds outside of the training set. Our review suggested that 13 studies performed external validation using this approach, and only five of these studies evaluated performance using at least 10 chemicals. The predictive gene sets derived from the work ranges from one to 1663 genes. Many factors contribute to which genes are identified as predictive of genotoxicity based on a particular dataset, including the training set of chemicals on which the gene set was derived and the cell-culture model, but also the top concentration used and the time point post-exposure used to identify responsive genes. Examination of Table 3.1 reveals a large amount of variability in the latter two factors across the studies. Thus, despite a significant amount of work in this area, inconsistencies in the approaches and models used in the studies, and other factors that contribute to developing, refining and testing the predictive gene set, have not yielded consistent findings and a broadly accepted gene expression

Table 3.1 Summary of predictive *in vitro* toxicogenomics studies in mammalian cells in culture, including details outlining the choice of cell line, study design and performance characteristics of the genomic biomarker.[a,b]

Publication	Cell culture	Training compounds	Exposure (time/ concentration)	Technology	Predictive gene signature	Predictive accuracy
Rieswijk *et al.* (2015)[94]	Primary mouse hepatocytes (male C57BL/6 mice)	3 GTX carcinogens/3 NGTX carcinogens	24 h (mRNA) and 48 h (miRNA and mRNA)/80% viability by MTT assay at 48 h	Affymetrix Mouse Genome 430 2.0 and miRCURY Locked-Nuclei Acid Array	264 genes, 6 pathways (24 h) and 456 genes, 3 miRNAs, 27 pathways (48 h)	Discriminative genes sets, miRNA (48 h only) and pathways distinguished GTX and NGTX carcinogens; predictive accuracy not reported
Li *et al.* (2015)[82]	Human TK6 cells	13 GTX/15 NGTX; 3 for validation: 1 GTX, 1 NGTX, 1 irrelevant positive	4 h exposure/dose optimization strategy by RT-qPCR	Agilent Human 4 × 44K	65 genes	100%
Buick *et al.* (2015)[79,89]	Human TK6 cells + rat liver S9; Human HepaRG cells (external dataset)	2 GTX/2 NGTX; validation in the presence of rat liver S9	4 h, 8 h, 24 h/3 concentrations based on RS by flow cytometry and RT-qPCR	Agilent SurePrint Human 8 × 60K	65 genes	100% at 8 h
Doktorova *et al.* (2014)[90]	Human HepaRG cells	10 GTX/10 NGTX hepatocarcinogens/10 NC	24 h and 72 h/ IC_{10} by MTT	Affymetrix Human Genome U133 Plus 2.0	440 genes (24 h) and 353 genes (72 h); top 30 significant pathways	Up to 88% (interlaboratory reproducibility 20–35% overlap of the AUC)

Lee *et al.* (2014)[86,88]	Human HepG2 cells (external dataset)	5 GTX/8 NGTX; 22 for validation: 11 GTX/11 NGTX	12 h, 24 h, 48 h for training set and 24 h and 48 h for test set/80% viability by MTT assay at 72 h or maximum dose of 2 mM in absence of cytotoxicity	Affymetrix Human Genome U133 Plus 2.0	6 gene expression signatures (gene sets); 2 identified that most accurately discriminated between GTX and NGTX	>90% for training set and >70% for the test set
Sakai *et al.* (2014)[133]	Human TK6 cells	8 genotoxins (6 DNA-reactive/2 DNA non-reactive); 12 for validation	4 h/50% RCG after a 20 h recovery (24 h)	Affymetrix Human Genome U133A/TaqMan Gene Expression Assays	2 genes	Not reported
Sakai *et al.* (2014)[83]	Human TK6 cells	8 genotoxins (6 DNA-reactive/2 DNA non-reactive); 14 for validation	4 h/41–53% reduction in RCG by cell counting after a 20 h recovery (24 h)	Affymetrix Human Genome U133A/TaqMan Gene Expression Assays	1 gene (out of 103 selected probes)	Not reported
Doktorova *et al.* (2013)[89]	Rat HepsC and HepsT cells; human hES-Hep, HepG2, HepaRG cells	5 GTX/5 NGTX carcinogens/5 NC	72 h/IC_{10} by MTT	Affymetrix Rat Genome 230 2.0 (HepC, HepsT); Human Genome U133 Plus 2.0 (hES-Hep, HepG2, HepaRG)	Top 62 discriminating pathways for 3 toxicity classes; 13 for GTX and 11 for NGTX (deregulated genes differed, but pathways overlapped)	HepaRG had lowest misclassification rate followed by hES-Hep, HepG2, HepsT and HepsC

Table 3.1 (*Continued*)

Publication	Cell culture	Training compounds	Exposure (time/concentration)	Technology	Predictive gene signature	Predictive accuracy
Keuhner *et al.* (2013)[80]	Human TK6 cells	2 DNA-damaging clastogens/2 aneugens; 1 for validation	4 h and 24 h/3 concentrations by MTS, PD (low = no cytotoxicity and high = 50% cytotoxic at 24 h)	Affymetrix Human Genome U133 Plus 2.0	1663 genes for clastogens and 363 for anuegens	Closer relationship of formaldehyde with clastogens than aneugens; 100% predicted test probability
Doktorova *et al.* (2012)[95]	Primary rat hepatocytes (HepsC) from male Sprague Dawley rats compared to *in vivo* rat data from Wistar rats at 1, 3, 7, and/or 14 days	3 GTX/3 NGTX carcinogens/2 NCs	24 h and 72 h/IC_{10} by MTT assay	Affymetrix Rat 230 2.0 (*in vitro* studies) and Affymetrix RAE230A (*in vivo* studies)/TaqMan RT-qPCR	Not reported (genes provided in supplementary table)	50% (correctly classified 4 out of 8 compounds: 3 false negatives and 1 false positive); clear correlation between the *in vivo* and *in vitro* data
Hochstenbach *et al.* (2012)[29]	Human peripheral blood mononuclear cells + rat liver S9	8 GTX/3 NGTX carcinogens	20 h/2 concentrations selected caused <10% cytotoxicity by trypan blue exclusion	Agilent Human 4 × 44K	78 genes, 8 pathways (GTX) and 26 genes, 30 pathways (NGTX)	Not reported
Magkoufopoulou *et al.* (2012)[88]	Human HepG2 cells	24 GTX/10 NGTX; 17 GTX/11 NGTX for validation	12 h, 24 h, 48 h/80% viability by MTT assay at 72 h or maximum dose of 2 mM in absence of cytotoxicity	Affymetrix Human Genome U133 Plus 2.0	33 genes for Ames positive/22 genes for Ames negative	Best prediction achieved by separating Ames positive from Ames negative at 24 h (89% accuracy and 91% specificity)

Boehme *et al.* (2011)[84]	Human HepG2 cells + rat liver S9	3 mutagenic/ 3 promutagenic/ 2 NGTX; 1 for validation	24 h (6 h treatment + 18 h recovery for S9) and 48 h/ non-cytotoxic doses (average of ~90% viability, lower than IC_{50}) based on literature, in-house data and P53 protein activation	Illumina Sentrix HumanRef-8 V2 BeadChip	91 genes for promutagenic compounds	100%
Magkoufopoulou *et al.* (2011)[134]	Human HepG2 cells	2 GTX/3 NGTX/ 3 FP-GTX	12 h, 24 h, 48 h/ 80% viability by MTT assay after 72 h	Affymetrix Human Genome U133 Plus 2.0	10 genes	100% correct prediction with a probability of 90% (2 out of 3 FP-GTX were correctly classified)
Yildirimman *et al.* (2011)[91]	Human hES-Hep002 cells	5 GTX/5 NGTX carcinogens/5 NC	72 h/IC_{10} concentrations by MTT assay	Affymetrix Human Genome U133 Plus 2.0/ TaqMan Gene Expression Assays	592 genes/ 37 pathways	Cross-validation accuracy of >95%; pathway scoring increases the performance of chemical classification
Jennen *et al.* (2010)[85]	Human HepG2 and HepaRG cells	2 GTX/3 NGTX carcinogens	12 h and 48 h/80% viability by MTT assay after 72 h	Affymetrix Human Genome U133 Plus 2.0	54/18 genes (for HepG2 at 12 h/48 h); 27/35 genes (for HepaRG at 12 h/48 h)	100%/100% (prediction/ probability for HepG2 at 12 h and 48 h); 100%/89% and 86%/100% (prediction/ probability for HepaRG at 12 h and 48 h, respectively)

Table 3.1 (*Continued*)

Publication	Cell culture	Training compounds	Exposure (time/concentration)	Technology	Predictive gene signature	Predictive accuracy
Mathijs *et al.* (2010)[93]	Primary mouse hepatocytes (from male C57BL/6 mice)	5 GTX/5 FP-GTX; validated with 2 GTX/3 FP-GTX	24 h and 48 h/ ~80% viability by MTT assay after 24 h	Affymetrix Mouse Genome 430 2.0	215 classifiers (24 h); 220 classifiers (48 h)	>80% prediction (100% correct classification at 24 h and 80% at 48 h)
Rohrbeck *et al.* (2010)[135]	Mouse BALB/c 3T3 cells (untrans-formed and transformed)	4 GTX carcinogens	24 h and 120 h (repeated treatment); 72 h + 21 day recovery (cell transformation assay)/IC_{20} by MTS assay and 2 additional concentrations	Affymetrix Mouse Genome 430 2.0/ RT-qPCR	41 genes (14 genes involved in tumorigenesis)	Gene signature identified in short-term (24 h and 120 h) untransformed and transformed cells (3 weeks); demonstrated robustness of predictive power (predictive accuracy not reported)
Ellinger-Ziegelbauer *et al.* (2009)[98]	Human TK6 cells	4 GTX (direct- and indirect-acting)	4 h and 24 h (4 h exposure + 20 h recovery)/3 concentrations, top one was 50% cytotoxic by growth inhibition by Coulter counter at 24 h	Targeted TaqMan Low Density Arrays (TQcards) and TaqMan Assays-on-Demand (47 genes selected from published data)	40-gene panel	Confirms the potential of toxicogenomics analysis for the investigation of GTX mechanisms (4 h is more optimal for mechanism of damage); good interlaboratory reproducibility (prediction accuracy not reported)

Mathijs *et al.* (2009)[92]	Primary mouse hepatocytes (from male C57BL/6 mice)	2 GTX/2 NGTX; 2 GTX for validation	12 h, 24 h, 36 h, 48 h/not reported (GTX doses same as those in Mathijs *et al.* (2009))	Affymetrix Mouse Genome 430 2.0	15 genes (24 h), 16 (36 h), 26 (48 h) for training set; 19 (24 h) and 22 (48 h) correlating with γH2AX	> 80% test probability for training set
Le Fevre *et al.* (2007)[81]	Human TK6 cells	11 anticancer drugs (DNA-reactive and non-DNA reactive); 3 anticancer drugs for validation	4 h and 24 h (4 h exposure + 20 h recovery)/3 concentrations (low = non-cytotoxic/GTX; medium = GTX but not cytotoxic; high = GTX and cytotoxic – 50% reduction seen by cell counting)	Affymetrix Human Genome U133A Plus 2.0/ RT-qPCR for 4 class-specific marker genes	28 marker genes	Not reported; compounds clustered based on MoA (drug class) at high concentration at 4 h time point
Amundson *et al.* (2005)[48]	Human TK6 and NH32 cells	3 non-DNA damaging/2 oxidative stressors/ 2 types of ionizing radiation/5 DNA-damaging agents	4 h/1 concentration (no notable decrease in cell viability by trypan blue exclusion)	7668 element cDNA microarrays	346 genes that differentiated among 4 categories of chemicals; 16 gene signature that discriminated p53 status among genotoxins	Not reported

Table 3.1 (*Continued*)

Publication	Cell culture	Training compounds	Exposure (time/concentration)	Technology	Predictive gene signature	Predictive accuracy
Caba *et al.* (2005)[64]	*Saccharomyces cerevisiae*	3 DNA-reactive/2 DNA non-reactive	120 min at 30 °C/equitoxic doses that produced 4–15% clonogenic survival relative to vehicle-treated controls	Affymetrix Yeast Genome S98	19 genes (11 DNA-reactive genes)	11 candidate genes appear to respond exclusively to GTX stress (not cytotoxic stress); predictive accuracy not reported
Kim *et al.* (2005)[68]	Mouse L5178Y cells	2 GTX/2 NGTX carcinogens and 2 GTX/2 NGTX non-carcinogens	24 h/90% viability by trypan blue dye exclusion/5 mg mL^{-1} in absence of cytotoxicity or based on the literature (highest mutant frequencies with minimal cytotoxicity)	TwinChip Mouse-7.4K Digital Genomics cDNA Microarray/RT-qPCR	10 genes for GTX agents/7 genes for carcinogenic agents	Not reported
van Delft *et al.* (2005)[72,73]	Human HepG2 cells (used data from van Delft *et al.* (2004)[72])	16 (GTX and NGTX carcinogens); 6 GTX and NGTX carcinogens for validation (2 chemicals overlap both training and test set)	24 h/< 25% cytotoxic by MTT assay	PHASE-1 Microarray Human-600	Varied from 14 to 31 for 4 supervised clustering methods and 3 different filtering approaches; 27 genes were selected at least 3 times	4 methods of supervised clustering methods tested: Pearson, NSC, KNN, WV with 3 different filtering approaches. None of the methods tested is superior to the others

Dickinson *et al.* (2004)[65]	Mouse L5178Y cells	1 GTX/1 NGTX (DNA-reactive/ non-DNA-reactive)	4 h and 24 h (4 h exposure + 20 h recovery)	Affymetrix Murine Genome U74Av2	71 genes for cisplatin and 55 genes for sodium chloride (14 genes in common); gene expression clearly distinguished between DNA-reactive and non-reactive chemicals	Not reported (no signature gene set presented)
Hu *et al.* (2004)[66]	Mouse L5178Y cells	3 direct-acting GTX/ 3 indirect-acting GTX	4 h and 24 h (4 h exposure + 20 h recovery)/ concentrations from non-cytotoxic/GTX to clearly GTX with 50–90% cytotoxicity by cell counting (Coulter) and micronuclei analysis	Affymetrix Mouse Genome U745A v2	43 genes (4 h) and 58 genes (24 h)	Robust, dose-responsive gene expression changes with discriminating power for direct *vs.* indirect genotoxins (predictive accuracy not reported)

Table 3.1 (*Continued*)

Publication	Cell culture	Training compounds	Exposure (time/concentration)	Technology	Predictive gene signature	Predictive accuracy
Islaih *et al.* (2004)[136]	Mouse L5178Y cells and human TK6 cells	2 GTX	4 h and 24 h (4 h exposure + 20 h recovery)/dose selections based on cytotoxicity (RSG) and *Tk* mutant frequency (low = 10–30% cytotoxic with weak induction of mutants; high = 50–80% cytotoxic and robust mutant induction)	Affymetrix Mouse Genome U74A v1 and Human Genome U95A Arrays	No distinctive gene expression changes identified	Unable to identify distinctive genes to differentiate 2 chemicals under investigation in mouse or human cells; different gene expression changes observed in mouse *vs.* human cells
van Delft *et al.* (2004)[72]	Human HepG2 cells	9 GTX/7 NGTX; validated with 3 GTX/3 NGTX carcinogens	24 h/<25% cytotoxic by MTT assay or 2 mM in the absence of cytotoxicity	PHASE-1 Microarray Human-600	20 most discriminating genes	81% correct classification for training set; 33% for validation set. By eliminating treatments and genes with marginal effects, this improved to 92%/100% and 94%/80%, respectively

Lee *et al.* (2003)[69]	Mouse L5178Y cells	3 GTX	24 h/60% cytotoxicity by cell proliferation (WST-1)	TwinChip Mouse-7.4K Digital Genomics cDNA Microarray	19 genes	Not reported
Seidel *et al.* (2003)[70]	Mouse L5178Y cells	2 DNA-damaging agents	4 h and 24 h (4 h exposure + 20 h recovery)/ 2 concentrations each based on pilot studies (low and high toxicity and mutant frequencies, respectively)	Atlas Mouse 1.2K (Clontech) Arrays	10 genes (bleomycin); 6 genes (hydrogen peroxide); 3 genes in common	3-gene signature could be a marker of exposure for DNA damage *via* free radicals; predictive accuracy not reported
Burczynski *et al.* (2000)[71]	Human HepG2 cells	3 compounds in 2 classes (cytotoxic anti-inflammatory drugs and DNA-damaging agents); validated with 86 toxic compounds	24 h/ED_{30} by reductase activity assay at 72 h	Not reported (<250-gene cDNA microarray)	7 of the 20 genes were common between the 2 analysis approaches used	Not reported

[a]The studies presented in this comparison table investigated the effects of chemical agents on gene expression and does not include those studies investigating radiation.

[b]Abbreviations: AUC: area under the curve; ED_{30}: 30% effective dose; FP-GTX: false-positive genotoxic; GTX: genotoxic; HepC: conventional primary rat hepatocytes; HepsT: trichostatin A-stabilized primary rat hepatocytes; hES-Hep: human embryonic stem cell-derived hepatocyte-like cells; IC_X: X% inhibitory concentration; MoA: mode of action; KNN: K-nearest neighbor analysis; MTS: 3-(4,5-dimethyl-2-yl)-5-(3-carboxymethoxyphenyl)-2-(4-sulfophenyl)-2*H*-tetrazolium; MTT: 3-(4, 5-dimethylthiazol-2-yl)-2, 5-diphenyltetrazolium bromide; NC: non-carcinogen; NGTX: non-genotoxic; NSC: nearest shrunken centroids analysis; PD: population doubling; RS: relative survival; RCG: relative cell growth; RSG: relative suspension growth; RT-qPCR: reverse transcriptase quantitative polymerase chain reaction; WV: weighted voting.

biomarker of genotoxicity. However, recognition of the limitations described above has led to several recent works that have produced much higher quality data for the purposes of developing a predictive signature for genotoxicity.

Recent work has applied much larger numbers of training and test chemicals than early studies to produce more robust predictors of genotoxicity. For example, Magkoufopoulou *et al.* (2012) trained their predictive assay using 34 genotoxic and non-genotoxic chemicals in human liver HepG2 cells, and then validated with an additional 28 chemicals, using three different genotoxicity prediction methods at 12, 24, and 48 hours.[88] They achieved the best accuracy by selecting predictive genes following the stratification of chemicals based on Ames test results at 24 hours. This large dataset was later leveraged by Lee *et al.* (2014) to demonstrate how knowledge of pathway perturbations can provide additional confirmation in support of a genotoxic mechanism (*i.e.* pathways involved in DNA damage response and repair were specifically enriched following exposure to genotoxic agents).[86] Yildirimman *et al.* (2011) also found that pathway scoring increases the performance of chemical classification of genotoxic carcinogens, non-genotoxic carcinogens, and non-carcinogens in human embryonic stem-cell-derived hepatocyte-like cells, as it provides information on how the affected genes are working together in response to a genotoxic insult.[91] Finally, interlaboratory reproducibility for gene expression signatures that differentiate genotoxic mechanisms has been investigated in human TK6 cells[98] and in human HepaRG cells,[90] with the data supporting reproducibility across laboratories and technical platforms.

HESI's Technical Committee on the Application of Genomics to Mechanism-Based Risk Assessment conducts projects to address key issues, challenges, and opportunities in implementing toxicogenomics data into regulatory decision making.[21,22] One focus of this committee is an extensive validation exercise that is being undertaken on a gene expression biomarker derived from human TK6 cells exposed to a reference set of 28 model compounds that are established as genotoxic or non-genotoxic.[42,79,82] A 65-gene classifier, that includes genes enriched in the p53 DNA damage-response pathway, known as TGx-28.65 (representing a 28-chemical training set, and comprised of 65 genes), was developed in TK6 cells. This signature (described in more detail in Chapter 2) was 100% accurate in predicting genotoxic and non-genotoxic chemicals in two validation exercises.[79,82] Importantly, a dose-setting protocol was developed to ensure consistency across experimental approaches.[82] TK6 cells (derived from a human lymphoblastoid cell) were selected for this purpose because they are an accepted model in genetic toxicology, express wild-type p53, and demonstrate a robust response to cellular stresses.[67,77,82,99] However, these cells lack the ability to express cytochrome P450s to metabolically activate chemicals, and thus do have some limitations.

Lastly, research efforts by Doktorova *et al.* (2012) demonstrated a clear correlation between *in vivo* and *in vitro* gene expression changes following exposure to genotoxic carcinogens, non-genotoxic carcinogens, and non-carcinogens in rats and primary rat hepatocytes.[95] Furthermore, these authors compared the transcriptomic responses of five different liver-based *in vitro* models derived from rat and human sources for the purposes of genotoxicity prediction, and determined that while all models tested were able to identify genotoxic carcinogens, the HepaRG cell line produced the lowest misclassification rate.[89] However, it is important to note that Jennen *et al.* (2010) found that HepaRG cells were a better suited *in vitro* liver model to interpret the biological effects of chemical exposure, whereas HepG2 cells were a more promising model for classification purposes using a toxicogenomics approach.[85] Since the ability to directly compare studies is hindered by differences in experimental design, it is difficult to determine the cause of opposing outcomes in predictive toxicogenomics studies. This highlights the need for standardized experimental conditions to determine the best path forward in the incorporation of transcriptomics data into a weight-of-evidence approach for genotoxicity prediction in risk assessment.

The studies described above each produced different gene sets (signatures) that predict outcome. This is unsurprising and will result from differences in cell culture systems, selection of chemicals used in the training sets (balance of genotoxic MoAs), chemical concentrations tested, and exposure time points, in addition to technical platforms, choice of statistical models, and data analysis. However, the signatures are generally consistent when considered at the pathway level. Indeed, predictive signatures for genotoxicity tend to be rich in p53-responsive genes, as well as genes involved in DNA repair, apoptosis, cell cycle progression, and signal transduction,[10] but the concordance of specific genes is variable between published genomic biomarkers.

In general, while there have been many targeted efforts by independent research groups to develop an *in vitro* predictive biomarker for genotoxicity testing, these gene signatures are not being used in genotoxicity risk assessments as they have not been rigorously tested across chemical classes, laboratories, or platforms. Ideally, a predictive biomarker should be qualified for use in risk assessment in order to increase its utility in weight of evidence approaches. In addition, the data and methodologies must be publicly available for broad use. Finally, it is absolutely critical that whatever system is applied is able to fully activate chemicals either through endogenous enzyme activity or through co-treatment with external metabolic activation systems. These comprise the aims of the HESI committee described above. Below, we expand on the final criterion in detail, as this has been a focus of research in our laboratory.

3.3.2 Integration of Metabolic Activation

Many chemicals must be converted into their reactive metabolites in order to produce a mutagenic effect *in vitro*.[100] A major shortcoming of many *in vitro* culture systems is that the cells inherently lack the appropriate metabolic activation to bioactivate the chemical under investigation. In order to minimize this deficiency, a metabolic activation system, typically rat liver S9 homogenates induced with Aroclor 1254 or phenobarbital/β-napthoflavone, are added to many *in vitro* mammalian-based test systems to provide a source of metabolizing enzymes, which may or may not produce similar metabolites as those produced *via* normal human liver metabolism.[34,101–103] It is recognized that S9 is variable in both protein content and enzyme activity (although much less so in inbred rodents *versus* humans), is toxic to cells, and leads to compound activation outside the cells, which means that the reactive metabolites must cross the plasma and nuclear membranes to exert their effects on DNA. The addition of S9 may produce artifacts due to higher levels of certain P450 enzymes, may lack the appropriate co-factors for detoxification, may cause non-target effects,[99] and does not precisely mimic the complex bioactivation and detoxification processes in humans. Thus, it is widely accepted that the addition of external rat liver S9 is an imperfect solution to metabolic incompetency issues. However, despite the known shortcomings, S9 has been used for many decades in genotoxicity bioassays, and provides the most accepted method of metabolic activation of chemicals for *in vitro* assays today.[34,101–103]

The HESI TGx-28.65 genotoxicity classifier described above was generated in human TK6 cells using 28 well-characterized genotoxic and non-genotoxic compounds that do not require metabolic activation, and was further validated using three additional chemicals in the absence of metabolic activation.[82] In order to broaden the utility of the predictive genotoxicity signature for chemicals requiring metabolic activation, the TGx-28.65 was tested in the presence of S9 for four additional chemicals using an external metabolic activation system, 1% phenobarbital/benzoflavone-induced rat liver S9.[79] The data suggest that the addition of external metabolic activation does not impede the ability of the predictive signature to classify genotoxicity in human TK6 cells. However, further testing of the TGx-28.65 classifier with an additional ten test chemicals in the presence of rat liver S9 accentuated some issues with the TK6(+S9) test system. For example, a known genotoxic agent, dimethylnitrosamine did not induce micronuclei, as expected, in the test system. Extensive testing of different concentrations of S9 (1–10%), different types of induction (Aroclor, ethanol, and phenobarbital/benzoflavone), and different mixtures of rat liver S9s in different proportions led to the additional 10 test chemicals being tested using 2% Aroclor-induced S9 in TK6 cells, except for furan, a chemical that requires CYP2E1 enzyme activity[104] (thus, 5% ethanol-induced rat liver S9 was used for furan[137]). This example highlights the difficulties that may be encountered when adding external metabolic activation to a test system that lacks inherent metabolic capabilities if a compound is not appropriately bioactivated, which could lead to false negatives when testing unknown chemicals. If the *in vitro* model system lacks metabolic capabilities, it is prudent to test different concentrations and types of metabolic activation systems to ensure appropriate metabolism of the compound. However, in order to better achieve metabolic activation of compounds *in vitro* that is relevant to human capabilities *in vivo*, it is preferable to use a human cell line with inherent metabolic capabilities and greater relevance to human metabolism *in vivo*.

As described above, the use of p53-proficient human cell lines with intact DNA repair capabilities and inherent metabolic capabilities is required to increase the human relevance of the test system while reducing the incidence of false test results.[41] Although the use of primary human hepatocytes (PHH) would most closely approximate human metabolic capabilities, as the liver is the primary organ for detoxification, PHHs are highly polymorphic in terms of metabolic enzyme expression, are more difficult to obtain, and are more limited in terms of growth activity, functional stability, and life span in culture.[105] HepG2 and HepaRG cells, both derived from human hepatocellular carcinomas, have been shown to be appropriate for both purposes.[106–108] These cell lines are proven to be a suitable tool for assessing *in vitro* genotoxicity endpoints, such as the *in vitro* comet and micronucleus assays. However, while both cell lines display inherent metabolic capabilities, HepaRG cells have been shown to express high levels of both phase I and phase II enzymes, in addition to nuclear receptors, and more closely approximate the metabolic capabilities of PHHs.[109–111] Furthermore, while both HepG2 cells and HepaRG cells have been shown to perform equally well in terms of toxicogenomics applications,[112] HepaRG cells have demonstrated some superiority over HepG2 cells in terms of their sensitivity and predictive accuracy for genotoxicity.[89] We evaluated whether the TGx-28.65 biomarker was able to accurately classify genotoxic and non-genotoxic chemicals in human HepaRG cells from the dataset produced by Doktorova *et al.* (2013) using Affymetrix DNA microarrays.[89] We found that the TGx-28.65 predictive gene signature also accurately classified chemicals for genotoxicity in these cells when tested using an external dataset generated with 15 prototypical compounds, including genotoxic carcinogens, non-genotoxic carcinogens, and non-carcinogens.[79,89] This analysis demonstrated that the TGx-28.65 predictive signature was reproducible across cell culture models (TK6 and HepaRG), laboratories, DNA microarray platforms (Affymetrix and Agilent DNA microarrays used in TGx-28.65 development), and across time points (72 h in HepaRG cells and 4 h in TK6 cells). Because HepaRG cells are metabolically more active and are p53 proficient, these cells provide an important model for genotoxicity research *in vitro*.

3.3.3 Additional Considerations for Experimental Design of Predictive Toxicogenomics Studies

As described above, there are many important considerations in the experimental design of *in vitro* toxicogenomics studies, such as choice of model system, number of chemicals used in the training and test sets, chemical selection, chemical concentration(s), exposure time(s), technical considerations, data analysis, and statistical tests. A well-designed *in vitro* toxicogenomics study is crucial for deriving high-quality gene expression data for mechanistic and predictive purposes.

The choice of cell line is a very important consideration when testing a chemical's potential toxicity *in vitro*, as well as adherence to good scientific practice for cell culture.[102,113] Some factors to consider include the cytogenetic stability of the cell model, choice of solvents for chemical suspension, culture conditions (pH and osmolality), origin of the cell line and passage number, p53 and DNA repair proficiency, metabolic capabilities, as well as confounding factors such as the presence of chemical or bacterial impurities or non-physiological conditions that may lead to artifacts or irrelevant positive results.[101,102,113] As described above, to maximize the human relevance of toxicogenomics assays and the ability to test environmental chemicals, most of which require metabolic activation,[100] it is preferable to choose stable, p53-proficient cells with inherent metabolic capabilities (*e.g.* HepaRG). These are also more highly differentiated cell lines that are more phenotypically and karyotypically stable than non-mammalian cell lines. These cell lines are adapted for use in standard genetic toxicity endpoints and more closely approximate primary human hepatocytes than other rodent cell lines.[41,105–109,111,112,114–116]

It is also critical to choose an appropriate top (highest) concentration for testing. The criteria for selection of the highest concentration is being revised and adopted in the OECD test guidelines for *in vitro* genetic toxicology as a result of discussions by experts at the international level. In the absence of excessive cytotoxicity and solubility issues, the revised maximum concentration is 10 mM, $2\ \mathrm{mg\ mL^{-1}}$, or $2\ \mu\mathrm{L\ mL^{-1}}$, whichever is lowest, or the lowest insoluble concentration for compounds with solubility issues.[76,77] Furthermore, for cytotoxic compounds, the top concentration should achieve $55 \pm 5\%$ cytotoxicity.[76,77] It would be wise to align the top concentration selection for *in vitro* toxicogenomics studies with those of the standard *in vitro* genetic toxicology test battery to allow for genotoxic and toxicogenomic endpoints to be analyzed from the same exposures (*e.g.* cells from the same exposure protocol can be used to determine gene mutation or clastogenicity, and to extract RNA for gene expression analysis). Ideally, each chemical should be tested over a range of concentrations and time points in multiple biological systems to ensure that the chemical's MoA is appropriately captured in the culture dish. It is also beneficial to anchor the gene expression data to cytotoxic and genotoxic endpoints as a means of validating the gene expression data at the cellular level.

Finally, standardized *in vitro* toxicogenomics protocols, data analysis, and statistical testing must be adhered to, and comprehensive validation must be performed across platforms, laboratories, and chemical classes. As DNA microarray technology has been in existence for almost 20 years,[117,118] standardized protocols for data analysis and interpretation of microarray data exist.[119,120] The data and analytical approaches must be made publicly available *via* deposition of genomics data in public repositories and detailed analytical descriptions in supplementary materials. It is also important to choose an appropriate classification algorithm when performing predictive toxicogenomics for compound classification. Indeed, reproducibility of predictive toxicogenomics results across technologies and between different research laboratories was identified as a threat to the use of genomic signatures of genotoxicity by the HESI IVGT committee.[53] One study compared four different supervised clustering methods to classify genotoxic and non-genotoxic carcinogens using gene expression microarray data derived from HepG2 cells. Clustering methods included Pearson correlation, nearest shrunken centroids analysis, K-nearest neighbor analysis, and weighted voting, with three different means of data filtration for each supervised clustering method. There was no clear supervised clustering method that outperformed the others in this example;[73] however, it is important to consider different classification algorithms and means of filtering the data in predictive toxicogenomics. The MicroArray Quality Control (MAQC)-II consortium has published several articles addressing common procedures for the development and validation of predictive toxicogenomics models for gene expression data.[121] They determined that the performance of the predictive model was largely determined by the endpoint being examined and the proficiency of the data analysts, and that models performed similarly using different methodologies to generate the model. The MAQC-II consortium drew several conclusions and recommendations that will be useful for regulatory purposes and we refer the reader to this document for further information.[121] Overall, standardization of predictive toxicogenomics tools and appropriate validation and qualification of genomic biomarkers will be essential if they are to be incorporated into human health risk assessment.

3.4 Summary and Conclusions

By integrating *in vitro* toxicogenomics for both mechanistic and predictive purposes into the current test strategy for genetic toxicology, it is possible to greatly reduce reliance on animal-intensive assays. Generally, chemical evaluation using *in vitro* approaches offers many advantages over animal-based testing, as they are less resource-intensive, more efficient, and more amenable to medium- to high-throughput platforms. *In vitro* approaches avoid potential ethical issues surrounding the use of laboratory animals, or at the very least reduce the number of animals required for chemical evaluation, and they directly address human relevance. *In vitro* toxicogenomics methodologies can provide a wealth of mechanistic insight into a chemical's MoA and pathway-based perturbations following a toxic insult, which can be used in a weight-of-evidence approach to genotoxicity testing and/or to classify chemicals based on their MoA. This type of information is lacking from the standard assays and thus could enhance current approaches.

In recent years, the use of high-throughput *in vitro* screening methods as the first tier of chemical screening has been emphasized,[14,15,122] with much interest surrounding the use of *in vitro* toxicogenomics signatures to predict adverse effects and MoA.[18,122] An important gap in current ToxCast screening capabilities is the lack of high-throughput screening approaches that are able to identify genotoxic carcinogens. Performance of current high-throughput methods in this area has been poor.[123,124] Thus, the availability of rapid genomic predictors of genotoxicity and carcinogenicity in human cells is essential to the success of such programs.

Currently, we contend that the addition of gene expression signatures to complement the standard genotoxicity testing strategy would generate rich mechanistic data for environmental chemicals and could simultaneously be used for predictive toxicogenomics purposes to classify compounds as genotoxic or non-genotoxic. In order to do so, clear technical guidelines need to be in place to guide the risk-assessment community on how to identify high-quality toxicogenomics data and how to best incorporate these data into their risk-assessment strategy. Bourdon-Lacombe *et al.* (2015) recently published a technical guide for the application of gene expression data into human health risk assessment for environmental chemicals;[125] additionally, a tiered testing strategy for incorporating toxicogenomics data into risk assessment has been proposed.[122] Many case studies have demonstrated how toxicogenomics data can be used in human health risk assessment and clearly depict the benefits of quantitative, mechanistic, and predictive toxicogenomics to complement the current toxicological methods.[104,126–129] In particular, a case study on the environmental carcinogen benzo[a]pyrene has shown how the TGx-28.65 biomarker might be used to inform human health risk assessment.[128]

Despite the promise described, there remain hurdles to overcome for routine *in vitro* toxicogenomics applications. Some of these issues are not unique to toxicogenomics (*e.g. in vitro* to *in vivo* extrapolation, and appropriate metabolic capabilities in a dish). The requirements for advanced statistics and bioinformatics that accompany big data, and the relevance of chemically induced pathways in cells to apical outcomes *in vivo* remain two large challenges. Although the cost of whole-genome profiling has decreased substantially over the past decade, the expense remains a barrier for many laboratories. In addition, practical higher-throughput methodologies for whole-genome analysis will greatly benefit the field. However, the current lack of validated, qualified predictive biomarkers remains the main challenge. In general, a publicly accessible, user-friendly source of validated signatures that can be applied to predict a diverse range of toxicities is required for routine use by regulators.

Technological advances in the field of predictive toxicogenomics continue to offer many advantages and challenges. The rise of next-generation sequencing means that gene expression analysis is no longer restricted to the probe sequences printed on a microarray slide or the limited dynamic range of microarrays, allowing an unrestricted, unbiased, and more quantitative scan of the whole genome. While many challenges with RNA sequencing (RNA-seq) exist, as they did for microarray a decade ago (*e.g.* standardization of protocols, bioinformatics challenges, and data analysis), the Sequencing Quality Control (SEQC) project, which is a multi-site cross-platform study[130] coordinated by the US Food and Drug Administration, has already demonstrated that RNA-seq is highly reproducible if data treatment and analysis are appropriate. It has already been shown that RNA-seq provides new insights into well-characterized genotoxic carcinogens, such as benzo[*a*]pyrene, by detecting 20% more gene expression changes over DNA microarrays, more insight into the perturbed biological processes, detection of alternative isoforms, and the ability to investigate allele-specific gene expression.[131] Moreover, the utility of legacy microarray data and established predictive biomarkers in the RNA-seq era has been demonstrated and confirms that RNA-seq-based predictive signatures can also be used in microarray data analysis.[132] The TGx-28.65 predictive gene signature is now being tested using RNA-seq technology to determine whether this genomic biomarker for genotoxicity can be expanded upon and/or refined using this powerful technology. Preliminary results suggest that the biomarker also works effectively on an RNA-seq platform. Developing predictive signatures based on robust DNA-damage-responsive genes that are conserved across species and tissues will increase the probability of success overall in this field.

In conclusion, it is evident that *in vitro* predictive toxicogenomics can offer a more efficient and economical complement to the current animal-based strategies. Although momentum is building, we remain at a proof-of-principle stage for *in vitro* predictive genotoxicity biomarkers. Focused, collaborative international efforts across government, academia, and industry, and qualified validation of biomarkers for use in chemical evaluation, are needed to move this field forward.

Acknowledgements

Health Canada research presented in this book chapter was supported by the Genomics Research and Development Initiative. The authors are thankful for helpful comments provided by Dr Reza Farmahin and Dr Sabina Halappanavar.

References

1. M. E. Meek and V. C. Armstrong, in *The Management of Industrial Chemicals in Canada, Risk Assessment of Chemicals: An Introduction*, ed. C. J. Van Leeuwen and T. G. Vermeire, Springer, Dordrecht, Netherlands, 2007, pp. 591–622.
2. G. Hess, Canada targets toxic chemicals, *Chem. Eng. News*, 2006, **84**, 13.
3. Council of Canadian Academies, *Integrating Emerging Technologies into Chemical Safety Assessment*, Council of Canadian Academies, Ottawa, ON, Canada, 2012.

4. National Research Council, in *Toxicity Testing in the 21st Century: A Vision and a Strategy*, ed. anonymous, The National Academies Press, Washington, D.C., 2007.
5. US EPA, *The U.S. Environmental Protection Agency's Strategic Plan for Evaluating the Toxicity of Chemicals*, Office of the Science Advisor, Science Policy Council, Washington, D.C., USA, 2009.
6. C. A. Afshari, H. K. Hamadeh and P. R. Bushel, The evolution of bioinformatics in toxicology: advancing toxicogenomics, *Toxicol. Sci.*, 2011, **120**(Suppl 1), S225–S237.
7. D. Krewski, M. Westphal, M. Al-Zoughool, M. C. Croteau and M. E. Andersen, New directions in toxicity testing, *Annu. Rev. Public Health*, 2011, **32**, 161–178.
8. A. K. Goetz, B. P. Singh, M. Battalora, J. M. Breier, J. P. Bailey, A. C. Chukwudebe and E. R. Janus, Current and future use of genomics data in toxicology: opportunities and challenges for regulatory applications, *Regul. Toxicol. Pharmacol.*, 2011, **61**, 141–153.
9. D. A. Keller, D. R. Juberg, N. Catlin, W. H. Farland, F. G. Hess, D. C. Wolf and N. G. Doerrer, Identification and characterization of adverse effects in 21st century toxicology, *Toxicol. Sci.*, 2012, **126**, 291–297.
10. M. D. Waters, M. Jackson and I. Lea, Characterizing and predicting carcinogenicity and mode of action using conventional and toxicogenomics methods, *Mutat. Res.*, 2010, **705**, 184–200.
11. M. R. Fielden and T. R. Zacharewski, Challenges and limitations of gene expression profiling in mechanistic and predictive toxicology, *Toxicol. Sci.*, 2001, **60**, 6–10.
12. E. F. Nuwaysir, M. Bittner, J. Trent, J. C. Barrett and C. A. Afshari, Microarrays and toxicology: the advent of toxicogenomics, *Mol. Carcinog.*, 1999, **24**, 153–159.
13. W. D. Pennie, J. D. Tugwood, G. J. Oliver and I. Kimber, The principles and practice of toxigenomics: applications and opportunities, *Toxicol. Sci.*, 2000, **54**, 277–283.
14. R. Kavlock, K. Chandler, K. Houck, S. Hunter, R. Judson, N. Kleinstreuer, T. Knudsen, M. Martin, S. Padilla, D. Reif, A. Richard, D. Rotroff, N. Sipes and D. Dix, Update on EPA's ToxCast program: Providing high throughput decision support tools for chemical risk management, *Chem. Res. Toxicol.*, 2012, **25**, 1287–1302.
15. A. D. Kligerman, R. R. Young, L. F. Stankowski Jr, K. Pant, T. Lawlor, M. J. Aardema and K. A. Houck, An evaluation of 25 selected ToxCast chemicals in medium-throughput assays to detect genotoxicity, *Environ. Mol. Mutagen.*, 2015, **56**(5), 468–476.
16. J. Lamb, E. D. Crawford, D. Peck, J. W. Modell, I. C. Blat, M. J. Wrobel, J. Lerner, J. P. Brunet, A. Subramanian, K. N. Ross, M. Reich, H. Hieronymus, G. Wei, S. A. Armstrong, S. J. Haggarty, P. A. Clemons, R. Wei, S. A. Carr, E. S. Lander and T. R. Golub, The Connectivity Map: using gene-expression signatures to connect small molecules, genes, and disease, *Science*, 2006, **313**, 1929–1935.

17. J. Lamb, The Connectivity Map: a new tool for biomedical research, *Nat. Rev. Cancer*, 2007, **7**, 54–60.
18. H. Zhu, J. Zhang, M. T. Kim, A. Boison, A. Sedykh and K. Moran, Big data in chemical toxicity research: the use of high-throughput screening assays to identify potential toxicants, *Chem. Res. Toxicol.*, 2014, **27**, 1643–1651.
19. N. Carmichael, M. Bausen, A. R. Boobis, S. M. Cohen, M. Embry, C. Fruijtier-Polloth, H. Greim, R. Lewis, M. E. Bette Meek, H. Mellor, C. Vickers and J. Doe, Using mode of action information to improve regulatory decision-making: an ECETOC/ILSI RF/HESI workshop overview, *Crit. Rev. Toxicol.*, 2011, **41**, 175–186.
20. R. Currie, Toxicogenomics: The challenges and opportunities to identify biomarkers, signatures and thresholds to support mode-of-action, *Mutat. Res.*, 2012, **746**, 97–103.
21. W. Pennie, S. D. Pettit and P. G. Lord, Toxicogenomics in risk assessment: an overview of an HESI collaborative research program, *Environ. Health Perspect.*, 2004, **112**, 417–419.
22. S. Pettit, S. A. des Etages, L. Mylecraine, R. Snyder, J. Fostel, R. T. Dunn 2nd, K. Haymes, M. Duval, J. Stevens, C. Afshari and A. Vickers, Current and future applications of toxicogenomics: Results summary of a survey from the HESI Genomics State of Science Subcommittee, *Environ. Health Perspect.*, 2010, **118**, 992–997.
23. J. Hrach, S. O. Mueller and P. Hewitt, Development of an *in vitro* liver toxicity prediction model based on longer term primary rat hepatocyte culture, *Toxicol. Lett.*, 2011, **206**, 189–196.
24. W. F. Van den Hof, M. L. Coonen, M. van Herwijnen, K. Brauers, W. K. Wodzig, J. H. van Delft and J. C. Kleinjans, Classification of hepatotoxicants using HepG2 cells: A proof of principle study, *Chem. Res. Toxicol.*, 2014, **27**, 433–442.
25. M. R. Fielden, B. P. Eynon, G. Natsoulis, K. Jarnagin, D. Banas and K. L. Kolaja, A gene expression signature that predicts the future onset of drug-induced renal tubular toxicity, *Toxicol. Pathol.*, 2005, **33**, 675–683.
26. Y. Minowa, C. Kondo, T. Uehara, Y. Morikawa, Y. Okuno, N. Nakatsu, A. Ono, T. Maruyama, I. Kato, J. Yamate, H. Yamada, Y. Ohno and T. Urushidani, Toxicogenomic multigene biomarker for predicting the future onset of proximal tubular injury in rats, *Toxicology*, 2012, **297**, 47–56.
27. J. L. A. Pennings, D. A. M. van Dartel, J. F. Robinson, T. E. Pronk and A. H. Piersma, Gene set assembly for quantitative prediction of developmental toxicity in the embryonic stem cell test, *Toxicology*, 2011, **284**, 63–71.
28. V. Chauhan, M. Howland and R. Wilkins, Identification of gene-based responses in human blood cells exposed to alpha particle radiation, *BMC Med. Genomics*, 2014, **7**, 43.
29. K. Hochstenbach, D. M. van Leeuwen, R. W. Gottschalk, H. Gmuender, S. B. Stolevik, U. C. Nygaard, M. Lovik, B. Granum, E. Namork,

H. van Loveren and J. H. van Delft, Transcriptomic fingerprints in human peripheral blood mononuclear cells indicative of genotoxic and non-genotoxic carcinogenic exposure, *Mutat. Res.*, 2012, **746**, 124–134.

30. R. Garcia-Escudero, A. B. Martinez-Cruz, M. Santos, C. Lorz, C. Segrelles, G. Garaulet, C. Saiz-Ladera, C. Costa, T. Buitrago-Perez, M. Duenas and J. M. Paramio, Gene expression profiling of mouse p53-deficient epidermal carcinoma defines molecular determinants of human cancer malignancy, *Mol. Cancer*, 2010, **9**, 193.
31. K. Z. Guyton, A. D. Kyle, J. Aubrecht, V. J. Cogliano, D. A. Eastmond, M. Jackson, N. Keshava, M. S. Sandy, B. Sonawane, L. Zhang, M. D. Waters and M. T. Smith, Improving prediction of chemical carcinogenicity by considering multiple mechanisms and applying toxicogenomic approaches, *Mutat. Res.*, 2009, **681**, 230–240.
32. J. R. Nevins and A. Potti, Mining gene expression profiles: Expression signatures as cancer phenotypes, *Nat. Rev. Genet.*, 2007, **8**, 601–609.
33. R. S. Thomas, T. M. O'Connell, L. Pluta, R. D. Wolfinger, L. Yang and T. J. Page, A comparison of transcriptomic and metabonomic technologies for identifying biomarkers predictive of two-year rodent cancer bioassays, *Toxicol. Sci.*, 2007, **96**, 40.
34. K. L. Dearfield, V. Thybaud, M. C. Cimino, L. Custer, A. Czich, J. S. Harvey, S. Hester, J. H. Kim, D. Kirkland, D. D. Levy, E. Lorge, M. M. Moore, G. Ouedraogo-Arras, M. Schuler, W. Suter, K. Sweder, K. Tarlo, J. van Benthem, F. van Goethem and K. L. Witt, Follow-up actions from positive results of *in vitro* genetic toxicity testing, *Environ. Mol. Mutagen.*, 2011, **52**, 177–204.
35. B. B. Gollapudi, G. E. Johnson, L. G. Hernandez, L. H. Pottenger, K. L. Dearfield, A. M. Jeffrey, E. Julien, J. H. Kim, D. P. Lovell, J. T. Macgregor, M. M. Moore, J. van Benthem, P. A. White, E. Zeiger and V. Thybaud, Quantitative approaches for assessing dose-response relationships in genetic toxicology studies, *Environ. Mol. Mutagen.*, 2013, **54**, 8–18.
36. V. Thybaud, M. Aardema, D. Casciano, V. Dellarco, M. R. Embry, B. B. Gollapudi, M. Hayashi, M. P. Holsapple, D. Jacobson-Kram, P. Kasper, J. T. MacGregor and R. Rees, Relevance and follow-up of positive results in *in vitro* genetic toxicity assays: an ILSI-HESI initiative, *Mutat. Res.*, 2007, **633**, 67–79.
37. FDA, in *Guidance for Industry and Other Stakeholders: Toxicological Principles for the Safety Assessment of Food Ingredients (Redbook 2000)*, ed. Anonymous, Center for Food Safety and Applied Nutrition (U.S. FDA), 2007 (revised).
38. ICH, *Guidance on Genotoxicity Testing and Data Interpretation for Pharmaceuticals Intended for Human Use*, S2(R1), 2011.
39. B. Mahadevan, R. D. Snyder, M. D. Waters, R. D. Benz, R. A. Kemper, R. R. Tice and A. M. Richard, Genetic toxicology in the 21st century: reflections and future directions, *Environ. Mol. Mutagen.*, 2011, **52**, 339–354.

40. L. Muller, Y. Kikuchi, G. Probst, L. Schechtman, H. Shimada, T. Sofuni and D. Tweats, ICH-harmonised guidances on genotoxicity testing of pharmaceuticals: evolution, reasoning and impact, *Mutat. Res.*, 1999, **436**, 195–225.
41. P. Fowler, K. Smith, J. Young, L. Jeffrey, D. Kirkland, S. Pfuhler and P. Carmichael, Reduction of misleading ("false") positive results in mammalian cell genotoxicity assays. I. Choice of cell type, *Mutat. Res.*, 2012, **742**, 11–25.
42. F. M. Goodsaid, S. Amur, J. Aubrecht, M. E. Burczynski, K. Carl, J. Catalano, R. Charlab, S. Close, C. Cornu-Artis, L. Essioux, A. J. Fornace Jr, L. Hinman, H. Hong, I. Hunt, D. Jacobson-Kram, A. Jawaid, D. Laurie, L. Lesko, H. H. Li, K. Lindpaintner, J. Mayne, P. Morrow, M. Papaluca-Amati, T. W. Robison, J. Roth, I. Schuppe-Koistinen, L. Shi, O. Spleiss, W. Tong, S. L. Truter, J. Vonderscher, A. Westelinck, L. Zhang and I. Zineh, Voluntary exploratory data submissions to the US FDA and the EMA: experience and impact, *Nat. Rev. Drug Discovery*, 2010, **9**, 435–445.
43. D. Kirkland, M. Aardema, L. Henderson and L. Muller, Evaluation of the ability of a battery of three *in vitro* genotoxicity tests to discriminate rodent carcinogens and non-carcinogens I. Sensitivity, specificity and relative predictivity, *Mutat. Res.*, 2005, **584**, 1–256.
44. M. S. Attene-Ramos, G. M. Nava, M. G. Muellner, E. D. Wagner, M. J. Plewa and H. R. Gaskins, DNA damage and toxicogenomic analyses of hydrogen sulfide in human intestinal epithelial FHs 74 Int cells, *Environ. Mol. Mutagen.*, 2010, **51**, 304–314.
45. S. M. Attia, S. F. Ahmad, K. M. Zoheir, S. A. Bakheet, G. K. Helal, A. R. Abd-Allah, N. O. Al-Harbi, K. A. Al-Hosaini and O. A. Al-Shabanah, Genotoxic evaluation of chloroacetonitrile in murine marrow cells and effects on DNA damage repair gene expressions, *Mutagenesis*, 2013, **29**, 55–62.
46. W. Ding, D. M. Petibone, J. R. Latendresse, M. G. Pearce, L. Muskhelishvili, G. A. White, C. W. Chang, R. A. Mittelstaedt, J. G. Shaddock, L. P. McDaniel, D. R. Doerge, S. M. Morris, M. E. Bishop, M. G. Manjanatha, A. Aidoo and R. H. Heflich, *In vivo* genotoxicity of furan in F344 rats at cancer bioassay doses, *Toxicol. Appl. Pharmacol.*, 2012, **261**, 164–171.
47. D. R. Geter, F. Zhang, M. R. Schisler, A. J. Wood, H. L. Kan, Y. C. Jeong, M. J. Bartels, L. McFadden and B. B. Gollapudi, Genetic damage, but limited evidence of oxidative stress markers in diethyl maleate-induced glutathione depleted mouse lymphoma L5178Y (TK(+/−)) cell cultures, *Toxicol. Mech. Methods*, 2012, **22**, 547–554.
48. S. A. Amundson, K. T. Do, L. Vinikoor, C. A. Koch-Paiz, M. L. Bittner, J. M. Trent, P. Meltzer and A. J. Fornace Jr, Stress-specific signatures: expression profiling of p53 wild-type and -null human cells, *Oncogene*, 2005, **24**, 4572–4579.

49. J. Aubrecht and E. Caba, Gene expression profile analysis: an emerging approach to investigate mechanisms of genotoxicity, *Pharmacogenomics*, 2005, **6**, 419–428.
50. H. Ellinger-Ziegelbauer, J. Aubrecht, J. C. Kleinjans and H. J. Ahr, Application of toxicogenomics to study mechanisms of genotoxicity and carcinogenicity, *Toxicol. Lett.*, 2009, **186**, 36–44.
51. R. K. Newton, M. Aardema and J. Aubrecht, The utility of DNA microarrays for characterizing genotoxicity, *Environ. Health Perspect.*, 2004, **112**, 420–422.
52. A. M. Lynch, J. C. Sasaki, R. Elespuru, D. Jacobson-Kram, V. Thybaud, M. De Boeck, M. J. Aardema, J. Aubrecht, R. D. Benz, S. D. Dertinger, G. R. Douglas, P. A. White, P. A. Escobar, A. Fornace Jr, M. Honma, R. T. Naven, J. F. Rusling, R. H. Schiestl, R. M. Walmsley, E. Yamamura, J. van Benthem and J. H. Kim, New and emerging technologies for genetic toxicity testing, *Environ. Mol. Mutagen.*, 2011, **52**, 205–223.
53. E. Zeiger, B. Gollapudi, M. J. Aardema, S. Auerbach, D. Boverhof, L. Custer, P. Dedon, M. Honma, S. Ishida, A. L. Kasinski, J. H. Kim, M. G. Manjanatha, J. Marlowe, S. Pfuhler, I. Pogribny, W. Slikker, L. F. Stankowski Jr, J. Y. Tanir, R. Tice, J. van Benthem, P. White, K. L. Witt and V. Thybaud, Opportunities to integrate new approaches in genetic toxicology: an ILSI-HESI workshop report, *Environ. Mol. Mutagen.*, 2015, **56**, 277–285.
54. X. Fan, E. K. Lobenhofer, M. Chen, W. Shi, J. Huang, J. Luo, J. Zhang, S. J. Walker, T. M. Chu, L. Li, R. Wolfinger, W. Bao, R. S. Paules, P. R. Bushel, J. Li, T. Shi, T. Nikolskaya, Y. Nikolsky, H. Hong, Y. Deng, Y. Cheng, H. Fang, L. Shi and W. Tong, Consistency of predictive signature genes and classifiers generated using different microarray platforms, *Pharmacogenomics J.*, 2010, **10**, 247–257.
55. X. Fan, L. Shao, H. Fang, W. Tong and Y. Cheng, Cross-platform comparison of microarray-based multiple-class prediction, *PLoS One*, 2011, **6**, e16067.
56. T. Heise, M. Schug, D. Storm, H. Ellinger-Ziegelbauer, H. J. Ahr, B. Hellwig, J. Rahnenfuhrer, A. Ghallab, G. Guenther, J. Sisnaiske, R. Reif, P. Godoy, H. Mielke, U. Gundert-Remy, A. Lampen, A. Oberemm and J. G. Hengstler, *In vitro – in vivo* correlation of gene expression alterations induced by liver carcinogens, *Curr. Med. Chem.*, 2012, **19**, 1721–1730.
57. W. Shi, M. Bessarabova, D. Dosymbekov, Z. Dezso, T. Nikolskaya, M. Dudoladova, T. Serebryiskaya, A. Bugrim, A. Guryanov, R. J. Brennan, R. Shah, J. Dopazo, M. Chen, Y. Deng, T. Shi, G. Jurman, C. Furlanello, R. S. Thomas, J. C. Corton, W. Tong, L. Shi and Y. Nikolsky, Functional analysis of multiple genomic signatures demonstrates that classification algorithms choose phenotype-related genes, *Pharmacogenomics J.*, 2010, **10**, 310–323.
58. H. Ellinger-Ziegelbauer, B. Stuart, B. Wahle, W. Bomann and H. J. Ahr, Characteristic expression profiles induced by genotoxic carcinogens in rat liver, *Toxicol. Sci.*, 2004, **77**, 19–34.

59. H. Ellinger-Ziegelbauer, B. Stuart, B. Wahle, W. Bomann and H. J. Ahr, Comparison of the expression profiles induced by genotoxic and nongenotoxic carcinogens in rat liver, *Mutat. Res.*, 2005, **575**, 61–84.
60. H. Ellinger-Ziegelbauer, H. Gmuender, A. Bandenburg and H. J. Ahr, Prediction of a carcinogenic potential of rat hepatocarcinogens using toxicogenomics analysis of short-term *in vivo* studies, *Mutat. Res.*, 2008, **637**, 23–39.
61. M. R. Fielden, A. Adai, R. T. Dunn 2nd, A. Olaharski, G. Searfoss, J. Sina, J. Aubrecht, E. Boitier, P. Nioi, S. Auerbach, D. Jacobson-Kram, N. Raghavan, Y. Yang, A. Kincaid, J. Sherlock, S. J. Chen, B. Car and Predictive Safety Testing Consortium, Carcinogenicity Working Group, Development and evaluation of a genomic signature for the prediction and mechanistic assessment of nongenotoxic hepatocarcinogens in the rat, *Toxicol. Sci.*, 2011, **124**, 54–74.
62. A. Y. Nie, M. McMillian, J. B. Parker, A. Leone, S. Bryant, L. Yieh, A. Bittner, J. Nelson, A. Carmen, J. Wan and P. G. Lord, Predictive toxicogenomics approaches reveal underlying molecular mechanisms of nongenotoxic carcinogenicity, *Mol. Carcinog.*, 2006, **45**, 914–933.
63. F. Yamada, K. Sumida, T. Uehara, Y. Morikawa, H. Yamada, T. Urushidani and Y. Ohno, Toxicogenomics discrimination of potential hepatocarcinogenicity of non-genotoxic compounds in rat liver, *J. Appl. Toxicol.*, 2013, **33**, 1284–1293.
64. E. Caba, D. A. Dickinson, G. R. Warnes and J. Aubrecht, Differentiating mechanisms of toxicity using global gene expression analysis in *Saccharomyces cerevisiae*, *Mutat. Res.*, 2005, **575**, 34–46.
65. D. A. Dickinson, G. R. Warnes, G. Quievryn, J. Messer, A. Zhitkovich, E. Rubitski and J. Aubrecht, Differentiation of DNA reactive and non-reactive genotoxic mechanisms using gene expression profile analysis, *Mutat. Res.*, 2004, **549**, 29–41.
66. T. Hu, D. P. Gibson, G. J. Carr, S. M. Torontali, J. P. Tiesman, J. G. Chaney and M. J. Aardema, Identification of a gene expression profile that discriminates indirect-acting genotoxins from direct-acting genotoxins, *Mutat. Res.*, 2004, **549**, 5–27.
67. M. Islaih, B. W. Halstead, I. A. Kadura, B. Li, J. L. Reid-Hubbard, L. Flick, J. L. Altizer, J. Thom Deahl, D. K. Monteith, R. K. Newton and D. E. Watson, Relationships between genomic, cell cycle, and mutagenic responses of TK6 cells exposed to DNA damaging chemicals, *Mutat. Res.*, 2005, **578**, 100–116.
68. J. Y. Kim, J. Kwon, J. E. Kim, W. S. Koh, M. K. Chung, S. Yoon, C. W. Song and M. Lee, Identification of potential biomarkers of genotoxicity and carcinogenicity in L5178Y mouse lymphoma cells by cDNA microarray analysis, *Environ. Mol. Mutagen.*, 2005, **45**, 80–89.
69. M. Lee, J. Kwon, S. N. Kim, J. E. Kim, W. S. Koh, E. J. Kim, M. K. Chung, S. S. Han and C. W. Song, cDNA microarray gene expression profiling of hydroxyurea, paclitaxel, and p-anisidine, genotoxic compounds with differing tumorigenicity results, *Environ. Mol. Mutagen.*, 2003, **42**, 91–97.

70. S. D. Seidel, H. L. Kan, W. T. Stott, M. R. Schisler and B. B. Gollapudi, Identification of transcriptome profiles for the DNA-damaging agents bleomycin and hydrogen peroxide in L5178Y mouse lymphoma cells, *Environ. Mol. Mutagen.*, 2003, **42**, 19–25.
71. M. E. Burczynski, M. McMillian, J. Ciervo, L. Li, J. B. Parker, R. T. Dunn 2nd, S. Hicken, S. Farr and M. D. Johnson, Toxicogenomics-based discrimination of toxic mechanism in HepG2 human hepatoma cells, *Toxicol. Sci.*, 2000, **58**, 399–415.
72. J. H. van Delft, E. van Agen, S. G. van Breda, M. H. Herwijnen, Y. C. Staal and J. C. Kleinjans, Discrimination of genotoxic from non-genotoxic carcinogens by gene expression profiling, *Carcinogenesis*, 2004, **25**, 1265–1276.
73. J. H. van Delft, E. van Agen, S. G. van Breda, M. H. Herwijnen, Y. C. Staal and J. C. Kleinjans, Comparison of supervised clustering methods to discriminate genotoxic from non-genotoxic carcinogens by gene expression profiling, *Mutat. Res.*, 2005, **575**, 17–33.
74. G. S. Akerman, B. A. Rosenzweig, O. E. Domon, C. A. Tsai, M. E. Bishop, L. J. McGarrity, J. T. Macgregor, F. D. Sistare, J. J. Chen and S. M. Morris, Alterations in gene expression profiles and the DNA-damage response in ionizing radiation-exposed TK6 cells, *Environ. Mol. Mutagen.*, 2005, **45**, 188–205.
75. M. Islaih, B. Li, I. A. Kadura, J. L. Reid-Hubbard, J. T. Deahl, J. L. Altizer, D. E. Watson and R. K. Newton, Comparison of gene expression changes induced in mouse and human cells treated with direct-acting mutagens, *Environ. Mol. Mutagen.*, 2004, **44**, 401–419.
76. OECD, *Test No. 473: In vitro Mammalian Chromosomal Aberration Test*, Organisation for Economic Co-operation and Development, 2014 (revised).
77. OECD, *Test No. 487: In vitro Mammalian Cell Micronucleus Test*, Organisation for Economic Co-operation and Development, 2014 (revised).
78. Z. G. Liu, R. Baskaran, E. T. Lea-Chou, L. D. Wood, Y. Chen, M. Karin and J. Y. Wang, Three distinct signalling responses by murine fibroblasts to genotoxic stress, *Nature*, 1996, **384**, 273–276.
79. J. K. Buick, I. Moffat, A. Williams, C. D. Swartz, L. Recio, D. R. Hyduke, H. H. Li, A. J. Fornace Jr, J. Aubrecht and C. L. Yauk, Integration of metabolic activation with a predictive toxicogenomics signature to classify genotoxic *versus* nongenotoxic chemicals in human TK6 cells, *Environ. Mol. Mutagen.*, 2015, **56**, 520–534.
80. S. Kuehner, K. Holzmann and G. Speit, Characterization of formaldehyde's genotoxic mode of action by gene expression analysis in TK6 cells, *Arch. Toxicol.*, 2013, **87**, 1999–2012.
81. A. C. Le Fevre, E. Boitier, J. P. Marchandeau, A. Sarasin and V. Thybaud, Characterization of DNA reactive and non-DNA reactive anticancer drugs by gene expression profiling, *Mutat. Res.*, 2007, **619**, 16–29.
82. H. H. Li, D. R. Hyduke, R. Chen, P. Heard, C. L. Yauk, J. Aubrecht and A. J. Fornace Jr, Development of a toxicogenomics signature for

genotoxicity using a dose-optimization and informatics strategy in human cells, *Environ. Mol. Mutagen.*, 2015, **56**, 505–519.
83. R. Sakai, C. Kondo, H. Oka, H. Miyajima, K. Kubo and T. Uehara, Utilization of CDKN1A/p21 gene for class discrimination of DNA damage-induced clastogenicity, *Toxicology*, 2014, **315**, 8–16.
84. K. Boehme, Y. Dietz, P. Hewitt and S. O. Mueller, Genomic profiling uncovers a molecular pattern for toxicological characterization of mutagens and promutagens *in vitro*, *Toxicol. Sci.*, 2011, **122**, 185–197.
85. D. G. Jennen, C. Magkoufopoulou, H. B. Ketelslegers, M. H. van Herwijnen, J. C. Kleinjans and J. H. van Delft, Comparison of HepG2 and HepaRG by whole-genome gene expression analysis for the purpose of chemical hazard identification, *Toxicol. Sci.*, 2010, **115**, 66–79.
86. W. J. Lee, S. C. Kim, S. J. Lee, J. Lee, J. H. Park, K. S. Yu, J. Lim and S. W. Kwon, Investigating the different mechanisms of genotoxic and non-genotoxic carcinogens by a gene set analysis, *PLoS One*, 2014, **9**, e86700.
87. C. Magkoufopoulou, S. M. Claessen, D. G. Jennen, J. C. Kleinjans and J. H. van Delft, Comparison of phenotypic and transcriptomic effects of false-positive genotoxins, true genotoxins and non-genotoxins using HepG2 cells, *Mutagenesis*, 2011, **26**, 593–604.
88. C. Magkoufopoulou, S. M. Claessen, M. Tsamou, D. G. Jennen, J. C. Kleinjans and J. H. van Delft, A transcriptomics-based in vitro assay for predicting chemical genotoxicity *in vivo*, *Carcinogenesis*, 2012, **33**, 1421–1429.
89. T. Y. Doktorova, R. Yildirimman, M. Vinken, M. Vilardell, T. Vanhaecke, H. Gmuender, R. Bort, G. Brolen, G. Holmgren, R. Li, C. Chesne, J. van Delft, J. Kleinjans, J. Castell, P. Bjorquist, R. Herwig and V. Rogiers, Transcriptomic responses generated by hepatocarcinogens in a battery of liver-based *in vitro* models, *Carcinogenesis*, 2013, **34**, 1393–1402.
90. T. Y. Doktorova, R. Yildirimman, L. Ceelen, M. Vilardell, T. Vanhaecke, M. Vinken, G. Ates, A. Heymans, H. Gmuender, R. Bort, R. Corvi, P. Phrakonkham, R. Li, N. Mouchet, C. Chesne, J. van Delft, J. Kleinjans, J. Castell, R. Herwig and V. Rogiers, Testing chemical carcinogenicity by using a transcriptomics HepaRG-based model?, *EXCLI J.*, 2014, **13**, 623–637.
91. R. Yildirimman, G. Brolen, M. Vilardell, G. Eriksson, J. Synnergren, H. Gmuender, A. Kamburov, M. Ingelman-Sundberg, J. Castell, A. Lahoz, J. Kleinjans, J. van Delft, P. Bjorquist and R. Herwig, Human embryonic stem cell derived hepatocyte-like cells as a tool for *in vitro* hazard assessment of chemical carcinogenicity, *Toxicol. Sci.*, 2011, **124**, 278–290.
92. K. Mathijs, K. J. Brauers, D. G. Jennen, A. Boorsma, M. H. van Herwijnen, R. W. Gottschalk, J. C. Kleinjans and J. H. van Delft, Discrimination for genotoxic and nongenotoxic carcinogens by gene expression profiling in primary mouse hepatocytes improves with exposure time, *Toxicol. Sci.*, 2009, **112**, 374–384.

93. K. Mathijs, K. J. Brauers, D. G. Jennen, D. Lizarraga, J. C. Kleinjans and J. H. van Delft, Gene expression profiling in primary mouse hepatocytes discriminates true from false-positive genotoxic compounds, *Mutagenesis*, 2010, **25**, 561–568.
94. L. Rieswijk, K. J. Brauers, M. L. Coonen, S. G. van Breda, D. G. Jennen and J. C. Kleinjans, Evaluating microRNA profiles reveals discriminative responses following genotoxic or non-genotoxic carcinogen exposure in primary mouse hepatocytes, *Mutagenesis*, 2015, **30**, 771–784.
95. T. Y. Doktorova, H. Ellinger-Ziegelbauer, M. Vinken, T. Vanhaecke, J. van Delft, J. Kleinjans, H. J. Ahr and V. Rogiers, Comparison of hepatocarcinogen-induced gene expression profiles in conventional primary rat hepatocytes with *in vivo* rat liver, *Arch. Toxicol.*, 2012, **86**, 1399–1411.
96. Y. Yang, S. J. Abel, R. Ciurlionis and J. F. Waring, Development of a toxicogenomics *in vitro* assay for the efficient characterization of compounds, *Pharmacogenomics*, 2006, **7**, 177–186.
97. R. Josse, J. Dumont, A. Fautrel, M. A. Robin and A. Guillouzo, Identification of early target genes of aflatoxin B1 in human hepatocytes, inter-individual variability and comparison with other genotoxic compounds, *Toxicol. Appl. Pharmacol.*, 2012, **258**, 176–187.
98. H. Ellinger-Ziegelbauer, J. M. Fostel, C. Aruga, D. Bauer, E. Boitier, S. Deng, D. Dickinson, A. C. Le Fevre, A. J. Fornace Jr, O. Grenet, Y. Gu, J. C. Hoflack, M. Shiiyama, R. Smith, R. D. Snyder, C. Spire, G. Tanaka and J. Aubrecht, Characterization and interlaboratory comparison of a gene expression signature for differentiating genotoxic mechanisms, *Toxicol. Sci.*, 2009, **110**, 341–352.
99. L. Godderis, R. Thomas, A. E. Hubbard, A. M. Tabish, P. Hoet, L. Zhang, M. T. Smith, H. Veulemans and C. M. McHale, Effect of chemical mutagens and carcinogens on gene expression profiles in human TK6 cells, *PLoS One*, 2012, **7**, e39205.
100. T. Shimada, D. Kim, N. Murayama, K. Tanaka, S. Takenaka, L. D. Nagy, L. M. Folkman, M. K. Foroozesh, M. Komori, H. Yamazaki and F. P. Guengerich, Binding of Diverse Environmental Chemicals with Human Cytochromes P450 2A13, 2A6, and 1B1 and Enzyme Inhibition, *Chem. Res. Toxicol.*, 2013, **26**, 517–528.
101. D. J. Kirkland, M. Aardema, N. Banduhn, P. Carmichael, R. Fautz, J. R. Meunier and S. Pfuhler, *In vitro* approaches to develop weight of evidence (WoE) and mode of action (MoA) discussions with positive *in vitro* genotoxicity results, *Mutagenesis*, 2007, **22**, 161–175.
102. D. Kirkland, Improvements in the reliability of *in vitro* genotoxicity testing, *Expert Opin. Drug Metab. Toxicol.*, 2011, **7**, 1513–1520.
103. L. Recio, K. G. Shepard, L. G. Hernandez and G. L. Kedderis, Dose-response assessment of naphthalene-induced genotoxicity and glutathione detoxication in human TK6 lymphoblasts, *Toxicol. Sci.*, 2012, **126**, 405–412.

104. A. F. Jackson, A. Williams, L. Recio, M. D. Waters, I. B. Lambert and C. L. Yauk, Case study on the utility of hepatic global gene expression profiling in the risk assessment of the carcinogen furan, *Toxicol. Appl. Pharmacol.*, 2014, **274**, 63–77.
105. H. H. Gerets, K. Tilmant, B. Gerin, H. Chanteux, B. O. Depelchin, S. Dhalluin and F. A. Atienzar, Characterization of primary human hepatocytes, HepG2 cells, and HepaRG cells at the mRNA level and CYP activity in response to inducers and their predictivity for the detection of human hepatotoxins, *Cell Biol. Toxicol.*, 2012, **28**, 69–87.
106. R. Josse, C. Aninat, D. Glaise, J. Dumont, V. Fessard, F. Morel, J. M. Poul, C. Guguen-Guillouzo and A. Guillouzo, Long-term functional stability of human HepaRG hepatocytes and use for chronic toxicity and genotoxicity studies, *Drug Metab. Dispos.*, 2008, **36**, 1111–1118.
107. L. Le Hegarat, J. Dumont, R. Josse, S. Huet, R. Lanceleur, A. Mourot, J. M. Poul, C. Guguen-Guillouzo, A. Guillouzo and V. Fessard, Assessment of the genotoxic potential of indirect chemical mutagens in HepaRG cells by the comet and the cytokinesis-block micronucleus assays, *Mutagenesis*, 2010, **25**, 555–560.
108. I. Valentin-Severin, L. Le Hegarat, J. C. Lhuguenot, A. M. Le Bon and M. C. Chagnon, Use of HepG2 cell line for direct or indirect mutagens screening: comparative investigation between comet and micronucleus assays, *Mutat. Res.*, 2003, **536**, 79–90.
109. T. B. Andersson, K. P. Kanebratt and J. G. Kenna, The HepaRG cell line: a unique *in vitro* tool for understanding drug metabolism and toxicology in human, *Expert Opin. Drug Metab. Toxicol.*, 2012, **8**, 909–920.
110. C. Aninat, A. Piton, D. Glaise, T. Le Charpentier, S. Langouet, F. Morel, C. Guguen-Guillouzo and A. Guillouzo, Expression of cytochromes P450, conjugating enzymes and nuclear receptors in human hepatoma HepaRG cells, *Drug Metab. Dispos.*, 2006, **34**, 75–83.
111. S. N. Hart, Y. Li, K. Nakamoto, E. A. Subileau, D. Steen and X. B. Zhong, A comparison of whole genome gene expression profiles of HepaRG cells and HepG2 cells to primary human hepatocytes and human liver tissues, *Drug Metab. Dispos.*, 2010, **38**, 988–994.
112. M. J. Jetten, J. C. Kleinjans, S. M. Claessen, C. Chesne and J. H. van Delft, Baseline and genotoxic compound induced gene expression profiles in HepG2 and HepaRG compared to primary human hepatocytes, *Toxicol. In Vitro*, 2013, **27**, 2031–2040.
113. D. Kirkland, S. Pfuhler, D. Tweats, M. Aardema, R. Corvi, F. Darroudi, A. Elhajouji, H. Glatt, P. Hastwell, M. Hayashi, P. Kasper, S. Kirchner, A. Lynch, D. Marzin, D. Maurici, J. R. Meunier, L. Muller, G. Nohynek, J. Parry, E. Parry, V. Thybaud, R. Tice, J. van Benthem, P. Vanparys and P. White, How to reduce false positive results when undertaking in vitro genotoxicity testing and thus avoid unnecessary follow-up animal tests: Report of an ECVAM Workshop, *Mutat. Res.*, 2007, **628**, 31–55.

114. A. Guillouzo, A. Corlu, C. Aninat, D. Glaise, F. Morel and C. Guguen-Guillouzo, The human hepatoma HepaRG cells: a highly differentiated model for studies of liver metabolism and toxicity of xenobiotics, *Chem. Biol. Interact.*, 2007, **168**, 66–73.
115. R. Josse, A. Rogue, E. Lorge and A. Guillouzo, An adaptation of the human HepaRG cells to the *in vitro* micronucleus assay, *Mutagenesis*, 2012, **27**, 295–304.
116. S. Knasmuller, V. Mersch-Sundermann, S. Kevekordes, F. Darroudi, W. W. Huber, C. Hoelzl, J. Bichler and B. J. Majer, Use of human-derived liver cell lines for the detection of environmental and dietary genotoxicants; current state of knowledge, *Toxicology*, 2004, **198**, 315–328.
117. G. Ramsay, DNA chips: state-of-the art, *Nat. Biotechnol.*, 1998, **16**, 40–44.
118. M. Schena, D. Shalon, R. W. Davis and P. O. Brown, Quantitative monitoring of gene expression patterns with a complementary DNA microarray, *Science*, 1995, **270**, 467–470.
119. A. Brazma, P. Hingamp, J. Quackenbush, G. Sherlock, P. Spellman, C. Stoeckert, J. Aach, W. Ansorge, C. A. Ball, H. C. Causton, T. Gaasterland, P. Glenisson, F. C. Holstege, I. F. Kim, V. Markowitz, J. C. Matese, H. Parkinson, A. Robinson, U. Sarkans, S. Schulze-Kremer, J. Stewart, R. Taylor, J. Vilo and M. Vingron, Minimum information about a microarray experiment (MIAME)-toward standards for microarray data, *Nat. Genet.*, 2001, **29**, 365–371.
120. A. Brazma, Minimum Information About a Microarray Experiment (MIAME)–successes, failures, challenges, *Sci. World J.*, 2009, **9**, 420–423.
121. L Shi, G. Campbell, W. D. Jones, F. Campagne, Z. Wen, S. J. Walker, Z. Su, T. M. Chu, F. M. Goodsaid, L. Pusztai, J. D. Shaughnessy Jr, A. Oberthuer, R. S. Thomas, R. S. Paules, M. Fielden, B. Barlogie, W. Chen, P. Du, M. Fischer, C. Furlanello, B. D. Gallas, X. Ge, D. B. Megherbi, W. F. Symmans, M. D. Wang, J. Zhang, H. Bitter, B. Brors, P. R. Bushel, M. Bylesjo, M. Chen, J. Cheng, J. Cheng, J. Chou, T. S. Davison, M. Delorenzi, Y. Deng, V. Devanarayan, D. J. Dix, J. Dopazo, K. C. Dorff, F. Elloumi, J. Fan, S. Fan, X. Fan, H. Fang, N. Gonzaludo, K. R. Hess, H. Hong, J. Huan, R. A. Irizarry, R. Judson, D. Juraeva, S. Lababidi, C. G. Lambert, L. Li, Y. Li, Z. Li, S. M. Lin, G. Liu, E. K. Lobenhofer, J. Luo, W. Luo, M. N. McCall, Y. Nikolsky, G. A. Pennello, R. G. Perkins, R. Philip, V. Popovici, N. D. Price, F. Qian, A. Scherer, T. Shi, W. Shi, J. Sung, D. Thierry-Mieg, J. Thierry-Mieg, V. Thodima, J. Trygg, L. Vishnuvajjala, S. J. Wang, J. Wu, Y. Wu, Q. Xie, W. A. Yousef, L. Zhang, X. Zhang, S. Zhong, Y. Zhou, S. Zhu, D. Arasappan, W. Bao, A. B. Lucas, F. Berthold, R. J. Brennan, A. Buness, J. G. Catalano, C. Chang, R. Chen, Y. Cheng, J. Cui, W. Czika, F. Demichelis, X. Deng, D. Dosymbekov, R. Eils, Y. Feng, J. Fostel, S. Fulmer-Smentek, J. C. Fuscoe, L. Gatto, W. Ge, D. R. Goldstein, L. Guo, D. N. Halbert, J. Han, S. C. Harris, C. Hatzis, D. Herman,

J. Huang, R. V. Jensen, R. Jiang, C. D. Johnson, G. Jurman, Y. Kahlert, S. A. Khuder, M. Kohl, J. Li, L. Li, M. Li, Q. Z. Li, S. Li, Z. Li, J. Liu, Y. Liu, Z. Liu, L. Meng, M. Madera, F. Martinez-Murillo, I. Medina, J. Meehan, K. Miclaus, R. A. Moffitt, D. Montaner, P. Mukherjee, G. J. Mulligan, P. Neville, T. Nikolskaya, B. Ning, G. P. Page, J. Parker, R. M. Parry, X. Peng, R. L. Peterson, J. H. Phan, B. Quanz, Y. Ren, S. Riccadonna, A. H. Roter, F. W. Samuelson, M. M. Schumacher, J. D. Shambaugh, Q. Shi, R. Shippy, S. Si, A. Smalter, C. Sotiriou, M. Soukup, F. Staedtler, G. Steiner, T. H. Stokes, Q. Sun, P. Y. Tan, R. Tang, Z. Tezak, B. Thorn, M. Tsyganova, Y. Turpaz, S. C. Vega, R. Visintainer, J. von Frese, C. Wang, E. Wang, J. Wang, W. Wang, F. Westermann, J. C. Willey, M. Woods, S. Wu, N. Xiao, J. Xu, L. Xu, L. Yang, X. Zeng, J. Zhang, L. Zhang, M. Zhang, C. Zhao, R. K. Puri, U. Scherf, W. Tong, R. D. Wolfinger and MAQC Consortium, The MicroArray Quality Control (MAQC)-II study of common practices for the development and validation of microarray-based predictive models, *Nat. Biotechnol.*, 2010, **28**, 827–838.

122. R. S. Thomas, M. A. Philbert, S. S. Auerbach, B. A. Wetmore, M. J. Devito, I. Cote, J. C. Rowlands, M. P. Whelan, S. M. Hays, M. E. Andersen, M. E. Meek, L. W. Reiter, J. C. Lambert, H. J. Clewell 3rd, M. L. Stephens, Q. J. Zhao, S. C. Wesselkamper, L. Flowers, E. W. Carney, T. P. Pastoor, D. D. Petersen, C. L. Yauk and A. Nong, Incorporating new technologies into toxicity testing and risk assessment: moving from 21st century vision to a data-driven framework, *Toxicol. Sci.*, 2013, **136**, 4–18.

123. R. Benigni, Evaluation of the toxicity forecasting capability of EPA's ToxCast Phase I data: can ToxCast *in vitro* assays predict carcinogenicity?, *J. Environ. Sci. Health, Part C: Environ. Carcinog. Ecotoxicol. Rev.*, 2013, **31**, 201–212.

124. A. W. Knight, S. Little, K. Houck, D. Dix, R. Judson, A. Richard, N. McCarroll, G. Akerman, C. Yang, L. Birrell and R. M. Walmsley, Evaluation of high-throughput genotoxicity assays used in profiling the US EPA ToxCast chemicals, *Regul. Toxicol. Pharmacol.*, 2009, **55**, 188–199.

125. J. A. Bourdon-Lacombe, I. D. Moffat, M. Deveau, M. Husain, S. Auerbach, D. Krewski, R. S. Thomas, P. R. Bushel, A. Williams and C. L. Yauk, Technical guide for applications of gene expression profiling in human health risk assessment of environmental chemicals, *Regul. Toxicol. Pharmacol.*, 2015, **72**, 292–309.

126. Y. Adeleye, M. Andersen, R. Clewell, M. Davies, M. Dent, S. Edwards, P. Fowler, S. Malcomber, B. Nicol, A. Scott, S. Scott, B. Sun, C. Westmoreland, A. White, Q. Zhang and P. L. Carmichael, Implementing Toxicity Testing in the 21st Century (TT21C): Making safety decisions using toxicity pathways, and progress in a prototype risk assessment, *Toxicology*, 2015, **332**, 102–111.

127. N. L. Chepelev, I. D. Moffat, S. Labib, J. Bourdon-Lacombe, B. Kuo, J. K. Buick, F. Lemieux, A. I. Malik, S. Halappanavar, A. Williams and C. L. Yauk, Integrating toxicogenomics into human health risk

assessment: Lessons learned from the benzo[a]pyrene case study, *Crit. Rev. Toxicol.*, 2015, **45**, 44–52.

128. I. Moffat, N. L. Chepelev, S. Labib, J. Bourdon-Lacombe, B. Kuo, J. K. Buick, F. Lemieux, A. Williams, S. Halappanavar, A. I. Malik, M. Luijten, J. Aubrecht, D. R. Hyduke, A. J. Fornace Jr., C. D. Swartz, L. Recio and C. L. Yauk, Comparison of toxicogenomics and traditional approaches to inform mode of action and points of departure in human health risk assessment of benzo[a]pyrene in drinking water, *Crit. Rev. Toxicol.*, 2015, **45**, 1–43.
129. R. S. Thomas, H. J. Clewell 3rd, B. C. Allen, L. Yang, E. Healy and M. E. Andersen, Integrating pathway-based transcriptomic data into quantitative chemical risk assessment: a five chemical case study, *Mutat. Res.*, 2012, **746**, 135–143.
130. SEQC/MAQC-III Consortium, A comprehensive assessment of RNA-seq accuracy, reproducibility and information content by the Sequencing Quality Control Consortium, *Nat. Biotechnol.*, 2014, **32**, 903–914.
131. J. Van Delft, S. Gaj, M. Lienhard, M. W. Albrecht, A. Kirpiy, K. Brauers, S. Claessen, D. Lizarraga, H. Lehrach, R. Herwig and J. Kleinjans, RNA-Seq provides new insights in the transcriptome responses induced by the carcinogen benzo[a]pyrene, *Toxicol. Sci.*, 2012, **130**, 427–439.
132. Z. Su, H. Fang, H. Hong, L. Shi, W. Zhang, W. Zhang, Y. Zhang, Z. Dong, L. J. Lancashire, M. Bessarabova, X. Yang, B. Ning, B. Gong, J. Meehan, J. Xu, W. Ge, R. Perkins, M. Fischer and W. Tong, An investigation of biomarkers derived from legacy microarray data for their utility in the RNA-seq era, *Genome Biol.*, 2014, **15**, 523.
133. R. Sakai, Y. Morikawa, C. Kondo, H. Oka, H. Miyajima, R. Sakai and T. Uehara, Combinatorial measurement of CDKN1A/p21 and KIF20A expression for discrimination of DNA damage-induced clastogenicity, *Int. J. Mol. Sci.*, 2014, **15**, 17256–17269.
134. C. Magkoufopoulou, S. M. Claessen, D. G. Jennen, J. C. Kleinjans and J. H. van Delft, Comparison of phenotypic and transcriptomic effects of false-positive genotoxins, true genotoxins and non-genotoxins using HepG2 cells, *Mutagenesis*, 2011, **26**, 593–604.
135. A. Rohrbeck, G. Salinas, K. Maaser, J. Linge, S. Salovaara, R. Corvi and J. Borlak, Toxicogenomics applied to *in vitro* carcinogenicity testing with Balb/c 3T3 cells revealed a gene signature predictive of chemical carcinogens, *Toxicol. Sci.*, 2010, **118**, 31–41.
136. M. Islaih, B. Li, I. A. Kadura, J. L. Reid-Hubbard, J. T. Deahl, J. L. Altizer, D. E. Watson and R. K. Newton, Comparison of gene expression changes induced in mouse and human cells treated with direct-acting mutagens, *Environ. Mol. Mutagen.*, 2004, **44**, 401–419.
137. C. L. Yauk, J. K. Buick, A. Williams, C. D. Swartz, L. Recio, H. H. Li, A. J. Fornace, E. M. Thomson and J. Aubrecht, Application of the TGx-28.65 transcriptomic biomarker to classify genotoxic and non-genotoxic chemicals in human TK6 cells in the presence of rat liver S9, *Environ. Mol. Mutagen.*, DOI: 10.1002/em.22004.

CHAPTER 4

In vivo *Signatures of Genotoxic and Non-genotoxic Chemicals*

SCOTT S. AUERBACH

Toxicoinformatic Group, Biomolecular Screening Branch, Division of the National Toxicology Program, National Institute of Environmental Health Sciences, PO Box 12233, MD K2-17, Research Triangle Park NC 27709, USA
Email: auerbachs@niehs.nih.gov

4.1 Introduction

Chemicals can cause cancer through two broadly defined modes of action: genotoxic and non-genotoxic. Genotoxic chemicals induce DNA mutations through a few mechanisms (alkylating DNA, depletion of nucleotides through antimetabolite mechanisms or interfering with molecular processes directly related to DNA) that affect the integrity or fidelity of DNA, whereas non-genotoxic chemicals do not appear to directly impact DNA, but instead interfere with processes related to cell death or proliferation or cause the formation of reactive oxygen species that can then directly affect the DNA.[1] Overall, non-genotoxic mechanisms are much more diverse and perturb biological systems through many routes.[1] Differentiating between genotoxic and non-genotoxic modes of action *in vivo* is a common practice in toxicology, as there is general agreement that chemicals that act by these modes of action should be distinguished when performing cancer risk assessment.[2] Importantly, it is often the case that chemicals possess both modes of action, as many genotoxic chemicals can activate mechanisms associated with non-genotoxic modes of action and non-genotoxic chemicals

Issues in Toxicology No. 28
Toxicogenomics in Predictive Carcinogenicity
Edited by Michael D. Waters and Russell S. Thomas

Published by the Royal Society of Chemistry, www.rsc.org

can manifest secondary genotoxic effects that are related to excessive toxicity and oxidative stress. Ideally, researchers are able to distinguish between primary and secondary modes of action; however, due to the integrated nature of biology it is often not simple to achieve this goal.

Typically, the first approach to evaluating genotoxicity involves an *in silico* assessment. Due to the large amounts of genotoxicity data that are available, particularly for the Ames assay, and the known structural alerts,[3] structure activity response (SAR) models perform quite well.[4] When using these models there are a number of issues to consider when determining the reliability of the model's predictions, including a model's applicability domain, sensitivity, specificity and balanced accuracy (www.oecd.org/officialdocuments/publicdisplaydocumentpdf/?doclanguage=en&cote=env/jm/mono%282007%292). When working in an under-studied area of chemical space it is often the case that SAR models are of limited utility.

When the SAR analysis for genotoxicity fails to confidently classify compounds the next step is typically an *in vitro* and/or *in vivo* evaluation of genotoxicity. A large number of *in vitro* and *in vivo* assays have been developed that evaluate the genotoxic properties of chemicals. The best known of the *in vitro* assays is the bacteria reverse mutation assay, also known as the Ames assay. Other assays that can be performed both *in vitro* and *in vivo* include the sister chromatid exchange, chromosomal aberration, micronuclei and comet assays.[5] The erythrocyte micronucleus is the most common assay that is used for *in vivo* genotoxicity assessment. Another well-established technique for evaluating *in vitro* and *in vivo* genotoxicity is the comet assay, which allows for tissue-specific evaluation of genetic damage.[6] A variety of less commonly used approaches for detecting *in vivo* mutagenesis have also been developed or are under evaluation, such as the Big Blue mouse models that can evaluate tissue-specific mutagenesis[7] and the Pig-a assay which evaluates mutagen-induced changes in blood cell markers.[7] The added value of the *in vivo* assessments is the greater similarity of the model system to humans and the external dose to internal effect relationship that has the potential to inform quantitative risk assessment. The latter is of particular distinction to the *in vitro* assays because of the poor relationship between genotoxic potency in most *in vitro* systems compared to *in vivo* assays. The specificity of a number of the traditional genotoxicity assays has been called into question, as a review of many years of results indicated that these approaches produce a large number of false positives.[8,9] In addition, there have been a number of chemicals that are not identified as genotoxic in *in vitro* assays that have genotoxic effects *in vivo* (Ames-negative genotoxic chemicals).[10] Hence, while the existing regulatory genotoxicity assays have been invaluable to chemical risk assessment, there are clear blind spots that need to be addressed.

The development of genomic technology over the last 20 years has raised the potential for use of transcriptomics and omics modalities for the development of signatures that can classify modes of action, differentiate mechanisms of action, classify subtypes of pathology and predict pathological outcomes (such as cancer). Hence, the true power of genomics is not necessarily just the ability to differentiate between modes of action, but the potential resampling of the data from many perspectives that allows for one assay (a transcriptomic measurement) to serve as window into a broad landscape of biology well beyond the initial question discussed here. Most traditional assays for evaluating genotoxicity are robust tools to solely address genotoxicity, whereas whole-genome gene expression data are extremely information-rich and can be analyzed to determine many of the nuanced effects that a chemical has on a biological system. Many of the molecular level effects detected by omics-based approaches can be linked to specific biological outcomes that are of concern for public health and can be related to nuanced mechanisms of action that extend well beyond the binary classification of genotoxic or non-genotoxic. Often the cost of genomics exceeds that of more traditional measures of genotoxicity, but when considering the totality of information provided by genomics (well beyond genotoxicity classification) the researcher with the appropriate informatics skills can glean tremendous value from the data. Furthermore, the price of genomic analysis is dropping rapidly and it is reasonable to assume that the cost of performing genomic analysis may be less than that of traditional tests in the near future.

This chapter discusses the application of genomics to the *in vivo* characterization of genotoxicity. It is often the case that genotoxicity is evaluated prior to carcinogenicity and therefore the focus of this chapter is purely on genotoxicity independent of the carcinogenicity. Where possible, results from human studies are considered; however, most results are from studies of species used in the pre-clinical setting, specifically rats and to a lesser extent mice. Furthermore, most of the results are focused on a small subset of organs that are common targets of genotoxic chemicals. This is simply because of the paucity of data in most organs and tissues. In cases where there are limited data, anecdotal support for signatures in under-studied tissues is highlighted. A focus is placed primarily on signatures of genotoxicity with the assumption that agents that do not affect genes in this signature by default would be classified as non-genotoxic. The reason for this approach, as noted above, is that the signatures of most chemicals with genotoxic properties fit into a couple of reasonably well-defined, highly interrelated classes that are clearly identifiable by transcriptomic approaches, whereas chemicals with non-genotoxic properties tend to be much more diverse in their biological effects, hence a strong coherency in transcriptomic data tends to be challenging to identify. Furthermore, it is likely that the patterns that indicate non-genotoxic effects vary quite considerably across tissues and species due to differences in receptor expression and how receptor-driven, non-genotoxic processes interact with the biological systems that vary across tissues and species.[11,12] It is likely that the non-genotoxic chemicals will need to be sub-divided into groupings related to their molecular initiating events [*e.g.* peroxisome proliferator-activated receptor (PPAR)α and aryl hydrocarbon receptor] in order to begin to identify coherent signatures, and even in if this is achieved, it is likely, as noted, that these signatures will diverge significantly between species. In the final part of the chapter various aspects of study design are reviewed, which will help the end user of the signatures most effectively implement them in their genotoxicity assessments.

A sizable amount of the results presented here come from meta-analysis/re-analysis of data from published (and some unpublished) data sets that have been munged and integrated in the NextBio analysis platform. Hence, while the original papers from which the data were derived will be cited, much of the results discussed will be from secondary analysis in NextBio, wherein multiple data sets derived from different manuscripts have been subjected to an integrated meta-analysis to come to the conclusions presented here.

4.2 General Signature of Genotoxicity

After reviewing a number of manuscripts to identify genomic signatures that differentiate between genotoxic and non-genotoxic chemicals, it became clear that there is strong similarly between genotoxic signatures across studies. This "general" signature of genotoxicity (GSG) that manifests in many tissues and across different species is the focus of this chapter (Table 4.1). The genes in this signature were formulated through mining of existing data and review of the literature. Pieces of this signature can be found in many papers that have evaluated *in vivo* and *in vitro* genotoxic perturbations. This signature tends to be somewhat restrictive and it is therefore plausible that this signature could be expanded to include a number of additional genes that are functionally related to those contained in the signature. The genes in the signature tend to be up-regulated following genotoxic challenge and have differing degrees of conservation of response across different tissues and species. Pathway analysis of these genes shows, not surprisingly, a strong enrichment for p53 signaling (Figure 4.1). The core components of the ataxia telangiectasia mutated (ATM)–p53 signaling pathway that are present in the signature comprise a well-characterized response to DNA damage (Figure 4.2). Specifically, following DNA damage, p53 is activated and induces expression of MGMT, CCNG1, CDKN1A, BAX and BTG2 (see Table 4.1). MGMT facilitates the repair of alkylated DNA.[13] CCNG1 binds to MDM2 which prevents the degradation of p53.[14] CDKN1A, BAX and BTG2 block cell cycle progression or, in cases of severe DNA damage, induce apoptosis.[15] It is important to note that components of the p53 pathway, such as those contained in the GSG, will respond to overtly cytotoxic doses of chemicals that are non-genotoxic by traditional measures. This is likely from secondary effects of the chemical that occur during excessive tissue damage (*i.e.* after administration of necrotizing doses of a chemical). It should be noted that there is a small subset of down-regulated genes that have been identified by a number of groups, but less consistently so than those that were included in the GSG. These genes are FHIT (fragile histidine triad) and WWOX (WW domain containing oxidoreductase) which are located at common chromosomal fragile sites in cancer.[16]

Table 4.1 Genes associated with the "general" signature of genotoxicity.

Symbol	EntrezGene ID	Description[a]	Notes
ADAM8	101	ADAM metallopeptidase domain 8	This gene encodes a member of the ADAM (a disintegrin and metalloprotease domain) family. Members of this family are membrane-anchored proteins structurally related to snake venom disintegrins, and have been implicated in a variety of biological processes involving cell–cell and cell–matrix interactions, including fertilization, muscle development and neurogenesis. The protein encoded by this gene may be involved in cell adhesion during neurodegeneration, and it is thought to be a target for allergic respiratory diseases, including asthma. Alternative splicing results in multiple transcript variants.
FAS	355	Fas (TNF receptor superfamily, member 6)	The protein encoded by this gene is a member of the tumor necrosis factor (TNF) receptor superfamily. This receptor contains a death domain. It has been shown to play a central role in the physiological regulation of programmed cell death, and has been implicated in the pathogenesis of various malignancies and diseases of the immune system. The interaction of this receptor with its ligand allows the formation of a death-inducing signaling complex that includes Fas-associated death domain protein (FADD), caspase 8 and caspase 10. The autoproteolytic processing of the caspases in the complex triggers a downstream caspase cascade, and leads to apoptosis. This receptor has been also shown to activate nuclear factor (NF)-kappaB, mitogen-activated protein kinase (MAPK)3/extracellular signal-regulated kinase (ERK)1, and MAPK8/c-Jun N-terminal kinase (JNK), and is found to be involved in transducing the proliferating signals in normal diploid fibroblast and T-cells. Several alternatively spliced transcript variants have been described, some of which are candidates for nonsense-mediated mRNA decay (NMD). The isoforms lacking the transmembrane domain may negatively regulate the apoptosis mediated by the full-length isoform.

Table 4.1 (*Continued*)

Symbol	EntrezGene ID	Description[a]	Notes
ATF3	467	Activating transcription factor 3	This gene encodes a member of the mammalian activation transcription factor/cAMP responsive element-binding (CREB) protein family of transcription factors. This gene is induced by a variety of signals, including many of those encountered by cancer cells, and is involved in the complex process of cellular stress response. Multiple transcript variants encoding different isoforms have been found for this gene. It is possible that alternative splicing of this gene may be physiologically important in the regulation of target genes.
BAX	581	BCL2-associated X protein	The protein encoded by this gene belongs to the B-cell CLL/lymphoma (BCL)2 protein family. BCL2 family members form hetero- or homodimers and act as anti- or pro-apoptotic regulators that are involved in a wide variety of cellular activities. This protein forms a heterodimer with BCL2, and functions as an apoptotic activator. This protein is reported to interact with, and increase the opening of, the mitochondrial voltage-dependent anion channel, which leads to the loss in membrane potential and the release of cytochrome c. The expression of this gene is regulated by the tumor suppressor p53 and has been shown to be involved in p53-mediated apoptosis. Multiple alternatively spliced transcript variants, which encode different isoforms, have been reported for this gene.
CCNG1	900	Cyclin G1	The eukaryotic cell cycle is governed by cyclin-dependent protein kinases (CDKs) whose activities are regulated by cyclins and CDK inhibitors. The protein encoded by this gene is a member of the cyclin family and contains the cyclin box. The encoded protein lacks the protein destabilizing (PEST) sequence that is present in other family members. Transcriptional activation of this gene can be induced by tumor protein p53. Two transcript variants encoding the same protein have been identified for this gene.

CDKN1A	1026	Cyclin-dependent kinase inhibitor 1A (p21, Cip1)	This gene encodes a potent cyclin dependent kinase inhibitor. The encoded protein binds to and inhibits the activity of cyclin-dependent kinase2 or cyclin-dependent kinase4 complexes, and thus functions as a regulator of cell cycle progression at G1. The expression of this gene is tightly controlled by the tumor suppressor protein p53, through which this protein mediates the p53-dependent cell cycle G1 phase arrest in response to a variety of stress stimuli. This protein can interact with proliferating cell nuclear antigen, a DNA polymerase accessory factor, and plays a regulatory role in S-phase DNA replication and DNA damage repair. This protein was reported to be specifically cleaved by CASP(apoptosis-related cysteine peptidase)3-like caspases, which thus leads to a dramatic activation of cyclin-dependent kinase2, and may be instrumental in the execution of apoptosis following caspase activation. Mice that lack this gene have the ability to regenerate damaged or missing tissue. Multiple alternatively spliced variants have been found for this gene.
DDB2	1643	Damage-specific DNA binding protein 2, 48 kDa	This gene encodes a protein that is necessary for the repair of ultraviolet light-damaged DNA. This protein is the smaller subunit of a heterodimeric protein complex that participates in nucleotide excision repair, and this complex mediates the ubiquitylation of histones H3 and H4, which facilitates the cellular response to DNA damage. This subunit appears to be required for DNA binding. Mutations in this gene cause xeroderma pigmentosum complementation group E, a recessive disease that is characterized by an increased sensitivity to UV light and a high predisposition for skin cancer development, in some cases accompanied by neurological abnormalities. Two transcript variants encoding different isoforms have been found for this gene.

Table 4.1 (*Continued*)

Symbol	EntrezGene ID	Description[a]	Notes
EGR1	1958	Early growth response 1	The protein encoded by this gene belongs to the EGR family of C2H2-type zinc-finger proteins. It is a nuclear protein and functions as a transcriptional regulator. The products of target genes it activates are required for differentitation and mitogenesis. Studies suggest that this is a cancer-suppressor gene.
FDXR	2232	Ferredoxin reductase	This gene encodes a mitochondrial flavoprotein that initiates electron transport for cytochromes P450 receiving electrons from NADPH. Multiple alternatively spliced transcript variants have been found for this gene.
FOSB	2354	FBJ murine osteosarcoma viral oncogene homolog B	The Fos gene family consists of four members: FOS, FOSB, FOSL1 and FOSL2. These genes encode leucine zipper proteins that can dimerize with proteins of the JUN family, thereby forming the transcription factor complex AP-1. As such, the FOS proteins have been implicated as regulators of cell proliferation, differentiation, and transformation. Alternatively spliced transcript variants encoding different isoforms have been found for this gene.
MDM2	4193	Mdm2, p53 E3 ubiquitin protein ligase homolog (mouse)	This gene encodes a nuclear-localized E3 ubiquitin ligase. The encoded protein can promote tumor formation by targeting tumor suppressor proteins, such as p53, for proteasomal degradation. This gene is itself transcriptionally regulated by p53. Overexpression or amplification of this locus is detected in a variety of different cancers. There is a pseudogene for this gene on chromosome 2. Alternative splicing results in a multitude of transcript variants, many of which may be expressed only in tumor cells.
MGMT	4255	*O*-6-methylguanine-DNA methyltransferase	Alkylating agents are potent carcinogens that can result in cell death, mutation and cancer. The protein encoded by this gene is a DNA repair protein that is involved in cellular defense against mutagenesis and toxicity from alkylating agents. The protein

			catalyzes transfer of methyl groups from O^6-alkylguanine and other methylated moieties of the DNA to its own molecule, which repairs the toxic lesions. Methylation of the genes promoter has been associated with several cancer types, including colorectal cancer, lung cancer, lymphoma and glioblastoma.
PLN	5350	Phospholamban	The protein encoded by this gene is found as a pentamer and is a major substrate for the cAMP-dependent protein kinase in cardiac muscle. The encoded protein is an inhibitor of cardiac muscle sarcoplasmic reticulum Ca^{2+}-ATPase in the unphosphorylated state, but inhibition is relieved upon phosphorylation of the protein. The subsequent activation of the Ca^{2+} pump leads to enhanced muscle relaxation rates, thereby contributing to the inotropic response elicited in heart by beta-agonists. The encoded protein is a key regulator of cardiac diastolic function. Mutations in this gene are a cause of inherited human dilated cardiomyopathy with refractory congestive heart failure.
TUBB2A	7280	Tubulin, beta 2A class IIa	Microtubules, key participants in processes such as mitosis and intracellular transport, are composed of heterodimers of alpha- and beta-tubulins. The protein encoded by this gene is a beta-tubulin. Defects in this gene are associated with complex cortical dysplasia with other brain malformations-5. Two transcript variants encoding distinct isoforms have been found for this gene.
BTG2	7832	BTG family, member 2	The protein encoded by this gene is a member of the BTG/Tob family. This family has structurally related proteins that appear to have antiproliferative properties. This encoded protein is involved in the regulation of the G1/S transition of the cell cycle.
GDF15	9518	Growth differentiation factor 15	The protein encoded by this gene belongs to the transforming growth factor (TGF)-beta family. The protein is expressed in a broad range of cell types, acts as a pleiotropic cytokine and is involved in the stress reponse program of cells after cellular injury. Increased protein levels are associated with disease states such as tissue hypoxia, inflammation, acute injury and oxidative stress.

Table 4.1 (*Continued*)

Symbol	EntrezGene ID	Description[a]	Notes
TP53I3	9540	Tumor protein p53 inducible protein 3	The protein encoded by this gene is similar to oxidoreductases, which are enzymes involved in cellular responses to oxidative stresses and irradiation. This gene is induced by the tumor suppressor p53 and is thought to be involved in p53-mediated cell death. It contains a p53 consensus binding site in its promoter region and a downstream pentanucleotide microsatellite sequence. p53 has been shown to transcriptionally activate this gene by interacting with the downstream pentanucleotide microsatellite sequence. The microsatellite is polymorphic, with a varying number of pentanucleotide repeats directly correlated with the extent of transcriptional activation by p53. It has been suggested that the microsatellite polymorphism may be associated with differential susceptibility to cancer. Alternatively spliced transcript variants encoding different isoforms have been found for this gene.
PLK2	10769	Polo-like kinase 2	The protein encoded by this gene is a member of the polo family of serine/threonine protein kinases that have a role in normal cell division. This gene is most abundantly expressed in testis, spleen and fetal tissues, and its expression is inducible by serum, suggesting that it may also play an important role in cells undergoing rapid cell division. Alternatively spliced transcript variants encoding different isoforms have been found for this gene.
PHLDA3	23612	Pleckstrin homology-like domain, family A, member 3	p53/TP53-regulated repressor of Akt/AKT1 signaling. Represses AKT1 by preventing AKT1-binding to membrane lipids, thereby inhibiting AKT1 translocation to the cellular membrane and activation. Contributes to p53/TP53-dependent apoptosis by repressing AKT1 activity. Its direct transcription regulation by p53/TP53 may explain how p53/TP53 can negatively regulate AKT1. May act as a tumor suppressor.

ATP6V1D	51382	ATPase, H + transporting, lysosomal 34kDa, V1 subunit D	This gene encodes a component of vacuolar (V)-ATPase, a multisubunit enzyme that mediates acidification of eukaryotic intracellular organelles. V-ATPase-dependent organelle acidification is necessary for such intracellular processes as protein sorting, zymogen activation, receptor-mediated endocytosis and synaptic vesicle proton gradient generation. V-ATPase is composed of a cytosolic V1 domain and a transmembrane V0 domain. The V1 domain consists of three A and three B subunits, two G subunits plus the C, D, E, F and H subunits. The V1 domain contains the ATP catalytic site. The V0 domain consists of five different subunits: a, c, c′, c″ and d. Additional isoforms of many of the V1 and V0 subunit proteins are encoded by multiple genes or alternatively spliced transcript variants. This gene encodes the V1 domain D subunit protein.
EDA2R	60401	Ectodysplasin A2 receptor	EDA-A1 and EDA-A2 are two isoforms of ectodysplasin that are encoded by the anhidrotic ectodermal dysplasia (EDA) gene. Mutations in EDA give rise to a clinical syndrome characterized by loss of hair, sweat glands and teeth. The protein encoded by this gene specifically binds to EDA-A2 isoform. This protein is a type III transmembrane protein of the tumor necrosis factor (TNF) receptor superfamily, and contains three cysteine-rich repeats and a single transmembrane domain, but lacks an N-terminal signal peptide. Alternatively spliced transcript variants have been found for this gene.
GAS5	60674	Growth arrest-specific 5 (non-protein coding)	This gene produces a spliced long non-coding RNA and is a member of the 5′ terminal oligo-pyrimidine class of genes. It is a small nucleolar RNA host gene, containing multiple C/D box snoRNA genes in its introns. Part of the secondary RNA structure of the encoded transcript mimics glucocorticoid response element (GRE) which means it can bind to the DNA binding domain of the glucocorticoid receptor (nuclear receptor subfamily 3, group C, member 1). This action blocks the glucocorticoid receptor from being activated and thereby stops it from regulating the transcription of its target genes.

Table 4.1 (*Continued*)

Symbol	EntrezGene ID	Description[a]	Notes
			This transcript is also thought to regulate the transcriptional activity of other receptors, such as androgen, progesterone and mineralocorticoid receptors, that can bind to its GRE mimic region. Multiple functions have been associated with this transcript, including cellular growth arrest and apoptosis. It has also been identified as a potential tumor suppressor, with its down-regulation associated with cancer in multiple different tissues.
AEN	64782	Apoptosis enhancing nuclease	Exonuclease with activity against single- and double-stranded DNA and RNA. Mediates p53-induced apoptosis. When induced by p53 following DNA damage, digests double-stranded DNA to form single-stranded DNA and amplifies DNA damage signals, leading to enhancement of apoptosis.
NHEJ1	79840	Nonhomologous end-joining factor 1	Double-strand breaks in DNA result from genotoxic stresses and are among the most damaging of DNA lesions. This gene encodes a DNA repair factor essential for the nonhomologous end-joining pathway, which preferentially mediates repair of double-stranded breaks. Mutations in this gene cause different kinds of severe combined immunodeficiency disorders.
TP53INP1	94241	Tumor protein p53 inducible nuclear protein 1	Antiproliferative and proapoptotic protein involved in the cell stress response which acts as a dual regulator of transcription and autophagy. Acts as a positive regulator of autophagy. In response to cellular stress or activation of autophagy, relocates to autophagosomes where it interacts with autophagosome-associated proteins GABARAP, GABARAPL1/L2, MAP1LC3A/B/C and regulates autophagy. Acts as an antioxidant and plays a major role in p53/TP53-driven oxidative stress response. Possesses both a p53/TP53-independent intracellular reactive oxygen species regulatory function and a p53/TP53-dependent transcription regulatory function. Positively regulates p53/TP53 and p73/TP73 and stimulates their capacity to induce apoptosis

			and regulate cell cycle. In response to double-strand DNA breaks, promotes p53/TP53 phosphorylation on Ser-46 and subsequent apoptosis. Acts as a tumor suppressor by inducing cell death by an autophagy and caspase-dependent mechanism. Can reduce cell migration by regulating the expression of SPARC [secreted protein, acidic, cysteine-rich (osteonectin)].
DDIT4L	115265	DNA-damage-inducible transcript 4-like	Inhibits cell growth by regulating the target of rapamycin (TOR) signaling pathway upstream of the TSC1-TSC2 complex and downstream of AKT1.
ZMAT3	64393	Zinc finger, matrin-type	This gene encodes a protein containing three zinc finger domains and a nuclear localization signal. The mRNA and the protein of this gene are upregulated by wildtype p53 and overexpression of this gene inhibits tumor cell growth, suggesting that this gene may have a role in the p53-dependent growth regulatory pathway. Alternative splicing of this gene results in two transcript variants encoding two isoforms differing in only one amino acid.
RRM2B	50484	Ribonucleotide reductase M2 B (TP53 inducible)	This gene encodes the small subunit of a p53-inducible ribonucleotide reductase. This heterotetrameric enzyme catalyzes the conversion of ribonucleoside diphosphates to deoxyribonucleoside diphosphates. The product of this reaction is necessary for DNA synthesis. Mutations in this gene have been associated with autosomal recessive mitochondrial DNA depletion syndrome, autosomal dominant progressive external ophthalmoplegia-5 and mitochondrial neurogastrointestinal encephalopathy. Alternatively spliced transcript variants have been described.
GADD45A	1647	Growth arrest and DNA damage inducible alpha	This gene is a member of a group of genes whose transcript levels are increased following stressful growth arrest conditions and treatment with DNA-damaging agents. The protein encoded by this gene responds to environmental stresses by mediating activation of the p38/JNK pathway *via* MTK1/MEKK4 kinase. The DNA damage-induced transcription of this gene is mediated by both p53-dependent and -independent mechanisms. Alternatively spliced transcript variants encoding distinct isoforms have been found for this gene.

Table 4.1 (*Continued*)

Symbol	EntrezGene ID	Description[a]	Notes
ENC1	8507	Ectodermal-neural cortex 1 (with BTB domain)	This gene encodes a member of the kelch-related family of actin-binding proteins. The encoded protein plays a role in the oxidative stress response as a regulator of the transcription factor Nrf2, and expression of this gene may play a role in malignant transformation. Alternatively spliced transcript variants encoding multiple isoforms have been observed for this gene.
TYMS	7298	Thymidylate synthetase	Thymidylate synthase catalyzes the methylation of deoxyuridylate to deoxythymidylate using 5,10-methylenetetrahydrofolate (methylene-THF) as a cofactor. This function maintains the dTMP (thymidine-5-prime monophosphate) pool critical for DNA replication and repair. The enzyme has been of interest as a target for cancer chemotherapeutic agents. It is considered to be the primary site of action for 5-fluorouracil, 5-fluoro-2-prime-deoxyuridine, and some folate analogs. Expression of this gene and that of a naturally occuring antisense transcript rTSalpha (GeneID:55556) vary inversely when cell-growth progresses from late-log to plateau phase.
SESN2	83667	Sestrin	This gene encodes a member of the sestrin family of PA26-related proteins. The encoded protein may function in the regulation of cell growth and survival. This protein may be involved in cellular response to different stress conditions.

[a]The gene descriptions are taken from the human Entrez Gene (www.ncbi.nlm.nih.gov/gene/) entry or from UniProt (www.uniprot.org/uniprot/) where an Entrez Gene description was not available.

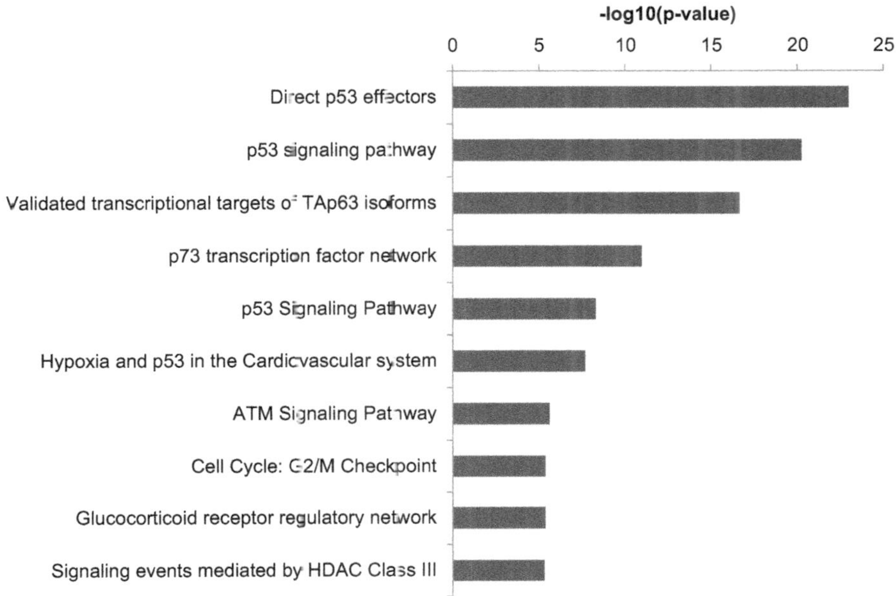

Figure 4.1 Pathways that are enriched in by the genes contained in the general signature of genotoxicity. The enrichment analysis was performed using the NextBio pathway analysis tool using annotation from the Molecular Signatures Database (http://software.broadinstitute.org/gsea/msigdb).

With a small fraction of genotoxic treatments (in most cases the treatments are intravenously delivered chemotherapeutics), there appears to be a second signature of genotoxicity that co-manifests or manifests in the absence of the GSG across multiple tissues. The signature is unpublished, but is clearly present in the DrugMatrix data across a number of tissues. There are four genes in this signature all of which are down-regulated by genotoxic chemical exposure: HBA1, HBB, ALAS2, and AHSP (see Table 4.2). All four of these transcripts are found at high levels in whole blood (Figure 4.3) and comprise four out of 13 genes in the hemoglobin chaperone pathway. Furthermore, these genes are contained in the leukopenia signature from DrugMatrix. Generally this signature is more pronounced in tissues that have a relatively high degree of blood perfusion. With such observations it is reasonable to hypothesize that the chemicals that manifest this signature in different organs may not be having a direct effect on the organ being evaluated, but instead are having an effect on the blood that is flowing through those organs. In other words, the transcripts are decreasing because the fraction of the pool of RNA from blood in the target organ is decreased, which then manifests in the microarray assessment as down-regulation of these blood-specific transcripts. In many ways this signature (Table 4.2) is likely an indicator of the effect of a genotoxic chemical on a target organ (bone marrow) that is commonly affected by chemotherapy due to its high rate of proliferation. Not surprisingly, this signature identifies some false-positive chemicals that are non-genotoxic bone marrow suppressants (*e.g.* leflutamide).

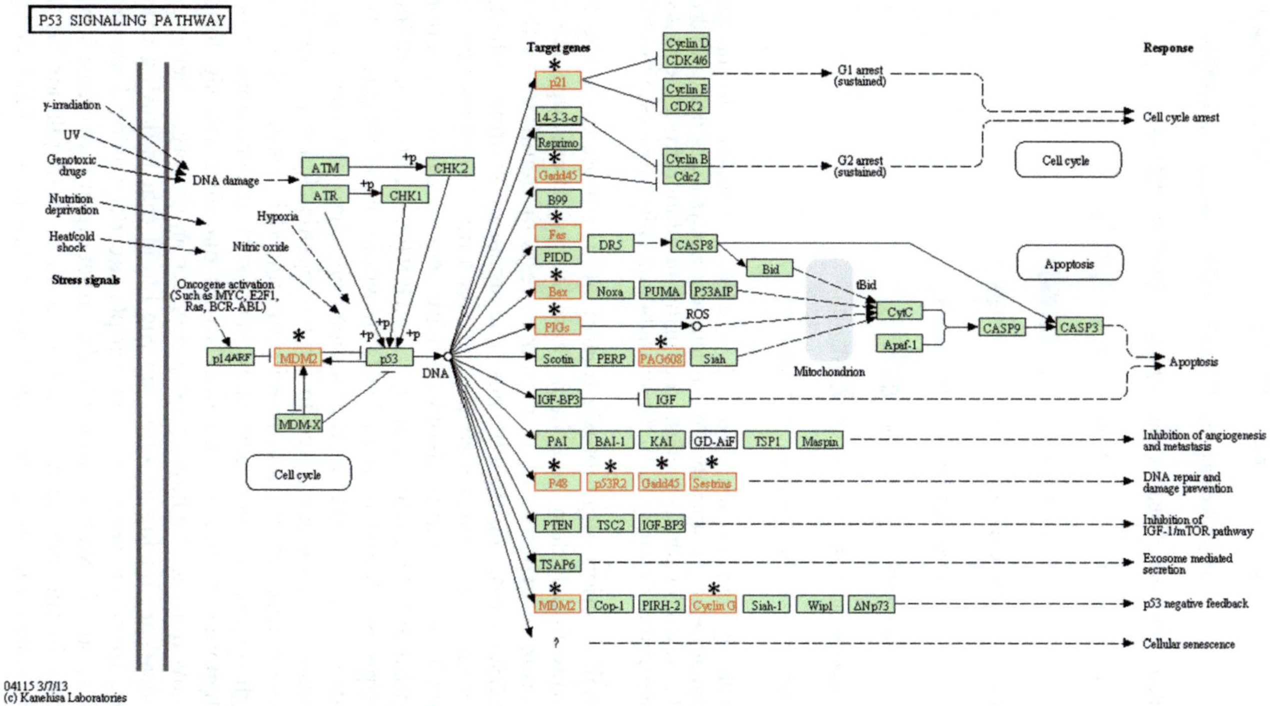

Figure 4.2 The genes from the general signature of genotoxicity were painted (*) onto the KEGG P53 signaling pathway. Notably the DNA repair and damage prevention subcomponent of the pathway is highly enriched. p21: CDKN1A; PIGs: TP53I3; PAG608: ZMAT3; P48: DDB2; p53R2: RRM2B. Refer to Table 4.1 for abbreviations.

Table 4.2 Bone marrow suppression signature genes.

Symbol	EntrezGeneID	Description[a]
HBA1	3039	The human alpha globin gene cluster located on chromosome 16 spans ~30 kb and includes seven loci: 5′- zeta - pseudozeta - mu - pseudoalpha-1 - alpha-2 - alpha-1 - theta - 3′. The alpha-2 (HBA2) and alpha-1 (HBA1) coding sequences are identical. These genes differ slightly over the 5′ untranslated regions and the introns, but they differ significantly over the 3′ untranslated regions. Two alpha chains plus two beta chains constitute HbA, which in normal adult life comprises about 97% of the total hemoglobin; alpha chains combine with delta chains to constitute HbA-2, which with HbF (fetal hemoglobin) makes up the remaining 3% of adult hemoglobin. Alpha-thalassemias result from deletions of each of the alpha genes as well as deletions of both HBA2 and HBA1; some nondeletion alpha-thalassemias have also been reported
HBB	3043	The alpha (HBA) and beta (HBB) loci determine the structure of the two types of polypeptide chains in adult hemoglobin, HbA. The normal adult hemoglobin tetramer consists of two alpha chains and two beta chains. Mutant beta globin causes sickle cell anemia. Absence of beta chain causes beta-zero-thalassemia. Reduced amounts of detectable beta globin causes beta-plus-thalassemia. The order of the genes in the beta-globin cluster is 5′-epsilon - gamma-G - gamma-A - delta - beta-3′
ALAS2	212	The product of this gene specifies an erythroid-specific mitochondrially located enzyme. The encoded protein catalyzes the first step in the heme biosynthetic pathway. Defects in this gene cause X-linked pyridoxine-responsive sideroblastic anemia. Alternatively spliced transcript variants encoding different isoforms have been identified
AHSP	51327	Acts as a chaperone to prevent the harmful aggregation of alpha-hemoglobin during normal erythroid cell development. Specifically protects free alpha-hemoglobin from precipitation. It is predicted to modulate pathological states of alpha-hemoglobin excess such as beta-thalassemia

[a]The gene descriptions are taken from the human Entrez Gene (www.ncbi.nlm.nih.gov/gene/) entry or from UniProt (www.uniprot.org/uniprot/) where an Entrez Gene description was not available.

This four-gene signature is referred to in this chapter as the bone marrow suppression signature (BSS). In the following sections the manifestation of the GSG and BSS in a variety of organ systems and species are reviewed.

4.3 Liver

The liver is a primary target for chemical toxicity for three primary reasons: (1) it sees a high concentration of chemical because chemicals absorbed from the gastrointestinal tract pass through the liver before entering the

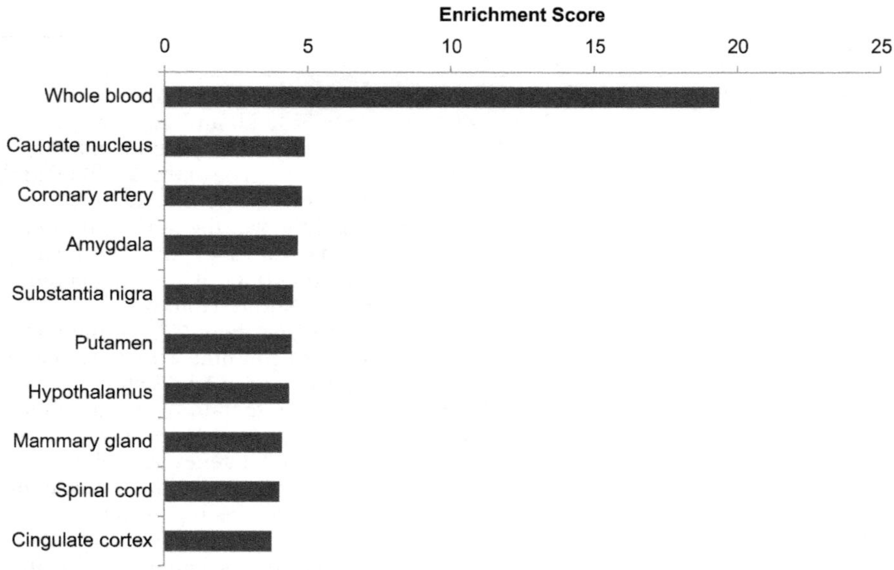

Figure 4.3 The bone marrow suppression signature genes are highly expressed in whole blood. The enrichment score reflects the relative expression of the four genes in the bone marrow suppression relative to the median of the RNA levels across all tissues.

circulation; (2) its bioactivation capacity is much greater than most other organs and tissues; and (3) its ability to bioconcentrate substances due to active transport. Due to the high sensitivity of the liver it is often the organ that determines the no effect level in toxicity studies and therefore often determines the point of departure used in a risk assessment. For these reasons, in additional to the relative homogeneity of the organ it has been a popular target for toxicogenomic characterization. In addition to the wide variety of studies that have evaluated a handful of chemicals in the liver using toxicogenomics there have been two systematically generated gene expression compendia that have evaluated a large number of chemicals in the liver. The compendium data sets are referred to as DrugMatrix (https://ntp.niehs.nih.gov/drugmatrix/index.html) and TG-Gates.[17] These and other data sets have been mined extensively to identify mode-of-action signatures (*e.g.* genotoxicity) and signatures for other endpoints such as pathology. The liver therefore provides an ideal place to start for reviewing the toxicogenomic effects of genotoxic and non-genotoxic chemicals in an *in vivo* setting.

4.3.1 Rat

A number of studies have sought to identify a signature of genotoxicity in rat liver[17–26] and there are some consistent findings across studies. The genes from the GSG that are most consistently induced in rat liver following genotoxic chemical exposure are CCNG1, NHEJ1, ADAM8, BAX, AEN, PLK2,

BTG2, MGMT, CDKN1A, MDM2, TP53INP1, ATP6V1D, ZMAT3, GDF15, FAS, PHLDA3, TYMS, NRCAM (neural cell adhesion molecule), ENC1 and PLN (see Table 4.1). As an example, the responses of the GSG genes in liver following repeated doses of aflatoxin b1 are shown in Figure 4.4. Additional genes that are induced by genotoxic chemical treatment in rat liver, but are not more broadly responsive in other tissues include LAMA5 (laminin alpha 5), MYBL1 (v-myb avian myeloblastosis viral oncogene homolog-like 1), JAM3 (junctional adhesion molecule 3), and BTG3 (BTG family member 3). Scoring of the GSG in against the DrugMatrix samples shows that the GSG is most strongly induced in rat liver by genotoxic treatments. The GSG does identify compounds that are not classically considered genotoxicants, such as carbon tetrachloride, thioacetamide and methapyrilene. The likely reason for these false positives is that they elicit a necrotizing response which produces a secondary genotoxic effect, similar to what is observed in many genotoxicity assessments when excessively toxic doses are used.[27] Example responses of

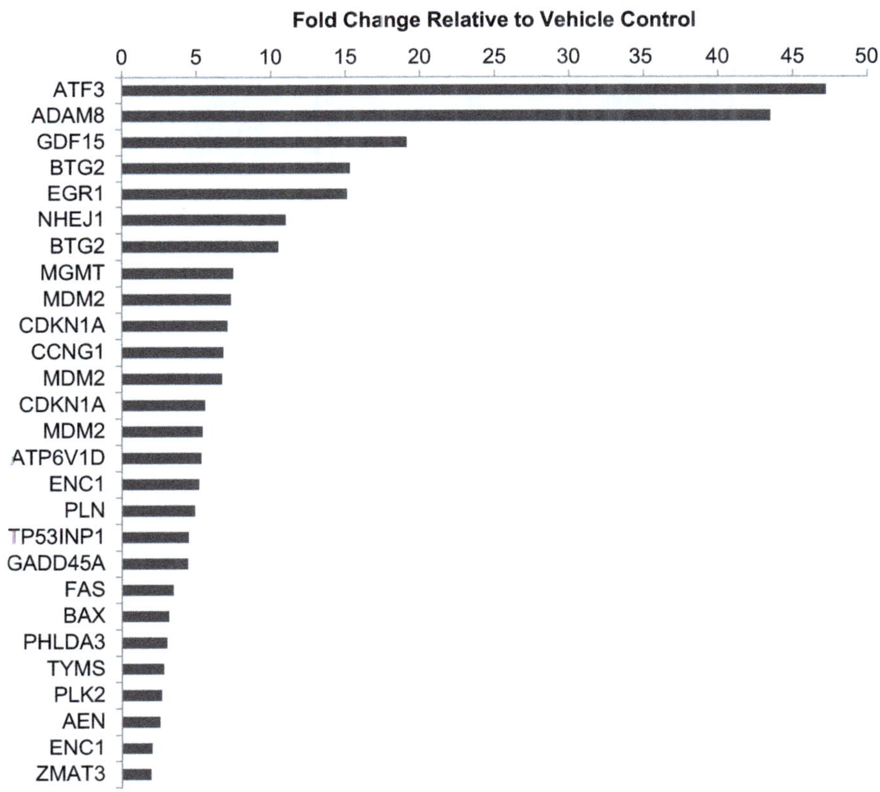

Figure 4.4 The response of genes from the general signature of genotoxicity following 5 days of oral dosing with 0.3 mg kg^{-1} of aflatoxin b1 (genotoxic fungal metabolite). Some genes are represented by more than one probe and this is reflected in the duplicated gene symbols. The data used in this figure are from DrugMatrix.

Figure 4.5 Example scoring of the general signature of genotoxicity (GSG) in NextBio. Alfatoxin B1 (A, genotoxic); 2-acetylaminofluorene (B, genotoxic); thioacetamide (C, cytotoxic but not typically considered genotoxic); and valproic acid (D, non-genotoxic) are representative treatments in which the GSG signature was affected. Both genotoxic treatments caused a strong induction of the GSG; however, so did the thioacetamide. The non-genotoxic agent caused and overall down-regulation of the signature. *P*-values reported in this figure are based on the Fisher exact test.

the GSG to a select set of prototype chemical treatments are shown in Figure 4.5.

Evaluation of the BSS signature in rat liver indicates that strongest effectors of this signature are genotoxic chemotherapeutics that are administered intravenously (IV) for at least 3 days (Figure 4.6). One striking observation is that a sizable number of the treatments that elicit a response from this signature have limited to no effect on the GSG. This is likely a byproduct of dose-limiting toxicity to the bone marrow that is in part caused by IV administration which bypasses the liver and has more direct effect on the bone marrow. That is, IV administration of these agents has little to no genotoxic effect in the liver at dose levels where there is overt toxicity in the bone marrow. The agents that had the greatest selective effects on this signature are the antimetabolites and the anthracyclines. Other chemicals such as that are not considered genotoxic, but can have effects on the cell cycle, such as the kinase inhibitor imatinib, cause a decrease in expression of the BSS; however, in most cases the effect on expression is less pronounced than that of the genotoxic agents.

There have been a number of papers that have sought to identify a signature for non-genotoxic effects in rat liver. Overall, the signatures of non-genotoxic chemical effects are not nearly as robust as those for genotoxic chemicals. The genes in these signatures tend to be associated with nuclear receptor signaling in the liver. The pathways that tend to be over-represented by the up-regulated genes in these signatures relate to xenobiotic metabolism, proteasome and to a lesser extent cell cycle and apoptosis. The pathways associated with down-regulated genes tend to be those associated with basic liver function such as complement and amino acid metabolism.

4.3.2 Mouse

A limited number of studies have sought to identify signatures from mouse liver that identify carcinogenic properties of chemicals.[15,17,28–33] The genes from the GSG that exhibit the strongest association with genotoxic chemical treatment in these studies are PHLDA3, CCNG1, EDA2R, ATPV1D, CDKN1A, GDF15, AEN, BAX, MDM2, DDIT4L, TRP53INP1, PLK2, MGMT, BTG2, ATF3, ENC1, and ZMAT3. Additional induced genes that are not contained in the GSG are GRIA3 (glutamate receptor, ionotropic, AMPA 3) and TOP2A (DNA topoisomerase 2-alpha). A review of existing studies did not identify the BSS in mice treated with genotoxic chemicals. This is likely due to the chemicals that have been selected for study in mice (*i.e.* they target the liver) and the route of administration that was used for most studies (oral). Similar to the rat, the GSG identifies some non-genotoxic chemicals, such as thioacetamide and (less confidently) WY-14 643, as a genotoxic drug, most likely for the same reasons noted for rat liver.

As with rat liver, the signature genes that are associated with non-genotoxic effects in mouse liver vary across studies. The pathways tend to be affected by non-genotoxic chemical exposure in mouse liver are related to cell cycle, apoptosis and xenobiotic metabolism.

Figure 4.6 Example scoring of the bone marrow suppression signature (BSS) in NextBio. Vinblastine (A, genotoxic); albendazole (B, genotoxic); epirubicin (C, cytotoxic but not typically considered genotoxic); and imatinib (D, non-genotoxic) are representative treatments in which the BSS signature was affected. The three genotoxic treatments caused a strong down-regulation of the BSS; however, so did the imatinib. *P*-values reported in this figure are based on the Fisher exact test.

4.3.3 Human

A search of NextBio did not identify studies using *in vivo* liver samples where chemical treatment was applied. Evaluation of the GSG across the NextBio database indicated it is activated by hepatitis B or C infection of human hepatocytes in a chimeric mouse (www.ncbi.nlm.nih.gov/projects/geo/query/acc.cgi?acc=GSE52752). There are many *in vitro* studies evaluating human primary hepatocytes and transformed liver cell lines. The most pronounced activation of the GSG is seen in a number of studies using HepG2 cells. The response of the GSG in the HepG2 studies tend to be disproportionately in response to genotoxic chemical treatment; however, very high doses of non-genotoxic compounds (*i.e.* those that were necrogenic) also induced the GSG.[34–36]

4.4 Kidney

Like the liver, the kidney is a common target organ in toxicity assessments. This is due to the high rate of blood flow which delivers a large amount of chemical to the organ, the active transport process that can bioconentrate chemicals and the presence of xenobiotic metabolizing enzymes (albeit less than the liver) that can bioactivate chemicals.

4.4.1 Rat

Despite the availability of data (DrugMatrix and TG-Gates) there has not been a systematic development of signatures that differentiate genotoxic from non-genotoxic agents. This is likely in part due to the more limited number of chemicals that cause a carcinogenic response in the kidney when compared to the liver. Mining the data that is available in NextBio clearly indicates that genotoxic agents produce a very similar response in the kidney as they do in liver. Specifically, a large number of the GSG genes are up-regulated following genotoxic chemical treatment. The genes that exhibit the highest degree of positive association with genotoxic chemical exposure in rat kidney are very similar to genes that were identified in liver in both rat and mouse, including CDKN1A, PHLDA3, CCNG1, BTG2, MDM2, AEN, TP53INP1, PLK2, MGMT, FAS1, GADD45A, ATP6V1D, ADAM8, TYMS, BAX, GDF15 and ENC1 (see Table 4.1). As an example, the response of the GSG genes in kidney to repeated doses of doxorubicin are shown in Figure 4.7. Other genes that appear to be consistently up-regulated in the kidney following genotoxic chemical treatment included ABCB1B (ATP-binding cassette 1B), JAM3, MAFF (v-maf avian musculoaponeurotic fibrosarcoma oncogene homolog F), RAPGEF5 (Rap guanine nucleotide exchange factor 5) and ARL4A (ADP ribosylation factor like GTPase 4A). As in the liver, there are number of non-genotoxic kidney toxicants that strongly induce the GSG when given at excessively toxic dose levels; these include furosemide, gentamicin and bacitracin. A review of the expression profiles of these chemicals shows a strong induction of HAVCR1 (also known as Kim-1), a biomarker of kidney toxicity, supporting the presumption that the activation of the

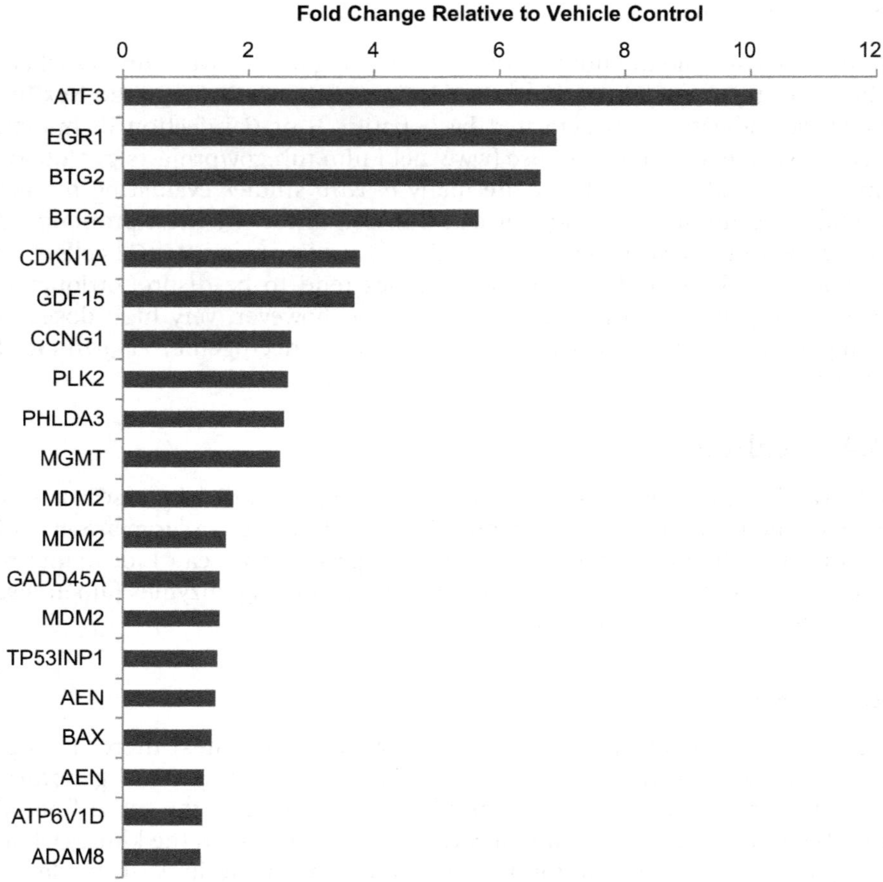

Figure 4.7 The response of genes from the general signature of genotoxicity in kidney following 3 days of intravenous dosing with 3.25 mg kg^{-1} of doxorubicin (genotoxic chemotherapeutic). Some genes are represented by more than one probe and this is reflected in duplicated gene symbols. The data used in this figure are from DrugMatrix.

genotoxicity signature may be related to overt renal toxicity. As with rat liver, the BSS is strongly affected by many of the same genotoxic chemical treatments in rat kidney. Also consistent with the liver, many of the same non-genotoxic chemicals that have immunosuppressive activity (*e.g.* leflutamide) or are kinase inhibitors that affect the cell cycle (*e.g.* imatinib) also cause a down-regulation of the four genes that are in the BSS.

Using NextBio in combination with the DrugMatrix and TG-Gates, a gene expression data signature for non-genotoxic activity was formulated. As with the rat liver non-genotoxic signature, the effects on specific genes were inconsistent and were easily biased based on the imbalanced inclusion of chemicals that act by different molecular initiating events. There were some pathways that were consistently up-regulated by non-genotoxic chemical treatments, including fatty acid metabolism and to a lesser extent immune signaling and ion transport, or down-regulated, such as cell cycle and to a considerably lesser extent steroid hormone function. Overall there is likely limited utility to the non-genotoxic carcinogen signature in kidney. It is likely that the best approach would be to evaluate the genotoxicity signature and to assume if it is not affected that a chemical is acting through a non-genotoxic mechanism.

4.4.2 Mouse

A systematic comparison of genotoxic and non-genotoxic chemical effects in mouse kidney has not been performed, hence only anecdotal evidence can be provided in support of the manifestation of the GSG and BSS signatures that can be formulated at this point. There have been a number of studies of the effect of individual genotoxic chemical treatments (cisplatin, ionizing radiation and doxorubicin) on mouse kidney.[21,37,38] Using NextBio to mine these data identified a number of genes from the GSG that are induced by these treatments, including CDKN1A, CCNG1, ZMAT3, PHLDA3, PLK2, MDM2, GDF15, TP53INP1, DDIT4L, BAX and EGR1 (see Table 4.1). In addition, genes not in the GSG that were induced by these genotoxic treatments included EPHX1 (epoxide hydrolase 1), and EXOC4 (exocyst complex component 4). Review of additional studies of the effect of high-dose vancomycin treatment and acute rhabdomyolysis on mouse kidney indicated that non-genotoxic chemicals administered at levels that produce frank toxicity in the kidney can produce a strong positive response in the GSG, as was highlighted with rat kidney.[22,39] Evaluation of the BSS showed that cisplatin-treated mouse kidneys that showed induction of the GSG (above) also showed marked down-regulated ALAS2; however, other genes in the BSS were unaffected.

4.4.3 Human

A search of NextBio did not identify studies using *in vivo* kidney samples where chemical treatment was applied. A review of other *in vivo* data from human kidneys did indicate that reperfusion following kidney transplant that can lead to tissue damage induces a weak response in the GSG.[40] In addition, *in vitro* studies of primary podocytes exhibit a relatively strong activation of the GSG when exposed to HAMLET (human alpha-lactalbumin made lethal to tumors), a compound that triggers a p38-mediated death indicating that the signature is associated with cell death/cytotoxicity which, as already noted, is often associated with DNA damage (www.ncbi.nlm.nih.gov/projects/geo/query/acc.cgi?acc=GSE23772).

4.5 Heart

The heart has relatively high energy requirements due to the need to maintain cyclical ion flux and muscle contraction. This makes the heart susceptible to chemicals that accentuate the energy requirements of these processes and those that interfere with energy production (*e.g.* mitochondrial toxicants). It is typically not a target in carcinogenicity studies, most likely due to its large post-mitotic state. Despite the relatively low cancer susceptibility of the heart it is responsive to number of genotoxic and non-genotoxic chemicals. The most notable class of genotoxic agents that affect the heart are the anthracycline class of drugs (*e.g.* doxorubicin), which can cause cardiomyopathy.[41]

4.5.1 Rat

Despite data being available from DrugMatrix, no systematic analysis of the data has been performed to identify a signature that differentiates responses from genotoxic and non-genotoxic compounds. Using NextBio the GSG was evaluated using the heart gene expression data from DrugMatrix. The top scoring chemical treatments were daunorubicin, epirubicin, mitoxantrone and idarubicin. The genes from the GSG that were most consistently affected across a collection of genotoxic chemical treatments in heart include AEN, PHLDA3, ZMAT3, CDKN1A, MGMT, MDM2, TYMS, BAX, SESN2 ENC1 and TP53INP1 (see Table 4.1). As an example, the response of the GSG genes in heart to repeated doses of doxorubicin are shown in Figure 4.8. In addition, genotoxic chemical treatment caused a strong down-regulation of the BSS signature in heart. This is not surprising, as the heart is a highly perfused organ and therefore the effect of blood flow would be expected to contribute significantly to the transcriptome measurements. As with other tissues, false positive effects produced by non-genotoxic chemicals on these signatures can be seen with chemicals that cause a cytotoxic response in the measured tissue or suppress bone marrow. This effect is seen within the DrugMatrix data and in other studies where myocardial necrosis was induced by a non-genotoxic agent (www.ncbi.nlm.nih.gov/geo/query/acc.cgi?acc=GSE7999) and non-genotoxic immune suppressants (*e.g.* leflutamide) caused down-regulation of the BSS.

4.5.2 Mouse

As there has been no systematic assessment of mouse heart following exposure to a variety of chemicals there is no way to identify a signature reflective of a broad mode of action such as genotoxicity. Furthermore, there are no existing compendium data sets to formulate a more robust evaluation of the GSG. Using NextBio two studies were identified that evaluated the effect of doxorubicin and etoposide on mouse heart gene expression.[42,43] The meta-analysis of these studies indicated that the genes from the GSG that are most strongly affected in mouse heart following genotoxic chemical treatment are PHLDA3, CDKN1A, CCNG1, PLK2, EDA2R, MDM2 and ENC1. Other treatments that caused an induction of the GSG in mouse heart included over-expression of CELF1, which causes myotonic dystrophy,[44]

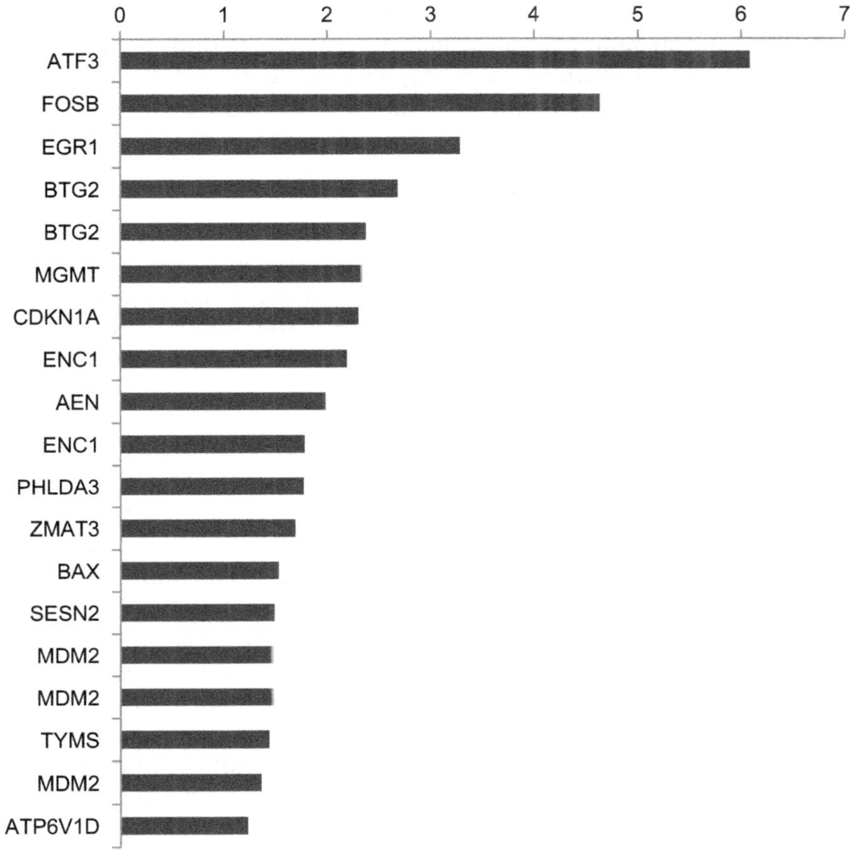

Figure 4.8 The response of genes from the general signature of genotoxicity in heart following 3 days of intravenous dosing with 3.25 mg kg^{-1} of doxorubicin (genotoxic chemotherapeutic). Some genes are represented by more than one probe and this is reflected in duplicated gene symbols. The data used in this figure are from DrugMatrix.

and knockout of TFR1 (transferrin receptor 1), which regulates iron homeostasis (www.ncbi.nlm.nih.gov/projects/geo/query/acc.cgi?acc=GSE68745) and myocardial infarction, suggesting, as was seen in rats, that the GSG can be induced by significant non-genotoxic adverse effects.[30] There was very limited evidence for an effect on the BSS in mouse heart. A single study of heart gene expression following doxorubin (known cardio- and genotoxicant) treatment in C57BL/6J mice showed a strong down-regulation of the orthologues of the BSS genes from rat (GSE59672, unpublished).

4.5.3 Human

A search of NextBio did not identify studies that evaluated *in vivo* gene expression in the heart following genotoxic chemical treatment. One study that did evaluate chemical exposure looked at the effects of propofol and sevoflurane (both non-genotoxic) in which a strong induction of the GSG was observed.[25] It is uncertain why this response was observed. Other scenarios where the GSG was induced include following coronary revascularization[45] and in cardiomyopathy.[46]

4.6 Skeletal Muscle

Like heart, skeletal muscle is not typically a target in cancer studies; however, unlike heart it does not typically respond toxicologically to genotoxic chemicals, in part because it does not have the same constant energy demand. Overall, the response to genotoxic chemicals would be expected to be weak in skeletal muscle.

4.6.1 Rat

Unlike liver, kidney and heart (to a lesser extent), the data set for deriving signatures is underdeveloped for skeletal muscle. This underdevelopment of data is particularly the case for genotoxic agents. With the limited data from DrugMatrix it can be concluded that chemicals that induce muscle damage, such as 3-hydroxy-3-methyl-glutaryl (HMG) CoA reductase inhibitors and PPARα activators, cause the GSG to be weakly activated. Other studies of the statins have observed a similar induction of this signature (www.ncbi.nlm.nih.gov/geo/query/acc.cgi?acc=GSE4418). In addition, a model of myasthenia gravis shows induction of the GSG. These findings support the idea that chemicals or biological states that elicit notable pathology in muscle can lead to an induction of the GSG.

4.6.2 Mouse

As with rat there are a very limited number of studies of chemical-associated effects in mouse muscle. One study of chemical-associated effects evaluated a high dose of a pan-PPAR activator (enough to produce increases in blood markers of muscle damage), which showed a strong activation of the GSG.[47] In addition, other studies that manifest skeletal muscle damage also saw induction of the GSG.[22,48]

4.6.3 Human

As with rats and mice, the human studies of skeletal muscle that most strongly induce the GSG are related to muscle damage.[49–51]

4.7 Bone Marrow and Blood

The bone marrow and blood are common targets of genotoxic agents, primarily because of their high rate of proliferation. This is the justification for using the *in vivo* erythrocyte micronucleus assay as a standard screen for identifying genotoxic agents. The phenotypes associated with chemical effects on the bone marrow and blood have been extensively characterized across many species, in particular humans, due to common side effects of bone marrow suppression during chemotherapy.

4.7.1 Rat

No publications have systematically derived a signature that differentiates genotoxic from non-genotoxic effects in rat bone marrow. There have been a limited number of studies that have evaluated single chemicals[52–54] and large amount of data is available from the DrugMatrix database; however, all these data are from an early-generation microarray that has limited genome coverage. Using a NextBio meta-analysis the genes from the GSG that are most consistently and strongly induced by genotoxic chemical exposure in bone marrow were found to be CDKN1A, PHLDA3, PLK2, CCNG1, AEN, ENC1, ATF3, GDF15, BAX, PLN, MDM2, SESN2 and TP53INP1. Other treatments that activate the GSG response in bone marrow-derived cells include biomaterial implants[55] an regeneration of bone marrow following physical ablation.[56] In addition to bone marrow gene expression, expression from peripheral blood mononuclear cells following doxorubicin exposure has been assessed. A clear induction of the GSG was observed (www.ncbi.nlm.nih.gov/projects/geo/query/acc.cgi?acc=GSE37260). Evaluation of the bone marrow suppression signature using the DrugMatrix bone marrow gene expression shows strong down-regulation of the signature genes ALAS2 and AHSP (the two genes from the signature that were measure in bone marrow with early-generation Codelink microarrays) by genotoxic chemicals.

4.7.2 Mouse

A limited number of independent studies have evaluated the effect of genotoxic exposures on blood and bone marrow gene expression in mice. Notable induction of the GSG was observed in two independent studies that used ionizing radiation to elicit expression in peripheral blood.[57,58] The genes from the signature that were most strongly affected by ionizing radiation include CDKN1A, PHLDA3, CCNG1, ZMAT3, BAX, MGMT and FAS. A parallel down-regulation of the BSS was also observed in these studies. A look across the NextBio database revealed that the GSG was also strongly induced by bacterial infection,[59,60] suggesting that inflammation can activate some of the same signaling pathways as radiation in peripheral blood.

4.7.3 Human

There have been numerous studies that have evaluated the effect of radiation exposure in humans on peripheral blood and lymphocyte gene expression (Figure 4.9).[61–64] The genes from the GSG that are most consistently affected

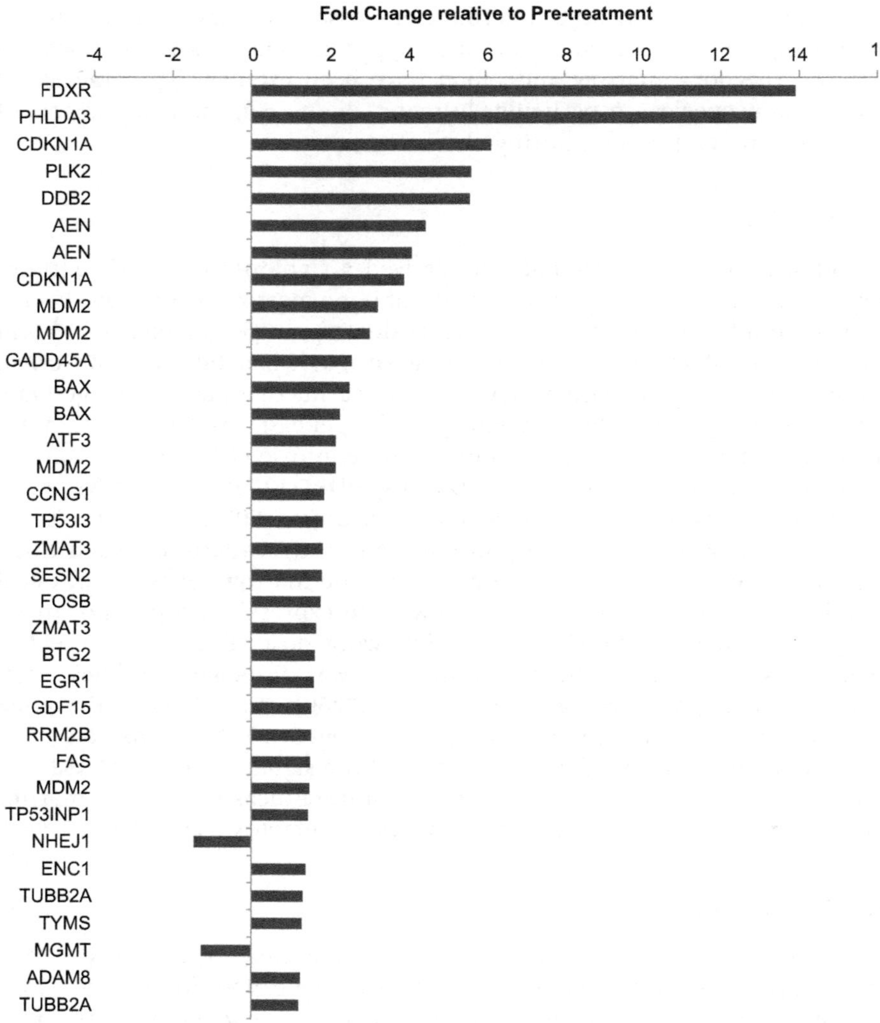

Figure 4.9 The response of genes from the general signature of genotoxicity (GSG) in whole blood following 4 hours of whole-body radiation. Some genes are represented by more than one probe and this is reflected in duplicated gene symbols. The data used in this figure are from www.ncbi.nlm.nih.gov/projects/geo/query/acc.cgi?acc=GSE20162.

by radiation exposure across these studies include FDXR, PHLDA3, AEN, DDB2, CDKN1A, GADD45A, MDM2, ZMAT3, CCNG1, RRM2B, FAS, PLK2, FOSB (FBJ murine osteosarcoma viral oncogene homolog B) and TP53I3. No studies related to chemical treatment were available to evaluate the behavior of the GSG. As with the mouse, review of the blood gene expression data in NextBio identified a number of infection-related gene expression studies where the GSG was induced.[65] A review of the BSS indicated that ionizing radiation had a limited effect on this signature in human blood. This is likely because this signature is a reflection of cell number, which was likely balanced in the above ionizing radiation studies.

4.8 Spleen

The spleen is a heavily perfused lymphoid organ that is often indirectly affected by toxicity in the blood system. The spleen's primary function is removal of damaged red blood cells; hence an increase in the rate of red blood cell damage often leads to congestion and enlargement of the spleen. Enlargement causes it to be more sensitive to hemorrhage following physical trauma.

4.8.1 Rat

No publications have described genotoxicity signatures in spleen; however, a large amount of data (early generation Codelink microarray data) exists in the DrugMatrix database that can be mined to identify signatures. Mining the DrugMatrix bone marrow gene expression data identified a number of genes from the GSG that are induced by genotoxic treatment in other tissues, including CDKN1A, ENC1 and ATF3. Notably, non-genotoxic treatments that produce immune suppression and pathological manifestation in the spleen also caused induction of the GSG signature, suggesting that the GSG is responsive to treatments that produce significant pathology and secondary DNA damage in the spleem, as seen in other tissues. An evaluation of the BSS demonstrated a down-regulation of the signature by both genotoxic agents and non-genotoxic immune suppressants (*e.g.* leflutamide) in rat spleen.

4.8.2 Mouse

Very few data are available on genotoxic or non-genotoxic chemical effects in the spleen of mice. One study that employed ionizing radiation observed the induction of a number of genes from the GSG in spleen, including PHLDA3, CDKN1A, DDIT4L, PLK2, EDA2R, SESN2, CCNG1, AEN, MGMT, ZMAT3, BAX, FAS, MDM2, ENC1, GDF15, TRP53INP1, ADAM8, BTG2 and FDXR.[66] As was seen in blood, the GSG in the spleen is responsive to infectious agents.[67,68]

4.8.3 Human

No data were identified from *in vivo* exposure to chemicals that evaluated spleen or other components of the lymphatic system gene expression. However, a study of TK6 spherocytic spleen cells shows a strong induction of the GSG in response to a number of genotoxic agents, suggesting that the response of the GSG in human spleen following exposure to genotoxic agents is plausible.[69]

4.9 Other Tissues

4.9.1 Rat

Studies of respiratory system tissues following 28 days of cigarette exposure, bleomycin treatment, high doses of inhaled formaldehyde or sulfur mustard exposure (all genotoxic exposures) showed induction of the GSG, with CDKN1A, CCNG1, GDF15, PLN, MDM2, PHLDA3, ATF3, ENC1, BAX, ATF3, BTG2 and ATP6V1D being consistently induced across the different studies.[11,70–72] Induction of the GSG was observed in the colon after exposure to the genotoxic agent azoxymethane, with GDF15, PLK2, PHLDA3, AEN, CCNG1 and MDM2 being the most strongly affected signature genes.[12]

4.9.2 Mouse

A relatively strong induction of the GSG was observed in the colon and jejunum following whole-body radiation in the mouse.[73] Over half of the genes from the GSG were induced by ionizing radiation, with PHLDA3, CCNG1, GDF15, DDIT4L, PLK2, EDA2R and BAX being most strongly and consistently induced. A similar, albeit weaker, response of the GSG signature was seen in mammary tissue following exposure to ionizing radiation, with CCNG1, RRM2B, CDKN1A, GDF15 and BTG2 being some of the most consistently induced genes.[74] Induction of the GSG was observed in lung following various genotoxic treatments including hyperoxia, radiation, urethane, benzo(*a*)pyrene, doxorubicin and bleomycin exposure.[42,75–80] The GSG genes that were most consistently affected were CDKN1A, CCNG1, PHLDA3, GDF15, SENS2, TP53INP1, ZMAT3, BAX, DDIT4L, AEN, MDM2 and EDA2R.

4.9.3 Human

Studies that evaluated responses to genotoxic exposure in healthy human tissue were primarily limited to blood and are discussed above. However, one study was identified that evaluated oral mucosa gene expression in cancer patients receiving cytarabine and daunorubicin chemotherapy.[81] In this study a weak but clear induction of the GSG signature was observed with ZMAT3, MDM2 and DDB2 being the most strongly induced genes (Figure 4.10). Another study of primary tissue following genotoxic exposure was performed on skin using ultraviolet radiation.[82] In this study a clear induction of the GSG was observed with AEN, MDM2, GADD45A and GDF15 having the most consistent response to ultraviolet irradiation. Finally, a couple of studies have assessed the response of tumors to genotoxic chemical treatments where the GSG was induced, including rectal cancer (radiation), breast cancer (cyclophosphamide, gemcitabine, paclitaxel and docetaxel) and chemo-/radiotherapy of squamous cell carcinoma of the cervix.[83–85]

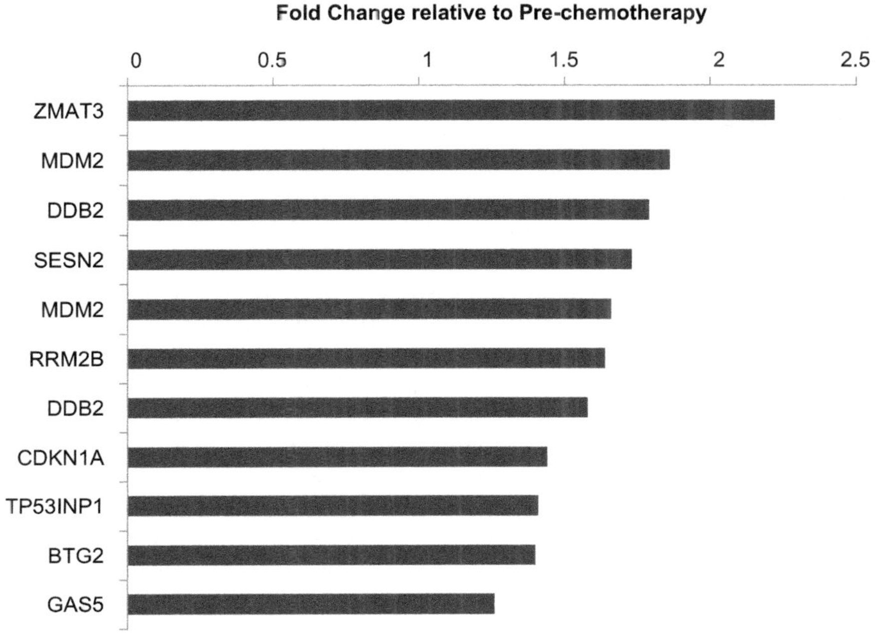

Figure 4.10 The response of genes from the general signature of genotoxicity (GSG) in buccal mucosa of chemotherapy patients 2 days after cytaribine + daunorubicin chemotherapy patients. The data used in this figure are from www.ncbi.nlm.nih.gov/projects/geo/query/acc.cgi?acc=GSE10746.

4.10 Study Design

4.10.1 Dose

Dose selection can have a significant effect on the generation and application of genomic signatures such as the GSG. If the animals are under-dosed there is often not enough signal to confidently classify the treatment. Alternatively, if the animals are over-dosed, not only is the study of questionable relevance to an eventual carcinogenicity study, but it will often cause a hypertoxic response that is non-specific (*i.e.* related to acute high-grade pathology) and is likely in part related to significant effects on body weight and starvation because the animals are in a moribund state. If a subchronic study has been performed prior to selecting a dose it is likely that a small multiple (×2–3) of the maximum tolerated dose (MTD) identified in the subchronic study can be used in the genomic study. The longer the duration of the genomic study the lower the multiple of the MTD should be used. If no toxicity data on the chemical exist it is likely that a short (5 day) range-finding study can be performed to determine an MTD based on body weight and clinical signs that can be used for a genomic study. When selecting the MTD it is critical to avoid doses where there is clear, overt toxicity.

Ultimately, as with any toxicological assessment, it is central that the animals are challenged, but not overtly sick.

4.10.2 Duration

Longer durations of exposure (*i.e.* repeated dose) tend to produce more coherent and reproducible responses in gene expression and are more confidently classified with signatures such as the GSG. Single-dose studies of short duration often have a significant amount of animal-to-animal variability. The greater noise seen in the shorter-term studies may in part be a sudden activation of stress-related signaling. This signaling likely settles after adaptation takes place and more specific gene expression related to a chemical's mechanism of action emerges. Another issue to consider is whether there is desire to relate the response of the signature to dose in order to compare it to the dose response of known toxicities. In the case of bioaccumalative or genotoxic chemicals the dose–response curves tend to shift to the left (*i.e.* greater potency) with increased duration. With bioaccumulative chemicals this is due to the accumulation of the chemical in tissue. With genotoxic chemicals it is likely to be due to the accumulation of mutations which lead to increasing numbers of cells activating the p53 pathway. This effect is particularly striking in rat liver when potent genotoxicants (such as the nitrosamines) are studied. In order to elicit a response from these chemicals in short-term studies it is often the case that many multiples of the carcinogenic dose need to be administered to elicit a clear gene expression response. However, after a longer period of exposure (*e.g.* 14–28 days) the responses begin to occur at doses that more closely approximate the points of departure for cancer. This is all in contrast to chemicals that act through non-genotoxic mechanisms (*e.g.* receptor-mediated) that can have rather robust responses in gene expression well below the carcinogenic dose after a single administration. Mixed modes of action can present a challenge: short-term exposures manifest robust changes in gene expression that are related to non-genotoxic effects, and then with longer durations of exposure the genotoxic signature, such as the GSG, begin to emerge. Overall, if dose selection is performed correctly then short-term repeat dose (5–7 days) studies are likely adequate for identifying a genotoxic hazard using a signature such as the GSG, or if a more quantitative assessment is desired then longer term (*e.g.* 14–90 days) studies should be performed.

4.10.3 Tissue Selection

Ideally, the tissue that is selected for genomic analysis is either the known or predicted target for carcinogenic/toxic effects. It is often the case that this is unknown or cannot be confidently predicted. If this is the case then selection should be based on historical data, *i.e.* which tissues are most commonly affected by chemical treatment. If the oral route (gavage) is used then it is often the liver and to a lesser extent the kidneys that are affected, unless of course the agent is a contact irritant, in which case the stomach or other parts of the gastrointestinal tract are likely to be a major target. If inhalation or lung instillation is used then the lung is the most likely target. Often a careful review of the literature of chemicals from the same class as the test article and a structural activity analysis can be very informative in selecting an appropriate target tissue.

4.11 Conclusion

At the beginning of this chapter an agglomerative general signature for response to genotoxic chemicals exposure was presented. This signature reflected a synthesis of number of studies, primarily in mice and rats, that sought to identify a signature that differentiates between genotoxic and non-genotoxic chemicals. A specific signature for non-genotoxic effects has not been formulated because there is no one coherent signature, and describing each of these signatures would collectively include everything that is not DNA damage, which is well beyond the scope of this chapter. The biological characterization of the GSG demonstrated it was strongly associated with activation of the p53 signaling pathway. Evaluation of the GSG across multiple tissues, species and treatments demonstrated its broad applicability. The GSG did have some limitations in that it was often activated by significant tissue damage or inflammation. A second signature related to genotoxicity is discussed which reflected an effect of chemicals on bone marrow, a common target of genotoxic chemicals. This signature is most strongly affected by chemicals that produced dose-limiting toxicities in the bone marrow and at a cellular level most likely reflect a decreased abundance of blood cells in the measured organs that is a byproduct of a reduced hematocrit. The BSS does have specificity challenges similar to the GSG, in that it also detects chemicals that are non-genotoxic bone marrow suppressants, such as leflutamide. Finally, a review of critical toxicogenomic study design parameters highlight the need to consider dose level, tissue selection and study duration.

Acknowledgements

The author would like to thank B. Alex Merrick and Richard Paules for critical review.

References

1. R. Benigni, C. Bossa and O. Tcheremenskaia, Nongenotoxic carcinogenicity of chemicals: Mechanisms of action and early recognition through a new set of structural alerts, *Chem. Rev.*, 2013, **113**(5), 2940–2957.
2. H. M. Bolt *et al.*, Carcinogenicity categorization of chemicals-new aspects to be considered in a European perspective, *Toxicol. Lett.*, 2004, **151**(1), 29–41.

3. J. Ashby and R. W. Tennant, Definitive relationships among chemical structure, carcinogenicity and mutagenicity for 301 chemicals tested by the U.S. NTP, *Mutat. Res.*, 1991, **257**(3), 229–306.
4. A. Hillebrecht *et al.*, Comparative evaluation of in silico systems for ames test mutagenicity prediction: scope and limitations, *Chem. Res. Toxicol.*, 2011, **24**(6), 843–854.
5. Expert Working Group, I.C.H., *Guidance on Genotoxicity Testing and Data Interpretation for Pharmaceuticals Intended for Human Use*. In International Conference on Harmonisation of Technical Requirements for Registration of Pharmaceuticals for Human Use, 2011.
6. A. R. Collins, The comet assay for DNA damage and repair: principles, applications, and limitations, *Mol. Biotechnol.*, 2004, **26**(3), 249–261.
7. S. W. Dean *et al.*, Transgenic mouse mutation assay systems can play an important role in regulatory mutagenicity testing in vivo for the detection of site-of-contact mutagens, *Mutagenesis*, 1999, **14**(1), 141–151.
8. D. Kirkland *et al.*, Evaluation of the ability of a battery of three in vitro genotoxicity tests to discriminate rodent carcinogens and non-carcinogens I. Sensitivity, specificity and relative predictivity, *Mutat. Res.*, 2005, **584**(1–2), 1–256.
9. D. Kirkland *et al.*, Evaluation of the ability of a battery of three in vitro genotoxicity tests to discriminate rodent carcinogens and non-carcinogens II. Further analysis of mammalian cell results, relative predictivity and tumour profiles, *Mutat. Res.*, 2006, **608**(1), 29–42.
10. M. D. Waters, M. Jackson and I. Lea, Characterizing and predicting carcinogenicity and mode of action using conventional and toxicogenomics methods, *Mutat. Res., Rev. Mutat. Res.*, 2011, **705**(3), 184–200.
11. U. Kogel *et al.*, A 28-day rat inhalation study with an integrated molecular toxicology endpoint demonstrates reduced exposure effects for a prototypic modified risk tobacco product compared with conventional cigarettes, *Food Chem. Toxicol.*, 2014, **68**, 204–217.
12. C. A. Kerr *et al.*, Genomic homeostasis is dysregulated in favour of apoptosis in the colonic epithelium of the azoxymethane treated rat, *BMC Physiol.*, 2013, **13**, 2.
13. B. Kaina *et al.*, MGMT: key node in the battle against genotoxicity, carcinogenicity and apoptosis induced by alkylating agents, *DNA Repair*, 2007, **6**(8), 1079–1099.
14. L. Zhao *et al.*, Cyclin G1 has growth inhibitory activity linked to the ARF-Mdm2-p53 and pRb tumor suppressor pathways, *Mol. Cancer Res.*, 2003, **1**(3), 195–206.
15. K. W. Kohn, Molecular interaction map of the mammalian cell cycle control and DNA repair systems, *Mol. Biol. Cell*, 1999, **10**(8), 2703–2734.
16. D. Iliopoulos *et al.*, Roles of FHIT and WWOX fragile genes in cancer, *Cancer Lett.*, 2006, **232**(1), 27–36.
17. S. J. Lee *et al.*, Distinguishing between genotoxic and non-genotoxic hepatocarcinogens by gene expression profiling and bioinformatic pathway analysis, *Sci. Rep.*, 2013, **3**, 2783.

18. H. Ellinger-Ziegelbauer *et al.*, Prediction of a carcinogenic potential of rat hepatocarcinogens using toxicogenomics analysis of short-term in vivo studies, *Mutat. Res.*, 2008, **637**(1–2), 23–39.
19. H. Ellinger-Ziegelbauer *et al.*, Characteristic expression profiles induced by genotoxic carcinogens in rat liver, *Toxicol. Sci.*, 2004, **77**(1), 19–34.
20. H. Ellinger-Ziegelbauer *et al.*, Comparison of the expression profiles induced by genotoxic and nongenotoxic carcinogens in rat liver, *Mutat. Res.*, 2005, **575**(1–2), 61–84.
21. H. Si *et al.*, Human and murine kidneys show gender- and species-specific gene expression differences in response to injury, *PLoS One*, 2009, **4**(3), e4802.
22. M. Fahling *et al.*, Tubular von Hippel-Lindau knockout protects against rhabdomyolysis-induced AKI, *J. Am. Soc. Nephrol.*, 2013, **24**(11), 1806–1819.
23. D. Gusenleitner *et al.*, Genomic models of short-term exposure accurately predict long-term chemical carcinogenicity and identify putative mechanisms of action, *PLoS One*, 2014, **9**(7), e102579.
24. H. Matsumoto *et al.*, Discrimination of carcinogens by hepatic transcript profiling in rats following 28-day administration, *Cancer Inf.*, 2009, 7, 253–269.
25. E. Lucchinetti *et al.*, Gene regulatory control of myocardial energy metabolism predicts postoperative cardiac function in patients undergoing off-pump coronary artery bypass graft surgery: inhalational versus intravenous anesthetics, *Anesthesiology*, 2007, **106**(3), 444–457.
26. K. Nakayama *et al.*, Differences in gene expression profiles in the liver between carcinogenic and non-carcinogenic isomers of compounds given to rats in a 28-day repeat-dose toxicity study, *Toxicol. Appl. Pharmacol.*, 2006, **217**(3), 299–307.
27. P. Fowler *et al.*, Reduction of misleading ("false") positive results in mammalian cell genotoxicity assays. I. Choice of cell type, *Mutat. Res.*, 2012, **742**(1–2), 11–25.
28. J. Eichner *et al.*, A Toxicogenomic Approach for the Prediction of Murine Hepatocarcinogenesis Using Ensemble Feature Selection, *PLoS One*, 2013, **8**(9), e73938.
29. J. P. M. Melis *et al.*, In vivo murine hepatic microRNA and mRNA expression signatures predicting the (non-)genotoxic carcinogenic potential of chemicals, *Arch. Toxicol.*, 2014, **88**(4), 1023–1034.
30. O. Tarnavski *et al.*, Mouse cardiac surgery: comprehensive techniques for the generation of mouse models of human diseases and their application for genomic studies, *Physiol. Genomics*, 2004, **16**(3), 349–360.
31. T. Watanabe *et al.*, Discrimination of genotoxic and non-genotoxic hepatocarcinogens by statistical analysis based on gene expression profiling in the mouse liver as determined by quantitative real-time PCR, *Mutat. Res.,– Genet. Toxicol. Environ. Mutagen.*, 2012, **747**(2), 164–175.
32. M. J. Jonker *et al.*, Finding transcriptomics biomarkers for in vivo identification of (non-)genotoxic carcinogens using wild-type and Xpa/p53 mutant mouse models, *Carcinogenesis*, 2009, **30**(10), 1805–1812.

33. N. Kossler *et al.*, Identification of specific mRNA signatures as fingerprints for carcinogenesis in mice induced by genotoxic and nongenotoxic hepatocarcinogens, *Toxicol. Sci.*, 2015, **143**(2), 277–295.
34. W. F. Van den Hof *et al.*, Classification of hepatotoxicants using HepG2 cells: A proof of principle study, *Chem. Res. Toxicol.*, 2014, **27**(3), 433–442.
35. C. Magkoufopoulou *et al.*, A transcriptomics-based in vitro assay for predicting chemical genotoxicity in vivo, *Carcinogenesis*, 2012, **33**(7), 1421–1429.
36. C. Magkoufopoulou *et al.*, Comparison of phenotypic and transcriptomic effects of false-positive genotoxins, true genotoxins and non-genotoxins using HepG2 cells, *Mutagenesis*, 2011, **26**(5), 593–604.
37. E. Schuler *et al.*, Time- and dose rate-related effects of internal (177)Lu exposure on gene expression in mouse kidney tissue, *Nucl. Med. Biol.*, 2014, **41**(10), 825–832.
38. M. S. Joo *et al.*, miR-125b transcriptionally increased by Nrf2 inhibits AhR repressor, which protects kidney from cisplatin-induced injury, *Cell Death Dis.*, 2013, **4**, e899.
39. C. Dieterich *et al.*, Gene expression analysis reveals new possible mechanisms of vancomycin-induced nephrotoxicity and identifies gene markers candidates, *Toxicol. Sci.*, 2009, **107**(1), 258–269.
40. J. Damman *et al.*, Hypoxia and Complement-and-Coagulation Pathways in the Deceased Organ Donor as the Major Target for Intervention to Improve Renal Allograft Outcome, *Transplantation*, 2015, **99**(6), 1293–1300.
41. K. Chatterjee *et al.*, Doxorubicin cardiomyopathy, *Cardiology*, 2010, **115**(2), 155–162.
42. B. Pang *et al.*, Drug-induced histone eviction from open chromatin contributes to the chemotherapeutic effects of doxorubicin, *Nat. Commun*, 2013, **4**, 1908.
43. S. Zhang *et al.*, Identification of the molecular basis of doxorubicin-induced cardiotoxicity, *Nat. Med.*, 2012, **18**(11), 1639–1642.
44. J. Giudice *et al.*, Alternative splicing regulates vesicular trafficking genes in cardiomyocytes during postnatal heart development, *Nat. Commun.*, 2014, **5**, 3603.
45. M. T. Ghorbel *et al.*, Off-pump coronary artery bypass surgery is associated with fewer gene expression changes in the human myocardium in comparison with on-pump surgery, *Physiol. Genomics*, 2010, **42**(1), 67–75.
46. K. C. Yang *et al.*, Deep RNA sequencing reveals dynamic regulation of myocardial noncoding RNAs in failing human heart and remodeling with mechanical circulatory support, *Circulation*, 2014, **129**(9), 1009–1021.
47. C. M. Lee *et al.*, Simultaneous non-negative matrix factorization for multiple large scale gene expression datasets in toxicology, *PLoS One*, 2012, **7**(12), e48238.
48. G. L. Warren *et al.*, Mechanisms of skeletal muscle injury and repair revealed by gene expression studies in mouse models, *J. Physiol.*, 2007, **582**(Pt 2), 825–841.

49. J. D. Crane *et al.*, Massage therapy attenuates inflammatory signaling after exercise-induced muscle damage, *Sci. Transl. Med.*, 2012, **4**(119), 119ra13.
50. S. A. Greenberg *et al.*, Interferon-alpha/beta-mediated innate immune mechanisms in dermatomyositis, *Ann. Neurol*, 2005, **57**(5), 664–678.
51. M. Bakay *et al.*, Nuclear envelope dystrophies show a transcriptional fingerprint suggesting disruption of Rb-MyoD pathways in muscle regeneration, *Brain*, 2006, **129**(Pt 4), 996–1013.
52. M. Davis *et al.*, Toxicogenomics profiling of bone marrow from rats treated with topotecan in combination with oxaliplatin: a mechanistic strategy to inform combination toxicity, *Front. Genet.*, 2015, **6**, 14.
53. B. Faiola *et al.*, Gene expression profile in bone marrow and hematopoietic stem cells in mice exposed to inhaled benzene, *Mutat. Res.*, 2004, **549**(1–2), 195–212.
54. S. E. Boley *et al.*, p53 heterozygosity alters the mRNA expression of p53 target genes in the bone marrow in response to inhaled benzene, *Toxicol. Sci.*, 2002, **66**(2), 209–215.
55. L. Mesure *et al.*, Gene expression study of monocytes/macrophages during early foreign body reaction and identification of potential precursors of myofibroblasts, *PLoS One*, 2010, **5**(9), e12949.
56. J. K. Wise *et al.*, Temporal gene expression profiling during rat femoral marrow ablation-induced intramembranous bone regeneration, *PLoS One*, 2010, **5**(10), e12987.
57. J. Lucas *et al.*, A translatable predictor of human radiation exposure, *PLoS One*, 2014, **9**(9), e107897.
58. D. R. Hyduke *et al.*, Identifying radiation exposure biomarkers from mouse blood transcriptome, *Int. J. Bioinf. Res. Appl.*, 2013, **9**(4), 365–385.
59. N. V. Johnson *et al.*, Haplotype Association Mapping Identifies a Candidate Gene Region in Mice Infected With Staphylococcus aureus, *G3: Genes, Genomes, Genet.*, 2012, **2**(6), 693–700.
60. S. H. Ahn *et al.*, Gene expression-based classifiers identify *Staphylococcus aureus* infection in mice and humans, *PLoS One*, 2013, **8**(1), e48979.
61. C. Mayer *et al.*, A radiation-induced gene expression signature as a tool to predict acute radiotherapy-induced adverse side effects, *Cancer Lett.*, 2011, **302**(1), 20–28.
62. C. Girardi *et al.*, Analysis of miRNA and mRNA expression profiles highlights alterations in ionizing radiation response of human lymphocytes under modeled microgravity, *PLoS One*, 2012, **7**(2), e31293.
63. T. Templin *et al.*, Radiation-induced micro-RNA expression changes in peripheral blood cells of radiotherapy patients, *Int. J. Radiat. Oncol., Biol., Phys.*, 2011, **80**(2), 549–557.
64. S. Paul *et al.*, Prediction of in vivo radiation dose status in radiotherapy patients using ex vivo and in vivo gene expression signatures, *Radiat. Res.*, 2011, **175**(3), 257–265.

65. M. A. Cazalis *et al.*, Early and dynamic changes in gene expression in septic shock patients: a genome-wide approach, *Intensive Care Med. Exp.*, 2014, **2**(1), 20.
66. C. Tonelli *et al.*, Genome-wide analysis of p53 transcriptional programs in B cells upon exposure to genotoxic stress in vivo, *Oncotarget*, 2015, **6**(28), 24611–24626.
67. C. Lachance *et al.*, Exacerbated type II interferon response drives hypervirulence and toxic shock by an emergent epidemic strain of Streptococcus suis, *Infect. Immun.*, 2013, **81**(6), 1928–1939.
68. J. Wu *et al.*, Strain-specific innate immune signaling pathways determine malaria parasitemia dynamics and host mortality, *Proc. Natl. Acad. Sci. U. S. A.*, 2014, **111**(4), E511–E520.
69. S. Kuehner, K. Holzmann and G. Speit, Characterization of formaldehyde's genotoxic mode of action by gene expression analysis in TK6 cells, *Arch. Toxicol.*, 2013, **87**(11), 1999–2012.
70. X. Yu *et al.*, A system-based approach to interpret dose- and time-dependent microarray data: quantitative integration of gene ontology analysis for risk assessment, *Toxicol. Sci.*, 2006, **92**(2), 560–577.
71. Y. Bauer *et al.*, A novel genomic signature with translational significance for human idiopathic pulmonary fibrosis, *Am. J. Respir. Cell Mol. Biol.*, 2015, **52**(2), 217–231.
72. R. S. Thomas *et al.*, A method to integrate benchmark dose estimates with genomic data to assess the functional effects of chemical exposure, *Toxicol. Sci.*, 2007, **98**(1), 240–248.
73. R. Kawashima *et al.*, Interleukin-13 damages intestinal mucosa via TWEAK and Fn14 in mice-a pathway associated with ulcerative colitis, *Gastroenterology*, 2011, **141**(6), 2119–2129 e8.
74. A. M. Snijders *et al.*, Genetic differences in transcript responses to low-dose ionizing radiation identify tissue functions associated with breast cancer susceptibility, *PLoS One*, 2012, **7**(10), e45394.
75. H. Y. Cho *et al.*, Targeted deletion of nrf2 impairs lung development and oxidant injury in neonatal mice, *Antioxid. Redox Signaling*, 2012, **17**(8), 1066–1082.
76. T. Moritake *et al.*, Strain-dependent damage in mouse lung after carbon ion irradiation, *Int. J. Radiat. Oncol., Biol., Phys.*, 2012, **84**(1), e95–e102.
77. M. D. Franco *et al.*, Transcriptome of normal lung distinguishes mouse lines with different susceptibility to inflammation and to lung tumorigenesis, *Cancer Lett.*, 2010, **294**(2), 187–194.
78. S. Halappanavar *et al.*, Pulmonary gene and microRNA expression changes in mice exposed to benzo(a)pyrene by oral gavage, *Toxicology*, 2011, **285**(3), 133–141.
79. R. Peng *et al.*, Bleomycin induces molecular changes directly relevant to idiopathic pulmonary fibrosis: a model for "active" disease, *PLoS One*, 2013, **8**(4), e59348.

80. T. Oga *et al.*, Prostaglandin F(2alpha) receptor signaling facilitates bleomycin-induced pulmonary fibrosis independently of transforming growth factor-beta, *Nat. Med.*, 2009, **15**(12), 1426–1430.
81. J. L. Mougeot *et al.*, Microarray analyses of oral punch biopsies from acute myeloid leukemia (AML) patients treated with chemotherapy, *Oral Surg. Oral Med. Oral Pathol. Oral Radiol. Endod.*, 2011, **112**(4), 446–452.
82. A. Tewari *et al.*, Upregulation of MMP12 and its activity by UVA1 in human skin: potential implications for photoaging, *J. Invest. Dermatol.*, 2014, **134**(10), 2598–2609.
83. S. Supiot *et al.*, Early dynamic transcriptomic changes during pre-operative radiotherapy in patients with rectal cancer: a feasibility study, *World J. Gastroenterol.*, 2013, **19**(21), 3249–3254.
84. L. Vera-Ramirez *et al.*, Transcriptional shift identifies a set of genes driving breast cancer chemoresistance, *PLoS One*, 2013, **8**(1), e53983.
85. M. Iwakawa *et al.*, The radiation-induced cell-death signaling pathway is activated by concurrent use of cisplatin in sequential biopsy specimens from patients with cervical cancer, *Cancer Biol. Ther.*, 2007, **6**(6), 905–911.

CHAPTER 5

Transcriptomic Dose–Response Analysis for Mode of Action and Risk Assessment

RUSSELL S. THOMAS*[a] AND MICHAEL D. WATERS[b]

[a] National Center for Computational Toxicology, U.S. Environmental Protection Agency, 109 T.W. Alexander Drive, Room D110D, Mail Code: D143-02, Research Triangle Park, NC 27711, USA; [b] Integrated Laboratory Systems (ILS), Inc., PO Box 13501, Research Triangle Park, NC 27709, USA

*Email: Thomas.Russell@epa.gov

5.1 Introduction

The field of "toxicogenomics" emerged nearly 15 years ago with a promise that combining the fields of toxicology and genomics would efficiently provide a mechanistic understanding of harmful physical and chemical agents that was not readily possible with conventional toxicological approaches.[1] At that time, both the technology and analysis tools were relatively immature. However, over the past decade and a half the technology has evolved to become a standard experimental and diagnostic tool. Among the technologies available, gene expression microarrays now provide the routine capability to reliably and reproducibly survey the entire transcriptome at detection thresholds around 1 pM.[2–4] In comparison with other technologies, microarrays have advantages in speed and ease of sample preparation, low per-sample cost, and well-established protocols and methods for data

Issues in Toxicology No. 28
Toxicogenomics in Predictive Carcinogenicity
Edited by Michael D. Waters and Russell S. Thomas

Published by the Royal Society of Chemistry, www.rsc.org

normalization and analysis.[2] For microarray content, gene sequences are available across a wide range of toxicologically relevant species including mouse, rat, pig, rabbit, dog, rhesus monkey, chimpanzee, chicken, cat, and human allowing the evaluation of cross-species conservation of expression patterns following chemical perturbation.

As an alternative to microarrays, whole-transcriptome, next-generation RNA sequencing (RNA-seq) offers a method for estimating transcript abundance in gene expression studies.[5] RNA-seq has the potential to overcome many of the limitations associated with microarrays as it does not rely on predetermined probe sequences for expression measurements, is not subject to the limitations of fluorescence and saturation of binding sites on an array, and is based on simple counting of reads that can be reliably aligned to a reference sequence. As count data, RNA-seq has effectively no limit to the dynamic range of signal detection, and, in theory, it can provide a higher degree of accuracy and precision in estimating relative expression levels. However, RNA-seq data also have potential challenges. Reliable quantification of expression levels appears to be highly dependent on read depth, with low transcript abundance characterized by high variance. In addition, RNA-seq libraries contain inherent biases resulting from the preparation process, and the variance of structure as a function of expression level is fundamentally different to that associated with microarrays.[5,6] Additionally, methods for normalization and statistical analysis of RNA-seq data are less mature, and no established best practices exist for RNA-seq data analyses.[5]

Technological advances in microarray and sequencing technologies have increased the ability to comprehensively assess transcriptional changes resulting from chemical and physical stressors. To be incorporated into human health risk assessment, transcriptomic technology must either focus on at least one of the four steps in the process (hazard, dose–response, exposure, and risk management) or in a broader context, provide inferences as to the mode of action (MoA), which can be used to refine an assessment. Many of the early studies applying gene expression technology in toxicology used relatively high, single doses of a chemical agent with a minimum number of experimental replicates in an attempt to predict hazard or identify MoA at these doses. The utility of gene expression data for this purpose has been demonstrated,[7–14] but is not the subject of this chapter. More recently, gene expression data have been applied to dose–response assessment to provide points of departure (PoD) for expression changes in individual genes, pathways, and cellular processes.[5,15,16] A PoD is the dose–response point that marks the beginning of a low-dose extrapolation. This point can be the lower bound on a dose for an estimated incidence or change in response level from a dose–response model (*e.g.*, benchmark dose, BMD), or a no observed adverse effect level (NOAEL) or lowest observed adverse effect level (LOAEL). The examination of dose-related changes allows more robust anchoring to pathological effects and the evaluation of dose-dependent changes in MoA.

5.2 Traditional Statistical Methods for Analyzing Transcriptomic Dose–Response Data

The majority of transcriptomic dose–response studies have been analyzed using a combination of traditional statistical tools that include analysis of variance (ANOVA) followed by pair-wise comparisons between dose groups and the associated control.[17,18] ANOVA modeling identifies genes that are significantly altered across dose and pair-wise comparison identifies genes that are significantly altered between a specific dose and the vehicle control or between doses. Genes that are identified as statistically significant may be subjected to additional fold-change cut-offs. Functional interpretation of the gene expression changes are typically performed using pathway or biological process enrichment analysis, which identify pathways or processes that have more or less significantly altered genes than expected by chance.[19] The pathway- or process-related changes can then be evaluated in a MoA context to identify key events associated with adverse phenotypes as a function of dose and time.

Analyzing transcriptomic dose–response studies using traditional statistical analysis methods is analogous to earlier methods used to define LOAELs or NOAELs for toxicological responses. The LOAEL is typically defined as the first dose producing a biologically significant change in the frequency or severity of adverse effects, while the NOAEL is the highest exposure dose at which there are no biologically significant effects; typically the dose preceding the LOAEL. The limitation of this approach is that dose spacing and experimental sample size can have a large impact on the final NOAEL and LOAEL and the approach does not account for variability in the estimate of the dose response or the slope of the dose–response curve. Similar limitations apply when using this approach for analyzing transcriptomic dose–response studies.

5.3 A Benchmark Dose Method for Analyzing Transcriptomic Data

For traditional toxicological endpoints, the BMD method was introduced in the 1980s to overcome the limitations of the NOAEL and LOAEL approach.[20] The BMD method fits a statistical model to the dose–response data and identifies a dose that causes a defined change in response. The application of the BMD method provides numerous advantages including better utilization of dose–response information, more appropriate reflection of experimental sample sizes, and the lack of constraint to experimental doses.[21] The BMD method is the standard approach for assessing the dose–response behavior of chemicals as they relate to adverse endpoints and is commonly applied to identify PoD values used in the derivation of standards for human health effects (Figure 5.1).[22,23]

To leverage many of the same advantages of the BMD approach, the BMD methodology has been integrated into gene expression analysis workflow for

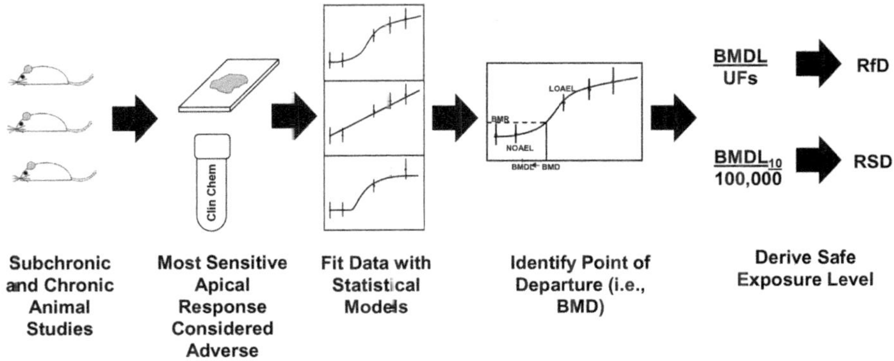

Figure 5.1 A flow chart outlining the dose–response assessment workflow for noncancer and cancer apical endpoints using benchmark dose (BMD). BMDL: 95% lower confidence limit of the BMD; BMR: benchmark response; LOAEL: lowest observed adverse effect level; NOAEL: no observed adverse effect level; RfD: reference dose; RSD: risk-specific dose; UF: uncertainty factor.

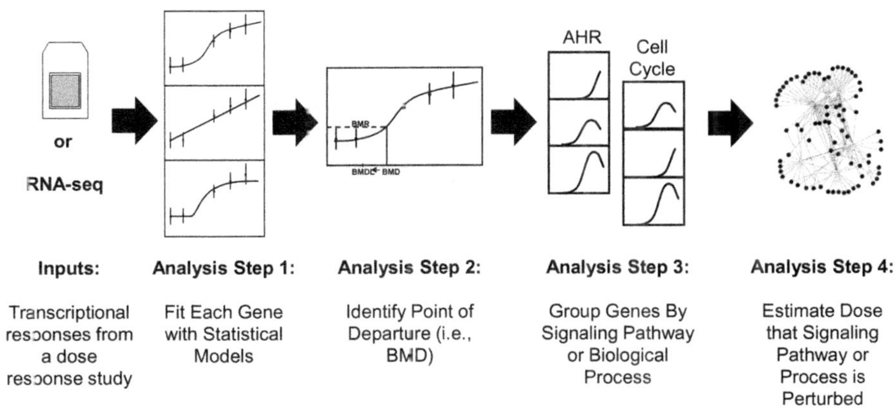

Figure 5.2 A flow chart outlining the transcriptomic benchmark dose (BMD) approach. Step 1 involves initial data processing and dose–response modeling. Step 2 involves BMD analysis. Step 3 groups genes by signaling pathway or biological function. Step 4 estimates the dose at which the pathway or process is perturbed. AHR: aryl hydrocarbon receptor; BMDL: 95% lower confidence limit of the BMD; BMR: benchmark response.

both microarray and RNA-seq data (Figure 5.2).[5,24] The first step in the workflow consists of collecting, normalizing, and $\log_2$ transforming the gene expression data. The workflow may then include a one-way ANOVA to identify all genes with significant dose–response behavior in order to reduce the computational requirements in the subsequent statistical modeling; however, this step is not always included. The probes or probe sets for the microarray data or genes for RNA-seq data are then fit as continuous data to a series of four different dose response models: linear, 2° polynomial,

3° polynomial, and power models. A Hill model is sometimes included if the chemical is suspected of operating by a receptor-mediated mechanism; however, including the Hill model can lead to unrealistically low BMD values if the lower asymptote is not fully anchored.[25] In this case, an additional filter may be applied to minimize the impact of the genes demonstrating the phenomena. Each model is run assuming constant variance and the benchmark response (BMR) factor is set to 1.349 multiplied by the standard deviation in the control samples. By definition, the BMR is the change in the response over the background level that results in an adverse effect. A normal distribution is assumed for control samples. It is assumed, *a priori*, that the transcriptional response could occur in either tail, with a 1% change of that occurring in the absence of exposure (0.5% in each tail). A BMR of 1.349 is the amount required to shift the mean transcriptional response of the control distribution such that the treated distribution contains 11% in a single tail, *i.e.* a 10% increase over the assumed background rate of response.

After model fitting, a single model is selected using a nested likelihood ratio test between the linear and the various polynomial models and a nominal *p*-value cut-off ($p < 0.05$).[26] If the power and Hill models are incorporated into the analysis, the Akaike information criterion is used to select between the best polynomial model, the power model, and/or Hill model.[27] A BMD and benchmark dose lower confidence limit (BMDL) is calculated for the selected model resulting in a BMD and BMDL value for each probe or probe set on the microarray or each gene with RNA-seq. Following model selection, unreliable probe sets for microarrays or genes for RNA-seq are usually removed. The filtering consists of removing probe sets or genes with a BMD value greater than the highest dose and those with a low-fit *p*-value. The filtering ensures that probe sets or genes with extrapolated BMD values or poor fits do not bias aggregated BMD estimates at the gene, pathway, or process level. For microarray data that utilize probes or probe sets, the BMDs and BMDLs for the individual probes and probe sets are averaged to obtain a single BMD and BMDL value for each gene.

To allow interpretation of the gene expression data in a broader context, the gene expression data are typically aggregated using one of several different public or commercial databases. These databases include, among others, the Gene Ontology (GO) Consortium database, the Kyoto Encyclopedia of Genes and Genomes (KEGG) biochemical and signaling pathways database, and the Ingenuity pathway and gene function databases. The individual genes are grouped according to function, signaling pathway, or disease and the associated BMD and BMDL values are summarized to represent the general behavior of the group as a function of dose. In most cases, the mean or median BMD and BMDL are sufficient to capture the general behavior of the group. There has been some debate as to whether the minimum BMD should be used to represent the pathway or process PoD since it is the most conservative. However, one needs to consider whether the minimum BMD derived from the effect on a single gene provides a realistic estimate of the dose that elicits a complex, adverse response such as apoptosis or cytoskeletal reorganization. When viewed from this perspective, use of the minimum BMD appears to be less scientifically sound. External stressors or perturbations cause coordinate changes in the expression of families of genes, many within the same functional class, suggesting that complex biological processes are not regulated by a changes in a single gene.[28–30] Furthermore, complex biological systems are inherently robust against small perturbations due to compensatory mechanisms that limit the effect of changes in a single gene.[31,32]

5.4 Application of Transcriptomic Dose–Response Analysis to MoA Assessment

Quantitative chemical risk assessment has historically relied heavily on identifying hazards through a suite of animal-based studies at relatively high doses, selecting the most sensitive, biologically significant pathological effect, and setting reference values based on this response with limited understanding of how the chemical causes toxicity or whether the effects observed at high doses are relevant to effects at environmentally realistic doses. In an effort to address these challenges and move towards chemical-specific risk assessments, the US Environmental Protection Agency (EPA) introduced MoA into its guidelines for carcinogen risk assessment.[33] The guidelines distinguish MoA from the more detailed understanding of mechanism and define MoA as the "sequence of key events and processes, starting with interaction of an agent with a cell, proceeding through operational and anatomical changes, and resulting in cancer formation."[33] The identification and use of MoA allows for better estimates of the shape of the dose–response curve in the low-dose region, dose-dependent transitions in MoA, assessment of cross-species relevance, identification of potentially susceptible populations, better estimates of intra- and inter-individual variability, and reduced uncertainty.

The utilization of transcriptomic dose–response analysis in the derivation of BMD, and within the MoA framework, can occur on several levels. The most obvious is the ability to correlate expression changes in key signaling pathways with adverse responses as a function of dose and use this concordance as part of the weight of evidence. The dose-dependent changes in expression may be identified using traditional statistical analysis approaches outlined in Section 5.2 or using transcriptomic BMD methods outlined in Section 5.3. The EPA guidelines permit the use of data on obligatory precursor events to inform the nature of the dose response below the concentrations at which the toxic sequelae are observed.[33] The identification of the key signaling events can be confirmed using orthogonal data or genetic knockdown or knockout approaches *in vitro* or *in vivo.* It has also been suggested that the dose–response component of a risk assessment could be based on quantitation of molecular endpoints, or "bioindicators" of response, selected on the basis of their association with obligatory precursor events.[34] These molecular endpoints or bioindicators of response can include gene expression changes in key cell signaling pathways.

Ideally, the transcriptomic dose–response studies would be performed as a function of time to relate temporal changes in the BMD values with the adverse responses and help identify transcriptional changes that may either resolve or become exacerbated over time. To address cross-species differences, dose–response transcriptomic studies could also be run in multiple species. For a large number of chemicals, significant cross-species differences have been observed in toxicological responses. The cross-species differences in transcriptomic dose–response changes allow identification of cell signaling pathways or processes that are critical to the adverse response and separation from those that may be adaptive. As part of an integrated risk assessment strategy, traditional pharmacokinetic (PK) or physiologically-based pharmacokinetic (PBPK) modeling should be performed to characterize target tissue doses.[35]

5.4.1 Case Study: Formaldehyde Exposure in Rat Nasal Epithelium

One example of the application of transcriptomic dose–response analysis in identifying and characterizing the dose–response behavior of key events in the MoA can be found in the investigation of transcriptional alterations following nasal formaldehyde exposure.[36–38] In an early transcriptomic study on formaldehyde, rats were exposed *via* whole-body inhalation to different doses of formaldehyde (0, 0.7, 2, 6, and 15 ppm) for 6 h, 5 days, or 19 days (6 h per day, 5 days per week; 15 ppm exposure at 6 h time point only). Nasal epithelial tissue was isolated from the region of the rat nose with high tumor incidence (level II region)[39] and subjected to gene expression analysis ($n = 4$–8 per dose per time point). Cell proliferation was examined using bromodeoxyuridine (BrdU) labeling at the 5 day time point. The study compared formaldehyde-induced transcriptional changes in nasal epithelium with corresponding histopathological changes at or below a proposed threshold for carcinogenicity (6 ppm) for exposures of up to 3 weeks in duration.

In a dose context, the transcriptional changes did not prove to be more sensitive than apical histological responses. This result was counter to commonly held thinking at the time. Using traditional statistical approaches, no significant gene changes were observed at the lowest concentration (0.7 ppm) and very few were observed at 2 ppm. Transcriptomic BMD analysis was performed and the gene-based BMD values were grouped based on GO biological processes and cellular components. Among the biological processes affected, protein transport processes and specific aspects of the inflammatory response were the most sensitive GO processes with BMD values between 1.5 and 2.5 ppm. GO categories related to cell proliferation and DNA damage had stable mean BMD values ranging from 6 to 8 ppm.

These values were consistent with the observed increase in (1) cell proliferation as measured by BrdU;[36] (2) the $BMD_{10\%}$ for nasal tumor formation in the rodent bioassay of 6.40 ppm;[40] and (3) the increased formation of DNA:protein cross-links around 6 ppm.[41] The cellular component GO categories with the lowest BMDs ranging from 1.5 to 4 ppm were associated with the cell membrane, extracellular components, maintenance of cell architecture, and anchorage to cell membrane. These low-concentration effects and their persistence across time indicate that the initial tissue responses to inhaled formaldehyde likely arise from extracellular reactivity of formaldehyde at or in the vicinity of the cell membrane.

The transcriptional changes in the formaldehyde study qualitatively supported a MoA that has dose-dependent contributions of extracellular and intracellular concentrations of formaldehyde at various inhaled concentrations. The cell has higher intracellular glutathione (GSH) and formaldehyde dehydrogenase than the extracellular space. At low concentrations of formaldehyde, a concentration gradient exists with higher concentrations in the extracellular spaces resulting in greater sensitivity to transcriptional changes. This transition appears to occur between 0.7 and 2 ppm. As formaldehyde concentrations increase and the detoxification processes become overwhelmed, the inter-compartmental differences in concentration will be lower and intracellular targets would become increasingly important. This dose-dependent transition likely occurs between 2 and 6 ppm. At higher concentrations, more extensive cellular damage occurs including DNA damage, cell death, and regenerative proliferation eventually leading to increased risk of tumors.

In a further study on formaldehyde by Andersen *et al.*,[37] concentration and exposure duration transitions in its MoA were examined which combined PK modeling for tissue formaldehyde acetal (FAcetal) and GSH, histopathology, and gene expression in nasal epithelium. Rats were exposed to 0, 0.7, 2, 6, 10, or 15 ppm formaldehyde 6 h per day for 1, 4, or 13 weeks. Nasal epithelial tissue was isolated from the region of the rat nose with high tumor incidence (level II region)[39] and subjected to gene expression analysis ($n = 4$–8 per dose per time point).

At the lower concentrations of formaldehyde, changes in a few stress-response genes were observed with transcriptomic BMD values of $\sim$1 ppm at each of the three exposure durations. These genes included Hmox1 and Maff, which are involved in the oxidative stress response. In addition, other stress-response genes were observed, including those that involve GSH, *i.e.*, Slc7a11 and Srxn1. Slc7a11 transports cystine (a cysteine disulfide) into cells in exchange for glutamate/cysteine. From the PK model, little change would be expected in intracellular FAcetal or GSH at 0.7 ppm. At 2.0 ppm, the PK model indicated that some small changes could be expected in GSH, which is consistent with the transcriptomic BMD values of the stress-response genes.

At tumorigenic concentrations of formaldehyde ($\geq$6 ppm), gene expression changes showed enrichment of pathways involved in cell cycle, DNA repair, and apoptosis. From the PK model, intracellular FAcetal is expected to increase more sharply with formaldehyde exposures >3 ppm. The transcriptional changes were consistent with replication stress due to reactivity of FAcetal with various cellular components; cells are activating pathways to ensure successful mitosis in an environment with enhanced replication. The tissue changes in the region from 6 to 15 ppm included cell proliferation, erosion, and necrosis that was more prominent at 1 week and increased severity of squamous metaplasia at the longer durations and higher exposure concentrations.

Apart from the DNA damage and cell proliferation responses, a group of downregulated genes was associated with Wnt and transforming growth factor (TGF)-β signaling pathways at the 4 week time point. The downregulated genes include Lrp6, Nlk, Rarg, Sfrp1, Sox2, and Tcf7l1, which are negative regulators of the Wnt pathway. With TGF-β, downregulated genes are positive regulators (Bmpr2, Crebbp, and Zfyve9), and decreases in these gene products would diminish pathway function. The coordinate change in Wnt and TGF-β is consistent with activation of the stem cell niche.[42] Stem cell niches in the upper airways are associated with basal cell populations that proliferate after epithelial cell damage.[43,44]

The transcriptomic results suggest a composite MoA for formaldehyde tumorigenicity. At early time points, cell cycle control and DNA repair are activated at transcriptomic BMDs in the 3–5 ppm range with inflammation at higher concentrations. At 4 weeks, tissues undergo remodeling with activation of the stem cell niche. By 13 weeks, there were more pervasive changes in gene expression, including activation of extracellular matrix and Notch signaling, and a shift of the cell cycle and DNA repair BMDs to lower values compared with the 4 week response. The results demonstrate the value of integrating PK modeling, histology, and transcriptomic studies across dose and time for establishing the MoA of a chemical.

5.4.2 Case Study: β-chloroprene in Mouse and Rat Lung

To examine how transcriptomic dose–response analysis could be used across species to identify the MoA of a toxicant, Thomas *et al.*[45] used β-chloroprene (2-chloro-1,3-butadiene) as a model chemical. In a chronic cancer bioassay, β-chloroprene produced a significant increase in female B6C3F1/Crl mouse lung tumors at the lowest exposure concentration of 12.8 ppm, whereas a small, but not statistically significant increase occurred in female F344/NCrl rats only at the highest exposure concentration of 80 ppm.[45] It is known that the metabolism of chloroprene results in the generation of reactive epoxides, and the rate of overall chloroprene metabolism is highly species dependent.[46] In the study by Thomas *et al.*, female mice were exposed to 0, 0.3, 3, 13, and 90 ppm chloroprene while female rats were exposed to 0, 5, 30, 90, and 200 ppm chloroprene. The exposure concentrations were selected to span those used for the cancer bioassay in the mouse and extend the concentration range in the rat. Both the mice and rats were treated for 5 days

(5 days of exposure) and 19 days (15 exposure days). Gene expression measurements were performed on the lungs and analyzed using both traditional statistical methods and the transcriptomic BMD approach.

In the female mouse lung, no genes showed significant changes at the lowest concentrations. At the remaining concentrations, the number of significant genes increased with concentration at both the 5 and 15 day time points. In the female rat lung, no significant changes in the expression of any genes were observed either at the 5 ppm concentration at the 5 or 15 day time point or at the 30 ppm concentration at the 5 day time point. The remaining concentrations showed a significant number of altered genes at both time points. In the enrichment analysis, a total of 18 pathways were enriched in the mouse lung at carcinogenic doses (≥12.8 ppm). One of the enriched pathways in the mouse lung was also enriched in the rat lung at noncarcinogenic doses (≤80 ppm), leaving 17 pathways as potential candidates for involvement in the MoA. When analyzed using transcriptomic BMD methods, a total of 11 signaling pathways had BMD values that were similar to the BMD value for chloroprene-induced lung tumors in the mouse. However, when these pathways were filtered to remove those with a median BMD value in the rat <80 ppm, only four pathways remained.

For the pathways identified using either the enrichment or BMD approaches, only two overlapped. These included GSH metabolism and methionine–cysteine–glutamate metabolism. The ratios of the BMD values across species for these pathways were similar to the ratio of the BMD values for lung tumors. The transcriptional perturbation of the glutathione pathway is consistent with the generation of reactive metabolites as a key event in the MoA for chloroprene-induced lung tumors.

Interestingly, the cross-species transcriptomic BMD ratio for the glutathione pathway was more similar to the differences in the tumor response than the previously proposed target tissue dose surrogate using a PBPK model.[47] The previous PBPK model was developed to account for total chloroprene oxidation in the basal state and did not include parameters for detoxification by epoxide hydrolase or induction of biotransformation enzymes resulting from treatment. The closer correspondence of the transcriptional changes with the tumor response was likely due to their reflection of the overall balance between metabolic activation and detoxification reactions.

5.4.3 Case Study: Naphthalene in the Rat Nasal and Olfactory Epithelium

Naphthalene is a volatile aromatic compound present in petroleum-based fuels and is a registered pesticide for moth larvae, leading to widespread human exposures. Concerns regarding the potential for health effects from human exposures to naphthalene were engendered by studies showing that chronic exposure to naphthalene vapor produced toxicity in the lungs of mice and in the nasal epithelium of both mice and rats.[48,49] In a B6C3F1 mouse 2 year bioassay,[48,50] a statistically significantly increased incidence of pulmonary alveolar/bronchiolar adenoma (a benign lesion) was observed in females exposed to naphthalene vapor at 30 ppm but not 10 ppm; for male mice the observed increase did not reach statistical significance. However, it has been suggested that this type of mouse lung tumor develops through a species-specific MoA based on metabolic and physiological susceptibility.[51] There were no nasal tumors observed in either sex of mice.

The toxicologic and carcinogenic potential of naphthalene was also studied in Fischer 344 (F344) rats exposed by inhalation to naphthalene vapor (0, 10, 30, or 60 ppm) for 2 years.[49,52] A statistically significant increase in neuroblastomas, a rare tumor type, was observed in the nasal olfactory epithelium of female rats at 60 ppm; increases in olfactory epithelial neuroblastomas in male rats were not statistically significant. In the same study, dose-dependent, statistically-significant increases in adenomas were observed beginning at 10 ppm in the respiratory epithelium of male rats; increases in respiratory epithelial adenomas in female rats were not statistically significant. Due to the rarity of the tumors observed in this study, and the fact that there were statistically significantly increases at the lowest exposure concentration, this study has resulted in a re-evaluation of the potential risks of naphthalene exposures.

In both sexes of rats, exposure to naphthalene causes increases in the incidences of various non-neoplastic nasal lesions.[53] Long *et al.*[53] found evidence of cellular injury and accelerated cell replication that was considered suggestive of a cytotoxic MoA. There is also strong evidence that the nasal effects of naphthalene result from the production of toxic metabolites,[53,54] including naphthalene epoxides and naphthoquinones. Napthalene is metabolized by P450 (Cyp) enzymes present in both respiratory and olfactory nasal tissues, particularly by Cyp2f4 in the rat.[55] However, the possible contribution of direct genotoxicity to naphthalene carcinogenicity could not be completely ruled out.[53]

A transcriptomic dose–response study was performed by Clewell *et al.*[56] to investigate the concentration response for gene expression changes in the nasal epithelium of F344 rats in order to elucidate the MoA for naphthalene toxicity/carcinogenicity and to determine potential PoD for risk assessment. Male and female Fischer 344 rats were exposed to naphthalene vapors at 0 (controls), 0.1, 1, 10, and 30 ppm for 6 h per day, 5 days per week, over a 90 day period. Following exposure, the respiratory epithelium and olfactory epithelium from the nasal cavity were dissected separately and used to perform gene expression microarray analysis.

Using a traditional statistical analysis approach, only a few significant gene expression changes were observed in the olfactory or respiratory epithelium of either sex at the lowest concentration (0.1 ppm). At the 1.0 ppm concentration, there was limited evidence of an oxidative stress response in the respiratory epithelium, but not in the olfactory epithelium. This finding is supported by the histopathology performed on animals from the same exposures;[57] there were no histopathological findings at the 0.1 ppm or

1.0 ppm exposure concentrations apart from minimal evidence of hyperplasia in the respiratory epithelium at 1.0 ppm. In contrast, significant enrichment in genes involved in oxidative stress, cell proliferation, DNA damage response, and GSH regulation was observed at bioassay concentrations of 10 and 30 ppm.

Using a transcriptomic BMD analysis, the female olfactory epithelium (the sex/site with the highest incidences of neuroblastomas in the National Toxicology Program (NTP) bioassay) showed BMD values for pathways related to GSH homeostasis, inflammation, and proliferation between 8 and 16 ppm. By comparison, the most sensitive pathways in the female olfactory epithelium had BMD values of approximately 6 ppm. In the male respiratory epithelium (the gender/site with the highest incidences of adenomas in the NTP bioassay), the BMD values for pathways related to GSH homeostasis, inflammation, and proliferation ranged between 0.5 and 0.9 ppm. The most sensitive pathway in the male respiratory epithelium had a BMD value of approximately 0.4 ppm.

The transcriptional evidence from this study is consistent with the histopathological findings in the NTP bioassay of cellular damage and hyper-proliferation, and supports the proposed cytotoxicity MoA.[53] Due to the substantial overlap of genes involved in cell cycle control and DNA damage response, it was not possible to draw a conclusion regarding the possible contribution of direct genotoxicity to the carcinogenicity of naphthalene using transcriptional data alone; however, it was apparent from the concentration response of pathway alterations that there is a dose-dependent transition in the MoA for naphthalene between 1.0 and 10 ppm. The transcriptional changes in GSH homeostasis, inflammation, and proliferation between 8 and 16 ppm in the female olfactory epithelium and between 0.5 and 0.9 ppm in the male respiratory epithelium are consistent with GSH depletion, oxidative stress, inflammation, and proliferative signaling as precursor events leading to increased olfactory epithelial neuroblastomas in female rats at 60 ppm and respiratory epithelial adenomas in male rats at 10 ppm.

5.4.4 Case Study: Fenofibrate and Methapyrilene for Receptor-mediated Nongenotoxic Carcinogenesis

According to EPA risk assessment guidelines, a threshold response can be assumed for a nongenotoxic carcinogen if a threshold-based MoA can be described.[33] However, this can be challenging because multiple MoAs and key events may exist for carcinogenicity and oftentimes extensive research is required to elucidate the MoA. Fenofibrate and methapyrilene are two prototypical nongenotoxic rodent carcinogens, one with a well-established MoA (fenofibrate), and the other where the MoA is still being elucidated (methapyrilene). The MoA for fenofibrate tumorigenesis involves activation of the peroxisome proliferator-activated receptor alpha (PPARA) receptor, proliferation of peroxisomes, expression of genes involved in lipid metabolism, oxidative stress resulting from excessive mitochondrial oxidation of fatty acids, with resulting DNA damage, and perturbation of the balance between cell proliferation and apoptosis, ultimately leading to an increased formation of tumors.[58,59] The MoA for methapyrilene tumorigenesis is not clear, but tumorigenicity appears to be related to a sustained proliferative response as an adaptation to chronic periportal injury and cell death.[60,61]

In a report by Bercu *et al.*,[16] a microarray study was performed to evaluate the ability of transcriptomic dose–response analysis to identify the MoA of fenofibrate and methapyrilene and assess the dose concordance of transcriptomic and histopathological responses. Female rats were exposed to fenofibrate in two daily oral gavage doses of 30 or 1000 mg kg^{-1} per day. Methapyrilene was delivered by oral gavage at 10 and 100 mg kg^{-1} per day over a period of 1, 3, and 7 days. Gene expression measurements were performed on the livers and analyzed using both a traditional statistical analysis and the transcriptomic BMD approach.

Fenofibrate dosing significantly altered signals for a relatively small number of genes over the two doses. The KEGG pathways most affected following fenofibrate exposure were largely related to PPARA activation and included fatty acid metabolism, PPAR signaling pathway, and valine, leucine, and isoleucine degradation. The transcriptomic BMD values for these pathways ranged from 5.0 to 5.9 mg kg^{-1} per day. The transcriptomic BMD values were consistent with functional measurements of peroxisomal β-oxidation that had a BMD of 6 mg kg^{-1} per day, but were lower than the BMD for hepatocellular carcinomas (57 mg kg^{-1} per day).

The number of genes affected by methapyrilene was relatively large and increased over time. The enrichment of KEGG pathways was similarly time dependent. Among enriched KEGG pathways, only one, ribosomal protein production, was changed at all time points. Other significant pathways following 7 days of exposure were related to metabolism, such as P450 metabolism, PPAR signaling pathway, arachidonic acid metabolism, cysteine metabolism, and androgen/estrogen metabolism. Although the MoA was difficult to infer from the data collected, the pathway enrichment results are consistent with sequential key events resulting in cellular and tissue changes over time. The transcriptomic BMD values for KEGG pathways following 7 days of exposure, the time point at which liver injury was observed,[62] ranged from 29 to 92 mg kg^{-1} per day. The BMD for hepatocellular carcinomas was approximately 7 mg kg^{-1} per day, which was more consistent with the most sensitive pathways involving disruption of steroid hormone metabolism. Overall, the results show good dose concordance between and histopathological and short-term transcriptomic responses for nongenotoxic carcinogens.

5.5 Applications of Transcriptomic Dose–Response Analysis to Assess Cross-species Extrapolation

Quantitative chemical risk assessment has relied heavily on the use of default uncertainty factors to assure the protection of public health in the absence of chemical-specific knowledge. One of the commonly used default uncertainty factors adjusts for potential cross-species differences in sensitivity. However, once the critical signaling pathways or biological processes are identified through *in vivo* transcriptomic studies, *in vitro* studies using primary human cells can be performed to evaluate the human relevance and relative sensitivity.

2,3,7,8-Tetrachlorodibenzo-*p*-dioxin (TCDD) is the prototype chemical for the large class of halogenated aromatic hydrocarbons and TCDD is known to produce an array of biochemical and physiological effects in laboratory animals, which includes cancer, endocrine alterations, immunological effects, and developmental defects.[63,64] The biochemical and physiological responses to TCDD vary significantly across species, which confounds the issue of extrapolating the responses to humans.[65–69] To assess the potential cross-species extrapolation associated with TCDD-induced liver toxicity, a transcriptomics approach was used to qualitatively and quantitatively compare the gene expression changes in human and rat primary hepatocytes exposed to TCDD.[25,70] Hepatocytes from five individual female rats and five individual human females were exposed for 24 h to 11 concentrations of TCDD ranging from 0.00001 to 100 nM and a vehicle control. Gene expression changes were analyzed using whole-genome microarrays and the measurements analyzed for the common orthologs shared between rats and humans using traditional statistical analysis and transcriptomic BMD approaches.

Using traditional statistical analysis approaches, the number of differentially expressed orthologs at any concentration was approximately three times higher in the rat than in human hepatocytes. In comparing the cross-species changes in expression, approximately 33% of the differentially expressed human orthologs were also altered in the rat hepatocytes while only 10% of the differentially expressed rat orthologs were altered in the human hepatocytes. Similar differences were observed at the pathway level using enrichment analysis. The commonly enriched pathways amounted to approximately 24% of the total enriched pathways in human hepatocytes and 22% of the total enriched pathways in rats. Among the pathways that were enriched in both species, the aryl hydrocarbon receptor (AHR) signaling pathway and a number of nuclear receptor pathways that lead to the induction of metabolic enzymes were shared. The enrichment in metabolic-related function suggested cross-species conservation in AHR-nuclear receptor cross-talk and the regulation of associated metabolic enzymes. Other shared pathways included stellate cell activation and fibrosis, epithelial-to-mesenchymal transition, cellular uptake by macropinocytosis, and cytoskeletal remodeling. In contrast, a variety of pathways were enriched in a single species, including some involved in growth factor signaling, apoptosis, energy metabolism, and the immune response. The results suggest that the function of AHR may have diverged across species in regulating these pathways, which could have implications for cross-species extrapolation of toxicity.

Transcriptomic BMD analysis of gene expression changes was used to evaluate cross-species differences in potency among differentially expressed orthologs and enriched pathways. Among the battery of AHR-regulated genes, cross-species differences in potency ranged from 2.41-fold for CYP1A2 to 6.96-fold for TIPARP, with human hepatocytes having higher BMD values for each ortholog. Across all orthologs, an average 18-fold difference in transcriptional BMD values was observed, with the rat more sensitive than the human. Similar cross-species differences in pathway BMD values were observed. One challenge using this approach is that few of the short-term transcriptional changes can be directly associated with any of the downstream adverse effects of TCDD since only a handful of AHR target genes have been directly linked to effects at the level of the cell or tissue. As a result, a cross-species comparison on the level of the transcriptome will potentially contain transcriptional changes associated with both adaptive and adverse effects. Despite this fact, the data from the study suggest that human hepatocytes responded qualitatively and quantitatively different than rat hepatocytes at both the gene and pathway level. Human hepatocytes were less sensitive than rat hepatocytes, while the discordantly affected cellular pathways may provide important clues to the MoA for TCDD-induced hepatotoxicity. Based on a report by the National Research Council, conserved responses may differ in magnitude between species, but if the differences are consistent, then appropriate adjustment factors can be derived for cross-species inferences.[71] The results also support the application of transcriptomic dose–response analysis to assess cross-species extrapolation in MoA and potency.

5.6 On the use of Toxicity Pathways in Toxicity Testing, Transcriptional BMD Analysis, and Risk Assessment

The traditional approach for estimating noncancer and cancer reference values in quantitative chemical risk assessment is time- and resource-intensive. The extent and nature of the studies required under the traditional approach has limited the number of chemicals with published risk assessments. In an early study by Thomas *et al.*,[72] the traditional approach for estimating noncancer and cancer reference values was compared with a transcriptomic BMD approach. In the study, female mice were exposed for 13 weeks to multiple concentrations of five chemicals that were positive for lung and/or liver tumor formation in a 2 year cancer bioassay. These five chemicals were 1,4-dichlorobenzene, 1,2,3-trichloropropane, propylene glycol mono-*t*-butyl ether, naphthalene, and methylene chloride. Histological and organ weight changes were evaluated on the lungs and livers. The histological, organ weight changes, and the original tumor incidences in the original cancer bioassay were analyzed using standard BMD methods to identify noncancer and cancer points of departure, respectively. In parallel, gene expression analysis was performed on the target tissues and analyzed using the transcriptomic BMD approach. The gene-based BMD values were grouped based on GO biological processes.

The transcriptional BMD values for the most sensitive GO categories for each chemical were compared with BMD values for the cancer and noncancer apical endpoints. The transcriptional BMD values were generally less than or similar to the corresponding values for the apical endpoints. The one exception was for methylene chloride in the lung where the BMD value for tumor incidence (790.7 mg m^{-3}) was less than the most sensitive transcriptional BMD value (1137.1 mg m^{-3}). However, a direct comparison between transcriptional and apical BMD values is likely to be confounded by the maximum tolerated dose (MTD).[73,74] To account for this potential confounding, partial correlation coefficients were calculated between the transcriptional BMD values for the different GO categories and the noncancer and cancer apical BMD values when adjusted for the MTD. For the noncancer effects, the transcriptional BMD values for GO categories related to immune response, protein maturation, and phosphorylation were also prevalent among those with the highest positive partial correlation. For the cancer effects, the transcriptional BMD values for GO categories related to JNK signaling, immune response, mRNA processing, and cell communication were among those with the highest positive partial correlation. Over two dozen GO categories showed a significant partial correlation coefficient for both cancer and noncancer endpoints, suggesting that a relatively small subset of genes contained in these GO categories may potentially be used to estimate both cancer and noncancer BMD values. Interestingly, in this study methylene chloride showed limited histological changes in the lung at the 13 week time point suggesting that cancer-related points of departure could not be accurately estimated from subchronic histological endpoints alone. However, the use of transcriptomic data provided a more comprehensive view of cellular responses than traditional histology and resulted in a more robust basis for estimating the PoD. The observed relationship between BMD values for transcriptomic and apical endpoints was an early indication that they could potentially be used as PoDs in noncancer and cancer risk assessment.

The use of pathways as a basis for toxicity testing and chemical risk assessment was proposed by the National Research Council[75] as well as by the EPA and the US National Toxicology Program.[76] To address this pathway focus, a reanalysis of the five-chemical study was performed[77] and then extended to include additional chemicals, multiple time points, and other target organs and species.[78] In the extended study, rats were dosed with multiple doses of six additional chemicals for 5 days and 2, 4, and 13 weeks. Similar to the original study, target organs were analyzed for traditional histological and organ weight changes and transcriptional changes using microarrays. Histological and organ weight changes in this study and the tumor incidences in the original cancer bioassays were analyzed using BMD methods to identify noncancer and cancer points of departure.

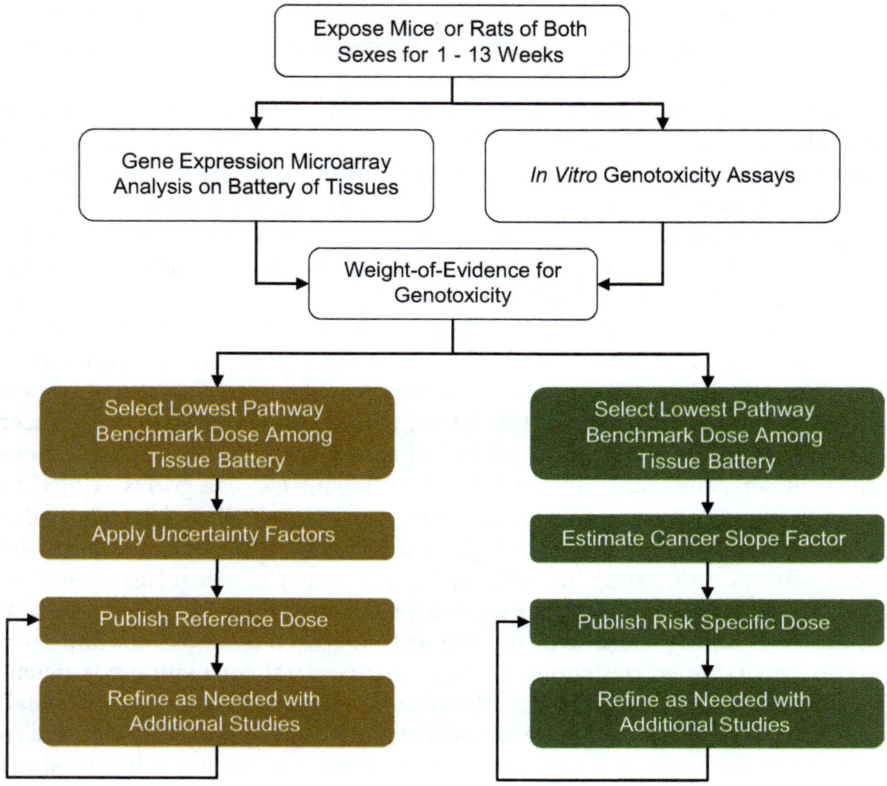

Figure 5.3 Flow chart outlining a framework for application of transcriptomic data to chemical risk assessment. The transcriptomic data are combined with traditional genotoxicity data to determine the potential for carcinogenic activity. Chemicals positive for genotoxicity are presumed to be carcinogenic and would proceed along the path associated with the green boxes. Chemicals negative for genotoxicity would be presumed to have a threshold and proceed down the noncancer path associated with the gold boxes.
Figure adapted from Thomas *et al.*[78]

The dose–response changes in gene expression were also analyzed using transcriptomic BMD methods and the responses grouped based on signaling pathways (Figure 5.3).

A comparison of transcriptional BMD values for the most sensitive pathway with BMD values for the noncancer and cancer apical endpoints showed a high degree of correlation at all time points. When the analysis included data from the earlier study, transcriptional BMD values for the most sensitive pathway were significantly correlated with noncancer ($r=0.827$, $p=0.0031$) and cancer-related ($r=0.940$, $p=0.0002$) BMD values at 13 weeks. The average ratio of apical-to-transcriptional BMD values was <2, suggesting that for the current chemicals, transcriptional perturbation did not occur at significantly lower doses than apical responses. The partial correlation coefficient between transcriptomic BMD values and the noncancer apical BMD values after accounting for the MTD was not significant, suggesting that the overall correlation between the noncancer apical BMD values and the pathway transcriptomic BMD values was equivalent to that observed with the MTD. However, the partial correlation coefficient between the transcriptomic BMD values and the cancer-related apical BMD values after accounting for the MTD was highly significant, suggesting that the overall correlation between the cancer-related apical BMD values and the transcriptomic BMD was greater than that observed with the MTD.

For most of the chemicals analyzed in this study, the most sensitive pathway was not consistent across time points. However, just as histopathological changes evolved over time in response to chemical treatment to reflect the temporal evolution of tissue injury, remodeling, and adaptation, the gene expression changes would also be expected to evolve to reflect this evolution. For example, a large number of the most sensitive pathways for trichlorobenzene at the 5 day time point were involved in cell cycle regulation which was consistent with histological data reporting an increase in mitotic nuclei only at this time point.[79] At the 2 week time point, a variety of metabolic pathways (*e.g.* amino acid, short chain fatty acids, nucleotide, and xenobiotic) were among the most sensitive for trichlorobenzene. Previous studies have reported treatment-related increases in the activities of many xenobiotic metabolizing enzymes in the liver.[80] Finally, at 13 weeks, the top two most sensitive pathways were involved in lipid metabolism, which was consistent with the histological appearance of hepatocyte vacuolization at this time point. Based on these studies, pathway-based transcriptomic data provide a similar correlation to the apical responses compared to the correlation observed using GO categories for biological processes,[72] and the study demonstrated that pathway-based transcriptional BMD values can also be used to predict BMD values based on apical endpoints.[78]

The practical application of these transcriptional BMD values was outlined in a framework for applying transcriptomic data to risk assessment.[78] For a chemical of interest, transcriptomic dose–response studies could be performed on untested chemicals at any single time point between 5 days and 13 weeks. The tests would be performed in mice and rats of both sexes. A battery of eight tissues that include those most frequently positive in rodent cancer bioassays (liver, lung, mammary gland, stomach, vascular system, kidney, hematopoietic system, and urinary bladder) could then be collected. In a previous analysis, these eight tissues cover 92% and 82% of targets for all mouse and rat carcinogens, respectively.[81] Additional tissues that include brain, the developing fetus, and gonadal tissue could be added for assessing neurological, developmental, and reproductive effects. Gene expression analysis on these tissues would allow the estimation of pathway-transcriptomic BMD and BMDL values. The lowest transcriptomic BMD value across all analyzed tissues could be used to estimate a PoD. In parallel, assays for genotoxicity could further expand the utility of the transcriptomic

BMD values. Alternatively, the transcriptomic data could be used to predict genotoxicity.[82] A weight-of-evidence analysis could then be used to determine the genotoxic potential of the chemical and inform decisions regarding the use of the transcriptomic BMD values for cancer-related slope factors or noncancer reference values. Although the most sensitive pathway may not mechanistically relate to the adverse effect, the results of the study demonstrate that basing the PoD on the most sensitive pathway would be generally protective until the key pathways are identified.

The application of transcriptomics within this framework poses both significant benefits and challenges. Importantly, the framework provides a relatively rapid and cost-effective means to provide provisional PoDs for chemicals lacking published reference values. The provisional transcriptional PoDs could be easily incorporated into the second tier of EPA's NexGen risk assessment program.[83] The second tier was designed to use high- and medium-throughput bioassay data for decision making on chemicals with limited traditional data. Similar to the proposal by Cote *et al.*,[83] if or when more data are collected, the provisional transcriptomic PoDs could either be updated or transitioned to a higher risk assessment tier. Among the obvious challenges, the proposed framework would require a shift from the current hazard-based labeling paradigm that relies on apical responses to a paradigm where the most sensitive adverse effect is not known, but the regulatory dose is set based on the lack of transcriptional perturbation. The latter paradigm has been referred to as a "region of safety". Concerns that may accompany this shift in focus may be whether the pathway-based transcriptomic BMD values are adequately protective across a wide range of chemicals and the difficulty in associating a transcriptional perturbation with actual risk. Although both concerns are valid, it is generally accepted that adverse phenotypic responses would not occur without direct or indirect alterations in transcriptional programs. In addition, for many chemicals, a PoD based on transcriptional perturbation is preferable to having no PoD at all. Additional concerns may include whether a PoD based on transcriptional perturbation is overly protective, especially when a given perturbation may not have a direct link to an adverse effect, and whether "no transcriptional perturbation" could become the new default. Based on the chemicals we have studied, concern that transcriptional perturbation would be overly protective may be unfounded, because the transcriptional BMD values for the most sensitive pathway were, on average, within a factor of two of both the noncancer- and cancer-related apical BMD values. However, this cannot be ruled out for a broader set of chemicals and mechanisms. The concern that this may potentially establish a new default is valid. However, traditional animal studies could be performed if it is believed that the "no transcriptional perturbation" level is too conservative.

The proposed framework outlines a practical, near-term application of transcriptomic data to chemical risk assessment. Over the longer term, better methods for interpreting transcriptomic data in the context of MoA and adverse endpoints will become available and should be utilized.

In addition, the transcriptomic data may be combined with other emerging technologies such as high-throughput *in vitro* screening to identify molecular initiating events and assemble the likely sequence of key events in the MoA. The combined use of these technologies for identifying MoAs will ultimately lead to more rapid and economical risk assessments while also helping to extrapolate across species and understand human relevance. In summary, the Thomas *et al.* study[78] demonstrated that pathway-based transcriptomic data can be practically applied to chemical risk assessment. Notably, the transcriptomic changes predicted both subchronic, noncancer PoDs as well as chronic, cancer-related PoDs, thereby providing a possible test method to increase the number of chemicals with published risk assessments.

5.7 Comparison of Traditional Risk Assessment Approaches with Those Applying Transcriptomics

Although the Thomas *et al.* studies[72,77,78] showed good correlation between transcriptomic BMD values and BMD values based on traditional histological responses and provide a practical framework for utilizing transcriptomic dose response studies in human health risk assessment, an in-depth case study was needed to more broadly compare traditional and transcriptomic assisted risk assessment approaches. Recently, Chepelev *et al.*[84] performed such a study involving exposure to benzo[*a*]pyrene (BaP) in drinking water. The intent was not to evaluate the safety of drinking water but rather to compare two alternative risk assessment methodologies: transcriptomics-informed and transcriptomics-only risk asessment. Transcriptomics data from human cell cultures and mice exposed to BaP were used to provide additional insight into the MoA of BaP and to enable derivation of tissue-specific PoDs. As expected, gene expression analysis confirmed that BaP is genotoxic in mice and facilitated the development of a definitive MoA. A high degree of consistency in perturbed pathways was observed between human lymphoblastoid TK6 cells and animal tissues. PoDs were similar for both traditional and transcriptional approaches (liver 1.2 *vs.* 1.0 mg kg^{-1} body weight per day; lungs 0.8 *vs.* 3.7 mg kg^{-1} body weight per day; forestomach 0.5 *vs.* 7.4 mg kg^{-1} body weight per day). The "transcriptomics-only" risk assessment approach provided useful information in a data-poor environment (*i.e.* in the absence of apical toxicology data). The case study revealed design flaws in traditional transcriptomic experiments and suggested how transcriptomics might identify gaps in established approaches as well as new testing strategies. The authors argued that transcriptomic analysis may be an efficient tool in hazard identification and MoA determination as well as in further evaluation of multi-tissue carcinogenic potential and potency.

A major concern moving forward is the breadth of target tissues that must be surveyed for cancer and noncancer effects. In the Chepelev *et al.*[84] case study, BaP adducts and mutations were detected in multiple tissues raising concerns that important effects in nontarget tissues could be missed. As transcriptomic approaches are becoming more high-throughput and less expensive (see Chapter 2 and Li *et al.*[82]), it is conceivable that all tissues collected in repeat-dose animal studies can be transcript-profiled. Thus, in future case studies the selection of appropriate chemicals with diverse MoAs and the sampling of tissues encompassing a wide array of potential toxic effects may serve to demonstrate the utility of transcriptomics in other aspects of human health risk assessment.

In a companion paper, Moffat *et al.*[85] focused on (1) obtaining biologically relevant data readily suitable for establishing a MoA; (2) examining the human relevance of a MoA derived from animal testing; and (3) proposing appropriate quantitative estimates for risk assessment. The authors described a strategy whereby transcriptomics can become an integral tool in risk assessment, especially when anchored by apical endpoints to increase confidence in a proposed MoA. It was acknowledged that the genotoxicity of BaP would be detected by standard *in vitro* tests. In contrast, transcriptomics does not indicate that a mutation has occurred, but instead indicates that a DNA damage response has been initiated, which is consistent with the presence of genotoxicity in a tissue. More importantly, transcriptomics provides essential details on the genotoxic MoA (*i.e.* activation of AHR, metabolism, induction of p53, and DNA damage response, in addition to the effects on cell cycle and signaling pathways). Futhermore, comparative transcriptomics provides insight into the relevance of the MoA to humans and can inform the risk assessment approach. Thus, transcriptomics data can support a positive Ames test or other conventional assay that addresses the ability of the compound to reach the target tissue, be activated, and interact with DNA *in vivo.* Through the analysis of multiple tissues, it offers insight into tissue-specificity, distribution, and metabolism. Moreover, the analysis reveals nongenotoxic carcinogenic (and noncarcinogenic) MoAs that would be missed through application of conventional genetic toxicology assays alone.

The latter point is illustrated by the case study of Jackson *et al.*,[86] also from the Yauk group, wherein furan, a hepatocarcinogen, induces cytotoxicity followed by regenerative proliferation. Although detailed genomics information (activation of certain DNA-damage response pathways) is currently not very informative in the absence of other information, development of specific genomic biomarkers for DNA damage will be invaluable in MoA-based, mechanistic approaches to toxicity testing[82] (see chapter 2 by Li *et al.*). As mentioned previously, a tiered testing paradigm has been recently proposed whereby first order margins of exposure are calculated using data from *in vitro* high-throughput screening assays.[87] Given the similarity between the PoDs derived by traditional and transcriptomics-based approaches,[84,86] it is proposed that PoDs derived from transcriptomics data can also be used to inform regulatory guidance on margins of exposure (historically used for screening of nongenotoxic carcinogens and noncancer effects) or maximum acceptable concentrations. Exposure information and uncertainties associated with hazard and dose response are obviously agent-specific. In summary, the two case studies[84,86] illustrate the utility of transcriptomics to inform MoA and PoD determination in risk assessment and provide examples of transcriptomics-informed and transcriptomics-only risk assessments.

5.8 Significance of Integrating Transcriptomic Data into Risk Assessment

There are multiple reasons to increase efficiency in the risk assessment process and incorporate modern technology in evaluating the potential health effects of chemicals. First, the number of chemicals for which definitive risk assessments have been performed, compared with the number of chemicals to which humans are exposed in the environment is diminishingly small. Second, chemicals without published risk assessment values are not considered quantitatively in the overall hazard index calculation when evaluating contaminated sites where multiple environmental chemicals exist.[88] Therefore, these chemicals are treated as not posing any hazard, leading to a potential underestimation of actual hazard at any given contaminated site. Third, the economic costs, time, and animal numbers required to perform the traditional toxicity tests that underlie current risk assessments limit the number of chemicals that can be evaluated.[75] The resulting studies cannot keep pace with the number of chemicals introduced into commerce or released into the environment.[89] These difficulties lead to a lack of data on which to base risk assessment values. Finally, the use of traditional high-dose animal studies and a simple linear extrapolation to environmental doses does not incorporate potential dose-dependent changes in mechanism that occur for many chemicals.[90]

The financial costs of environmental regulations are estimated to be hundreds of billions of dollars per year and are increasing over time.[91] There is little doubt that the use of high-dose animal studies together with many of the default assumptions associated with chemical risk assessments has contributed significantly to these costs. With the maturation of new technologies and the EPA risk assessment guidelines, there exists a unique opportunity to incorporate unique data-driven approaches for performing MoA-based chemical risk assessments. The technological advances include the capability to perform a genome-wide survey of transcriptional states of a cell or tissue following chemical exposure. The key advantage to applying the technology to risk assessment is both the sensitivity and comprehensive examination of the direct and indirect molecular changes. For direct effects, these may include transcriptional changes resulting from chemical binding to ligand-activated cellular receptors, such as those observed following exposure to the AHR ligand TCDD.[92] Indirect effects may include processes such as cellular remodeling following chronic damage (*e.g.* metaplasia observed in nasal epithelium following formaldehyde exposure).[93] In both cases, the underlying transcriptional program would be invariably different. In many cases, the transcriptional changes would identify potential key events in the MoA and, at a minimum, the changes would provide "bioindicators" that have been widely discussed as potential qualitative and quantitative surrogates for the tumor response.[94] When performed in dose–response studies, the genome-wide measurements of transcriptional changes can provide both critical effect levels (*e.g.* BMD) at which these alterations occur and an experimental basis for a risk assessment.

At present, risk assessments can be developed based on key precursor events related to MoA. These assessments would be primarily developed when the gene categories affected at the lowest doses represent mechanisms of action different from DNA damage and repair. In the longer term, genomic dose–response evaluations might be examined to show groups of interacting processes that are required for cancer at high doses and the absence of activation of these pathways at lower doses. Dose-dependent transitions in MoA, illustrated by hierarchical oxidative stress[95,96] may represent the normal manner in which the body deals with reactive stressors. The National Research Council report on toxicity testing[75] discusses progressive levels of perturbation, from responses below thresholds, early adaptive responses for dealing with reactive stressors, on to overt toxicity and perhaps carcinogenicity.[75] Further development of mechanistic dose–response models should provide a much broader range of risk assessment applications than simple linear low-dose extrapolations or margins of exposure calculations as discussed in this chapter. The next-generation suite of computational models will require these same genomic dose–response tools and other "omic" tools to describe dose-dependent activation of transcriptional networks.[97,98]

References

1. E. F. Nuwaysir, M. Bittner, J. Trent, J. C. Barrett and C. A. Afshari, Microarrays and toxicology: the advent of toxicogenomics, *Mol. Carcinog.*, 1999, **24**, 153–159.
2. L. Shi, L. H. Reid, W. D. Jones, R. Shippy, J. A. Warrington, S. C. Baker, P. J. Collins, F. de Longueville, E. S. Kawasaki, K. Y. Lee, Y. Luo, Y. A. Sun, J. C. Willey, R. A. Setterquist, G. M. Fischer, W. Tong, Y. P. Dragan, D. J. Dix, F. W. Frueh, F. M. Goodsaid, D. Herman, R. V. Jensen, C. D. Johnson, E. K. Lobenhofer, R. K. Puri, U. Schrf, J. Thierry-Mieg, C. Wang, M. Wilson, P. K. Wolber, L. Zhang, S. Amur, W. Bao, C. C. Barbacioru, A. B. Lucas, V. Bertholet, C. Boysen, B. Bromley, D. Brown, A. Brunner, R. Canales, X. M. Cao, T. A. Cebula, J. J. Chen, J. Cheng, T. M. Chu, E. Chudin, J. Corson, J. C. Corton, L. J. Croner, C. Davies, T. S. Davison, G. Delenstarr, X. Deng, D. Dorris, A. C. Eklund, X. H. Fan, H. Fang, S. Fulmer-Smentek, J. C. Fuscoe, K. Gallagher, W. Ge, L. Guo, X. Guo, J. Hager, P. K. Haje, J. Han, T. Han, H. C. Harbottle, S. C. Harris, E. Hatchwell, C. A. Hauser, S. Hester,

H. Hong, P. Hurban, S. A. Jackson, H. Ji, C. R. Knight, W. P. Kuo, J. E. LeClerc, S. Levy, Q. Z. Li, C. Liu, Y. Liu, M. J. Lombardi, Y. Ma, S. R. Magnuson, B. Maqsodi, T. McDaniel, N. Mei, O. Myklebost, B. Ning, N. Novoradovskaya, M. S. Orr, T. W. Osborn, A. Papallo, T. A. Patterson, R. G. Perkins, E. H. Peters, R. Peterson, K. L. Philips, P. S. Pine, L. Pusztai, F. Qian, H. Ren, M. Rosen, B. A. Rosenzweig, R. R. Samaha, M. Schena, G. P. Schroth, S. Shchegrova, D. D. Smith, F. Staedtler, Z. Su, H. Sun, Z. Szallasi, Z. Tezak, D. Thierry-Mieg, K. L. Thompson, I. Tikhonova, Y. Turpaz, B. Vallanat, C. Van, S. J. Walker, S. J. Wang, Y. Wang, R. Wolfinger, A. Wong, J. Wu, C. Xiao, Q. Xie, J. Xu, W. Yang, L. Zhang, S. Zhong, Y. Zong and W. Slikker Jr., The MicroArray Quality Control (MAQC) project shows inter- and intraplatform reproducibility of gene expression measurements, *Nat. Biotechnol.*, 2006, **24**, 1151–1161.

3. R. Shippy, S. Fulmer-Smentek, R. V. Jensen, W. D. Jones, P. K. Wolber, C. D. Johnson, P. S. Pine, C. Boysen, X. Guo, E. Chudin, Y. A. Sun, J. C. Willey, J. Thierry-Mieg, D. Thierry-Mieg, R. A. Setterquist, M. Wilson, A. B. Lucas, N. Novoradovskaya, A. Papallo, Y. Turpaz, S. C. Baker, J. A. Warrington, L. Shi and D. Herman, Using RNA sample titrations to assess microarray platform performance and normalization techniques, *Nat. Biotechnol.*, 2006, **24**, 1123–1131.
4. D. Abdueva, M. R. Wing, B. Schaub and T. J. Triche, Experimental comparison and evaluation of the Affymetrix exon and U133Plus2 GeneChip arrays, *PLoS One*, 2007, **2**, e913.
5. M. B. Black, B. B. Parks, L. Pluta, T. M. Chu, B. C. Allen, R. D. Wolfinger and R. S. Thomas, Comparison of microarrays and RNA-seq for gene expression analyses of dose-response experiments, *Toxicol. Sci.*, 2014, **137**, 385–403.
6. N. F. Lahens, I. H. Kavakli, R. Zhang, K. Hayer, M. B. Black, H. Dueck, A. Pizarro, J. Kim, R. Irizarry, R. S. Thomas, G. R. Grant and J. B. Hogenesch, IVT-seq reveals extreme bias in RNA sequencing, *Genome Biol.*, 2014, **15**, R86.
7. R. S. Thomas, T. M. O'Connell, L. Pluta, R. D. Wolfinger, L. Yang and T. J. Page, A comparison of transcriptomic and metabonomic technologies for identifying biomarkers predictive of two-year rodent cancer bioassays, *Toxicol. Sci.*, 2007, **96**, 40–46.
8. R. S. Thomas, L. Pluta, L. Yang and T. A. Halsey, Application of genomic biomarkers to predict increased lung tumor incidence in 2-year rodent cancer bioassays, *Toxicol. Sci.*, 2007, **97**, 55–64.
9. N. Zidek, J. Hellmann, P. J. Kramer and P. G. Hewitt, Acute hepatotoxicity: a predictive model based on focused Illumina microarrays, *Toxicol. Sci.*, 2007, **99**, 289–302.
10. M. R. Fielden, B. P. Eynon, G. Natsoulis, K. Jarnagin, D. Banas and K. L. Kolaja, A gene expression signature that predicts the future onset of drug-induced renal tubular toxicity, *Toxicol. Pathol.*, 2005, **33**, 675–683.

11. M. D. Waters, M. Jackson and I. Lea, Characterizing and predicting carcinogenicity and mode of action using conventional and toxicogenomics methods, *Mutat. Res.*, 2010, **705**, 184–200.
12. G. P. Daston and J. M. Naciff, Predicting developmental toxicity through toxicogenomics, *Birth Defects Res., Part C*, 2010, **90**, 110–117.
13. S. S. Auerbach, R. R. Shah, D. Mav, C. S. Smith, N. J. Walker, M. K. Vallant, G. A. Boorman and R. D. Irwin, Predicting the hepatocarcinogenic potential of alkenylbenzene flavoring agents using toxicogenomics and machine learning, *Toxicol. Appl. Pharmacol.*, 2010, **243**, 300–314.
14. T. Uehara, M. Hirode, A. Ono, N. Kiyosawa, K. Omura, T. Shimizu, Y. Mizukawa, T. Miyagishima, T. Nagao and T. Urushidani, A toxicogenomics approach for early assessment of potential non-genotoxic hepatocarcinogenicity of chemicals in rats, *Toxicology*, 2008, **250**, 15–26.
15. T. W. Simon, S. S. Simons Jr., R. J. Preston, A. R. Boobis, S. M. Cohen, N. G. Doerrer, P. A. Fenner-Crisp, T. S. McMullin, C. A. McQueen, J. C. Rowlands and R. D.-R. Subteam, The use of mode of action information in risk assessment: quantitative key events/dose-response framework for modeling the dose-response for key events, *Crit. Rev. Toxicol.*, 2014, **44**(Suppl. 3), 17–43.
16. J. P. Bercu, R. A. Jolly, K. M. Flagella, T. K. Baker, P. Romero and J. L. Stevens, Toxicogenomics and cancer risk assessment: a framework for key event analysis and dose-response assessment for nongenotoxic carcinogens, *Regul. Toxicol. Pharmacol.*, 2010, **58**, 369–381.
17. G. K. Smyth, in *Bioinformatics and Computational Biology Solutions using R and Bioconductor*, ed. R. Gentleman, V. Carey, S. Dudoit, R. A. Irizarry and W. Huber, Springer, New York, Editon edn., 2005, pp. 397–420.
18. M. K. Kerr and G. A. Churchill, Statistical design and the analysis of gene expression microarray data, *Genet. Res.*, 2001, **77**, 123–128.
19. W. Huang da, B. T. Sherman and R. A. Lempicki, Systematic and integrative analysis of large gene lists using DAVID bioinformatics resources, *Nat. Protoc.*, 2009, **4**, 44–57.
20. K. S. Crump, A new method for determining allowable daily intakes, *Fundam. Appl. Toxicol.*, 1984, **4**, 854–871.
21. A. F. Filipsson, S. Sand, J. Nilsson and K. Victorin, The benchmark dose method–review of available models, and recommendations for application in health risk assessment, *Crit. Rev. Toxicol*, 2003, **33**, 505–542.
22. EPA, *The Use of the Benchmark Dose Approach in Health Risk Assessment* EPA/630/R-94/007, Office of Research and Development, U.S. Environmental Protection Agency, Washington, D.C., 1995.
23. D. R. Mattison and J. D. Sandler, Summary of the workshop on issues in risk assessment: quantitative methods for developmental toxicology, *Risk Anal.*, 1994, **14**, 595–604.
24. R. S. Thomas, B. C. Allen, A. Nong, L. Yang, E. Bermudez, H. J. Clewell, 3rd and M. E. Andersen, A method to integrate benchmark dose

estimates with genomic data to assess the functional effects of chemical exposure, *Toxicol. Sci.*, 2007, **98**, 240–248.
25. M. B. Black, R. A. Budinsky, A. Dombkowski, D. Cukovic, E. L. LeCluyse, S. S. Ferguson, R. S. Thomas and J. C. Rowlands, Cross-species comparisons of transcriptomic alterations in human and rat primary hepatocytes exposed to 2,3,7,8-tetrachlorodibenzo-p-dioxin, *Toxicol. Sci.*, 2012, **127**, 199–215.
26. D. Posada and T. R. Buckley, Model selection and model averaging in phylogenetics: advantages of akaike information criterion and bayesian approaches over likelihood ratio tests, *Syst. Biol.*, 2004, **53**, 793–808.
27. H. Akaike, 2nd International Symposium on Information Theory, 1973.
28. T. R. Hughes, M. J. Marton, A. R. Jones, C. J. Roberts, R. Stoughton, C. D. Armour, H. A. Bennett, E. Coffey, H. Dai, Y. D. He, M. J. Kidd, A. M. King, M. R. Meyer, D. Slade, P. Y. Lum, S. B. Stepaniants, D. D. Shoemaker, D. Gachotte, K. Chakraburtty, J. Simon, M. Bard and S. H. Friend, Functional discovery *via* a compendium of expression profiles, *Cell*, 2000, **102**, 109–126.
29. R. A. Currie, G. Orphanides and J. G. Moggs, Mapping molecular responses to xenoestrogens through Gene Ontology and pathway analysis of toxicogenomic data, *Reprod. Toxicol.*, 2005, **20**, 433–440.
30. J. I. Murray, M. L. Whitfield, N. D. Trinklein, R. M. Myers, P. O. Brown and D. Botstein, Diverse and specific gene expression responses to stresses in cultured human cells, *Mol. Biol. Cell*, 2004, **15**, 2361–2374.
31. H. Kitano, Biological robustness, *Nat. Rev. Genet.*, 2004, **5**, 826–837.
32. H. Kitano and K. Oda, Robustness trade-offs and host-microbial symbiosis in the immune system, *Mol. Syst. Biol.*, 2006, **2**, 2006 0022.
33. EPA, *Guidelines for Carcinogen Risk Assessment* EPA/630/P-03/001F, U.S. Environmental Protection Agency, Washington, D.C., 2005.
34. R. J. Preston, Quantitation of molecular endpoints for the dose-response component of cancer risk assessment, *Toxicol. Pathol.*, 2002, **30**, 112–116.
35. H. J. Clewell, 3rd, M. E. Andersen and H. A. Barton, A consistent approach for the application of pharmacokinetic modeling in cancer and noncancer risk assessment, *Environ. Health Perspect.*, 2002, **110**, 85–93.
36. M. E. Andersen, H. J. Clewell, 3rd, E. Bermudez, G. A. Willson and R. S. Thomas, Genomic signatures and dose-dependent transitions in nasal epithelial responses to inhaled formaldehyde in the rat, *Toxicol. Sci.*, 2008, **105**, 368–383.
37. M. E. Andersen, H. J. Clewell, 3rd, E. Bermudez, D. E. Dodd, G. A. Willson, J. L. Campbell and R. S. Thomas, Formaldehyde: integrating dosimetry, cytotoxicity, and genomics to understand dose-dependent transitions for an endogenous compound, *Toxicol. Sci.*, 2010, **118**, 716–731.
38. X. Ye, Z. Ji, C. Wei, C. M. McHale, S. Ding, R. Thomas, X. Yang and L. Zhang, Inhaled formaldehyde induces DNA-protein crosslinks and

oxidative stress in bone marrow and other distant organs of exposed mice, *Environ. Mol. Mutagen.*, 2013, **54**, 705–718.
39. T. M. Monticello, J. A. Swenberg, E. A. Gross, J. R. Leininger, J. S. Kimbell, S. Seilkop, T. B. Starr, J. E. Gibson and K. T. Morgan, Correlation of regional and nonlinear formaldehyde-induced nasal cancer with proliferating populations of cells, *Cancer Res.*, 1996, **56**, 1012–1022.
40. P. M. Schlosser, P. D. Lilly, R. B. Conolly, D. B. Janszen and J. S. Kimbell, Benchmark dose risk assessment for formaldehyde using airflow modeling and a single-compartment, DNA-protein cross-link dosimetry model to estimate human equivalent doses, *Risk Anal.*, 2003, **23**, 473–487.
41. M. Casanova, K. T. Morgan, E. A. Gross, O. R. Moss and H. A. Heck, DNA-protein cross-links and cell replication at specific sites in the nose of F344 rats exposed subchronically to formaldehyde, *Fundam. Appl. Toxicol.*, 1994, **23**, 525–536.
42. L. Li and H. Clevers, Coexistence of quiescent and active adult stem cells in mammals, *Science*, 2010, **327**, 542–545.
43. M. R. Alison, Stem cells and cancer in the aerodigestive tract, *Eur. J. Cancer*, 2009, **45**(Suppl. 1), 175–185.
44. J. C. Snyder, R. M. Teisanu and B. R. Stripp, Endogenous lung stem cells and contribution to disease, *J. Pathol.*, 2009, **217**, 254–264.
45. R. S. Thomas, M. W. Himmelstein, H. J. Clewell, 3rd, Y. Yang, E. Healy, M. B. Black and M. E. Andersen, Cross-species transcriptomic analysis of mouse and rat lung exposed to chloroprene, *Toxicol. Sci.*, 2013, **131**, 629–640.
46. M. W. Himmelstein, S. C. Carpenter and P. M. Hinderliter, Kinetic modeling of beta-chloroprene metabolism: I. In vitro rates in liver and lung tissue fractions from mice, rats, hamsters, and humans, *Toxicol. Sci.*, 2004, **79**, 18–27.
47. M. W. Himmelstein, S. C. Carpenter, M. V. Evans, P. M. Hinderliter and E. M. Kenyon, Kinetic modeling of beta-chloroprene metabolism: II. The application of physiologically based modeling for cancer dose response analysis, *Toxicol. Sci.*, 2004, **79**, 28–37.
48. N. T. P. (NTP), Toxicology and carcinogenesis studies of naphthalene in B6C3F1mice (inhalation studies), *Technical Report Series No 410. NTP. Research Triangle Park, NC.*, 1992.
49. N. T. P. (NTP), Toxicology and carcinogenesis studies of naphthalene in F344/N rats, *U.S. Department of Health and Human Services National Toxicology Program*, Washington, D.C., 2000.
50. K. M. Abdo, S. L. Eustis, M. McDonald, M. P. Jokinen, B. Adkins and J. K. Haseman, Naphthalene: a respiratory tract toxicant and carcinogen for mice, *Inhalation Toxicol.*, 1992, **4**, 393–409.
51. G. Cruzan, J. Bus, M. Banton, R. Gingell and G. Carlson, Mouse specific lung tumors from CYP2F2-mediated cytotoxic metabolism: an endpoint/toxic response where data from multiple chemicals converge to support a mode of action, *Regul. Toxicol. Pharmacol.*, 2009, **55**, 205–218.

52. K. M. Abdo, S. Grumbein, B. J. Chou and R. Herbert, Toxicity and carcinogenicity study in F344 rats following 2 years of whole-body exposure to naphthalene vapors, *Inhalation Toxicol.*, 2001, **13**, 931–950.
53. P. H. Long, R. A. Herbert, J. C. Peckham, S. L. Grumbein, C. C. Shackelford and K. Abdo, Morphology of nasal lesions in F344/N rats following chronic inhalation exposure to naphthalene vapors, *Toxicol. Pathol.*, 2003, **31**, 655–664.
54. K. T. Bogen, J. M. Benson, G. S. Yost, J. B. Morris, A. R. Dahl, H. J. Clewell, 3rd, K. Krishnan and C. J. Omiecinski, Naphthalene metabolism in relation to target tissue anatomy, physiology, cytotoxicity and tumorigenic mechanism of action, *Regul. Toxicol. Pharmacol.*, 2008, **51**, S27–36.
55. R. M. Baldwin, M. A. Shultz and A. R. Buckpitt, Bioactivation of the pulmonary toxicants naphthalene and 1-nitronaphthalene by rat CYP2F4, *J. Pharmacol. Exp. Ther.*, 2005, **312**, 857–865.
56. H. J. Clewell, A. Efremenko, J. L. Campbell, D. E. Dodd and R. S. Thomas, Transcriptional responses in the rat nasal epithelium following subchronic inhalation of naphthalene vapor, *Toxicol. Appl. Pharmacol.*, 2014, **280**, 78–85.
57. D. E. Dodd, B. A. Wong, E. A. Gross and R. A. Miller, Nasal epithelial lesions in F344 rats following a 90-day inhalation exposure to naphthalene, *Inhalation Toxicol.*, 2012, **24**, 70–79.
58. J. E. Klaunig, M. A. Babich, K. P. Baetcke, J. C. Cook, J. C. Corton, R. M. David, J. G. DeLuca, D. Y. Lai, R. H. McKee, J. M. Peters, R. A. Roberts and P. A. Fenner-Crisp, PPARalpha agonist-induced rodent tumors: modes of action and human relevance, *Crit. Rev. Toxicol.*, 2003, **33**, 655–780.
59. D. Y. Lai, Rodent carcinogenicity of peroxisome proliferators and issues on human relevance, *J. Environ. Sci. Health, Part C: Environ. Carcinog. Ecotoxicol. Rev.*, 2004, **22**, 37–55.
60. A. E. Mercer, S. L. Regan, C. M. Hirst, E. E. Graham, D. J. Antoine, C. A. Benson, D. P. Williams, J. Foster, J. G. Kenna and B. K. Park, Functional and toxicological consequences of metabolic bioactivation of methapyrilene *via* thiophene S-oxidation: Induction of cell defence, apoptosis and hepatic necrosis, *Toxicol. Appl. Pharmacol.*, 2009, **239**, 297–305.
61. M. L. Cunningham, L. L. Pippin, N. L. Anderson and M. L. Wenk, The hepatocarcinogen methapyrilene but not the analog pyrilamine induces sustained hepatocellular replication and protein alterations in F344 rats in a 13-week feed study, *Toxicol. Appl. Pharmacol.*, 1995, **131**, 216–223.
62. J. F. Waring, R. G. Ulrich, N. Flint, D. Morfitt, A. Kalkuhl, F. Staedtler, M. Lawton, J. M. Beekman and L. Suter, Interlaboratory evaluation of rat hepatic gene expression changes induced by methapyrilene, *Environ. Health Perspect.*, 2004, **112**, 439–448.
63. L. S. Birnbaum, The mechanism of dioxin toxicity: relationship to risk assessment, *Environ. Health Perspect.*, 1994, **102**(Suppl. 9), 157–167.

64. R. Pohjanvirta and J. Tuomisto, Short-term toxicity of 2,3,7,8-tetrachlorodibenzo-p-dioxin in laboratory animals: effects, mechanisms, and animal models, *Pharmacol. Rev.*, 1994, **46**, 483–549.
65. A. B. Okey, An aryl hydrocarbon receptor odyssey to the shores of toxicology: the Deichmann Lecture, International Congress of Toxicology-XI, *Toxicol. Sci.*, 2007, **98**, 5–38.
66. R. A. Budinsky, E. L. LeCluyse, S. S. Ferguson, J. C. Rowlands and T. Simon, Human and rat primary hepatocyte CYP1A1 and 1A2 induction with 2,3,7,8-tetrachlorodibenzo-p-dioxin, 2,3,7,8-tetrachlorodibenzofuran, and 2,3,4,7,8-pentachlorodibenzofuran, *Toxicol. Sci.*, 2010, **118**, 224–235.
67. E. A. Carlson, C. McCulloch, A. Koganti, S. B. Goodwin, T. R. Sutter and J. B. Silkworth, Divergent transcriptomic responses to aryl hydrocarbon receptor agonists between rat and human primary hepatocytes, *Toxicol. Sci.*, 2009, **112**, 257–272.
68. C. A. Flaveny, I. A. Murray and G. H. Perdew, Differential gene regulation by the human and mouse aryl hydrocarbon receptor, *Toxicol. Sci.*, 2010, **114**, 217–225.
69. P. C. Boutros, R. Yan, I. D. Moffat, R. Pohjanvirta and A. B. Okey, Transcriptomic responses to 2,3,7,8-tetrachlorodibenzo-p-dioxin (TCDD) in liver: comparison of rat and mouse, *BMC Genomics*, 2008, **9**, 419.
70. M. J. Aardema and J. T. MacGregor, Toxicology and genetic toxicology in the new era of "toxicogenomics": impact of "-omics" technologies, *Mutat. Res.*, 2002, **499**, 13–25.
71. NRC, *Applications of Toxicogenomic Technologies to Predictive Toxicology and Risk Assessment*, National Academy of Sciences Press, Washington, D.C., 2007.
72. R. S. Thomas, H. J. Clewell, 3rd, B. C. Allen, S. C. Wesselkamper, N. C. Wang, J. C. Lambert, J. K. Hess-Wilson, Q. J. Zhao and M. E. Andersen, Application of transcriptional benchmark dose values in quantitative cancer and noncancer risk assessment, *Toxicol. Sci.*, 2011, **120**, 194–205.
73. M. Razzaghi and D. W. Gaylor, On the correlation coefficient between the TD50 and the MTD, *Risk Anal.*, 1996, **16**, 107–113.
74. D. Krewski, D. W. Gaylor, A. P. Soms and M. Szyszkowicz, An overview of the report: correlation between carcinogenic potency and the maximum tolerated dose: implications for risk assessment, *Risk Anal.*, 1993, **13**, 383–398.
75. NRC, *Toxicity Testing in the 21st Century: A Vision and a Strategy*, National Research Council of the National Academies, Washington, D.C., 2007.
76. F. S. Collins, G. M. Gray and J. R. Bucher, Toxicology. Transforming environmental health protection, *Science*, 2008, **319**, 906–907.
77. R. S. Thomas, H. J. Clewell, 3rd, B. C. Allen, L. Yang, E. Healy and M. E. Andersen, Integrating pathway-based transcriptomic data into quantitative chemical risk assessment: a five chemical case study, *Mutat. Res.*, 2012, **746**, 135–143.

78. R. S. Thomas, S. C. Wesselkamper, N. C. Wang, Q. J. Zhao, D. D. Petersen, J. C. Lambert, I. Cote, L. Yang, E. Healy, M. B. Black, H. J. Clewell, 3rd, B. C. Allen and M. E. Andersen, Temporal concordance between apical and transcriptional points of departure for chemical risk assessment, *Toxicol. Sci.*, 2013, **134**, 180–194.
79. D. E. Dodd, L. J. Pluta, M. A. Sochaski, K. A. Funk and R. S. Thomas, Subchronic hepatotoxicity evaluation of 1,2,4-tribromobenzene in Sprague-Dawley rats, *Int. J. Toxicol*, 2012, **31**, 250–256.
80. G. P. Carlson and R. G. Tardiff, Effect of 1,4-dibromobenzene and 1,2,4-tribromobenzene on xenobiotic metabolism, *Toxicol. Appl. Pharmacol.*, 1977, **42**, 189–196.
81. L. S. Gold, N. B. Manley, T. H. Slone and J. M. Ward, Compendium of chemical carcinogens by target organ: results of chronic bioassays in rats, mice, hamsters, dogs, and monkeys, *Toxicol. Pathol.*, 2001, **29**, 639–652.
82. H. H. Li, D. R. Hyduke, R. Chen, P. Heard, C. L. Yauk, J. Aubrecht and A. J. Fornace Jr., Development of a toxicogenomics signature for genotoxicity using a dose-optimization and informatics strategy in human cells, *Environ. Mol. Mutagen.*, 2015, **56**(6), 505–519.
83. I. Cote, P. T. Anastas, L. S. Birnbaum, R. M. Clark, D. J. Dix, S. W. Edwards and P. W. Preuss, Advancing the next generation of health risk assessment, *Environ. Health Perspect.*, 2012, **120**, 1499–1502.
84. N. L. Chepelev, I. D. Moffat, S. Labib, J. Bourdon-Lacombe, B. Kuo, J. K. Buick, F. Lemieux, A. I. Malik, S. Halappanavar, A. Williams and C. L. Yauk, Integrating toxicogenomics into human health risk assessment: Lessons learned from the benzo[a]pyrene case study, *Crit. Rev. Toxicol.*, 2015, **45**, 44–52.
85. I. Moffat, N. L. Chepelev, S. Labib, J. Bourdon-Lacombe, B. Kuo, J. K. Buick, F. Lemieux, A. Williams, S. Halappanavar, A. I. Malik, M. Luijten, J. Aubrecht, D. R. Hyduke, A. J. Fornace Jr., C. D. Swartz, L. Recio and C. L. Yauk, Comparison of toxicogenomics and traditional approaches to inform mode of action and points of departure in human health risk assessment of benzo[a]pyrene in drinking water, *Crit. Rev. Toxicol.*, 2015, **45**, 1–43.
86. A. F. Jackson, A. Williams, L. Recio, M. D. Waters, I. B. Lambert and C. L. Yauk, Case study on the utility of hepatic global gene expression profiling in the risk assessment of the carcinogen furan, *Toxicol. Appl. Pharmacol.*, 2014, **274**, 63–77.
87. R. S. Thomas, M. A. Philbert, S. S. Auerbach, B. A. Wetmore, M. J. Devito, I. Cote, J. C. Rowlands, M. P. Whelan, S. M. Hays, M. E. Andersen, M. E. Meek, L. W. Reiter, J. C. Lambert, H. J. Clewell, 3rd, M. L. Stephens, Q. J. Zhao, S. C. Wesselkamper, L. Flowers, E. W. Carney, T. P. Pastoor, D. D. Petersen, C. L. Yauk and A. Nong, Incorporating new technologies into toxicity testing and risk assessment: moving from 21st century vision to a data-driven framework, *Toxicol. Sci.*, 2013, **136**, 4–18.

88. EPA, *Risk Assessment Guidance for Superfund Volume I: Human Health Evaluation Manual (Part A)*, U.S. Environmental Protection Agency, Washington, D.C., 1989.
89. R. Judson, A. Richard, D. J. Dix, K. Houck, M. Martin, R. Kavlock, V. Dellarco, T. Henry, T. Holderman, P. Sayre, S. Tan, T. Carpenter and E. Smith, The toxicity data landscape for environmental chemicals, *Environ. Health Perspect.*, 2009, **117**, 685–695.
90. W. Slikker Jr., M. E. Andersen, M. S. Bogdanffy, J. S. Bus, S. D. Cohen, R. B. Conolly, R. M. David, N. G. Doerrer, D. C. Dorman, D. W. Gaylor, D. Hattis, J. M. Rogers, R. W. Setzer, J. A. Swenberg and K. Wallace, Dose-dependent transitions in mechanisms of toxicity: case studies, *Toxicol. Appl. Pharmacol.*, 2004, **201**, 226–294.
91. R. B. Conolly, B. D. Beck and J. I. Goodman, Stimulating research to improve the scientific basis of risk assessment, *Toxicol. Sci.*, 1999, **49**, 1–4.
92. S. Safe, Molecular biology of the Ah receptor and its role in carcinogenesis, *Toxicol. Lett.*, 2001, **120**, 1–7.
93. W. D. Kerns, K. L. Pavkov, D. J. Donofrio, E. J. Gralla and J. A. Swenberg, Carcinogenicity of formaldehyde in rats and mice after long-term inhalation exposure, *Cancer Res.*, 1983, **43**, 4382–4392.
94. R. J. Preston, Cancer risk assessment for 1,3-butadiene: data integration opportunities, *Chem.–Biol. Interact.*, 2007, **166**, 150–155.
95. G. G. Xiao, M. Wang, N. Li, J. A. Loo and A. E. Nel, Use of proteomics to demonstrate a hierarchical oxidative stress response to diesel exhaust particle chemicals in a macrophage cell line, *J. Biol. Chem.*, 2003, **278**, 50781–50790.
96. A. Nel, T. Xia, L. Madler and N. Li, Toxic potential of materials at the nanolevel, *Science*, 2006, **311**, 622–627.
97. U. Alon, Network motifs: theory and experimental approaches, *Nat. Rev. Genet*, 2007, **8**, 450–461.
98. U. Alon, *An Introduction to Systems Biology: Design Principles of Biological Circuits*, Chapman & Hall, Boca Raton, FL, 2006.

CHAPTER 6

Using Transcriptomics to Evaluate Thresholds in Genotoxicity Dose–Response

PATRICK D. McMULLEN,[a] SALIL PENDSE,[a] YEYEJIDE ADELEYE,[b] PAUL L. CARMICHAEL,[b] MELVIN E. ANDERSEN[a] AND REBECCA A. CLEWELL*[c]

[a] The Hamner Institutes for Health Sciences, Research Triangle Park, NC, USA; [b] Unilever, Safety and Environmental Assurance Centre, UK; [c] ScitoVation, 6 Davis Drive, Research Triangle Park, NC 27709, USA
*Email: rclewell@scitovation.com

6.1 Introduction

6.1.1 Challenges in Assessing the Safety of Potential Genotoxicants

Quantitative understanding of the dose–response relationship for chemical toxicity underpins risk assessment. Yet, extrapolation from high-dose animal studies to lower doses at which human exposures occur remain one of the greater challenges in interpreting toxicity studies. Tools such as benchmark dose (BMD) model the data, predict the more likely shape of the curve, and, because of their ability to take multiple dose levels into account, often improve analyses over the classical no observed adverse effect level or lowest adverse effect level (LOAEL) approaches. However, for genotoxic compounds, these tools are not widely used. The default assumption is a

Issues in Toxicology No. 28
Toxicogenomics in Predictive Carcinogenicity
Edited by Michael D. Waters and Russell S. Thomas

Published by the Royal Society of Chemistry, www.rsc.org

low-dose linear response, wherein genotoxic agents are considered to induce DNA damage proportional to the administered dose below the lowest tested dose.[1] This practice is based on the assumption that even a single molecule of a DNA reactive chemical may cause a mutation and thereby increase the risk for cancer.[2,3] However, this assumption does not consider nonlinearities resulting from biokinetic processes and cellular defense mechanisms.

In recent years, several genotoxic chemicals have been reported to produce threshold-shaped dose–response curves for mutation and genotoxicity assays, both *in vivo* and *in vitro*.[4–8] These data point to the possibility that the shape of the dose–response curve may be more complex than the default assumption of linearity—a fact that could dramatically change estimated points of departure for chemicals identified as potential genotoxicants. However, there is much debate surrounding the possibility of threshold-shaped dose–response curves. Some concerns are a result of uncertainty in sensitivity of the assays or the fact the few studies account for metabolism (bioactivation or bioinactivation). Other criticisms center on statistical analyses and the ability to fit a linear curve to the available data. Nonetheless, it is clear that statistical methods alone cannot either prove or disprove the existence of a threshold. Rather, to truly define the shape of the dose–response curves, we must look to the underlying biology and develop targeted experiments that identify and measure the key processes governing cellular response at low doses.

Based on the overwhelming consensus of the biological literature, a biological pathway for DNA damage can be constructed as follows: (1) DNA damage occurs through chemical interaction with DNA or off-target effects in the cell that lead to DNA lesions; (2) DNA damage is recognized by sensor kinases such as ATM, ATR, or DNA-PK; (3) these kinases activate a variety of response proteins including p53 and MDM2 *via* phosphorylation; (4) phosphorylation of p53 and MDM2 disrupts the p53–MDM2 complex in the cytosol and allows translocation of p53 to the nucleus; which (5) leads to both assembly of DNA repair machinery and tetramerization and transcriptional activation of p53 (Figure 6.1). Transcriptional responses of p53—the so-called guardian of the genome—include activation of cell cycle arrest, DNA repair, and—in the event of irreparable damage–apoptosis. To the extent that this representation of cellular response to DNA damage is accurate, the transcriptional response to DNA-damaging compounds should inform the success or failure of the cell to respond to DNA damage stress, and thereby affect the shape of the dose–response curve.

Gene expression changes have proven to be reasonable predictors for results of classical apical endpoints *in vivo* (*i.e.* the 2 year rodent bioassay).[9] Affordable, high-throughput transcriptomic technologies have increased our ability to evaluate cellular response to chemical stress at reasonable cost. Taking advantage of this capability, we evaluated the relationship between transcriptional response and the shape of the dose–response curves for genotoxicity using several DNA damaging chemicals *in vitro*. For these studies, whole-genome transcriptomic responses were evaluated using gene

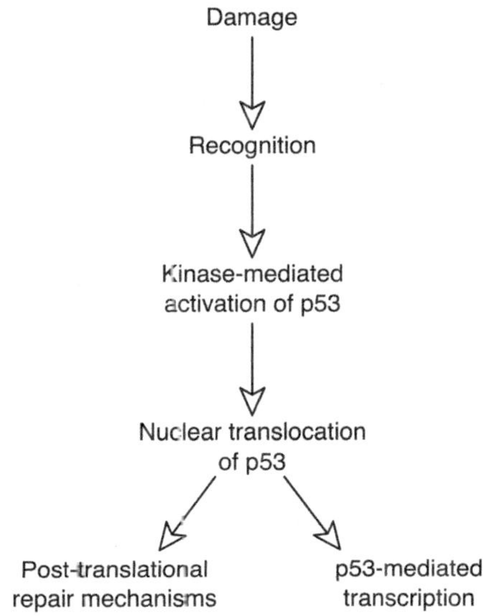

Figure 6.1 Generalized DNA damage repair pathway. Damage sensors, signal transducers, and repair machinery depend on the nature of the damage, but this structure is common across many types of DNA damage.

array technologies, and micronuclei (whole or fragments of DNA lost during replication) were used as a measure of permanent (irreparable) DNA damage. Micronuclei are associated with increased DNA damage and the Organisation for Economic Co-operation and Development guideline micronucleus *in vitro* and *in vivo* assays are widely used for genotoxicity screening efforts.[10] Prototype compounds were selected to represent different mechanisms of DNA damage, including alkylation and adduct formation, oxidative damage, topoisomerase II inhibition, DNA cross-linking, and damage similar to γ-irradiation (Table 6.1). Two approaches were taken to evaluating the role of transcriptional response in preventing micronucleus induction: (1) comparison of gene signatures for all of the prototype compounds at a single chemical dose that led to a similar activation of the p53-DNA damage pathway (*i.e.*, 1.5-fold increase in total p53); and (2) in-depth dose–response evaluation for a subset of chemicals. The goal of these efforts was to determine if these transcriptional pathways were responsible for maintaining homeostasis at low levels of DNA damage.

6.1.2 Understanding the Nature of the Threshold-shaped Dose–Response Curve

Several possible methodological explanations exist for the apparent threshold in micronucleus response. The micronucleus assay is often criticized for its lack of sensitivity at low doses. This observation, coupled with the nonzero background rate of micronucleus formation in most cell types, makes it

Table 6.1 Compounds evaluated for micronucleus and transcriptomic response.

Chemical	Type of damage	Primary repair mechanisms
Bromate	Oxidative damage	Base excision repair
Hydrogen peroxide	Oxidative damage	Base excision repair
Quercetin	Oxidative damage	Base excision repair
N-ethyl-*N*-nitrosourea	DNA adducts	Base excision repair
Glycidamide	DNA adducts	Base excision repair
Methyl methanesulfonate	DNA adducts	Base excision repair
Curcumin	Oxidative damage (topoisomerase II inhibition)	Base excision repair/non-homologous end joining
Etoposide	Topoisomerase II inhibition	Non-homologous end joining
Mitomycin C	Cross-linking	Homologous recombination?
Neocarzinostatin	DNA-specific oxidative damage	Non-homologous end joining

difficult to distinguish small effects in micronucleus frequency. To confound this further, cell-based assays are subject to kinetic effects arising from degradation and non-specific binding in culture, along with effects from the cells themselves (*i.e.*, metabolism and clearance).[11] As such, the nominal concentration used for the assay may differ substantially from the concentration in the cell. If this difference is nonlinear, it can lead to the appearance of a threshold response. These processes, which may in fact lead to a threshold-shaped response curve do not support an altered approach to risk assessment practices for genotoxic compounds because they do not require a true nonlinearity in the concentration–response relationship. Rather, they represent either a lack of understanding of the dose or a failure of the assay to obtain required sensitivity. As such, they are often cited as reasons for exercising caution when interpreting data that appear to have a threshold-like shape. Further complicating this issue is the fact that differentiating between linear and threshold concentration–response functions *via* statistics is confounded by the high degree of variability in the micronucleus assay.

Beyond explanations of thresholds that center on methodological limitations, however, are the biologically driven explanations for how a cell might maintain homeostasis at low levels of genotoxic stress, leading to concentration–response relationships with real thresholds. Many stress-responsive pathways share a common architecture that involves a sensor coupled to a transcription factor.[12] The transcriptional response includes machinery for adapting to the stress and repairing its consequences. Genotoxic damage is detected by sensors, including ATR for adduct formation and single strand breaks, and ATM for double-strand breaks. These kinases drive phosphorylation and transcriptional activation of p53 leading to protective cellular responses.[13] This adaptive response constitutes a negative feedback that leads to the triage and repair of DNA damage. It is plausible that this response has a finite capacity for repair, which would lead to qualitatively different response for damage levels that saturate repair machinery.

Additionally, post-translational processes from existing repair machinery could produce a threshold dose response. Cells respond to genotoxic stress on the order of seconds or minutes by forming DNA repair centers.[14–19] These post-translational processes (*i.e.* processes that do not require a transcriptional response) have been described for many indirect mechanisms of DNA damage (alkylation, adduct formation, and oxidative stress). This mechanism—a rapid post-translational response with limited capacity coupled with a slower transcriptional response—can lead to a systems-control phenomenon known as *perfect adaptation*.[20,21] These control mechanisms, if tuned correctly, produce a true threshold response in which there is no change in permanent damage throughout a region of concentration.[22]

6.2 Methods for Evaluating Low-dose Genotoxicity

The 2007 National Research Council report, *Toxicity Testing in the 21st Century: a Vision and a Strategy*, proposed that better understanding of the structure and dynamics of toxicity pathways could lead to better risk-assessments. Following this approach, dose–response behaviors for *in vitro* model systems could be predicted based on the structure of the cellular circuitry the dynamics of these circuits. These computational systems biology pathway models are potentially powerful tool for characterizing the concentration–response relationship for genotoxicants at low doses.

In the absence of biologically grounded models that describe and predict concentration response, we have to rely on statistical models evaluating the empirical data sets. These models, however, suffer from experimental variability, measurement errors, and other stochastic factors that complicate the process of evaluating the shape of the response curve near experimental control or background levels. Essentially, the issue is characterization of the slope of the concentration–response function near the origin. A positive slope near zero indicates that an incremental increase in concentration above background levels leads to an incremental increase in response. This behavior can be consistent with the assumption of low-dose linearity. A zero-slope function implies that there is a range of concentrations where there is exactly zero response—that there is no change in response with increasing concentration. If the assay results serve as a basis for extrapolation to a human risk-assessment, this latter behavior is consistent with the low-dose nonlinear model commonly used to evaluate non-genotoxic carcinogens. The third possibility—a negative slope near as concentration nears zero—is representative of hormesis/non-monotonicity.[24]

Lutz and Lutz proposed a pragmatic approach for evaluating whether dose–response curves have thresholds.[25] Here, two alternative mathematical models are parameterized and evaluated against the data. The first option is a linear model. The second model option has a threshold dose, below which the concentration response is flat, and above which the concentration is linear. This type of piecewise linear function is also known as a hockey-stick model. The two models are compared using the F-test, a statistic that balances goodness-of-fit against the number of parameters.

While the Lutz model represents an important step forward in creating a standardized test for thresholds, there remain a number of challenges in interpreting the results of the Lutz model for toxicological data. First, the result can depend strongly on the statistical power of the experiment (namely, experimental variability and number of replicates). Second, the resulting *p*-value is a direct comparison between the linear and threshold models. If neither of these models captures the concentration–response relationship, the evaluation of which is a better fit is immaterial. If the data are better described by a higher-order polynomial or a Hill function, the Lutz framework (in its current implementation) is oblivious to this. If the higher concentrations used to formulate the concentration–response curve cause cytotoxicity (for *in vitro* models) or systemic toxicity (for *in vivo* models), these processes need to be captured since they can confound the analysis. Concentrations causing nonspecific effects really need to be omitted. For these reasons, the authors provide the following word of caution in the application of the statistical model: "If the hockey stick model fits the data significantly better than linearity, the threshold-like appearance of the dose–response curve will have to be corroborated by mechanistic considerations and experimental testing of the respective hypothesis."[25] In the same vein, the lack of a statistical significance for the threshold model does not prove a linear concentration–response relationship. Thus, the only way to move toward a better understanding of concentration response is by developing a more in depth quantitative understanding of the underlying biology for DNA damage response.

6.3 Micronucleus Dose Response

The micronucleus assay is a widely used screening tool for identifying potential genotoxic compounds. Micronuclei are formed when pieces of a chromosome or entire chromosomes are not incorporated into the daughter nuclei during mitosis.[26] The presence of micronuclei becomes a measure of permanent (irreparable) damage to the DNA. The relative ease and high-throughput capability of this assay compared with other mutation and genotoxicity endpoints has led to widespread use in screening.[10] However, the assay measures rare events, which inevitably leads to concerns about the sensitivity at low doses and the interpretation of threshold-shaped dose–response curves. Furthermore, while the micronucleus assay is highly sensitive to double-strand break agents and spindle poisons, it may not be as sensitive to other forms of DNA damage—such as alkylation—that require comparatively more damage to cause micronuclei formation. Thus, we again face issues of biological *vs.* methodological thresholds.

In our studies, we attempted to determine whether the apparent threshold concentration–response curves seen in the micronucleus assay could be explained by the biology of the DNA damage response signaling pathway. We evaluated micronucleus induction for several compounds that induce different types of DNA damage (Table 6.1) in human cells that express wild-type p53 (HT-1080). These cells are commonly used to evaluate the p53 network and, in our laboratory, we have shown that they demonstrate appropriate p53 transcriptional response to DNA damaging compounds, including cell cycle arrest and apoptosis.[8,27] Both the magnitude of response and the shape of the dose–response curve varied across the compounds (Figure 6.2).

The potential for threshold-like behavior in the micronuclei response curves was evaluated using a the Lutz and Lutz model to test whether a threshold model is more likely than a linear model.[25] While this analysis does not "prove" threshold behavior, a p-value <0.05 indicates that the dose–response curve is nonlinear and is likely to exhibit threshold behavior. Chemicals with statistically significant deviations from linearity were curcumin, hydrogen peroxide, neocarzinostatin, methyl methanesulfonate, and quercetin. Below, we describe our approaches to identifying potential transcriptional mechanisms that could provide the biological underpinning for the observed threshold-shaped concentration–response curves.

6.4 Comparing Gene Signatures for Chemicals Causing Different Types of DNA Damage

Genotoxic stressors fall into distinct classes based on the type of damage and the mechanism by which specific chemicals cause DNA damage. The varying nature of the DNA damage requires the cell to mount different types of cytoprotective responses. For example, ultraviolet radiation causes thymine dimers that require removal *via* nucleotide excision repair. Ionizing radiation causes double-strand breaks that require a qualitatively different repair mechanism: nonhomologous end joining. Since we have these disparate DNA repair responses, we would expect that different damage mechanisms should lead to different patterns of gene expression, with agents causing damage by similar mechanisms producing a more conserved transcriptomic response.

Toxicogenomic responses have been used successfully to categorize responses to developmental toxicants,[28–30] and a number of approaches exist for evaluating similarities and differences in toxicogenomics responses across chemical groups.[31] Here, we considered the transcriptomic response to a set of mechanistically diverse genotoxicants, covering (1) reactive oxygen species generation; (2) adduct formation; (3) double-strand breaks; and (4) cross-linking (Table 6.1). We determined equipotent concentrations of the compounds by characterizing the degree to which each compound induced total p53 concentrations in HT1080 cells. We then used gene expression microarrays ($n = 4–5$ for each compound) to assess gene

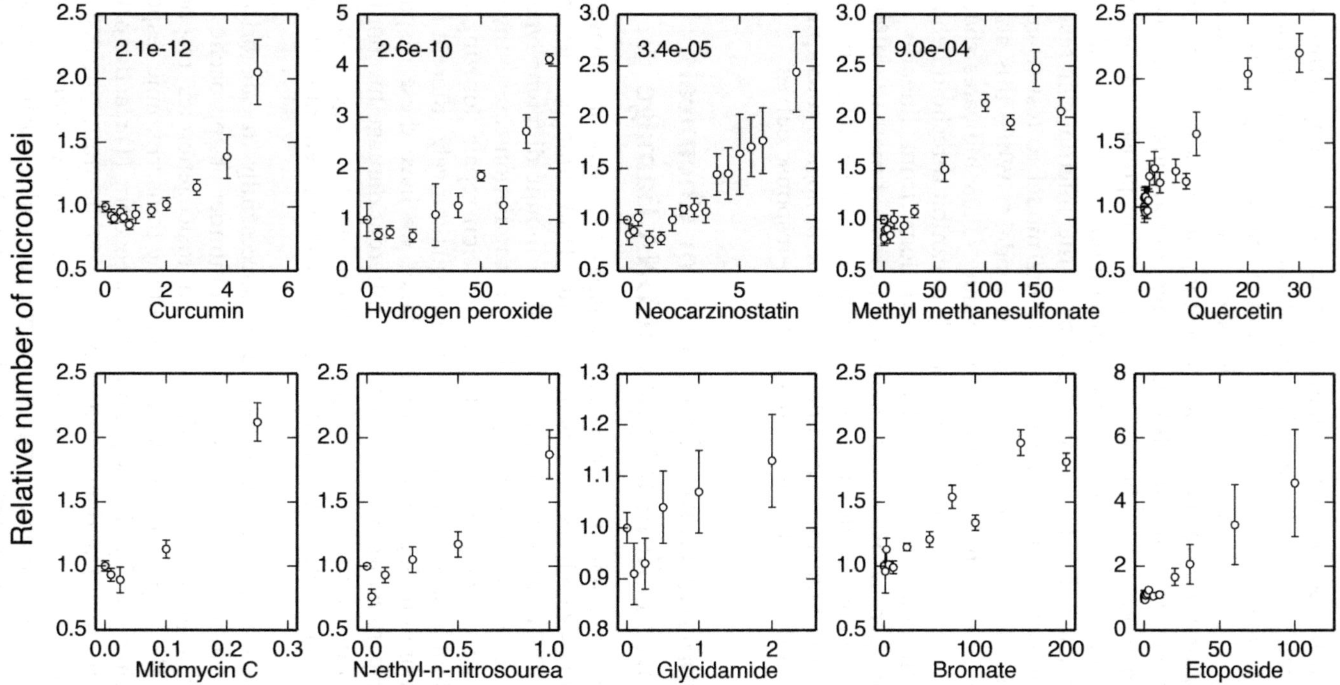

Figure 6.2 Micronucleus response to prototypical DNA-damaging compounds. The numbers of micronuclei per parent cell relative to vehicle controls (mean ± SEM) are plotted against compound concentration. Each concentration response was tested against the Lutz threshold model.[25] Shaded plots denote concentration response data for which the threshold model is a better fit than the linear model, significant at the indicated *p*-value.

expression changes. These data, along with a comprehensive description of our materials and methods, are publically available in the National Center for Biotechnology Information Gene Expression Omnibus (GSE74725).

A number of teams have identified suites of genes that serve as transcriptional biomarkers of genotoxicity.[9,13,32–35] We previously established a core response of damage mechanism-independent genes. This core set is composed of genes that (1) are transcriptionally regulated through direct binding by p53; and (2) are common to the transcriptional response to etoposide, methyl methanesufonate, and quercetin—compounds that cause damage by three different mechanisms (Figure 6.3A). Functional analysis of these genes revealed expected Gene Ontology categories, including "regulation of apoptosis" and "response to DNA damage stimulus" (hypergeometric test; FDR-corrected $p<0.01$). We then mapped the categories significantly associated with the core response genes onto a network defined by the Gene Ontology structure to better understand the relationships between them (Figure 6.3B, see McMullen *et al.*[36] for details). Our hypothesis is that this core suite of p53-regulated genes is part of a universal DNA-repair response that does not depend on damage mechanism or repair processes. Other processes occur that are specific to the nature of the damage/chemical. These secondary pathways include machinery for addressing damage to cellular components other than DNA, such as oxidative stress, inflammation, and other off-target (*i.e.* non-p53-mediated) effects.

Toxicogenomics can be used to determine the mechanism of genotoxic damage and the type of repair processes implemented (see Chapters 3 and 4). Here, we plotted the response to the 10 compounds of several classes of genes (Figure 6.4). The expression change of the core p53 response is conserved across the 10 compounds. To quantify the similarity of this response across the compounds, we computed the correlation of the expression changes for each pair of compounds; the mean correlation coefficient for the core response is 0.71, indicating that this response is extremely similar across compounds with different mechanisms for DNA damage. To understand whether the genes in the core p53 response could be expanded to include genes with similar functions, we identified the genes differentially expressed in response to at least one compound, and considered the response for genes annotated with the same categories as the core response (Figure 6.4, "stress response/apoptosis"). The correlation across compounds was much lower for these genes (0.37), indicating aspects of the p53-mediated damage response that are specific to the type of damage. The correlation for the remaining differentially expressed was 0.37. Some of these transcriptional changes may be due to off-target reactivity.

Clustering algorithms can be used to identify groups of compounds that elicit similar gene expression profiles. Here, we used hierarchical clustering (cosine distance and complete linkage), applied over the entire transcriptome, to ascertain whether compounds with similar modes of action elicit similar responses. DNA adduct-forming compounds (*N*-ethyl-*N*-nitrosourea, methyl methanesufonate, and glycidamide) cluster together.

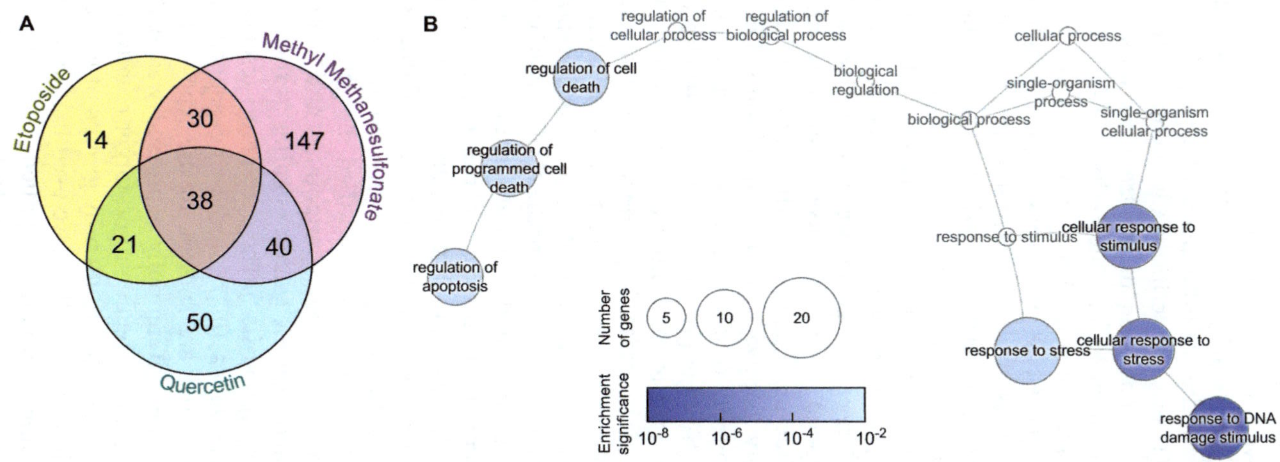

Figure 6.3 The core p53 response. Compounds with different mechanisms of DNA damage cause transcriptional regulation that involves both a core response agnostic to mechanism and a peripheral response specific to the compound—or perhaps the mode of action. (A) Here, we identified 38 genes that are differentially expressed in response to etoposide, quercetin, and methyl methanesulfonate. (B) We used overrepresentation analysis to identify Gene Ontology categories represented by the core response. The categories are organized according to the structure of the Gene Ontology.
Reprinted with permission.[8]

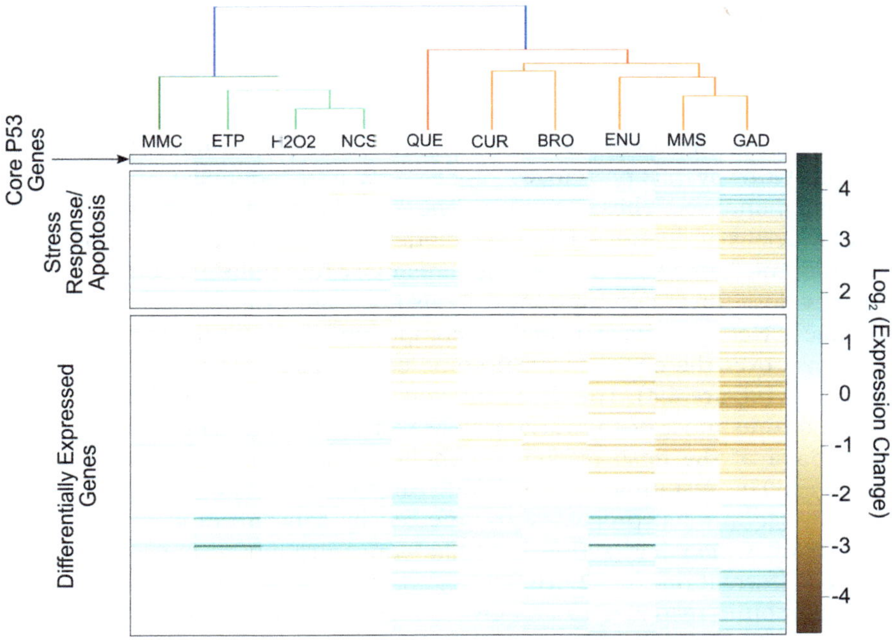

Figure 6.4 Genotoxic compounds with similar modes of action invoke similar gene expression responses. Gene expression changes for treatment of HT-1080 cells with 10 DNA damaging agents. Rows represent genes that fall into various categories pertaining to DNA damage response. The "core p53 genes" are a universal response that we defined previously (Figure 6.3A).[8] "Stress response/apoptosis" are genes annotated with the same Gene Ontology categories that are enriched in the core p53 genes (Figure 6.3B). The balance of the matrix is composed of genes that are differentially expressed in response to one or more of the compounds. Compounds have been organized according to their transcriptional responses across the entire genome by hierarchical clustering.

Oxidative stressors form two clades, with hydrogen peroxide and neocarzinostatin forming one group, and curcumin and bromate forming another. Interestingly, the changes in expression of genes related to the core response ("cell cycle/apoptosis") are more similar within mechanistic classes (*e.g.* glycidamide and methyl methanesulfonate) than outside of mechanistic classes (*e.g.* glycidamide and hydrogen peroxide). Our work highlighted the utility of using multiple "omic" data streams, as well as looking beyond simple classification of chemicals based on individual gene signatures, to evaluate and classify chemical activity by biological pathways. In this case, we eliminated off-target signals by focusing on genes directly regulated by the DNA damage responsive transcription factor p53 using published ChIP-seq data.[37–40] We then evaluated chemical similarity based on biological pathway rather than individual gene response, increasing confidence in the biological significance of the observed patterns.

6.5 Integrating Dose–Response Trends for Transcriptomic and Micronucleus Endpoints

In addition to performing transcriptomic dose–response analyses for several chemicals, we also evaluated dose–response data for various biomarkers of the p53 pathway (p53 protein, cell cycle arrest, and apoptosis). Six doses were evaluated and were chosen to represent those below and above the BMD for micronucleus induction. The Lutz model—used in Figure 6.2 to evaluate whether micronucleus concentration–response functions are consistent with the low-dose linear assumption—can also be used to assess gene expression changes. Here, we used principal component analysis to summarize the response over the entire transcriptome on a microarray into a single dose response (Figure 6.4). The first principal component, describing the vector that captures most of the variability across genes and conditions, captures the shape of the dose response without regard to the magnitude of the changes. The resulting "eigenvector" can be characterized to understand how transcriptional processes are affected by concentration. Interestingly, curcumin, which has a threshold-like concentration response for micronucleus formation ($p = 2\times10^{-12}$; Figure 6.2), also has a threshold-like toxicogenomic response ($p = 1\times10^{-16}$; Figure 6.5A). In contrast, etoposide has a linear concentration response for both micronucleus formation and gene expression (Figures 6.2 and 6.5B).

The correspondence of the dose–response behavior between these two endpoints suggested that the same upstream processes may be contributing to the concentration–response curve for transcriptional regulation and micronucleus formation. If this relationship could be better understood, the toxicogenomic endpoints might be suitable surrogates for genotoxic chemical risk assessment. However, more work is needed before this observation can be generalized: while we observed concordance between these two endpoints for some compounds, with others they were not in close agreement. For example, methyl methanesulfonate yielded a threshold-like concentration response for micronuclei, but the transcriptomic response could not be distinguished from linearity. Nonetheless, the similarity in the concentration–response curves for micronucleus and transcriptomic response is intriguing.

We further evaluated the role of transcriptional activation of p53 in micronucleus outcome by evaluating the in-depth dose response for several phenotypic cellular responses that require a transcriptional response, as well as measures of p53 activation—phospho-p53 (serine 15). Based on the conventional understanding of the ability of p53 to cause cell cycle arrest apoptosis to prevent permanent DNA damage, we expected that transcriptomic activation and other protective responses (cell cycle arrest, and apoptosis) would occur at lower doses. Only at doses that exceeded the ability of the cell to maintain cell cycle control would we expect to see increases in the measure of permanent (heritable) DNA damage. However, we consistently saw very different dose–response behaviors for these markers (Figure 6.6).

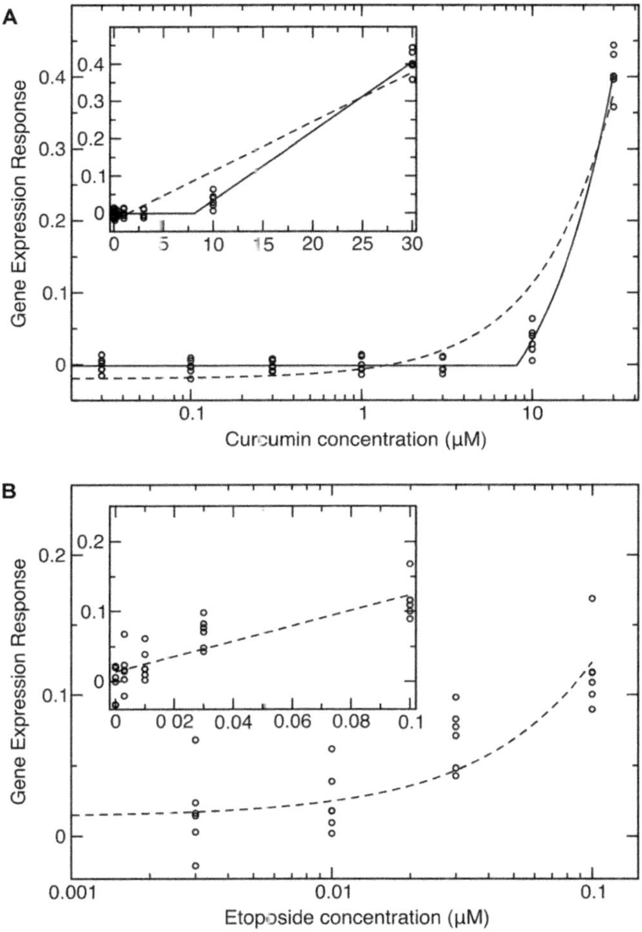

Figure 6.5 Concentration response of genome-wide expression changes can be summarized using principal component analysis. The first principal component captures the predominant source of response variance in the experiment; typically this reflects the concentration response. Here we plot the first principal component (arbitrary units) as a function of concentration for (A) curcumin, which has an apparent threshold in the micronucleus response, and (B) etoposide, which does not. The Lutz threshold model (solid trace) reflects the curcumin transcriptional response better than the linear model (dashed trace, $p = 1 \times 10^{-16}$, F-test). For etoposide, the concentration response predicted by the two models is similar, so the threshold model is not shown. Insets include the same data on a linear scale for comparison.

To compare the dose response for transcriptional activation to the cellular endpoints, we used BMD analysis. BMD estimates the point-of-departure for experimental data by fitting various concentration–response functions. The function that best reflects the data is used to analytically determine the

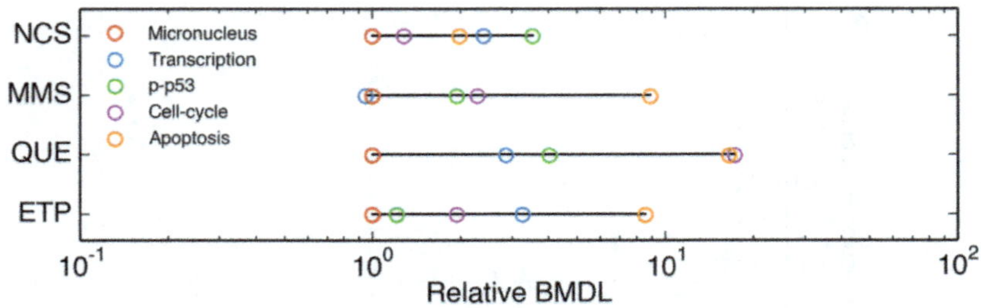

Figure 6.6 Relationship between the benchmark doses (BMD) for several DNA damage endpoints for neocarzinostatin (NCS), methyl methanesufonate (MMS), quercetin (QUE), and etoposide (ETP). Because effective concentrations of these compounds vary widely, BMDL (lower confidence limit of BMD) concentrations (horizontal axis) are expressed relative to the BMD for micronucleus formation. For all four compounds, micronucleus induction (red) is the most sensitive—or nearly so—endpoint. With the exception of MMS, coordinated transcriptional responses do not occur at concentrations that cause micronuclueus formation.

concentration that reflects an increase of one standard deviation above baseline. This approach has advantages over the traditional LOAEL technique, namely that it is not as dependent on the concentrations selected for the experiment (provided the concentration–response function can be adequately characterized). For protein and cellular endpoints, the Environmental Protection Agency's BMD software was used. For the transcriptomic data, we used the BMDExpress software, which calculates the BMD for each gene and determines the average BMD for each Gene Ontology biological process.[41]

In all cases, the most sensitive pathways for the chemicals were not related to DNA damage or p53 activation (Figure 6.6). The gene categories with the lowest BMDs were "androstenedione and testosterone biosynthesis and metabolism", "immune response", "immune response (inflammatory)", "cytoskeleton remodeling", and "development" for curcumin, etoposide, methyl methanesulfonate (MMS), neocarzinostatin, and quercetin, respectively (unpublished data).[8] Apoptotic, cell cycle and DNA damage categories were not observed until higher doses (Figure 6.6). With all compounds, except MMS, the BMD for transcriptional activation was greater than that for micronuclei formation. With MMS the two endpoints had similar BMDs. To our surprise, there were significant increases in micronuclei at doses that did not affect p53 transcriptional response. Regardless of the chemical or the type of DNA damage, the p53 transcriptional response does not appear to be the operative process in preventing induction of micronuclei.

The lack of sensitivity of the gene expression response as a marker of DNA damage has also been observed in other systems.[42] Gene expression changes in a mouse lymphoma cell line were minimal at concentrations that produced a three-fold increase in micronuclei in response to cisplatin[35] or an increase in mutation frequency in response to bleomycin, hydrogen peroxide, or MMS.[43,44] Concentrations of benzo(*a*)pyrene diol that caused increased adduct formation in TK6 human lyphoblast cells did not induce a robust gene expression response.[45] Low sensitivity of gene expression experiments may be the result of measuring the average response for an effect that is heterogenous across populations of cells.[46] Similar observations have been made *in vivo*. Comparison of BMDs calculated from transcriptomic and classical apical endpoints for cancer in rodents showed a high degree of correlation for several carcinogenic compounds.[47,48]

When evaluating these studies, it is important to distinguish initial DNA damage (or lesions) and the downstream measures of irreparable damage or heritable mutation. Although often described interchangeably under the term "genotoxicity endpoint", on the cellular level, these endpoints represent different cellular events. The initial lesion measures chemical reactivity with DNA. The permanent damage or heritable mutation indicates the phenotypic outcome, occurring in the absence of efficient cellular repair. In comparing transcriptional responses to genotoxic endpoints, we must be careful to treat these initial stressors differently from downstream responses. Nonetheless, the various lines of evidence, both *in vitro* and *in vivo*, indicate that transcriptional response is generally a higher dose response, occurring at concentrations equal to—or greater than—those required to lead to permanent (irreparable damage). Thus, we need to look beyond transcriptional responses to determine the cellular defense processes that can produce threshold-type dose–response curves.

6.6 Homeostasis, Transcriptional Regulation, and Post-translational Activity

Transcriptional responses help cells respond to perturbations, but are resource intensive and slow. Cells have post-translational mechanisms that maintain homeostasis in the face of transient stressors.[49] These post-translational processes use pre-existing proteins, ensuring rapid response to transient insults. The composite signaling motifs described previously–consisting of both post-translational modifications and transcriptional upregulation—are highly conserved stress-response motifs, including the DNA damage response. In response to genotoxicity, the cells rely on a variety of post-translational modifications to resolve the damage. Glutathione is responsible for reducing oxidative stressors before they have a chance to react with DNA bases.[50–52] The cellular response to formation of DNA adducts activates base excision repair pathways to repair abasic sites at the location of damage.[53–56] For compounds causing DNA cross-linking, either nucleotide excision repair or homologous recombination mitigates the damage. Both these repair pathways involve formation of large protein complexes at the site of damage that do not require transcriptional induction. These robust processes, working to repair DNA damage in the absence of synthesis of new protein, can explain the threshold-like responses of various genotoxicants at low doses.

Concentration–response for micronucleus and hypoxanthine (guanine) phosphoribosyltransferase mutations in cells treated with various alkylating agents have apparent threshold-shaped dose–response curves for several compounds, including methylnitrosourea.[4,57,58] One of the key post-translational responses to DNA alkylation is the removal of the O^6-alkylguanine adduct by O^6-alkylguanine DNA alkyltransferase (MGMT) through a SN2 reaction mechanism. Dose–response studies in wild-type cells and cells where MGMT was chemically inactivated showed the role of MGMT in creating the threshold-like dose–response behaviors. In the absence of MGMT, the point of departure was shifted left, indicating that the post-translational repair of the DNA adduct was important for preventing mutagenicity at low doses.

In our laboratory, we have focused on another post-translational response to DNA damage—double strand break repair through DNA repair centers. DNA repair centers—complexes of DNA repair, kinase and scaffold proteins—form at sites of double strand breaks. These repair centers respond to double strand breaks and repair the lesions without requiring transcriptional activation. By its nature, the process is highly efficient, but capacity limited. We evaluated the role of DNA repair centers in preventing micronucleus formation at low chemical concentrations and the potential for this post-translational response to induce threshold-type dose–response curves for double strand break agents. Our prototype compound, neocarzinostatin (NCS), is an engineered peptide, surrounding a chromophore.[59] When delivered to the DNA, NCS interacts with the DNA and causes a short burst of oxidative damage. Neocarzinostatin is often used as a γ-irradiation mimic *in vitro*.[60–62] We found that NCS produced a threshold-shaped dose–response curve for micronuclei (Figure 6.2).[63,64]

We evaluated the time- and concentration–response for DNA repair center formation following NCS treatment using confocal microscopy in cells stained for two components of the DNA repair center complex (p-H2AX and P53BP1). Double strand breaks were more rapidly resolved (measured as foci dissolution) at low concentrations (Figure 6.7). More importantly, the dose- and time–response indicated that post-translational repair processes prevent long-term damage at low doses. At higher doses that saturate these post-translation processes, there is activation of gene transcription. These doses correspond to those that cause formation of micronuclei. This transition from post-translational response (DNA repair center formation) to transcriptional response appears to represent a "tipping point" between adaptation and adversity (Figure 6.8).

The ability to measure DNA damage and repair at doses below those inducing micronuclei give us an unprecedented look at the cellular dynamics that define the shape of the micronucleus curve. DNA repair center formation at doses well below those where permanent damage (micronuclei) or transcriptional activation occur make clear that there are regions of dose where DNA damage and DNA repair are active before activation of these other processes. DNA repair center formation (a sentinel of DNA damage) was present at 10-fold lower concentrations than those leading

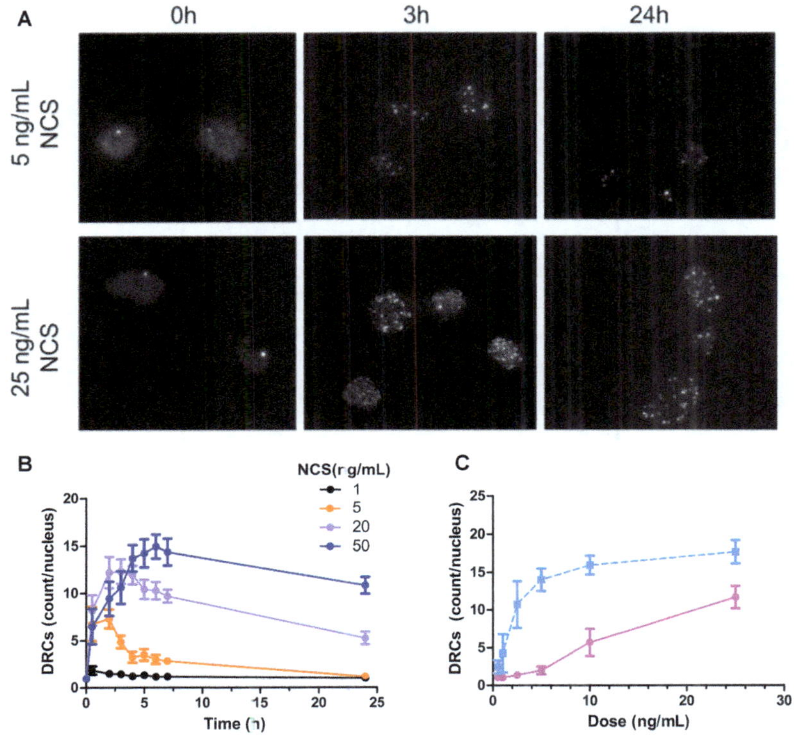

Figure 6.7 Time-course behaviors of DNA-repair centers serve to differentiate regions of adaptation from those showing overt adversity responses. (A) Images of DNA-repair center foci following treatment with neocarzinostatin (NCS). (B) Dose- and time-response for DNA-repair center foci (as foci per nucleus) following treatment with NCS. At lower exposures, foci resolve quickly. At higher, doses they persist out beyond 24 h. (C) Plots of foci remaining at 24 h (in purple) and total number of foci produced (in blue).
Reproduced with permission.[12]

to micronuclei or transcription. More importantly, the resolution of this damage through the post-translational formation of DNA repair centers appears to prevent conversion of transient breaks to permanent damage. These studies indicate that the threshold in micronuclei response is not due to chemical kinetics or assay sensitivity: it is a real biological phenomenon resulting from repair activity within the cell.

6.7 Suitability of Gene Expression Changes for Genotoxic Adversity

Our studies used HT-1080 cells, a human cancer cell with normal p53. Transformed cell lines, such as these cancer cells, may be missing key components of DNA damage sensing and repair machinery. Nonetheless,

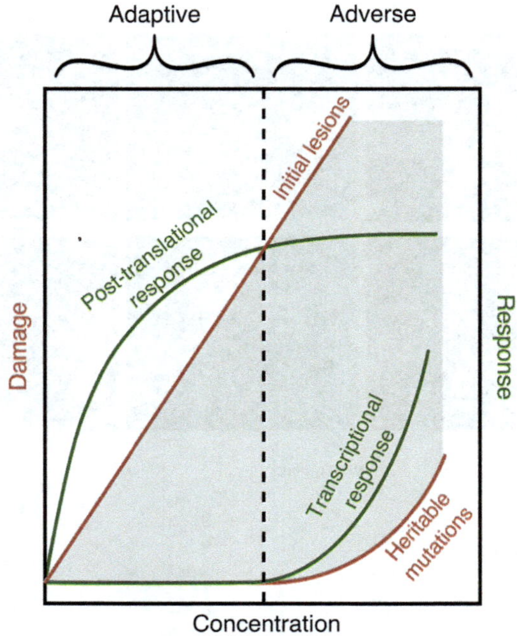

Figure 6.8 Interplay between DNA damage and response. Genotoxicants cause DNA damage (initial lesions) proportional to their concentration. At low concentrations, existing cellular machinery can repair the damage by forming DNA repair centers (post-translational response). This process constitutes an adaptive response, as it does not lead to an increase in permanent damage (heritable mutations). However, the post-translational response has a finite capacity. When concentrations are sufficiently high, unrepaired lesions can become permanent damage (adverse response). The transcriptional response to DNA damage is negligible at concentrations that do not yield fixed mutations (see Figure 6.6).

this work has illuminated the relative importance of different cellular behaviors in maintaining repair at low levels of DNA-damage. Transcriptional endpoints were useful for identifying modes of action for a variety of toxicological endpoints. For DNA-damaging compounds, gene expression studies differentiated between genotoxic and nongenotoxic compounds. More importantly, we found that gene expression assays may be useful in setting points of departure for mutation risk. Our work and that from other investigators indicate that cells use existing cellular machinery to handle small amounts of DNA damage.[4,8,19,23,49,57,58] Large-scale gene expression responses do not occur until the insult is sufficiently great to cause fixed DNA damage (Figure 6.6). Because the transcriptional responses are associated with permanent damage rather than the formation of initial lesions, statistically significant changes in gene expression likely serve more as a sensitive indicator of concentrations leading to mutagenicity than evidence of successful repair in sub-threshold regions for mutagenicity (Figure 6.8).

References

1. National Research Council, *Science and Decisions: Advancing Risk Assessment*, The National Academies Press, Washington, DC, 2009.
2. K. S. Crump, D. G. Hoel, C. H. Langley and R. Peto, Fundamental carcinogenic processes and their implications for low dose risk assessment, *Cancer Res.*, 1976, **36**, 2973–2979.
3. M. Crawford and R. Wilson, Low-Dose Linearity: The Rule or the Exception?, *Hum. Ecol. Risk Assess.*, 1996, **2**, 305–330.
4. G. E. Johnson, S. H. Doak, S. M. Griffiths, E. L. Quick, D. O. Skibinski, Z. M. Zair and G. J. Jenkins, Non-linear dose-response of DNA-reactive genotoxins: recommendations for data analysis, *Mutat. Res.*, 2009, **678**, 95–100.
5. L. H. Pottenger, M. R. Schisler, F. Zhang, M. J. Bartels, D. D. Fontaine, L. G. McFadden and B. B. Gollapudi, Dose-response and operational thresholds/NOAELs for *in vitro* mutagenic effects from DNA-reactive mutagens, MMS and MNU, *Mutat. Res.*, 2009, **678**, 138–147.
6. E. Gocke and L. Muller, *In vivo* studies in the mouse to define a threshold for the genotoxicity of EMS and ENU, *Mutat. Res.*, 2009, **678**, 101–107.
7. S. M. Bryce, J. Shi, J. Nicolette, M. Diehl, P. Sonders, S. Avlasevich, S. Raja, J. C. Bemis and S. D. Dertinger, High content flow cytometric micronucleus scoring method is applicable to attachment cell lines, *Environ. Mol. Mutagen.*, 2010, **51**, 260–266.
8. R. A. Clewell, B. Sun, Y. Adeleye, P. Carmichael, A. Efremenko, P. D. McMullen, S. Pendse, O. J. Trask, A. White and M. E. Andersen, Profiling dose-dependent activation of p53-mediated signaling pathways by chemicals with distinct mechanisms of DNA damage, *Toxicol. Sci.*, 2014, **142**, 56–73.
9. H. Ellinger-Ziegelbauer, H. Gmuender, A. Bandenburg and H. J. Ahr, Prediction of a carcinogenic potential of rat hepatocarcinogens using toxicogenomics analysis of short-term *in vivo* studies, *Mutat. Res.*, 2008, **637**, 23–39.
10. Organisation for Economic Co-operation and Development, in *OECD Guideline for the Testing of Chemicals, Section 4*, OECD Publishing, Editon edn, 2014.
11. F. A. Groothuis, M. B. Heringa, B. Nicol, J. L. Hermens, B. J. Blaauboer and N. I. Kramer, Dose metric considerations in *in vitro* assays to improve quantitative *in vitro-in vivo* dose extrapolations, *Toxicology*, 2015, **332**, 30–40.
12. Q. Zhang, S. Bhattacharya, J. Pi, R. Clewell, P. Carmichael and M. Andersen, Adaptive Posttranslational Control in Cellular Stress Response Pathways and its Relationship to Toxicity Testing and Safety Assessment, *Toxicol. Sci.*, 2015, **147**, 302–316.
13. H. H. Li, D. R. Hyduke, R. Chen, P. Heard, C. L. Yauk, J. Aubrecht and A. J. Fornace, Jr., Development of a toxicogenomics signature for

genotoxicity using a dose-optimization and informatics strategy in human cells, *Environ. Mol. Mutagen.*, 2015, **56**(6), 505–519.

14. T. Haaf, E. I. Golub, G. Reddy, C. M. Radding and D. C. Ward, Nuclear foci of mammalian Rad51 recombination protein in somatic cells after DNA damage and its localization in synaptonemal complexes, *Proc. Natl. Acad. Sci. U. S. A.*, 1995, **92**, 2298–2302.
15. R. S. Maser, K. J. Monsen, B. E. Nelms and J. H. Petrini, hMre11 and hRad50 nuclear foci are induced during the normal cellular response to DNA double-strand breaks, *Mol. Cell. Biol.*, 1997, **17**, 6087–6096.
16. B. E. Nelms, R. S. Maser, J. F. MacKay, M. G. Lagally and J. H. Petrini, *In situ* visualization of DNA double-strand break repair in human fibroblasts, *Science*, 1998, **280**, 590–592.
17. E. P. Rogakou, D. R. Pilch, A. H. Orr, V. S. Ivanova and W. M. Bonner, DNA double-stranded breaks induce histone H2AX phosphorylation on serine 139, *J. Biol. Chem.*, 1998, **273**, 5858–5868.
18. L. B. Schultz, N. H. Chehab, A. Malikzay and T. D. Halazonetis, p53 binding protein 1 (53BP1) is an early participant in the cellular response to DNA double-strand breaks, *J. Cell Biol.*, 2000, **151**, 1381–1390.
19. S. V. Costes, A. Boissiere, S. Ravani, R. Romano, B. Parvin and M. H. Barcellos-Hoff, Imaging features that discriminate between foci induced by high- and low-LET radiation in human fibroblasts, *Radiat. Res.*, 2006, **165**, 505–515.
20. D. Muzzey, C. A. Gomez-Uribe, J. T. Mettetal and A. van Oudenaarden, A systems-level analysis of perfect adaptation in yeast osmoregulation, *Cell*, 2009, **138**, 160–171.
21. A. Miermont, J. Uhlendorf, M. McClean and P. Hersen, The Dynamical Systems Properties of the HOG Signaling Cascade, *J. Signal Transduction*, 2011, **2011**, 930940.
22. Q. Zhang, S. Bhattacharya, R. B. Conolly, H. J. Clewell, N. E. Kaminski and M. E. Andersen, Molecular signaling network motifs provide a mechanistic basis for cellular threshold responses, *Environ. Health Perspect.*, 2014, **122**, 1261–1270.
23. T. Neumaier, J. Swenson, C. Pham, A. Polyzos, A. T. Lo, P. Yang, J. Dyball, A. Asaithamby, D. J. Chen, M. J. Bissell, S. Thalhammer and S. V. Costes, Evidence for formation of DNA repair centers and dose-response nonlinearity in human cells, *Proc. Natl. Acad. Sci. U. S. A.*, 2012, **109**, 443–448.
24. M. Kirsch-Volders, L. Gonzalez, P. Carmichael and D. Kirkland, Risk assessment of genotoxic mutagens with thresholds: a brief introduction, *Mutat. Res.*, 2009, **678**, 72–75.
25. W. K. Lutz and R. W. Lutz, Statistical model to estimate a threshold dose and its confidence limits for the analysis of sublinear dose-response relationships, exemplified for mutagenicity data, *Mutat. Res.*, 2009, **678**, 118–122.

26. S. Samanta and P. Dey, Micronucleus and its applications, *Diagn. Cytopathol.*, 2012, **40**, 84–90.
27. B. Sun, S. M. Ross, O. J. Trask, P. L. Carmichael, M. Dent, A. White, M. E. Andersen and R. A. Clewell, Assessing dose-dependent differences in DNA-damage, p53 response and genotoxicity for quercetin and curcumin, *Toxicology In Vitro*, 2013, **27**, 1877–1887.
28. P. T. Theunissen, J. L. Pennings, J. F. Robinson, S. M. Claessen, J. C. Kleinjans and A. H. Piersma, Time-response evaluation by transcriptomics of methylmercury effects on neural differentiation of murine embryonic stem cells, *Toxicol. Sci.*, 2011, **122**, 437–447.
29. D. A. van Dartel, J. L. Pennings, F. J. van Schooten and A. H. Piersma, Transcriptomics-based identification of developmental toxicants through their interference with cardiomyocyte differentiation of embryonic stem cells, *Toxicol. Appl. Pharmacol.*, 2010, **243**, 420–428.
30. S. A. Hermsen, T. E. Pronk, E. J. van den Brandhof, L. T. van der Ven and A. H. Piersma, Concentration-response analysis of differential gene expression in the zebrafish embryotoxicity test following flusilazole exposure, *Toxicol. Sci.*, 2012, **127**, 303–312.
31. D. A. van Dartel, J. L. Pennings, J. F. Robinson, J. C. Kleinjans and A. H. Piersma, Discriminating classes of developmental toxicants using gene expression profiling in the embryonic stem cell test, *Toxicol. Lett.*, 2011, **201**, 143–151.
32. K. Boehme, Y. Dietz, P. Hewitt and S. O. Mueller, Genomic profiling uncovers a molecular pattern for toxicological characterization of mutagens and promutagens *in vitro*, *Toxicol. Sci.*, 2011, **122**, 185–197.
33. M. Iida, C. H. Anna, J. Hartis, M. Bruno, B. Wetmore, J. R. Dubin, S. Sieber, L. Bennett, M. L. Cunningham, R. S. Paules, K. B. Tomer, C. D. Houle, A. B. Merrick, R. C. Sills and T. R. Devereux, Changes in global gene and protein expression during early mouse liver carcinogenesis induced by non-genotoxic model carcinogens oxazepam and Wyeth-14,643, *Carcinogenesis*, 2003, **24**, 757–770.
34. S. A. Amundson, M. Bittner, P. Meltzer, J. Trent and A. J. Fornace, Jr., Physiological function as regulation of large transcriptional programs: the cellular response to genotoxic stress, *Comp. Biochem. Physiol., Part B: Biochem. Mol. Biol.*, 2001, **129**, 703–710.
35. D. A. Dickinson, G. R. Warnes, G. Quievryn, J. Messer, A. Zhitkovich, E. Rubitski and J. Aubrecht, Differentiation of DNA reactive and non-reactive genotoxic mechanisms using gene expression profile analysis, *Mutat. Res.*, 2004, **549**, 29–41.
36. P. D. McMullen, S. Bhattacharya, C. G. Woods, B. Sun, K. Yarborough, S. M. Ross, M. E. Miller, M. T. McBride, E. L. LeCluyse, R. A. Clewell and M. E. Andersen, A map of the PPARalpha transcription regulatory network for primary human hepatocytes, *Chem. Biol. Interact.*, 2014, **209**, 14–24.

37. L. Smeenk, S. J. van Heeringen, M. Koeppel, B. Gilbert, E. Janssen-Megens, H. G. Stunnenberg and M. Lohrum, Role of p53 serine 46 in p53 target gene regulation, *PLoS One*, 2011, **6**, e17574.
38. P. Kapranov, S. E. Cawley, J. Drenkow, S. Bekiranov, R. L. Strausberg, S. P. Fodor and T. R. Gingeras, Large-scale transcriptional activity in chromosomes 21 and 22, *Science*, 2002, **296**, 916–919.
39. S. Cawley, S. Bekiranov, H. H. Ng, P. Kapranov, E. A. Sekinger, D. Kampa, A. Piccolboni, V. Sementchenko, J. Cheng, A. J. Williams, R. Wheeler, B. Wong, J. Drenkow, M. Yamanaka, S. Patel, S. Brubaker, H. Tammana, G. Helt, K. Struhl and T. R. Gingeras, Unbiased mapping of transcription factor binding sites along human chromosomes 21 and 22 points to widespread regulation of noncoding RNAs, *Cell*, 2004, **116**, 499–509.
40. O. Aksoy, A. Chicas, T. Zeng, Z. Zhao, M. McCurrach, X. Wang and S. W. Lowe, The atypical E2F family member E2F7 couples the p53 and RB pathways during cellular senescence, *Genes Dev.*, 2012, **26**, 1546–1557.
41. L. Yang, B. C. Allen and R. S. Thomas, BMDExpress: a software tool for the benchmark dose analyses of genomic data, *BMC Genomics*, 2007, **8**, 387.
42. R. K. Newton, M. Aardema and J. Aubrecht, The utility of DNA microarrays for characterizing genotoxicity, *Environ. Health Perspect.*, 2004, **112**, 420–422.
43. S. D. Seidel, H. L. Kan, W. T. Stott, M. R. Schisler and B. B. Gollapudi, Identification of transcriptome profiles for the DNA-damaging agents bleomycin and hydrogen peroxide in L5178Y mouse lymphoma cells, *Environ. Mol. Mutagen.*, 2003, **42**, 19–25.
44. S. D. Seidel, B. R. Sparrow, H. L. Kan, W. T. Stott, M. R. Schisler, V. A. Linscombe and B. B. Gollapudi, Profiles of gene expression changes in L5178Y mouse lymphoma cells treated with methyl methanesulfonate and sodium chloride, *Mutagenesis*, 2004, **19**, 195–201.
45. G. S. Akerman, B. A. Rosenzweig, O. E. Domon, L. J. McGarrity, L. R. Blankenship, C. A. Tsai, S. J. Culp, J. T. MacGregor, F. D. Sistare, J. J. Chen and S. M. Morris, Gene expression profiles and genetic damage in benzo(a)pyrene diol epoxide-exposed TK6 cells, *Mutat. Res.*, 2004, **549**, 43–64.
46. V. Thybaud, A. C. Le Fevre and E. Boitier, Application of toxicogenomics to genetic toxicology risk assessment, *Environ. Mol. Mutagen.*, 2007, **48**, 369–379.
47. R. S. Thomas, H. J. Clewell, 3rd, B. C. Allen, L. Yang, E. Healy and M. E. Andersen, Integrating pathway-based transcriptomic data into quantitative chemical risk assessment: a five chemical case study, *Mutat. Res.*, 2012, **746**, 135–143.
48. R. S. Thomas, S. C. Wesselkamper, N. C. Wang, Q. J. Zhao, D. D. Petersen, J. C. Lambert, I. Cote, L. Yang, E. Healy, M. B. Black, H. J. Clewell, 3rd, B. C. Allen and M. E. Andersen, Temporal concordance

between apical and transcriptional points of departure for chemical risk assessment, *Toxicol. Sci.*, 2013, **134**, 180–194.
49. Q. Zhang, S. Bhattacharya, J. Pi, R. Clewell, P. Carmichael and M. Andersen, Adaptive Posttranslational Control in Cellular Stress Response Pathways and its Relationship to Toxicity Testing and Safety Assessment, *Toxicol. Sci.*, 2015, **147**, 302–316.
50. Y. Miyamoto, Y. H. Koh, Y. S. Park, N. Fujiwara, H. Sakiyama, Y. Misonou, T. Ookawara, K. Suzuki, K. Honke and N. Taniguchi, Oxidative stress caused by inactivation of glutathione peroxidase and adaptive responses, *Biol. Chem.*, 2003, **384**, 567–574.
51. M. Valko, C. J. Rhodes, J. Moncol, M. Izakovic and M. Mazur, Free radicals, metals and antioxidants in oxidative stress-induced cancer, *Chem. Biol. Interact.*, 2006, **160**, 1–40.
52. M. Valko, D. Leibfritz, J. Moncol, M. T. Cronin, M. Mazur and J. Telser, Free radicals and antioxidants in normal physiological functions and human disease, *Int. J. Biochem. Cell Biol.*, 2007, **39**, 44–84.
53. B. A. Sokhansanj, G. R. Rodrigue, J. P. Fitch and D. M. Wilson, 3rd, A quantitative model of human DNA base excision repair. I. Mechanistic insights, *Nucleic Acids Res.*, 2002, **30**, 1817–1825.
54. T. Lindahl and R. D. Wood, Quality control by DNA repair, *Science*, 1999, **286**, 1897–1905.
55. T. Lindahl, B. Sedgwick, M. Sekiguchi and Y. Nakabeppu, Regulation and expression of the adaptive response to alkylating agents, *Annu. Rev. Biochem.*, 1988, **57**, 133–157.
56. M. D. Evans, M. Dizdaroglu and M. S. Cooke, Oxidative DNA damage and disease: induction, repair and significance, *Mutat. Res.*, 2004, **567**, 1–61.
57. S. H. Doak, K. Brusehafer, E. Dudley, E. Quick, G. Johnson, R. P. Newton and G. J. Jenkins, No-observed effect levels are associated with up-regulation of MGMT following MMS exposure, *Mutat. Res.*, 2008, **648**, 9–14.
58. S. H. Doak, G. J. Jenkins, G. E. Johnson, E. Quick, E. M. Parry and J. M. Parry, Mechanistic influences for mutation induction curves after exposure to DNA-reactive carcinogens, *Cancer Res.*, 2007, **67**, 3904–3911.
59. S. Kobayashi, M. Hori, G. X. Wang and M. Hirama, Formal total synthesis of neocarzinostatin chromophore, *J. Org. Chem.*, 2006, **71**, 636–644.
60. J. Ahn, M. Urist and C. Prives, Questioning the role of checkpoint kinase 2 in the p53 DNA damage response, *J. Biol. Chem.*, 2003, **278**, 20480–20489.
61. M. Ouchi and T. Ouchi, Distinct DNA damage determines differential phosphorylation of Chk2, *Cancer Biol. Ther.*, 2014, **15**, 1700–1704.
62. S. Sinha, R. Ghildiyal, V. S. Mehta and E. Sen, ATM-NFkappaB axis-driven TIGAR regulates sensitivity of glioma cells to radiomimetics in the presence of TNFalpha, *Cell Death Dis.*, 2013, **4**, e615.

63. R. A. Clewell, P. D. McMullen, Y. Adeleye, P. Carmichael and M. E. Andersen, in *Validating Alternative Methods for Toxicity Testing*, ed. C. Eskes and M. P. Whelan, Springer International Publishing AG, in press.
64. R. A. Clewell and M. E. Andersen, Approaches for characterizing threshold dose-response relationships for DNA-damage pathways involved in carcinogenicity *in vivo* and micronuclei formation *in vitro*, *Mutagenesis*, 2016, gev078.

CHAPTER 7

Dissecting Modes of Action of Non-genotoxic Carcinogens

MIRJAM M. SCHAAP,[a,b] JAN VAN BENTHEM,[a] MIRIAM N. JACOBS,[c] ANNAMARIA COLACCI,[d] ANNE S. KIENHUIS,[a] HARRY VAN STEEG[a,b] AND MIRJAM LUIJTEN*[a,b]

[a] Centre for Health Protection, National Institute for Public Health and the Environment (RIVM), Bilthoven, The Netherlands; [b] Department of Toxicogenetics, Leiden University Medical Center, Leiden, The Netherlands; [c] Centre for Radiation, Chemical and Environmental Hazards Centre, Public Health England, UK; [d] Center for Environmental Toxicology, Environmental Protection and Health Prevention Agency, Bologna, Italy
*Email: Mirjam.Luijten@rivm.nl

7.1 Introduction

7.1.1 Current Regulatory Requirements and Difficulties with Respect to Non-genotoxic Carcinogens

Carcinogenesis is a major toxicity endpoint in the safety evaluation of substances. It is a complex endpoint, characterized by a sequence of stages that involve initiation, promotion, and progression. Overall, regulatory requirements for carcinogenicity testing are quite similar across different regulatory and legislative frameworks of chemicals, but regional differences or sector differences may occur. The existing testing strategies mostly focus on genetic

Issues in Toxicology No. 28
Toxicogenomics in Predictive Carcinogenicity
Edited by Michael D. Waters and Russell S. Thomas

Published by the Royal Society of Chemistry, www.rsc.org

toxicity, as the majority of the known, probable or possible human carcinogens classified by the International Agency for Research on Cancer (IARC) have a genotoxic potential.[1] The first tier of the standard regulatory test battery generally consists of a set of well-established *in vitro* genotoxicity tests in bacterial and mammalian cells. A positive *in vitro* outcome triggers, in most regulatory frameworks, subsequent *in vivo* testing to confirm the positive *in vitro* finding. Dependent on the regulatory guidelines, a 2 year carcinogenicity study in rodents and/or a short-term carcinogenicity assay with transgenic mouse models may be required for substances with positive *in vivo* genotoxicity results. Because the carcinogenic potential in one species is not necessarily adaptable to another, studies are often required in two species.[2]

Although the existing testing strategies for cancer hazard assessment seem to be satisfactory, there are also important concerns. The primary strategy is aimed to demonstrate whether a substance has the potential to cause mutations or chromosomal damage, which, in turn, may lead to adverse health effects, including cancer. Consequently, substances that induce cancer *via* non-genotoxic mechanisms, *i.e.* the so-called non-genotoxic carcinogens, may remain unidentified. This lack of detection of non-genotoxic carcinogens may result in a substantial risk for society and the environment.[1]

7.1.2 What are Non-genotoxic Carcinogens?

Carcinogens can be categorized as either genotoxic or non-genotoxic, according to the mechanism(s) by which they induce tumors. Genotoxic carcinogens, also known as DNA-reactive carcinogens, are substances that directly interact with DNA, either as the parent substance or as reactive metabolite(s). Non-genotoxic carcinogens (NGTXC) are substances that act through distinct mechanisms without directly altering DNA, chromosome number or structure. Unlike genotoxic carcinogens, NGTXC can have diverse modes of action that are predominantly tissue and species specific. Examples of mechanisms of cancer induction by NGTXC include, but are not limited to, receptor-mediated endocrine modulation, non-receptor-mediated endocrine modulation, tumor promotion, induction of tissue-specific toxicity and inflammatory responses, immune suppression, inhibition of intercellular communication, and epigenetic alterations.[1,3–5] (See also Table 7.1, in which the major mechanisms are listed.) Additionally, there are several substances, such as carbon tetrachloride and some nickel compounds, that induce cancer *via* non-genotoxic mechanisms that are not fully understood.[1] The wide diversity in mechanism, the tissue specificity, and the absence of genotoxicity makes identification and characterization of NGTXC very challenging. Furthermore, NGTXC exhibit non-linear dose–response curves with a clear threshold. They may also exhibit temporal characteristics due to which the carcinogenesis process may be reversible at some early stages.

Table 7.1 Modes of action of non-genotoxic carcinogens.[a]

Key mechanism	Relevance of key mechanism(s)/ endpoints	Main relevant pathways and important genes involved	Related pathways/ cross-interactions	Endpoint/hallmark affected cross-reactions	Examples of major genes of the pathway(s) involved
Receptor binding and activation	Initiating event as part of hormone-mediated/ xenobiotic metabolism processes Angiogenesis receptors would be relevant, as the tyrosine phosphorylation patterns and the activation of major downstream signaling pathways are comparable[6]	WNT/catenin signaling pathway	ER, AR, PPARs RXR, RAR, VDR,[7] GR, FXR, AhR, ARNT	Metabolic enzyme induction Genetic instability Tumor microenvironment Cell cycle, Invasion and metastasis	AhR: modulation of heat shock proteins[8] PPARs: β catenin[9]
Cell proliferation	Early downstream event	p13k-AKT signaling pathway	VEGF signaling pathway MAPK signaling pathway	Cell proliferation Angiogenesis DNA repair	Ras, Raf-1, MEK, ERK
Gap junctional intercellular communication	Endpoint: block of dye transfer	Gap junction	MAPK signaling pathway	Apoptosis Differentiation; Tissue homeostasis	Ras, Raf-1, MEK, ERK, connexins
	Possibly connected with apoptosis and cell morphological transformation		Calcium signaling pathway	Sustained proliferative signaling	PLC, PKC
P450 induction	P450 futile cycling: CYP2E1 is involved in the oxygenation of	Oxidative stress	NADPH oxidase activity Oxidative stress	Deregulated metabolism	Phospho-Akt[10]

Table 7.1 (*Continued*)

Key mechanism	Relevance of key mechanism(s)/ endpoints	Main relevant pathways and important genes involved	Related pathways/ cross-interactions	Endpoint/hallmark affected cross-reactions	Examples of major genes of the pathway(s) involved
	difficult to oxidize substrates, generates ROS. Stabilization of CYP2E1 protein leads to prolonged burst of ROS production which may result in tissue necrosis, mutation and malignancy	Hepatic inflammation Xenobiotic metabolism			
	CYP1A1 AhR-mediated transduction	Mitochondrial transcription and mitochondrial stress signaling[11]	ER pathway cross-talk[12] MMPs Retinoic acid signaling[13]	Correlation with migratory processes and metastasis potential	E-cadherin, JNK, NEDD9
Induction of oxidative stress	ROS-triggered mitochondrial pathway inducing apoptosis[14] either directly through activation of gene transcriptional pathways or indirectly through hypomethylation	p53 signaling pathway	Mitochondrial malignancy	MDM family of p53-proteins (P21 and BCL2)	
			AhR		HMOX
			Caspase-3-dependent PKC-delta	Apoptosis	CASP3
		NADH-dependent signaling pathway	HIF-signaling pathway; Erb-2 signaling pathway	Angiogenesis HIF-1	VEGF, MMPs, IL8
				Inflammation Deregulated metabolism Senescence	CXCL10, CXCL11

Immuno-suppression	Chemokine signaling pathway	TGF-β signaling pathway	Transcription factors	Immunosuppression Angiogenesis	IL6[15]
Cell signaling pathways	Increased cell proliferation or selective cell death (apoptosis or necrosis). The release of calcium may result in the activation of kinases, such as PKC	Cyclic AMP signaling[16]	Calcium–calmodulin-activated pathway	Cytoskeleton restructuring (tumor microenvironment)	PKC, NFκB, AP-1
		Transcription factors, such as NFκB and AP-1, involved in the mediation of cell proliferation and apoptosis	P\13K-Akt;	Apoptosis	
		Changes in DNA methylation and hydroxymethylation patterns	Intracellular signal transducers such as nitric oxide	Necrosis	
Increased resistance to apoptotic cell death	Regulation of apoptosis, mainly by BCL2 family of mitochondria, death receptors and caspase network Intrinsic pathway: mitochondria-mediated apoptosis (initiated by *e.g.* high cytoplasmic calcium levels, ROS, xenobiotics)[17] Extrinsic pathway: all non-mitochondrial pathways	ERα, P53, ErbB-/HER-2 tyrosine kinase, ERK, MAPK, P16/P53, BCL2/P53, PPARα, gap junction intracellular communication, hypersecretion of luteinizing hormone Also ALK, VEGF, HER receptors, BRAF, Rho-associated protein kinase, fibroblast growth factor-9, cathepsins, cyclooxygenases, prostaglandins[18]	A process with many downstream pathways; cross-reaction with almost all the pathways indicated in this table.		Annexin V[19]
Pathogenic angiogenesis and neoangiogenesis	Endothelial cell activation in response to angiogenic factors.	Two key signaling pathways that interact: VEGF receptor- and TGF-mediated signaling	Calcium signaling pathway MAPK signaling pathway	Proliferation	Ras, Raf-1, ERK

Table 7.1 (*Continued*)

Key mechanism	Relevance of key mechanism(s)/ endpoints	Main relevant pathways and important genes involved	Related pathways/ cross-interactions	Endpoint/hallmark affected cross-reactions	Examples of major genes of the pathway(s) involved
	This ultimately leads to re-organization of endothelial cells to form tubules, which interconnect to form a network	pathways leading to tumor microenvironment and angiogenesis	Focal adhesion P13k-AKt bFGF MMPs and urokinase PDGF	Migration Cell survival Cell proliferation, migration and invasion	FAK CASP9, Bad MMP1 and MMP2, integrins, angiopoietin
Genetic instability	DNA repair pathways (combination with gene expression) DNA methylation	Caspase pathway, *e.g.* in colon carcinoma DCC[20]	Multiple cross-interactions, *e.g.* inflammation and inactivation of p53, to cell proliferation to invasion and metastasis	TGF-β signaling pathway	Caspase 9
Cellular senescence/ telomerase	Alteration of hTERT	hTERT pathway	ERG-mediated signaling IL6	Senescence	hTERT, p21
Metastasis (migration, intra- and extravasation, survival outside of original tissue)	The only hallmark definitively related to cancer; ability of cells to growth in semisolid medium; ability of cells to cross membrane barriers.	Leukocyte transendothelial pathway	p53 signaling MAPK signaling pathway VEGF signaling	Migration	Calmodulins

[a]Abbreviations: AhR: aryl hydrocarbon receptor; ALK: anaplastic lymphoma kinase; AP: activator protein; AR: androgen receptor; ARNT: aryl hydrocarbon receptor nuclear translocator; CYP: cytochrome P450; ER: endoplasmic reticulum; ERK: extracellular signal-regulated kinase; FXR: farnesoid X receptor; GR: glucocorticoid receptor; HER: human epidermal growth factor receptor; hTERT: human telomerase reverse transcriptase; IL: interleukin; MAPK: mitogen-activated protein kinase; MMP: matrix metalloproteinase; NADH: nicotinamide adenine dinucleotide; NADPH: nicotinamide adenine dinucleotide phosphate; NFκB: nuclear factor κB; PKC: protein kinase C; PLC: phospholipase C; PPAR: peroxisome proliferator-activated receptor; RAR: retinoic acid receptor; ROS: reactive oxygen species; RXR: retinoid X receptor; TGF: transforming growth factor; VDR: vitamin D receptor; VEGF: vascular endothelial growth factor.

7.1.3 Possible Methods for the Identification of Non-genotoxic Carcinogens

Of the known, probable or possible human carcinogens classified by the IARC, 12% (45 out of 371) appear to be non-genotoxic. For one-third of these substances, exposure is high enough to expect a significantly increased cancer risk.[1] This underlines the importance of alternative, preferably *in silico* or *in vitro* test systems predicting non-genotoxic carcinogenic features of a substance for which toxicity data are lacking. Due to the diversity of mechanisms of cancer induction, it is not feasible to assume that all NGTXC can be detected in a single test system. Rather, a panel of tailored test methods covering multiple biological traits of NGTXC will be required.

Several *in silico* and *in vitro* methods have been suggested for the detection of NGTXC. Among these (quantitative) structure–activity relationships are measurements of replicative DNA synthesis as an indicator of cell proliferation, measurements of inhibition of gap-junction intercellular communications, and the application of toxicogenomics. Since the focus of this book is on toxicogenomics, this is the subject of this chapter.

Toxicogenomics approaches to detect carcinogenic features of chemicals have been extensively investigated *in vivo* as well as *in vitro*.[6–12] Overall, these studies have yielded biologically relevant gene signatures, which may be employed to obtain insight into the possible mode of action of a (carcinogenic) substance or to predict the carcinogenicity of a substance. In contrast to toxicogenomics biomarkers for genotoxic carcinogens, identification of robust and reliable predictive biomarkers for NGTXC has proven troublesome, even *in vivo*. Although integration of different types of omics data, such as mRNA, miRNA, and protein expression profiles may enhance the robustness of these biomarkers,[13] we believe that focusing on mechanisms of cancer induction is a more promising strategy for identification of NGTXC. Recently, we demonstrated that application of toxicogenomics in *in vitro* systems such as primary mouse hepatocytes and mouse embryonic stem cells may be a useful approach to detect NGTXC.[14,15]

7.1.4 Comparison Approach

The method we developed is the so-called 'comparison approach'. This approach, outlined in Figure 7.1, involves an unsupervised analysis that is based on a clustering approach using gene set enrichment analysis (GSEA).[16] It makes use of limited gene sets, consisting of the most significantly up- and downregulated genes per chemical. For each chemical, a top 60 (top 30 up and top 30 down) of most significantly regulated genes, ranked according to T-statistics, is selected (see Schaap *et al.*[15] for details). The resulting gene sets for each chemical are pairwise compared, followed by calculation of scores. These scores are the sums of the absolute T-statistics values of the genes that are overlapping between two gene sets, *i.e.* 'hits' (see Figure 7.1). T-statistics is used as a basis for the scores because it provides

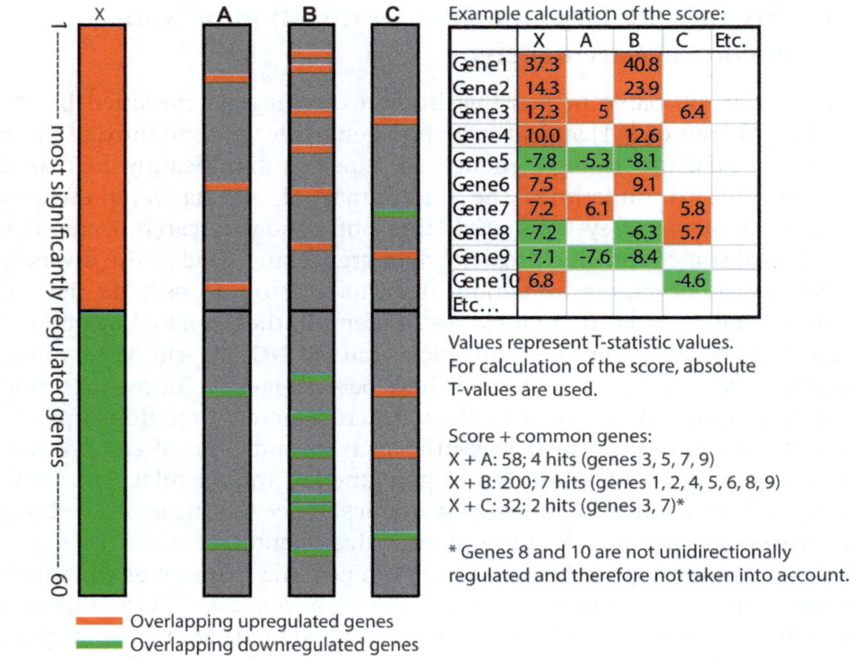

Figure 7.1 Outline of the comparison approach. Per chemical (X in this figure), the 60 most significantly regulated genes (top 30 up and top 30 down, ranked according to T-statistics) are compared to those of the other chemicals tested (indicated here as A, B, and C). For each combination of chemicals, the number of common genes (= hit) was determined (provided the direction of regulation was the same), and the sum of T-statistics (= score) was calculated. These two parameters were used to find the best match for the chemical of interest.
Reproduced from: Schaap *et al.*,[15] with kind permission from Springer Science and Business Media.

information on the direction as well as the significance of regulation of expressed genes. When developing this approach, we also tested whether the use of a higher number of transcripts, such as a top 100 or top 200, would alter or improve the outcome of the analysis. This appeared not to be the case. To distinguish between noise and most likely valuable results, cut-off criteria are applied: these have been arbitrarily set at 10 hits and a minimum score of 100.

Application of the comparison approach to gene expression data obtained from primary mouse hepatocytes and mouse embryonic stem cells upon exposure to a selected set of chemicals resulted in the recognition of NGTXC with diverse modes of action.[15] Using the data from primary mouse hepatocytes, we were able to categorize chemicals such as peroxisome proliferators, aryl hydrocarbon receptor agonists, metalloids and skin tumor promotors according to their modes of action. However, immunosuppressants such as cyclosporin A were only detected using gene expression data from mouse embryonic stem cells.[15] This highlights the need to use a combination of *in vitro* systems, to ensure that important aspects such as metabolic competence and expression of relevant cellular receptors, but also differences in cell cycle (*e.g.* dividing *versus* non-dividing cells) are covered. The accuracy of the comparison approach in detecting NGTXC may be further enhanced by applying other 'omics' approaches such as, for instance, (phospho)proteomics. In the studies performed so far, only a single concentration was tested for each substance. Testing multiple concentrations per substance may further strengthen the method. In spite of applying a concentration selection procedure, some of the substances tested showed considerable toxic responses while other substances did not. This suggests that testing of a (wide) range of concentrations may be advisable when using *in vitro* toxicogenomics approaches in order to obtain reliable substance-specific (transcriptional) responses.

7.2 Improving the Comparison Approach: A Case Study

7.2.1 Cyclosporin A and Tacrolimus as Model Compounds

Testing a range of concentrations (opposed to testing only a single concentration) for each substance may strengthen the comparison approach. We investigated this using cyclosporin A (CSA) and tacrolimus (FK506) as model compounds. In a previous study with the comparison approach, these two chemicals were not recognized as having similar modes of action when tested in primary mouse hepatocytes.[15] CSA and FK506 are cornerstone immunosuppressive agents administered to solid-organ-transplant recipients to prevent and treat allograft rejection.[17] Their best-documented mode of action is suppression of the immune system through inhibition of the phosphatase activity of calcineurin in T-cells. This prevents activation of members of the nuclear factor of activated T-cells family, affecting the transcription of cytokines important in the immune response. This effect is further enhanced due to involvement of calcineurin in the activation of nuclear factor-κB, which has a huge variety of target genes within T-cells.[17–19] Another route by which immune suppression is exerted is through inhibition of the mitogen-activated protein kinase (MAPK) pathways JNK and p38 pathways.[20] In addition, both chemicals have been reported to induce endoplasmic reticulum (ER) stress and consequent unfolded protein response (UPR), both in immune cells and cells of non-lymphoid lineages.[21,22] The ER stress response accompanies nearly all forms of acute and chronic liver disease, and has been shown to contribute to fibrotic remodeling and cholestasis.[23,24] Furthermore, ER stress has been linked to transforming growth factor (TGF)-β signaling.[25] Both CSA and FK506 stimulate the expression of TGF-β and other fibrogenic proteins, which is thought to be an important factor in the development of chronic nephrotoxicity observed upon long-term treatment with these substances.[26] Another important effect induced by both substances is decrease of cellular energy production due to mitochondrial dysfunction.[27–29] CSA is known to cause dose-dependent cholestasis in liver due to impaired biliary excretion of bile salts.[30,31] For FK506, only a few studies exist that suggest that FK506 may also induce cholestasis in rodents.[32–34] Systemic overproduction of TGF-β due to CSA exposure has been reported to increase tight junctional permeability in the liver, which may contribute to cholestasis.[35] CSA, but not FK506, has been shown to inhibit calcium-dependent mitochondrial permeability transition by binding to cyclophilin D, thereby preventing subsequent activation of the apoptosis cascade.[36] Because of the large overlap in modes of action, CSA and FK506 are interesting candidates for a case study to assess the added value of testing multiple concentrations when using the comparison approach.

7.2.2 Concentration Selection

Selecting adequate test concentrations is essential for any toxicity study, including microarray studies. For this case study, test concentrations for microarray analyses of primary mouse hepatocytes were selected based on cytotoxicity data. Cytotoxicity was evaluated using the MTT colorimetric assay according to Schaap *et al.*,[14] measured at 24 and 72 h after start of exposure of the cells. Primary mouse hepatocytes were exposed to the following concentration ranges: 0.3–30 μM CSA and 1–100 μM FK506. The results for the MTT measurements are depicted in Figure 7.2. Primary mouse hepatocytes exposed to CSA and FK506 clearly responded differently in terms of cytotoxicity (Figure 7.2A and B). FK506 has been shown to be more potent than CSA.[37,38] FK506 exposure did not induce significant cytotoxicity up to 40 μM, while exposure to 100 μM FK506 abruptly induced complete cell death. This effect was observed at both 24 h (Figure 7.2A) and 72 h (Figure 7.2B) after the start of the exposures. Compared to FK506, the dynamics of the cytotoxicity curves observed for CSA were much more gradual.

For the gene expression studies, four concentrations per chemical were tested. Concentrations were selected as follows: the top concentration was the concentration that induced a cytotoxicity of no more than 10%. The three additional concentrations were each three-fold lower than the preceding one. This resulted in the following test concentrations for microarray analyses: 0.3, 1.0, 3.0, and 10 μM of CSA; and 1.0, 3.0, 10, and 30 μM of FK506.

7.2.3 Microarray Analyses

Together with vehicle controls, each of the concentrations tested for CSA and FK506 was hybridized to HT MG-430 PM Array plates (Affymetrix), using four biological replicates. The number of differentially expressed genes (DEGs), calculated using an analysis of variance and based on a false discovery rate (FDR) <0.05, ranged from 1 to 5801 for CSA, and from 1 to 7342 for FK506. Principal component analysis (PCA) of the gene expression data, performed

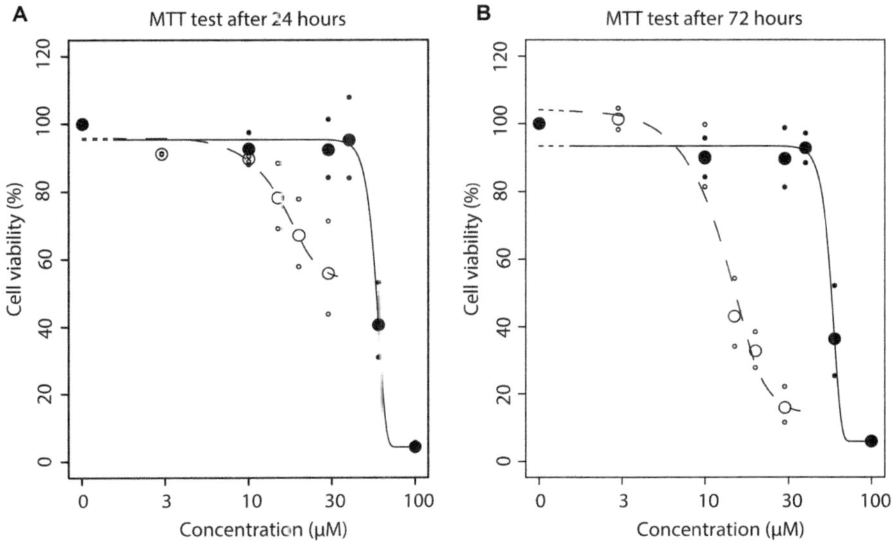

Figure 7.2 Cytotoxicity in primary mouse hepatocytes due to exposure to cyclosporin A (CSA) (open circles) and FK506 (closed circles) was evaluated by performing the MTT colorimetric assay at two time points: (A) immediately after 24 h of exposure; and (B) after an additional 48 h of culturing with normal serum-free culture medium.

using all genes as input, showed that the first principal component explains the largest part (69%) of the variation in the entire dataset (Figure 7.3). This variation was mainly caused by the highest concentrations tested for CSA and FK506. Despite their large effects on gene expression, these concentrations induced maximum cytotoxicity levels of 10% (after 72 h of exposure), a criterion commonly used in *in vitro* gene expression studies when testing single concentrations. The expression profiles obtained for the lower concentrations were regulated in the opposite direction, with the lowest concentrations being closest to the origin of the PCA plot. The second principal component, explaining about 15% of the variation, separates the responses induced by CSA and FK506. This implies that these chemicals, although having an overlap in mode of action, also differ in their response at the gene expression level.

7.2.4 Comparison Approach on a Concentration Range

The comparison approach (see Section 7.1.4 for details) was applied to the gene expression dataset obtained for different concentrations of CSA and FK506 tested in primary mouse hepatocytes. We hypothesized that by testing a range of concentrations the various mechanisms of action of each chemical as well as the overlap in mechanism of action between the two chemicals would be revealed. Using cut-off values of ≥10 hits and a minimum score

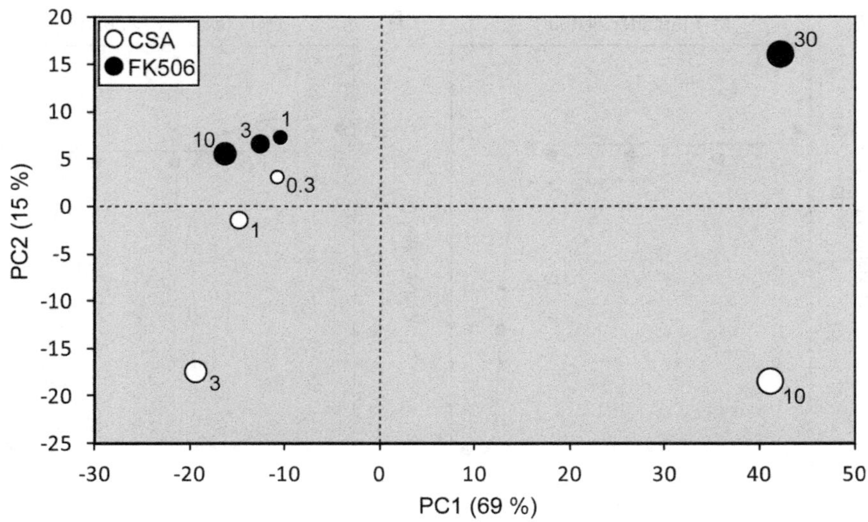

Figure 7.3 Principal component (PC) analysis of the gene expression values, using all available data. Open circles indicate cyclosporin A (CSA) and closed circles indicate FK506, whereby the size of the circle corresponds with the concentration tested, *i.e.* larger circles indicate higher concentrations and *vice versa.*

of 100, the comparison approach yielded an overlap for 3 μM CSA and 10 μM FK506 (score 144; 10 hits) (see Table 7.2). The strongest match in terms of score was observed for 10 μM CSA and 30 μM FK506 (score 147; 6 hits), although the criterion of ≥10 hits was not reached for this combination. The score associated with the match found for 10 μM CSA and 10 μM FK506 was somewhat lower but still convincing (score 116; 6 hits).

7.2.5 Biological Response at the Pathway Level

The gene expression profiles obtained for CSA and FK506 were studied in more detail in order to determine the biological relevance of the results from the comparison approach. For this, we evaluated the biological response in primary mouse hepatocytes upon exposure to CSA or FK506 by performing pathway analysis using MetaCore from GeneGo Inc. (http://lsresearch.thomsonreuters.com/pages/solutions/1/metacore). The resulting GeneGo pathways are listed in Tables 7.3 and 7.4. For each concentration tested, the top 1000 differentially expressed genes (based on the FDR) was used as input. Additionally, GSEA ($p<0.01$ and FDR<0.10, data not shown)[16] was performed to find supportive evidence for the pathways identified with MetaCore.

For FK506, the two lowest concentrations tested (*i.e.* 1.0 and 3.0 μM) induced only a minor response at the gene expression level (see Table 7.3). GSEA revealed downregulation of a few pathways related to the immune

Table 7.2 Outcome of comparison approach performed for cyclosporin A (CSA) and FK506.

	Match1	Score1	Hits1	Cut-off values met	Match2	Score2	Hits2	Cut off values met
CSA								
0.3 μM	3 μM FK506	32	2	0[a]	10 μM FK506	31	2	0[a]
1 μM	10 μM FK506	41	3	0[a]	3 μM FK506	37	3	0[a]
3 μM	10 μM FK506	144	10	2	3 μM FK506	80	6	0[a]
10 μM	30 μM FK506	147	6	1[b]	10 μM FK506	116	6	1[b]
FK506								
1 μM	3 μM CSA	45	3	0[a]	1 μM CSA	38	3	0[a]
3 μM	3 μM CSA	80	6	0[a]	1 μM CSA	39	3	0[a]
10 μM	3 μM CSA	144	10	2	10 μM CSA	116	6	1[b]
30 μM	10 μM CSA	147	6	1[b]	3 μM CSA	79	4	0[a]

[a]Cases where both cut-off values were not met (*i.e.* hits 10 and score 100).
[b]Cases where one of the two cut-off values was met.

response, but enriched pathways mainly involved induction of processes related to TGF-β receptor signaling. These effects were more pronounced in cells exposed to 10 μM FK506: a large number of TGF-β receptor-dependent pathways was upregulated. Most of these pathways relate to epithelial–mesenchymal transition (EMT), a process by which epithelial cells lose their cell polarity and cell–cell adhesion. Furthermore, multiple pathways involved in lipid metabolism, such as regulation of glucose and lipid metabolism *via* farnesoid X receptor (FXR) and fatty acid synthase activity in hepatocytes, were strongly downregulated (see Table 7.3). Pathways related to oxidative stress or suppression of the immune system were not abundantly overrepresented. The gene expression response observed for the highest concentration of FK506 (30 μM) was somewhat different from the lower concentrations tested. Instead of TGF-β receptor signaling, the majority of regulated pathways involved downregulation of amino acid, lipid and energy metabolism. Furthermore, glutathione metabolism was impaired, one of the factors known to play a role in the development of cholestasis.

The number of pathways significantly regulated by CSA was, in comparison to FK506, relatively limited (see Table 7.4), despite the fact that CSA induced a comparable number of DEGs. Like for FK506, analysis in MetaCore resulted in (virtually) no pathways for the two lowest concentrations of CSA (*i.e.* 0.3 and 1.0 μM). However, GSEA showed a large number of gene sets regulated by these concentrations. The lowest concentration of CSA induced downregulation of processes involved in fatty acid and glutathione metabolism, as well as a few pathways related to bile acid synthesis. To some extent, this was also observed for 1.0 μM CSA. Exposure to this concentration, however, mainly resulted in a clear effect on cell cycle, DNA replication, and DNA repair: numerous pathways were downregulated. This induction of cell cycle arrest was also found, although to a lesser extent, in cells exposed to 3.0 μM CSA, but not for 10 μM CSA. This finding may be explained by the

Table 7.3 Results pathway analyses for FK506. For each concentration a top 1000 genes was further analyzed for overrepresentation in MetaCore to identify significant pathways with a false discovery rate (FDR) < 0.05.[a]

GeneGo pathway maps	*p*-value (FDR < 0.05)	Count
1 μM FK506		
No significantly regulated pathways		
3 μM FK506		
Immune response_T-cell subsets: secreted signals	3.47E-03	6/25
Breast cancer (general schema)	3.38E-02	6/41
10 μM FK506		
Regulation of metabolism_Bile acids regulation of glucose and lipid metabolism *via* FXR	1.13E-05	10/37
Stimulation of TGF-β signaling in lung cancer	1.13E-05	11/48
Transcription_Sirtuin6 regulation and functions	1.71E-04	11/64
Regulation of lipid metabolism_Regulation of lipid metabolism *via* LXR, NF-Y and SREBP	8.905E-4	8/38
Adiponectin in pathogenesis of type 2 diabetes	9.888E-4	7/29
Development_Insulin, IGF-1 and TNF-α in brown adipocyte differentiation	9.888E-4	9/53
Cell adhesion_Plasmin signaling	2.880E-3	7/35
Cytoskeleton remodeling_TGF, WNT and cytoskeletal remodeling	2.920E-3	12/111
Development_Regulation of EMT	3.182E-3	9/64
Development_Role of IL-8 in angiogenesis	3.249E-3	9/65
Development_IGF-1 receptor signaling	3.623E-3	8/52
Immune response_CRTH2 signaling in Th2 cells	7.308E-3	7/44
Development_TGF-β-induction of EMT *via* ROS	7.308E-3	5/20
Immune response_TNF-R2 signaling pathways	7.438E-3	7/45
Putative pathways for stimulation of fat cell differentiation by bisphenol A	7.438E-3	6/32
Development_TGF-β-dependent induction of EMT *via* MAPK	9.001E-3	7/47
Development_TGF-β receptor signaling	1.221E-2	7/50
Development_Glucocorticoid receptor signaling	1.221E-2	5/24
Development_Oligodendrocyte differentiation from adult stem cells	1.221E-2	7/51
Development_Growth factors in regulation of oligodendrocyte precursor cell survival	1.221E-2	6/37
Development_β-adrenergic receptors in brown adipocyte differentiation	1.221E-2	6/37
Cell cycle_Regulation of G1/S transition (part 1)	1.354E-2	6/38
IGF family signaling in colorectal cancer	2.816E-2	7/60
Development_Transcription factors in segregation of hepatocytic lineage	2.816E-2	5/30
Cell adhesion_Gap junctions	2.816E-2	5/30
Multiple myeloma (general schema)	2.873E-2	4/18
Development_TGF-β-dependent induction of EMT *via* RhoA, PI3K and ILK.	2.922E-2	6/46
FGF signaling in pancreatic cancer	2.922E-2	6/46
Signal transduction_PTMs in IL-17-induced CIKS-independent signaling pathways	2.922E-2	6/46

Table 7.3 (*Continued*)

GeneGo pathway maps	p-value (FDR < 0.05)	Count
Regulation of lipid metabolism_Regulation of fatty acid synthase activity in hepatocytes	3.075E-2	4/19
Development_PDGF signaling *via* STATs and NF-κB	3.075E-2	5/32
Immune response_C3a signaling	3.321E-2	6/48
Normal and pathological TGF-β-mediated regulation of cell proliferation	3.332E-2	5/33
Development_Thromboxane A2 signaling pathway	3.385E-2	6/49
Cytoskeleton remodeling_Integrin outside-in signaling	3.385E-2	6/49
Immune response_C5a signaling	3.560E-2	6/50
Development_A2B receptor: action *via* G-protein α s	3.560E-2	6/50
Mucin expression in CF airways	3.978E-2	7/69
Regulation of lipid metabolism_Insulin regulation of fatty acid metabolism	4.107E-2	8/89
Inhibition of neutrophil migration by proresolving lipid mediators in COPD	4.107E-2	7/70
Development_Differentiation of white adipocytes	4.244E-2	6/53
Development_EGFR signaling pathway	4.244E-2	7/71
Reproduction_GnRH signaling	4.458E-2	7/72
Th17 cells in CF	4.458E-2	6/54
Immune response_Platelet activating factor/PTAFR pathway signaling	4.786E-2	6/55
Mechanism of action of DGAT1 inhibitors in obesity and diabetes mellitus, type II (variant 1)	4.874E-2	4/24
Triacylglycerol biosynthesis in obesity and diabetes mellitus, type II	4.874E-2	4/24
30 μM FK506		
Tryptophan metabolism/rodent version	8.77E-08	21/102
Leucine, isoleucine and valine metabolism/rodent version	1.68E-07	18/80
Tryptophan metabolism	1.68E-07	20/101
Leucine, isoleucine and valine metabolism p. 2	3.680E-6	16/78
Regulation of lipid metabolism_PPAR regulation of lipid metabolism	3.311E-5	11/42
Peroxisomal branched chain fatty acid oxidation	3.885E-5	15/83
Propionate metabolism p. 1	1.024E-4	10/39
Protein folding and maturation_Angiotensin system maturation\human version	2.085E-4	10/43
Butanoate metabolism	2.085E-4	12/63
Lysine metabolism/rodent version	2.085E-4	14/86
Propionate metabolism p. 2	2.882E-4	12/66
Protein folding and maturation_Angiotensin system maturation\rodent version	4.530E-4	10/48
Regulation of lipid metabolism_RXR-dependent regulation of lipid metabolism *via* PPAR, RAR and VDR	4.744E-4	8/30
Mitochondrial ketone bodies biosynthesis and metabolism	1.936E-3	7/27
Mitochondrial long chain fatty acid β-oxidation	2.180E-3	12/83
Fructose metabolism/rodent version	2.180E-3	12/83
Lysine metabolism	2.313E-3	12/84

Table 7.3 (*Continued*)

GeneGo pathway maps	*p*-value (FDR < 0.05)	Count
Fructose metabolism	2.659E-3	11/73
Phenylalanine metabolism/rodent version	4.366E-3	10/65
NAD metabolism	4.366E-3	14/119
Phenylalanine metabolism	5.165E-3	10/67
Regulation of metabolism_Bile acids regulation of glucose and lipid metabolism *via* FXR	1.029E-2	7/37
Regulation of lipid metabolism_Regulation of fatty acid synthesis: NLTP and EHHADH	1.043E-2	5/18
L-Arginine metabolism	1.154E-2	10/75
Ascorbate metabolism/rodent version	1.169E-2	6/28
Arginine metabolism/rodent version	1.310E-2	11/91
Transcription_Sirtuin6 regulation and functions	1.310E-2	9/64
Tyrosine metabolism p. 2 (melanin)	2.219E-2	10/83
Mitochondrial unsaturated fatty acid β-oxidation	2.652E-2	7/45
Glycolysis and gluconeogenesis p. 1	2.928E-2	7/46
Histidine-glutamate-glutamine and proline metabolism/rodent version	3.482E-2	12/120
Glycine, serine, cysteine and threonine metabolism/ rodent version	4.157E-2	12/123
Glutathione metabolism	4.590E-2	8/64
Glutathione metabolism/human version	4.917E-2	8/65

[a]Abbreviations: CF: cystic fibrosis; COPD: chronic obstructive pulmonary disease; DGAT: diglyceride acyltransferase; EGFR: epidermal growth factor receptor; EMT: epithelial-to-mesenchymal transition; FGF: fibroblast growth factor; FXR: farnesoid X receptor; GnRH: gonadotropin-releasing hormone; IGF: insulin-like growth factor; IL: interleukin; ILK: integrin-linked kinase; LXR: liver X receptor; MAPK: mitogen-activated protein kinase; NAD: nicotinamide adenine dinucleotide; NLTP: non-specific lipid transfer protein; NF: nuclear factor; PDGF: platelet-derived growth factor; PI3K: phosphoinositide 3-kinase; PPAR: peroxisome proliferator-activated receptor; PTAFR: platelet-activating factor receptor; PTM: post-translational modification; RAR: retinoic acid receptor; ROS: reactive oxygen species; RXR: retinoid X receptor; SREBP: sterol regulatory element-binding proteins; STAT: signal transducer and activator of transcription; TGF: transforming growth factor; TNF: tumor necrosis factor; VDR: vitamin D receptor.

fact that activation of the TGF-β signaling pathway induces both cell cycle arrest and apoptosis. The latter, however, is inhibited by CSA due to inhibition of mitochondrial permeability transition. For 3.0 and 10 μM CSA, we observed induction of ER stress and UPR in a dose-dependent manner, as identified by GSEA and MetaCore analysis (see Table 7.4). Exposure to 10 μM CSA also induced downregulation of a large number of pathways related to amino acid, lipid and glucose metabolism, including fatty acid oxidation in mitochondria and peroxisomes.

Taken together, FK506 and CSA induced in primary mouse hepatocytes partly an overlapping, and partly a different response. The overlap was mainly found at high concentrations and comprises downregulation of cell energy metabolism plus some indications of cholestasis. At lower concentrations, FK506 showed a dose-dependent induction of TGF-β signaling, while CSA specifically induced cell cycle arrest (possible TGF-β induced) and ER stress plus consequent UPR in a dose-dependent manner.

Table 7.4 Results pathway analyses for cyclosporin A (CSA). For each concentration the top 1000 genes were further analyzed for overrepresentation in MetaCore to identify significant pathways with a false detection rate (FDR) < 0.05.[a]

GeneGo pathway maps	*p*-value (FDR < 0.05)	Count
0.3 μM CSA		
Cell cycle_Chromosome condensation in prometaphase	7.22E-03	5/21
1 μM CSA		
No significantly regulated pathways		
3 μM CSA		
Transcription_Sirtuin6 regulation and functions	7.08E-04	11/64
Apoptosis and survival_Endoplasmic reticulum stress response pathway	7.08E-04	10/55
CFTR folding and maturation (normal and CF)	7.98E-03	6/24
Regulation of lipid metabolism_Regulation of lipid metabolism *via* LXR, NF-Y and SREBP	8.909E-3	7/38
Cell cycle_Regulation of G1/S transition (part 1)	8.909E-3	7/38
Immune response_Antigen presentation by MHC class I	1.017E-2	6/28
N-Glycan biosynthesis p1	1.275E-2	9/71
Cholesterol biosynthesis	4.414E-2	10/103
10 μM CSA		
Tryptophan metabolism/rodent version	6.98E-05	16/102
Tryptophan metabolism	1.80E-04	15/101
Mitochondrial ketone bodies biosynthesis and metabolism	2.38E-04	8/27
CFTR folding and maturation (normal and CF)	9.450E-4	7/24
Retinol metabolism/rodent version	3.989E-3	10/66
Leucine, isoleucine and valine metabolism/rodent version	3.989E-3	11/80
Apoptosis and survival_Endoplasmic reticulum stress response pathway	4.228E-3	9/55
Retinol metabolism	4.776E-3	10/70
DeltaF508-CFTR traffic/endoplasmic reticulum-to-Golgi in CF	7.963E-3	5/17
wtCFTR traffic/endoplasmic reticulum-to-Golgi (normal)	7.963E-3	5/17
Peroxisomal branched-chain fatty acid oxidation	1.353E-2	10/83
Mitochondrial long-chain fatty acid β-oxidation	1.353E-2	10/83
Transport_FXR-regulated cholesterol and bile acids cellular transport	1.382E-2	7/42
Linoleic acid/rodent version	1.487E-2	6/31
Leucine, isoleucine and valine metabolism p. 2	2.973E-2	9/78
Blood coagulation_Blood coagulation	4.652E-2	6/39

[a]Abbreviations: CF: cystic fibrosis; CFTR: cystic fibrosis transmembrane conductance regulator; FXR: farnesoid X receptor; LXR: liver X receptor; MHC: major histocompatibility complex; NF: nuclear factor; SREBP: sterol regulatory element-binding proteins; wt: wild-type.

7.2.6 Biological Relevance of Comparison Approach

To assess the biological relevance of the results from the comparison approach we studied the concentration–response curves of the genes that were used as input for the comparison approach. Since we aimed to interpret these genes in terms of their biological relevance, the non-annotated genes were excluded from this analysis. The 60 genes per concentration tested for each of the substances, including the level of regulation compared to controls, are given in supplementary Tables 7.1 (CSA) and 7.2 (FK506).[†]

Strikingly, there was a gene present in all gene sets for FK506 and nearly all gene sets for CSA: S100 calcium binding protein G (*S100g*). This gene, strongly downregulated in transcription upon exposure to FK506 or CSA, is a direct target of the vitamin D receptor (VDR). VDR is, next to pregnane X receptor (PXR) and constitutive androstane receptor (CAR), a member of the nuclear receptor family of transcription factors, which regulates steroid and xenobiotic metabolism by inducing the phase I cytochrome P450 (CYP) enzymes, phase II conjugating transferases, and the phase III transporters.[39] Both FK506 and CSA are not only metabolized by CYP3A enzymes, but have also been described as inhibitors of CYP3A.[40] In our study, various CYP3A enzymes (*Cyp3a11*, *Cyp3a25* and *Cyp3a44*) were indeed downregulated, especially by CSA. Another gene that was repeatedly found in the gene sets for CSA and FK506 is lymphocyte antigen 6 complex, locus D (*Ly6d*). Transcription of this gene was also decreased upon exposure to one of the chemicals. *Ly6d* has been reported to be involved in the differentiation of lymphoid progenitors into B cells.[41] This finding may therefore be considered indicative of immunosuppression.

The list for CSA contains genes encoding glutathione transferases and reductase (*Gsta2*, *Gsta4*, and *Gsr*), which play a major role in phase II metabolism. Each of these genes was downregulated, but not in a concentration-dependent manner. Upon xenobiotic exposure, rather the opposite would be expected. A decrease in expression of these genes, however, has also been found *in vivo* and has been associated with an increase in protein level.[42,43] A clear concentration-dependent decline in mRNA expression was observed for genes involved in cholestasis and lipid metabolism. These genes include the efflux transporter P-glycoprotein (*Abcb1a*), and carboxylesterase 1 (*Ces1*). CSA is known to inhibit P-glycoprotein, which is one of the ATP binding cassette transporters responsible for the secretion of bile components into the bile canaliculus.[44] Carboxylesterase 1 (*Ces1*) was also downregulated. *Ces1* is a direct target of the FXR, which is considered the master regulator of bile acid homeostasis, but also regulates lipid and glucose metabolism.[45] Exposure to hepatotoxicants such as CSA are known to impair the FXR signaling pathway.[30,46]

Downregulation of processes involved in cell cycle upon CSA exposure was identified with GSEA, but not with MetaCore analyses. Among the genes used for the comparison approach was *Mapk4*. It was downregulated at all concentrations, but not in a concentration-dependent manner. This decrease in expression may be considered indicative of the moderate response

[†]Electronic supplementary information available: DOI: 10.1039/9781782624059-XXX.

found with GSEA. Downregulation of fatty acid oxidation and glucose metabolism was reflected by a concentration-dependent decrease in transcription of cytochrome P450 enzymes (*Cyp2c55* and *Cyp4a10*), acyl-CoA dehydrogenase (*Acadl*) and fructose biphosphatase 2 (*Fbp2*). The main effect induced by CSA exposure, according to pathway analyses, was upregulation of processes involved in protein folding, ER stress and detection and transport of misfolded proteins. A large number of genes involved in these processes showed a (strong) concentration-dependent increase in transcription and were part of the gene sets used in the comparison approach. Examples include genes that encode the protein disulfide isomerases A3, A4, and A6 (*Pdia3/4/6*), enzymes involved in protein folding, and stress-modulated transcription factor p8 (*Nupr1*).[47] The expression of genes involved in a rescue unit in the ER membrane to stabilize membrane proteins in response to stress was also increased. This rescue unit comprises the stress-associated endoplasmic reticulum protein 1 (*Serp1*), other components of translocon (*Sec61b*, *Sec61a*), and molecular chaperons in the ER (*e.g.* calnexin; *Canx*).[48]

For FK506, the lists of 60 most significantly regulated genes (for each concentration) used for the comparison approach are given in supplementary Table 7.2.[‡] Exposure to FK506 induced expression of carboxylesterase 6 (*Ces6*) and member 1a4 of the solute carrier organic anion transporter family (*Slco1a4*), both target genes of PXR, one of the regulators of gene activation in response to xenobiotic stress.[49] Analogous to CSA, expression of the transporter P-glycoprotein (*Abcb1a*) was decreased. This effect is in concordance with an increased concentration-dependent transcription of cytochrome P450 enzyme *Cyp8b1* and cysteine sulfinic acid decarboxylase (*Csad*), which are repressed through FXR under normal conditions. Dysregulation of lipid metabolism was also reflected by regulation of peroxisome proliferator-activated receptor-α target genes vanin-1 (*Vnn*) and apolipoprotein A-V (*Apoa5*). Both gene products have been shown to lower plasma triglyceride levels.[50,51] However, these genes were regulated in opposite direction in the present study. This might be explained by the fact that vanin-1 exhibits a biphasic expression upon oxidative stress.[52] A decrease in fatty acid oxidation and glucose metabolism was reflected in a similar manner as for CSA: transcription of cytochrome P450 enzymes (*Cyp4a10*, and *Cyp4a31*), acetyl-CoA acyltransferase (*Acaa1b*), acyl-CoA dehydrogenase (*Acadl*), and fructose biphosphatase 2 (*Fbp2*) was decreased in a concentration-dependent fashion. Induction of TGF-β signaling, one of the main findings from the pathway analyses for FK506, was also represented in the gene sets. Expression of TGF-β receptor 1 (*Tgfbr1*) and insulin-like growth factor 1 (*Igf1*) was somewhat induced, whilst expression of stress-modulated transcription factor p8 (*Nupr1*) and *Fyn*, a proto-oncogene, was strongly enhanced. Fyn has been demonstrated to be involved in EMT, which is triggered by, among others, TGF-β.[53] *Nupr1* has been reported to be activated by TGF-β1 at the

[‡]Electronic supplementary information available: DOI: 10.1039/9781782624059-XXX.

transcriptional level, mediated by the SMAD proteins.[54] TGF-β1 is also known to increase the expression of plasminogen activator inhibitor-1, which in turn decreases plasminogen activator (*Plat*) expression.[55,56] Both FK06 and CSA substantially decreased expression of *Plat* in primary mouse hepatocytes. Transcription of ligands to the epidermal growth factor (EGF) receptor family, such as neuregulin 1 (*Nrg1*) and betacellulin (*Btc*), was also decreased. It is likely that the TGF-β and EGF signaling pathways interact and that both positive and negative regulations exist between the two pathways.[57]

The comparison approach performed for the case study resulted in a match between CSA and FK506 for three combinations (see Section 7.2.4): 3 μM CSA and 10 μM FK506 (10 hits), 10 μM CSA and 30 μM FK506 (6 hits), and 10 μM CSA and 10 μM FK506 (6 hits). The genes that contributed to a match for these concentrations are listed in Table 7.5. In total, there are 14 unique genes. An increasing degree of differential regulation (compared to vehicle controls) is indicated with increasing intensity in color. Table 7.5 nicely shows that each of these genes is regulated in a clear concentration-dependent manner. The highest concentrations tested (10 μM and 30 μM for CSA and FK506, respectively) had a major effect on transcription. For example, chemokines *Cxcl10*, *Ccl5*, and *Cxcl2* were strongly induced only upon exposure to either 10 μM CSA or 30 μM FK506. PCA analysis of the complete expression profiles also demonstrated that the highest concentrations tested were the main cause for the observed variation in gene expression (Figure 7.3). Although cytotoxicity levels for 10 μM CSA and 30 μM FK506 did not exceed 10% (Figure 7.2), it may be argued that the match for this combination is not based solely on overlap in specific mechanism of action, but in part also on more generic toxicity. Consequently, caution is warranted in cases where the comparison approach only yields a match for chemicals at (slightly) toxic concentrations.

7.3 Discussion and Future Prospects

Chemicals that are carcinogenic in a non-genotoxic manner may pose a substantial risk to humans and the environment. Adequate, preferably non-animal, test methods to detect this 'class' of chemicals in safety assessments are therefore in high demand. One of the promising approaches is to employ toxicogenomics in combination with suitable *in vitro* systems. In this chapter, we explored whether a toxicogenomics approach that focuses on mechanism of action could be improved. The nature of a toxicological response induced by a given chemical often depends on cell type and dose. The key research question was if this so-called 'comparison approach' would benefit from the use of a concentration range instead of a single concentration per chemical. The hypothesis was that the use of multiple concentrations would resolve issues regarding selecting adequate test concentrations and, in addition, would provide more insight into the mechanism(s) of action a given chemical may have. To this end, we performed a case study in primary mouse hepatocytes with CSA and FK506 as

Table 7.5 Genes that contributed to the match between cyclosporin A CSA and FK506. Differential expression values are indicated as fold changes compared to vehicle controls. An increasing degree of differential regulation is indicated with increasing intensity in color; red is downregulated, green is upregulated.

GeneSymbols	GeneENTREZ	0.3 µM CSA	1 µM CSA	3 µM CSA	10 µM CSA	1 µM FK506	3 µM FK506	10 µM FK506	30 µM FK506
Match: 3 µM CSA + 10 µM FK506 (10 genes)									
S100g	12309	0.42	0.43	0.42	0.39	0.35	0.36	0.39	0.35
Ly6d	17068	0.77	0.68	0.31	0.26	0.62	0.40	0.28	0.25
D17H6S56E-5	110956	0.91	0.80	0.51	0.26	0.93	0.84	0.62	0.60
Acadl	11363	0.94	0.88	0.75	0.64	1.00	0.93	0.70	0.71
Plat	18791	0.92	0.92	0.32	0.11	1.02	0.81	0.47	0.19
Ivl	16447	1.06	1.05	0.40	0.33	0.79	0.62	0.41	0.31
Abcb1a	18671	0.83	0.75	0.42	0.32	0.90	0.72	0.63	0.57
AU018778	234564	0.50	0.52	0.39	0.29	0.64	0.50	0.48	0.30
2010003K11Rik	69861	0.72	0.70	0.49	0.30	0.86	0.61	0.51	0.20
Fbp2	14120	0.98	0.72	0.50	0.32	1.04	0.91	0.61	0.33
Match: 10 µM CSA + 30 µM FK506 (6 genes)									
Plat	18791	0.92	0.92	0.32	0.11	1.02	0.81	0.47	0.19
Hmgcs2	15360	0.79	0.88	0.75	0.13	1.02	0.94	0.72	0.11
Aldh1a1	11668	0.89	0.85	0.64	0.24	0.91	0.80	0.74	0.26
Acaa1b	235674	0.72	0.66	0.60	0.16	0.82	0.67	0.50	0.21
2010003K11Rik	69861	0.72	0.70	0.49	0.30	0.86	0.61	0.51	0.20
Net1	56349	0.95	0.94	0.63	0.37	0.93	0.85	0.77	0.46
Match: 10 µM CSA + 10 µM FK506 (6 genes)									
D17H6S56E-5	110956	0.91	0.80	0.51	0.26	0.93	0.84	0.62	0.60
Ly6d	17068	0.77	0.68	0.31	0.26	0.62	0.40	0.28	0.25
Plat	18791	0.92	0.92	0.32	0.11	1.02	0.81	0.47	0.19
Acadl	11363	0.94	0.88	0.75	0.64	1.00	0.93	0.78	0.71
Acaa1b	235674	0.72	0.66	0.60	0.16	0.82	0.67	0.50	0.21
2010003K11Rik	69861	0.72	0.70	0.49	0.30	0.86	0.61	0.51	0.20

model compounds. Expression profiles were similar to such a degree that a match between the two model compounds was found for multiple concentrations tested, whilst such a match was absent in studies in which only single concentrations of test compounds were used. The gene sets on which the overlap in expression was based differed with different concentrations tested, thereby highlighting the fact that the two compounds may have multiple mechanisms of action. Furthermore, we demonstrated that the in the comparison approach applied gene sets reliably reflect the perturbed pathways and biological processes derived from the complete expression profiles. This confirms the applicability of the comparison approach to identify modes of action of chemicals with limited data. Based on these findings, we therefore conclude that using multiple concentrations in the comparison approach is of significant added value. Nevertheless, the robustness of this approach needs to be substantiated by testing concentration ranges of larger sets of chemicals.

In addition, issues such as optimizing the combination of *in vitro* systems used in the comparison approach need to be resolved. The case study presented here clearly showed that primary mouse hepatocytes do not cover all biological responses found *in vivo*. The main mode of action of CSA and FK506, *i.e.* immunosuppression, was not recognized with our approach, whilst this effect has been detected *in vivo* in the liver, at least for CSA.[30] This may be explained by the fact that the cultures used for this study mostly contain hepatocytes, and relatively few non-parenchymal liver cells. Consequently, cells of the adaptive (T- and B-lymphocytes) and innate (Kupffer cells and natural killer cells) immune system are highly underrepresented. Since each *in vitro* system has its shortcomings, a combination of multiple *in vitro* systems will be required to ensure coverage of a broad range of mechanisms.

In conclusion, the case study presented here demonstrates that testing multiple concentrations per chemical strongly enhances our method to detect mechanisms of action of NGTXC. Although the comparison approach does not provide specific information on which of these mechanisms is the main driver of the carcinogenic process, we consider this method a promising tool for the detection of NGTXC. It addresses the need to obtain detailed knowledge about the mechanisms of toxicity of a substance of interest, as propagated by the National Research Council in their report *Toxicity testing in the 21st century: a vision and strategy*. Furthermore, use of the comparison approach is not necessarily limited to carcinogens, but may be applicable to environmental and pharmaceutical chemicals in general. In this way, toxicogenomics will become a useful and feasible tool for the safety evaluation of chemicals.

References

1. L. G. Hernandez, H. van Steeg, M. Luijten and J. van Benthem, Mechanisms of non-genotoxic carcinogens and importance of a weight of evidence approach, *Mutat. Res.*, 2009, **682**, 94–109.

2. P. A. Marone, W. C. Hall and A. W. Hayes, Reassessing the two-year rodent carcinogenicity bioassay: a review of the applicability to human risk and current perspectives, *Regul. Toxicol. Pharmacol.*, 2014, **68**, 108–118.
3. M. D. Waters, M. Jackson and I. Lea, Characterizing and predicting carcinogenicity and mode of action using conventional and toxicogenomics methods, *Mutat. Res.*, 2010, **705**, 184–200.
4. D. Hattis, M. Chu, N. Rahmioglu, R. Goble, P. Verma, K. Hartman and M. Kozlak, A preliminary operational classification system for nonmutagenic modes of action for carcinogenesis, *Crit. Rev. Toxicol.*, 2009, **39**, 97–138.
5. R. L. Melnick, M. C. Kohn and C. J. Portier, Implications for risk assessment of suggested nongenotoxic mechanisms of chemical carcinogenesis, *Environ. Health Perspect.*, 1996, **104**(Suppl 1), 123–134.
6. H. Ellinger-Ziegelbauer, H. Gmuender, A. Bandenburg and H. J. Ahr, Prediction of a carcinogenic potential of rat hepatocarcinogens using toxicogenomics analysis of short-term in vivo studies, *Mutat. Res.*, 2008, **637**, 23–39.
7. M. R. Fielden, A. Adai, R. T. Dunn 2nd, A. Olaharski, G. Searfoss, J. Sina, J. Aubrecht, E. Boitier, P. Nioi, S. Auerbach, D. Jacobson-Kram, N. Raghavan, Y. Yang, A. Kincaid, J. Sherlock, S. J. Chen, B. Car and C. W. G. Predictive Safety Testing Consortium, Development and evaluation of a genomic signature for the prediction and mechanistic assessment of nongenotoxic hepatocarcinogens in the rat, *Toxicol. Sci.*, 2011, **124**, 54–74.
8. J. P. Melis, K. W. Derks, T. E. Pronk, P. Wackers, M. M. Schaap, E. Zwart, W. F. van Ijcken, M. J. Jonker, T. M. Breit, J. Pothof, H. van Steeg and M. Luijten, In vivo murine hepatic microRNA and mRNA expression signatures predicting the (non-)genotoxic carcinogenic potential of chemicals, *Arch. Toxicol.*, 2014, **88**, 1023–1034.
9. T. Watanabe, T. Suzuki, M. Natsume, M. Nakajima, K. Narumi, S. Hamada, T. Sakuma, A. Koeda, K. Oshida, Y. Miyamoto, A. Maeda, M. Hirayama, H. Sanada, H. Honda, W. Ohyama, E. Okada, Y. Fujiishi, S. Sutou, A. Tadakuma, Y. Ishikawa, M. Kido, R. Minamiguchi, I. Hanahara and C. Furihata, Discrimination of genotoxic and non-genotoxic hepatocarcinogens by statistical analysis based on gene expression profiling in the mouse liver as determined by quantitative real-time PCR, *Mutat. Res.*, 2012, **747**, 164–175.
10. T. Uehara, Y. Minowa, Y. Morikawa, C. Kondo, T. Maruyama, I. Kato, N. Nakatsu, Y. Igarashi, A. Ono, H. Hayashi, K. Mitsumori, H. Yamada, Y. Ohno and T. Urushidani, Prediction model of potential hepatocarcinogenicity of rat hepatocarcinogens using a large-scale toxicogenomics database, *Toxicol. Appl. Pharmacol.*, 2011, **255**, 297–306.
11. F. Yamada, K. Sumida, T. Uehara, Y. Morikawa, H. Yamada, T. Urushidani and Y. Ohno, Toxicogenomics discrimination of potential hepatocarcinogenicity of non-genotoxic compounds in rat liver, *J. Appl. Toxicol.*, 2013, **33**, 1284–1293.

12. K. Z. Guyton, A. D. Kyle, J. Aubrecht, V. J. Cogliano, D. A. Eastmond, M. Jackson, N. Keshava, M. S. Sandy, B. Sonawane, L. Zhang, M. D. Waters and M. T. Smith, Improving prediction of chemical carcinogenicity by considering multiple mechanisms and applying toxicogenomic approaches, *Mutat. Res.*, 2009, **681**, 230–240.
13. M. Romer, J. Eichner, U. Metzger, M. F. Templin, S. Plummer, H. Ellinger-Ziegelbauer and A. Zell, Cross-platform toxicogenomics for the prediction of non-genotoxic hepatocarcinogenesis in rat, *PLoS One*, 2014, **9**, e97640.
14. M. M. Schaap, E. P. Zwart, P. F. Wackers, I. Huijskens, B. van de Water, T. M. Breit, H. van Steeg, M. J. Jonker and M. Luijten, Dissecting modes of action of non-genotoxic carcinogens in primary mouse hepatocytes, *Arch. Toxicol.*, 2012, **86**, 1717–1727.
15. M. M. Schaap, P. F. Wackers, E. P. Zwart, I. Huijskens, M. J. Jonker, G. Hendriks, T. M. Breit, H. van Steeg, B. van de Water and M. Luijten, A novel toxicogenomics-based approach to categorize (non-)genotoxic carcinogens, *Arch. Toxicol.*, 2015, **89**, 2413–2427.
16. A. Subramanian, P. Tamayo, V. K. Mootha, S. Mukherjee, B. L. Ebert, M. A. Gillette, A. Paulovich, S. L. Pomeroy, T. R. Golub, E. S. Lander and J. P. Mesirov, Gene set enrichment analysis: a knowledge-based approach for interpreting genome-wide expression profiles, *Proc. Natl. Acad. Sci. U. S. A.*, 2005, **102**, 15545–15550.
17. J. M. Barbarino, C. E. Staatz, R. Venkataramanan, T. E. Klein and R. B. Altman, PharmGKB summary: cyclosporine and tacrolimus pathways, *Pharmacogenet. Genomics*, 2013, **23**, 563–585.
18. M. M. Hamawy, Molecular actions of calcineurin inhibitors, *Drug News Perspect.*, 2003, **16**, 277–282.
19. S. Matsuda and S. Koyasu, Mechanisms of action of cyclosporine, *Immunopharmacology*, 2000, **47**, 119–125.
20. S. Matsuda and S. Koyasu, Regulation of MAPK signaling pathways through immunophilin-ligand complex, *Curr. Top. Med. Chem.*, 2003, **3**, 1358–1367.
21. S. Du, N. Hiramatsu, K. Hayakawa, A. Kasai, M. Okamura, T. Huang, J. Yao, M. Takeda, I. Araki, N. Sawada, A. W. Paton, J. C. Paton and M. Kitamura, Suppression of NF-kappaB by cyclosporin a and tacrolimus (FK506) via induction of the C/EBP family: implication for unfolded protein response, *J. Immunol.*, 2009, **182**, 7201–7211.
22. M. Kitamura, Induction of the unfolded protein response by calcineurin inhibitors: a double-edged sword in renal transplantation, *Nephrol., Dial., Transplant.*, 2010, **25**, 6–9.
23. L. Dara, C. Ji and N. Kaplowitz, The contribution of endoplasmic reticulum stress to liver diseases, *Hepatology*, 2011, **53**, 1752–1763.
24. H. Tanjore, W. E. Lawson and T. S. Blackwell, Endoplasmic reticulum stress as a pro-fibrotic stimulus, *Biochim. Biophys. Acta*, 2013, **1832**, 940–947.

25. S. Lenna and M. Trojanowska, The role of endoplasmic reticulum stress and the unfolded protein response in fibrosis, *Curr. Opin. Rheumatol.*, 2012, **24**, 663–668.
26. A. Khanna, M. Plummer, C. Bromberek, B. Bresnahan and S. Hariharan, Expression of TGF-beta and fibrogenic genes in transplant recipients with tacrolimus and cyclosporine nephrotoxicity, *Kidney Int.*, 2002, **62**, 2257–2263.
27. A. Wolf, C. F. Trendelenburg, C. Diez-Fernandez, P. Prieto, S. Houy, W. E. Trommer and A. Cordier, Cyclosporine A-induced oxidative stress in rat hepatocytes, *J. Pharmacol. Exp. Ther.*, 1997, **280**, 1328–1334.
28. N. Simon, C. Morin, S. Urien, J. P. Tillement and B. Bruguerolle, Tacrolimus and sirolimus decrease oxidative phosphorylation of isolated rat kidney mitochondria, *Br. J. Pharmacol.*, 2003, **138**, 369–376.
29. U. Christians, S. Gottschalk, J. Miljus, C. Hainz, L. Z. Benet, D. Leibfritz and N. Serkova, Alterations in glucose metabolism by cyclosporine in rat brain slices link to oxidative stress: interactions with mTOR inhibitors, *Br. J. Pharmacol.*, 2004, **143**, 388–396.
30. A. S. Kienhuis, A. P. Vitins, J. L. Pennings, T. E. Pronk, E. N. Speksnijder, M. Roodbergen, J. H. van Delft, M. Luijten and L. T. van der Ven, Cyclosporine A treated in vitro models induce cholestasis response through comparison of phenotype-directed gene expression analysis of in vivo Cyclosporine A-induced cholestasis, *Toxicol. Lett.*, 2013, **221**, 225–236.
31. A. Sharanek, P. B. Azzi, H. Al-Attrache, C. C. Savary, L. Humbert, D. Rainteau, C. Guguen-Guillouzo and A. Guillouzo, Different dose-dependent mechanisms are involved in early cyclosporine a-induced cholestatic effects in hepaRG cells, *Toxicol. Sci.*, 2014, **141**, 244–253.
32. S. Sanchez-Campos, R. Lopez-Acebo, P. Gonzalez, J. M. Culebras, M. J. Tunon and J. Gonzalez-Gallego, Cholestasis and alterations of glutathione metabolism induced by tacrolimus (FK506) in the rat, *Transplantation*, 1998, **66**, 84–88.
33. K. Mizuta, E. Kobayashi, H. Uchida, A. Fujimura, H. Kawarasaki and K. Hashizume, Influence of tacrolimus on bile acid and lipid composition in continuously drained bile using a rat model. Comparative study with cyclosporine, *Transplant Int.*, 1999, **12**, 316–322.
34. K. Mizuta, E. Kobayashi, H. Uchida, S. Hishikawa and H. Kawarasaki, Increase of bile acid production by tacrolimus in the rat liver, *Transplant. Proc.*, 2003, **35**, 437–438.
35. L. Lora, E. Mazzon, D. Billington, C. Milanesi, R. Naccarato and D. Martines, Effects of cyclosporin A on paracellular and transcellular transport of horseradish peroxidase in perfused rat livers, *Dig. Dis. Sci.*, 1997, **42**, 514–521.
36. P. C. Waldmeier, J. J. Feldtrauer, T. Qian and J. J. Lemasters, Inhibition of the mitochondrial permeability transition by the nonimmunosuppressive cyclosporin derivative NIM811, *Mol. Pharmacol.*, 2002, **62**, 22–29.

37. T. Kino, H. Hatanaka, S. Miyata, N. Inamura, M. Nishiyama, T. Yajima, T. Goto, M. Okuhara, M. Kohsaka, H. Aoki *et al.*, FK-506, a novel immunosuppressant isolated from a Streptomyces. II. Immunosuppressive effect of FK-506 in vitro, *J. Antibiot.*, 1987, **40**, 1256–1265.
38. H. Takeuchi, K. Okuyama, O. Konno, Y. Jojima, I. Akashi, Y. Nakamura, H. Iwamoto, K. Hama, T. Iwahori, M. Uchiyama, T. Ashizawa, N. Matsuno, T. Nagao, T. Hirano and K. Oka, Optimal dose and target trough level in cyclosporine and tacrolimus conversion in renal transplantation as evaluated by lymphocyte drug sensitivity and pharmacokinetic parameters, *Transplant. Proc.*, 2005, **37**, 1745–1747.
39. Y. K. Seo, Y. T. Chung, S. Kim, I. Echchgadda, C. S. Song and B. Chatterjee, Xenobiotic- and vitamin D-responsive induction of the steroid/bile acid-sulfotransferase Sult2A1 in young and old mice: the role of a gene enhancer in the liver chromatin, *Gene*, 2007, **386**, 218–223.
40. R. Amundsen, A. Asberg, I. K. Ohm and H. Christensen, Cyclosporine A- and tacrolimus-mediated inhibition of CYP3A4 and CYP3A5 in vitro, *Drug Metab. Dispos.*, 2012, **40**, 655–661.
41. Q. Zhang, B. L. Esplin, R. Iida, K. P. Garrett, Z. L. Huang, K. L. Medina and P. W. Kincade, RAG-1 and Ly6D independently reflect progression in the B lymphoid lineage, *PLoS One*, 2013, **8**, e72397.
42. S. Bramow, P. Ott, F. Thomsen Nielsen, K. Bangert, N. Tygstrup and K. Dalhoff, Cholestasis and regulation of genes related to drug metabolism and biliary transport in rat liver following treatment with cyclosporine A and sirolimus (Rapamycin), *Pharmacol. Toxicol.*, 2001, **89**, 133–139.
43. A. Van Summeren, J. Renes, D. Lizarraga, F. G. Bouwman, J. P. Noben, J. H. van Delft, J. C. Kleinjans and E. C. Mariman, Screening for drug-induced hepatotoxicity in primary mouse hepatocytes using acetaminophen, amiodarone, and cyclosporin a as model compounds: an omics-guided approach, *OMICS*, 2013, **17**, 71–83.
44. M. Trauner and J. L. Boyer, Bile salt transporters: molecular characterization, function, and regulation, *Physiol. Rev.*, 2003, **83**, 633–671.
45. P. L. Jansen and F. G. Schaap, How sweet it is to activate FXR, *Hepatology*, 2014, **59**, 1665–1667.
46. S. Modica, R. M. Gadaleta and A. Moschetta, Deciphering the nuclear bile acid receptor FXR paradigm, *Nucl. Recept. Signaling*, 2010, **8**, e005.
47. J. Huang, N. Viswakarma, S. Yu, Y. Jia, L. Bai, A. Vluggens, M. Cherkaoui-Malki, M. Khan, I. Singh, G. Yang, M. S. Rao, J. Borensztajn and J. K. Reddy, Progressive endoplasmic reticulum stress contributes to hepatocarcinogenesis in fatty acyl-CoA oxidase 1-deficient mice, *Am. J. Pathol.*, 2011, **179**, 703–713.
48. A. Yamaguchi, O. Hori, D. M. Stern, E. Hartmann, S. Ogawa and M. Tohyama, Stress-associated endoplasmic reticulum protein 1 (SERP1)/Ribosome-associated membrane protein 4 (RAMP4) stabilizes membrane proteins during stress and facilitates subsequent glycosylation, *J. Cell Biol.*, 1999, **147**, 1195–1204.

49. C. Xu, X. Wang and J. L. Staudinger, Regulation of tissue-specific carboxylesterase expression by pregnane x receptor and constitutive androstane receptor, *Drug Metab. Dispos.*, 2009, **37**, 1539–1547.
50. J. A. van Diepen, P. A. Jansen, D. B. Ballak, A. Hijmans, G. J. Hooiveld, S. Rommelaere, F. Galland, P. Naquet, F. P. Rutjes, R. P. Mensink, P. Schrauwen, C. J. Tack, M. G. Netea, S. Kersten, J. Schalkwijk and R. Stienstra, PPAR-alpha dependent regulation of vanin-1 mediates hepatic lipid metabolism, *J. Hepatol.*, 2014, **61**, 366–372.
51. N. Vu-Dac, P. Gervois, H. Jakel, M. Nowak, E. Bauge, H. Dehondt, B. Staels, L. A. Pennacchio, E. M. Rubin, J. Fruchart-Najib and J. C. Fruchart, Apolipoprotein A5, a crucial determinant of plasma triglyceride levels, is highly responsive to peroxisome proliferator-activated receptor alpha activators, *J. Biol. Chem.*, 2003, **278**, 17982–17985.
52. C. Berruyer, F. M. Martin, R. Castellano, A. Macone, F. Malergue, S. Garrido-Urbani, V. Millet, J. Imbert, S. Dupre, G. Pitari, P. Naquet and F. Galland, Vanin-1-/- mice exhibit a glutathione-mediated tissue resistance to oxidative stress, *Mol. Cell. Biol.*, 2004, **24**, 7214–7224.
53. T. S. Gujral, M. Chan, L. Peshkin, P. K. Sorger, M. W. Kirschner and G. MacBeath, A noncanonical Frizzled2 pathway regulates epithelial-mesenchymal transition and metastasis, *Cell*, 2014, **159**, 844–856.
54. R. M. Pommier, J. Gout, D. F. Vincent, C. E. Cano, B. Kaniewski, S. Martel, J. Rodriguez, G. Fourel, U. Valcourt, J. C. Marie, J. L. Iovanna and L. Bartholin, The human NUPR1/P8 gene is transcriptionally activated by transforming growth factor beta via the SMAD signalling pathway, *Biochem. J.*, 2012, **445**, 285–293.
55. Y. Zhu, W. L. Yin, Y. F. Ba, L. Tian, Z. Q. Gu, M. S. Zhang and C. N. Zhong, Transforming growth factor-1 promotes the transcriptional activation of plasminogen activator inhibitor type 1 in carcinoma-associated fibroblasts, *Mol. Med. Rep.*, 2012, **6**, 1001–1005.
56. S. M. Wileman, N. A. Booth, N. Moore, B. Redmill, J. V. Forrester and R. M. Knott, Regulation of plasminogen activation by TGF-beta in cultured human retinal endothelial cells, *Br. J. Ophthalmol.*, 2000, **84**, 417–422.
57. M. Jia and S. Souchelnytstkyi, Comments on the cross-talk of TGFbeta and EGF in cancer, *Exp. Oncol.*, 2011, **33**, 170–173.

CHAPTER 8

Human Embryonic Stem Cells as Biological Models to Examine the Impact of Xenobiotics on the Genome and Epigenome

LESLIE RECIO

Integrated Laboratory Systems (ILS), Inc., PO Box 13507, Research Triangle Park, NC 27709, USA
Email: lrecio@ils-inc.com

8.1 Introduction

A number of human diseases are caused by environmental factors that can promote diseases by affecting changes in the genome or by affecting the epigenome, the regulatory network that controls the expression of genes. The epigenome includes all of the epigenetic mechanisms that regulate gene expression and controls the circumstances in which a gene is active or inactive. Although novel approaches which are integrating genomics technologies into animal and cell culture models to define cellular response pathways and mode-of-action are making an impact in toxicology, there has been a limited focus on the impact of environmental agents on the epigenome, a key biological interface between the environment, the cell, and the genome. Unlike the static genomic DNA, the epigenome is dynamic,

Issues in Toxicology No. 28
Toxicogenomics in Predictive Carcinogenicity
Edited by Michael D. Waters and Russell S. Thomas

Published by the Royal Society of Chemistry, www.rsc.org

poised to act in response to physiological cues or to environmental stressors in real time or remain latent for many years and responding decades later. The primary focus of *in vitro* test systems used in safety assessment has been on direct cellular and molecular targets, apical measures of cellular health status, and DNA damage and toxicogenomic responses with limited direct measures of epigenomic impacts.[1] Many of the analytical tools used to assess and track the role of the epigenome in response to physiological cues can now be used in the safety assessment of environmental agents, pharmaceuticals and pesticides. The National Institutes of Health (NIH) Roadmap Epigenomics Mapping Consortium is evaluating epigenetic tools and producing epigenomic reference maps of human tissues based on the hypothesis that the origins of health and susceptibility to disease are, in part, the result of epigenetic regulation of the genomic blueprint (see www.roadmapepigenomics.org/).

Although the causes of the vast majority of birth defects are uncertain, environmental exposures or pharmaceuticals are linked to approximately 3% of all birth defects.[2,3] Current biological assays accepted by regulatory agencies to assess the potential to cause developmental and reproductive toxicity involve the use of animal models. The animal testing requirements for developmental toxicity[4] calls for two animal studies requiring hundreds of rats and rabbits and costing $200–300 K per test substance. In certain cases, the two-generation reproduction toxicity study, which is among the most costly and time consuming of toxicology tests, using up to 3200 animals per substance, can be required.[5] Not only are these models costly and time consuming, but pronounced interspecies differences in response are known to occur with animal tests.[6] There is a clear need for a battery of human cell-based assays predictive of potential human reproductive and developmental toxicants to be used prior to the conduct of costly and time-consuming animal testing.

Human embryonic stem (hES) cell-based biological assays can be used to assess traditional measures of cellular health status and determine the impact of toxicants on the genome and epigenome that regulate stem cell biology. Biological assays to assess the impact on the hES cell epigenome of xenobiotic exposures and in safety assessment are needed, since virtually none exist. Dysfunction from xenobiotic exposures of the highly regulated epigenetic programs that maintain stemness (self-renewal and pluripotency) and differentiation can produce aberrantly programmed descendent daughter cells leading to spontaneous abortions, birth defects, neurological disorders, and other effects later in life as the organism responds to environmental and biochemical stimuli.[7] For the large number of compounds entering the environment and the need to assess the potential of reproductive and developmental toxicity in safety assessments, hES cell-based assays may be developed into a battery of medium-throughput *in vitro* toxicology assays that can be integrated with toxicogenomics- and epigenomics-based signature profiling to assess the disruption of the hES epigenome and predict the potential to cause reproductive and developmental toxicity.

There are an alarming number of substances entering the human environment that have been identified as potential developmental and reproductive toxicants in a broad spectrum of animal species that need to be examined in human-based systems.[8] hES cells offer a unique platform for developing bioassays aimed at assessing the impact of xenobiotics on the genome and epigenome that can be developed into short-term cell-based screening assays to predict reproductive and developmental toxicity. The large number of chemicals that need to be prioritized with respect to reproductive and developmental toxicity testing using definitive *in vivo* tests and the need to reduce animal use safety assessments necessitates the development and validation of human-relevant bioassays using known target cells. Furthermore, the recent European Commission prohibition banning the testing of finished cosmetic products and cosmetic ingredients on animals makes the reliance on non-animal test systems more urgent http://ec.europa.eu/growth/sectors/cosmetics/animal-testing/index_en.htm.

8.2 hES Cells as Biological Models in Toxicology

Stem cells have two unique biological properties that distinguish them from all other cell types used in toxicology: (1) self-renewal—they are undifferentiated cells that can proliferate while remaining in an undifferentiated state for long periods; and (2) pluripotency—under certain conditions/physiological cues, they can be induced to differentiate into cells with specialized tissue functions (Figure 8.1) (see http://stemcells.nih.gov/info/basics/Pages/Default.aspx).

Innovations in cell culture methods have provided the reagents (defined serum-free culture media) and technologies required to grow and maintain hES cells in an undifferentiated state. Under certain physiological or experimental conditions using specific growth factors, hES cells can be

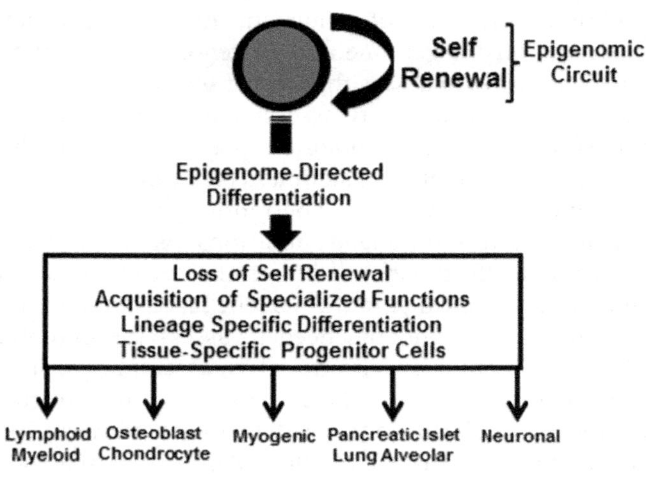

Figure 8.1 The properties of stem cells and regulation of stem cell biology by the epigenome.

induced to differentiate along specific lineages (*e.g.* neuronal lineage). hES cell differentiation involves a series of phenotypic and underlying molecular changes orchestrated to result in the progressive loss of "self-renewal" and the parallel acquisition of specialized functions of mature cells. The maintenance of pluripotency and the orchestration of hES cell differentiation along specific lineages is controlled by a regulatory circuit of transcription factors and networks of genes under the direction of the hES cell epigenome, specifically micro-(mi)RNAs and the more recently recognized long noncoding RNAs (lncRNAs).[9–11] Disruption of this regulatory circuit by exposures to certain xenobiotics can alter this evolutionarily conserved finely tuned process leading to loss of pluripotency or "stemness" and aberrant or aborted specific lineage differentiation during embryonic development, and later in life in adult tissue-specific progenitor cells.[12] The development of genomic and epigenomic signatures in hES cells predictive of developmental and reproductive toxicity potential has a scientific niche as part of a safety assessment test battery to assess potential reproductive and developmental toxicity.

There is a clear need in toxicology for human-relevant *in vitro* test systems using hES cell-based assays in toxicology. Since hES cells are at the origins of human development and are a unique biological and genomic landscape not represented in any other cell type used in toxicology. Furthermore, use of hES cells eliminates interspecies differences in genomic and epigenomic response networks and development of predictive biomarkers that can be used to assess the relevance of finding in rodents supporting robust human hazard identification. The use of hES cells and other stem cells as cellular and molecular toxicology platforms integrating traditional measures of toxicology with genomics and epigenomics profiling is consistent with the National Toxicology Program, US Environmental Protection Agency, and US Food and Drug Administration TOX21 efforts, where pathways of toxicity are defined using human *in vitro* cell systems combined with computational approaches for hazard identification.[13–15] Innovative, highly predictive medium-throughput cell culture models to predict potential hazards to humans are being used to reduce animal usage as evidenced by the TOX21 program and the continuing high attrition rate of drug candidates due to safety concerns.[16,17] The hES toxicity testing platform integrated with traditional measures of cellular health status, genomic, and epigenome biomarkers can provide a robust systems biology driven approach to *in vitro* hazard assessment of developmental and reproductive potential.

8.3 Developing hES Cells into a Toxicology Testing Platform

The unique properties and cell culture conditions for growth and maintenance of hES cells is a challenge in developing stem cell culture systems into robust reproducible *in vitro* toxicology assays. *In vitro* toxicologists, regulatory agencies, and investigators have long relied on well-established protocols that use decades-old established rodent and human cell lines that grow as monolayers attached to plastic or as suspension cultures, have rapid cell culture doubling times (8–14 h), carry chromosomal anomalies, and in certain cases carry mutations in critical biological response genes like *p53*, a central regulator of the DNA damage response.[18] Considerable effort and expertise is required in growing and maintaining hES cells such as the WA09 cell line from WiCell (www.wicell.org/). The WA09 cell line is a human blastocyst-derived, pluripotent cell line that remains in an undifferentiated state in defined media with a cell culture doubling time of 31–33 h, maintains a normal karyotype, requires strict cell culture conditions for growth in matrigel serum-free or feeder cells with media, and, most importantly, can withstand freeze/thaw media with protocols for recovery of hES with high viability.[19] The WA09 cell line is one of the original cell lines approved for NIH-funded research. hES cells do not express drug metabolizing enzymes, which is a significant issue of concern for all high-throughput testing systems.

The cell growth and maintenance of WA09 cells is challenging for the design of a toxicology/toxicogenomics-based medium-throughput assay and requires re-tooling and re-thinking of traditional single-exposure study designs. It is important to assess the impact of xenobiotic exposures on the genome and epigenome which regulate "stemness" within hES cells, as well as on differentiation along lineage-specific lines, since either effect can lead to aberrant programming resulting in potential developmental and reproductive hazards (Figure 8.2).

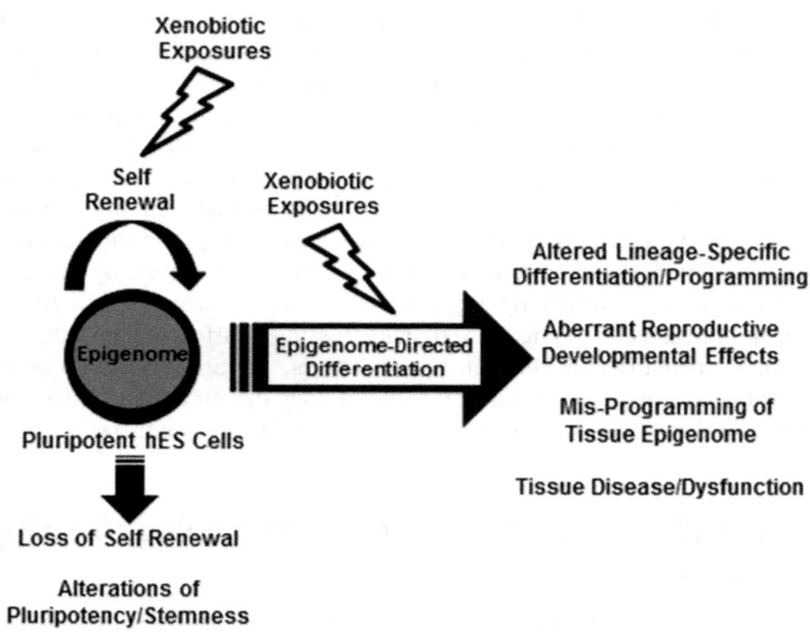

Figure 8.2 Xenobiotic exposure to human embryonic stem (hES) cells can disrupt the epigenomic circuity that maintains stemness and impact upon epigenome-directed lineage-specific differentiation.

Due to the long cell-culture doubling time and the desire to expose cells throughout all the phases of the cell cycle, unlike traditional *in vitro* toxicology testing which typically uses single exposures, experimental designs to assess cytotoxicity and metabolomic-based changes with hES cells use multiple days of repeat exposures.[20] These same designs are being adapted by our laboratory to examine the impact of teratogens and non-teratogens on the genome and epigenome of hES WA09 cells.

8.4 Stem Cells as a Biological Platform to Examine the Impact of Xenobiotics on the Genome and Epigenome

The maintenance of stemness and the differentiation of stem cells along specific lineages is regulated by networks of genes controlled by the stem cell epigenome, specifically miRNAs and lncRNAs.[9–11] Our laboratory is focused on developing *in vitro* assays integrating traditional measures of cellular health status with interrogation of the stem cell transcriptome response (protein-coding gene mRNAs, and the non-coding epigenome miRNA and lncRNAs) that regulates stem cell biology. Disruption of this highly conserved regulatory circuit by exposures to xenobiotics can alter finely tuned processes affecting stemness and pluripotency. Numerous epigenetic mechanisms regulating the expression of genes have been identified, and are interlinked, working together to regulate gene expression by affecting transcript stability, chromatin modifications and remodeling, recruitment of transcriptional cofactors, and nuclear organization in response to physiological mediators or xenobiotic stressors.

8.5 The miRNA and lncRNA Epigenome

The Encyclopedia of DNA Elements (ENCODE) project is compiling a comprehensive list of functional elements in the human genome that act at the protein and RNA levels, and regulatory elements that control how and when a gene is actively transcribed (www.genome.gov/encode/). Interrogation of the RNA content of a variety of human tissues and cell lines has revealed that, collectively, at least 75% of the genome is transcribed in one cell type or another.[21,22] The pervasive transcription of the cellular genome produces a broad array of RNA species, including ribosomal (r)RNAs, protein-encoding messenger (m)RNAs and a number of non-coding (nc)RNAs; protein-coding mRNAs account for less than 5% of the transcribed RNA species.[22] ncRNAs are defined as transcribed segments that produce RNA molecules that are not being translated into proteins.[23] While the pervasive transcription of the genome was once considered transcriptional noise and the functions of most ncRNAs was unknown, it is apparent that certain ncRNAs play critical roles as epigenetic regulators of gene expression for key biological processes controlling stem cell biology.[24]

ncRNAs interact with transcriptional complexes or mRNA species to indirectly regulate gene expression and protein translation and have been classified based on length. miRNAs are small (~22 nucleotide) ncRNA molecules that regulate gene expression and are particularly important in regulating stem cell "self-renewal", maintaining a pluripotent phenotype and stem cell differentiation by repressing the translation of specific mRNAs in stem cells and in differentiating daughter cells.[25,26] By base-pairing of miRNAs to the 3′ untranslated portion of a target mRNA or multiple target mRNAs, miRNAs downregulate protein translation. Since miRNA can bind to multiple discrete sequences, miRNAs can control networks of genes and pathways that produce an integrated phenotypic response to environmental stimuli (*e.g.* proliferation of specific immune cell lineages in response to inflammatory signals).[27] The significance of miRNA in the regulation of cellular response to cellular stressors and as circulating biomarkers of tissue injury has been reviewed by others.[28]

lncRNAs are larger ncRNA segments (>200 nucleotides) that are frequently associated with transcriptional complexes. lncRNAs are emerging as key epigenetic regulators of stem cell biology regulatory self-renewal and pluripotency. lncRNAs were initially discovered through their association with transcriptionally active chromatin complexes in stem cells.[24] Since they are frequently co-transcribed with nearby protein-coding genes, lncRNA transcription appears to be associated with their co-organization in the genome with transcriptional regulatory elements of protein-coding genes. lncRNAs function in a broad range of biological roles (see Rinn and Chang[24] for references), including interacting with chromatin complexes and affecting transcription; maintenance of DNA methylation/demethylation and transcriptional gene silencing; global p53-mediated DNA damage response; and stem cell pluripotency and differentiation.

Once referred to as the "dark matter" or "junk" DNA of the genome, lncRNAs are emerging as novel epigenomic gene regulators that are associated with a broad range of human disease including cancer, neurodegenerative disorders, and diabetes.[29,30] The essential role of lncRNAs in carcinogenesis can be illustrated by the fact that ~13% of the differentially expressed RNA transcripts in liver of mice exposed to the nongenotoxic carcinogen furan were lncRNAs; one of these (lncRNA p21) is known to have a role in the p53 DNA damage response and is a major regulator of UV-induced apoptosis in skin, a hallmark event of cancer.[31–33]

The pervasive transcription of the genome produces a broad range of RNAs with diverse functions beyond acting as a "messenger" between a gene and protein. A paradigm shift is emerging where novel RNA crosstalk is observed within gene regulatory networks mediated by ncRNAs. Aberrant RNA crosstalk within the epigenome has implications for human development and disease, and at present, this layer of gene expression regulation is not reflected in high-throughput toxicology assays. Interrogation of how the genome and epigenome (transcriptome) regulate hES biology and control hES cell response to xenobiotic toxicants is needed to advance predictive tools that can be used to examine compounds of unknown biological activity in toxicology safety assessments. Such information can be used to develop a better mechanistic understanding of the interaction of toxicants with the biologically unique hES cell genomic and epigenomic landscape. Identifying critical response genes and ncRNA networks and pathways can be used to construct transcriptome-based predictive biomarkers of epigenomic responses to xenobiotic exposures in hES cells, thus offering the potential to rapidly identify epigenetic toxicants that could affect embryonic developmental programs.[1,33]

8.6 Epigenomic Biomarkers and Toxicity Testing in the 21st Century

Cells respond to environmental stressors by activating the expression of genomic and epigenomic networks that activate biological and cellular response networks to act as cellular defense mechanisms. The integration of genomic tools with investigative toxicology test systems (referred to as toxicogenomics) is demonstrating that cellular and genomic responses to cytotoxic and genotoxic stress are based on the genetic and epigenetic constitution of the test system or individual and involve dose-dependent hierarchical activation of genetic pathways that can define the molecular mechanisms driving biological or phenotypic outcomes (phenotypic anchors). Although a number of studies have examined the impact of environmental agents on the cellular and tissue transcriptome, few studies have focused on integrating the tools of epigenomics into toxicology assessment. Many of the epigenomics tools used to examine stem cell biology and lineage-specific differentiation have matured sufficiently to be useful tools to address toxicological concerns regarding the interaction of environmental agents with the epigenome. The importance of hES cells in human development makes them a target cell population of concern and developing hES cell systems as toxicological tools for safety assessment and research should be a priority. The unique properties of hES cells integrated with toxicogenomics expression profiling of the genome and epigenome makes them a useful biological testing platform for toxicological studies that can evaluate the impact on the epigenome maintaining stemness and driving lineage-specific differentiation. This approach will reveal fundamental information on how the epigenome of stem cells responds to environmental agents that may also provide insights and future direction on how the dynamic epigenome of the developing embryo and stem and progenitor cells of organs may respond to environmental stressors.

Acknowledgements

The author is supported in part by National Institute of Environmental Health Sciences Small Business Innovation Research: R44ES022114.

References

1. M. D. Waters, M. Jackson and I. Lea, Characterizing and predicting carcinogenicity and mode of action using conventional and toxicogenomics methods, *Mutat. Res.*, 2010, **705**, 184–200.
2. A. Christianson, C. Howson and B. Modell, *March of Dimes Global Report of Birth Defects: The Hidden Toll of Dying and Disabled Children*, March of Dimes Birth Defects Foundation, White Plains, New York, 2006.
3. National Research Council; Committee on Developmental Toxicology, *Scientific Frontiers in Developmental Toxicology and Risk Assessment Board on Environmental Studies and Toxicology Commission on Life Sciences*, National Academies Press, Washington, DC, 2000.
4. OECD, *Test No. 414: Prenatal Development Toxicity Study*, OECD Publishing, Paris, 2001.
5. OECD, *Test No. 416: Two-Generation Reproduction Toxicity*, OECD Publishing, Paris, 2001.
6. J. L. Schardein, B. A. Schwetz and M. F. Kenel, Species sensitivities and prediction of teratogenic potential, *Environ. Health Perspect.*, 1985, **61**, 55–67.
7. D. C. Dolinoy and R. L. Jirtle, Environmental epigenomics in human health and disease, *Environ. Mol. Mutagen.*, 2008, **49**, 4–8.
8. T. H. Shepard, 2010, Catalog of Teratogenic Agents, Johns Hopkins University, Baltimore, MD, http://depts.washington.edu/terisdb/index.html.
9. M. Guttman, I. Amit, M. Garber, C. French, M. F. Lin, D. Feldser, M. Huarte, O. Zuk, B. W. Carey, J. P. Cassady, M. N. Cabili, R. Jaenisch, T. S. Mikkelsen, T. Jacks, N. Hacohen, B. E. Bernstein, M. Kellis, A. Regev, J. L. Rinn and E. S. Lander, Chromatin signature reveals over a thousand highly conserved large non-coding RNAs in mammals, *Nature*, 2009, **458**, 223–227.
10. T. R. Leonardo, H. L. Schultheisz, J. F. Loring and L. C. Laurent, The functions of microRNAs in pluripotency and reprogramming, *Nat. Cell Biol.*, 2012, **14**, 1114–1121.
11. M. Guttman, J. Donaghey, B. W. Carey, M. Garber, J. K. Grenier, G. Munson, G. Young, A. B. Lucas, R. Ach, L. Bruhn, X. Yang, I. Amit, A. Meissner, A. Regev, J. L. Rinn, D. E. Root and E. S. Lander, lincRNAs act in the circuitry controlling pluripotency and differentiation, *Nature*, 2011, **477**, 295–300.
12. I. Beerman and D. J. Rossi, Epigenetic Control of Stem Cell Potential during Homeostasis, Aging, and Disease, *Cell Stem Cell*, 2015, **16**, 613–625.
13. R. S. Thomas, M. A. Philbert, S. S. Auerbach, B. A. Wetmore, M. J. Devito, I. Cote, J. C. Rowlands, M. P. Whelan, S. M. Hays, M. E. Andersen, M. E. Meek, L. W. Reiter, J. C. Lambert, H. J. Clewell 3rd, M. L. Stephens, Q. J. Zhao, S. C. Wesselkamper, L. Flowers, E. W. Carney, T. P. Pastoor, D. D. Petersen, C. L. Yauk and A. Nong, Incorporating new technologies

into toxicity testing and risk assessment: moving from 21st century vision to a data-driven framework, *Toxicol. Sci.*, 2013, **136**, 4–18.
14. R. R. Tice, C. P. Austin, R. J. Kavlock and J. R. Bucher, Improving the human hazard characterization of chemicals: a Tox21 update, *Environ. Health Perspect*, 2013, **121**, 756–765.
15. National Research Council; Committee on Toxicity and Assessment of Environmental Agents, *Toxicity Testing in the 21st Century: A Vision and a Strategy*, National Academies Press, Washington, DC, 2007.
16. N. C. Kleinstreuer, J. Yang, E. L. Berg, T. B. Knudsen, A. M. Richard, M. T. Martin, D. M. Reif, R. S. Judson, M. Polokoff, D. J. Dix, R. J. Kavlock and K. A. Houck, Phenotypic screening of the ToxCast chemical library to classify toxic and therapeutic mechanisms, *Nat. Biotechnol.*, 2014, **32**, 583–591.
17. M. J. Waring, J. Arrowsmith, A. R. Leach, P. D. Leeson, S. Mandrell, R. M. Owen, G. Pairaudeau, W. D. Pennie, S. D. Pickett, J. Wang, O. Wallace and A. Weir, An analysis of the attrition of drug candidates from four major pharmaceutical companies, *Nat. Rev. Drug Discovery*, 2015, **14**, 475–486.
18. P. Fowler, R. Smith, K. Smith, J. Young, L. Jeffrey, P. Carmichael, D. Kirkland and S. Pfuhler, Reduction of misleading ("false") positive results in mammalian cell genotoxicity assays. III: sensitivity of human cell types to known genotoxic agents, *Mutat. Res., Genet. Toxicol. Environ. Mutagen.*, 2014, **767**, 28–36.
19. J. A. Thomson, J. Itskovitz-Eldor, S. S. Shapiro, M. A. Waknitz, J. J. Swiergiel, V. S. Marshall and J. M. Jones, Embryonic stem cell lines derived from human blastocysts, *Science*, 1998, **282**, 1145–1147.
20. J. A. Palmer, A. M. Smith, L. A. Egnash, K. R. Conard, P. R. West, R. E. Burrier, E. L. Donley and F. R. Kirchner, Establishment and assessment of a new human embryonic stem cell-based biomarker assay for developmental toxicity screening, *Birth Defects Res., Part B*, 2013, **98**, 343–363.
21. M. B. Clark, P. P. Amaral, F. J. Schlesinger, M. E. Dinger, R. J. Taft, J. L. Rinn, C. P. Ponting, P. F. Stadler, K. V. Morris, A. Morillon, J. S. Rozowsky, M. B. Gerstein, C. Wahlestedt, Y. Hayashizaki, P. Carninci, T. R. Gingeras and J. S. Mattick, The reality of pervasive transcription, *PLoS Biol.*, 2011, **9**, e1000625, discussion e1001102.
22. S. Djebali, C. A. Davis, A. Merkel, A. Dobin, T. Lassmann, A. Mortazavi, A. Tanzer, J. Lagarde, W. Lin, F. Schlesinger, C. Xue, G. K. Marinov, J. Khatun, B. A. Williams, C. Zaleski, J. Rozowsky, M. Roder, F. Kokocinski, R. F. Abdelhamid, T. Alioto, I. Antoshechkin, M. T. Baer, N. S. Bar, P. Batut, K. Bell, I. Bell, S. Chakrabortty, X. Chen, J. Chrast, J. Curado, T. Derrien, J. Drenkow, E. Dumais, J. Dumais, R. Duttagupta, E. Falconnet, M. Fastuca, K. Fejes-Toth, P. Ferreira, S. Foissac, M. J. Fullwood, H. Gao, D. Gonzalez, A. Gordon, H. Gunawardena, C. Howald, S. Jha, R. Johnson, P. Kapranov, B. King, C. Kingswood, O. J. Luo, E. Park, K. Persaud, J. B. Preall, P. Ribeca, B. Risk, D. Robyr,

M. Sammeth, L. Schaffer, L. H. See, A. Shahab, J. Skancke, A. M. Suzuki, H. Takahashi, H. Tilgner, D. Trout, N. Walters, H. Wang, J. Wrobel, Y. Yu, X. Ruan, Y. Hayashizaki, J. Harrow, M. Gerstein, T. Hubbard, A. Reymond, S. E. Antonarakis, G. Hannon, M. C. Giddings, Y. Ruan, B. Wold, P. Carninci, R. Guigo and T. R. Gingeras, Landscape of transcription in human cells, *Nature*, 2012, **489**, 101–108.

23. U. Munoz-Najar and J. M. Sedivy, Epigenetic control of aging, *Antioxid. Redox Signaling*, 2011, **14**, 241–259.
24. J. L. Rinn and H. Y. Chang, Genome regulation by long noncoding RNAs, *Annu. Rev. Biochem.*, 2012, **81**, 145–166.
25. W. T. Guo, X. W. Wang and Y. Wang, Micro-management of pluripotent stem cells, *Protein Cell*, 2014, **5**, 36–47.
26. V. Havelange and R. Garzon, MicroRNAs: emerging key regulators of hematopoiesis, *Am. J. Hematol.*, 2010, **85**, 935–942.
27. I. Alvarez-Garcia and E. A. Miska, MicroRNA functions in animal development and human disease, *Development*, 2005, **132**, 4653–4662.
28. O. F. Laterza, L. Lim, P. W. Garrett-Engele, K. Vlasakova, N. Muniappa, W. K. Tanaka, J. M. Johnson, J. F. Sina, T. L. Fare, F. D. Sistare and W. E. Glaab, Plasma MicroRNAs as sensitive and specific biomarkers of tissue injury, *Clin. Chem.*, 2009, **55**, 1977–1983.
29. T. Gutschner and S. Diederichs, The hallmarks of cancer: a long non-coding RNA point of view, *RNA Biol.*, 2012, **9**, 703–719.
30. O. Wapinski and H. Y. Chang, Long noncoding RNAs and human disease, *Trends Cell Biol.*, 2011, **21**, 354–361.
31. J. R. Hall, Z. J. Messenger, H. W. Tam, S. L. Phillips, L. Recio and R. C. Smart, Long noncoding RNA lincRNA-p21 is the major mediator of UVB-induced and p53-dependent apoptosis in keratinocytes, *Cell Death Dis.*, 2015, **6**, e1700.
32. L. Recio, S. L. Phillips, T. Maynor, M. Waters, A. F. Jackson and C. L. Yauk, Differential expression of long noncoding RNAs in the livers of female B6C3F1 mice exposed to the carcinogen furan, *Toxicol. Sci.*, 2013, **135**, 369–379.
33. E. A. Gibb, E. A. Vucic, K. S. Enfield, G. L. Stewart, K. M. Lonergan, J. Y. Kennett, D. D. Becker-Santos, C. E. MacAulay, S. Lam, C. J. Brown and W. L. Lam, Human cancer long non-coding RNA transcriptomes, *PLoS One*, 2011, **6**, e25915.

CHAPTER 9

Novel Data Streams in the Assessment of Mutagenicity and Carcinogenicity: Implications for Cancer Hazard Assessment

KATHRYN Z. GUYTON*[a] AND MICHAEL D. WATERS[b]

[a] Monographs Section, International Agency for Research on Cancer, 150, cours Albert Thomas, 69372 Lyon Cedex 08, France; [b] Integrated Laboratory Systems (ILS), Inc., PO Box 13501, Research Triangle Park, NC 27709, USA

*Email: GuytonK@iarc.fr

9.1 Integrating Across Multiple Data Streams to Reach Hazard Conclusions: Mechanistic Data can be Critical When Human Evidence is Less than Sufficient

Advances in cancer prevention have come from the identification of chemical hazards, stimulating policy actions to address them.[1–4] Carcinogenic hazard identification is an assessment of whether an agent causes cancer. Hazard identification does not predict the magnitude of cancer risks under

Issues in Toxicology No. 28
Toxicogenomics in Predictive Carcinogenicity
Edited by Michael D. Waters and Russell S. Thomas

Published by the Royal Society of Chemistry, www.rsc.org

specific conditions or in certain populations. The Monographs Programme of the International Agency for Research on Cancer (IARC) assesses carcinogenic hazards from chemical and biological agents, occupations, complex environmental exposures, and lifestyle factors. These evaluations provide an important first step in the international societal decision-making process to identify and control carcinogenic hazards. IARC Monograph evaluations are conducted according to published procedures for the identification, evaluation and integration of scientific evidence of cancer hazard (http://monographs.iarc.fr/ENG/Preamble/). These procedures have been in use for more than 40 years, and have been updated over the years.[5] According to these procedures, the IARC assembles international groups of scientists with a range of relevant scientific expertise (Working Groups) to review and assess the quality and strength of scientific evidence. These Working Group classify agents for cancer hazard, assessing the likelihood that the agents of concern pose a cancer hazard to humans.[6] The IARC has used this approach for over four decades, since the first Monograph meeting in 1972.[7] According to the IARC Preamble, classifications of carcinogenic hazard are into one of five categories, ranging from carcinogenic to humans to probably not carcinogenic to humans (Appendix 9.1).[8]

9.1.1 Introduction to the IARC Monographs Hazard Classification Process

The process for the preparation of an IARC Monograph is described clearly in the Preamble to the Monographs, which is published on the IARC Monographs website and as part of each Monograph. The IARC convenes advisory groups every 5 years to ensure that the topics of new Monographs reflect current research and public health priorities. These advisory groups comprise senior officials from national and international health agencies (who contribute information about carcinogens for which new evaluations would be useful) and senior scientists from cancer research institutes and academic organizations (who contribute information about cancer research in progress and anticipated publication dates). These advisory groups evaluate nominations solicited from individuals and organizations worldwide, recommending agents for first-time evaluation or for re-evaluation based on availability of significant new data. Thus the evaluation process is dynamic and reflects both new scientific results and emerging scientific understanding. For example, in 2013 the IARC sought nominations for agents to be evaluated in 2015–2019.[9] Researchers, health organizations, various stakeholders with different perspectives, and the public at large—a total of 49 people from 14 different countries—submitted nominations of 114 agents for consideration. In April 2014 the programme convened an advisory group of 21 senior scientists from 13 countries to prioritize agents by the extent of human exposure, suspicion of carcinogenicity, public health relevance, and public concern.[10]

Agents are selected for evaluation based on the prevalence and intensity of exposure (of both occupational groups and the general population) and availability of sufficient information for an evaluation of carcinogenicity, as well as advice from the IARC scientific council and advisory groups. Most evaluations concern specific compounds. Monographs are also published on occupations, industries or workplace factors (*e.g.* shift work, firefighting, painting, and leather tanning and processing). Individual chemicals to which exposures occur in these and other occupational settings have also been evaluated.

Once agents are selected for evaluation, expert Working Groups are assembled to review, evaluate, and integrate the available scientific evidence. The Working Groups then classify the agent as carcinogenic, probably carcinogenic, possibly carcinogenic, not classifiable, or probably not carcinogenic to humans (see Appendix 9.1). Working Group members are also responsible for writing the IARC Monograph, which must both review the available evidence and give the rationale for the Working Group's conclusions.

Members of the expert Working Groups are selected based on subject matter expertise as well as the absence of real or apparent conflicts of interest. Most Working Group candidates are identified from literature searches. Members are typically scientists who have conducted research relevant to the agent under review, but not necessarily on the specific agent. In addition to Working Group members, invited specialists, representatives of health agencies, stakeholder observers, and the IARC Secretariat also attend meetings. The list of potential meeting participants, including disclosure of relevant conflicts of interest, is posted on the IARC website several months in advance of the meeting (http://monographs.iarc.fr/ENG/Meetings/).

Importantly, only Working Group members conduct the actual evaluation.[11,12] IARC staff facilitate the evaluation process and ensure that the procedures described in the Preamble[8] are followed; however, they do not determine the outcomes.

IARC evaluations of carcinogenicity are based exclusively on publicly available information. The evidence comes from epidemiological studies, animal bioassays, pharmacokinetic studies, mechanistic experiments including short-term predictive tests, and surveys of human exposure. The evidence for cancer in humans and in experimental animals is reviewed systematically. The review is inclusive of scientific articles that have been published or accepted for publication in peer-reviewed scientific journals. Publicly available government reports are also included.

The initial assembly and organization of the literature is performed by IARC staff[12] according to the Preamble and further detailed in the instruction for authors.[13] The IARC conducts and documents literature searches (PubMed searches are documented using HAWCproject.org), incorporating relevant articles cited in prior IARC Monographs (if available) or identified through contact with investigators in the field. Working Group members are then assigned various authorship and peer review tasks.

They may refine existing or perform additional literature searches to identify any further publications that may be informative. Finally, relevant published or otherwise publicly available information may be identified in response to a public call for data before the Working Group meeting.

The Working Group meets for 8 days, during which the assembled draft documents are further reviewed, revised and summarized by four discipline-related subgroups. Specifically, the subgroups address evidence on exposure, cancer in humans, cancer in experimental animals, and mechanistic and related effects. All of the meeting participants, including the Working Group, invited specialists, stakeholder observers, representatives, and the IARC/World Health Organization secretariat, have access to the draft documents. The summaries are discussed within and across subgroups, and information from all disciplines is discussed in plenary sessions to inform the assignment of the agents to a specific carcinogenicity category based on the evaluation criteria.

Typically, several related agents are evaluated in a single 8 day meeting. After a full discussion of the scientific evidence, the Working Group members follow the published IARC procedures for evaluating and then integrating information from epidemiological studies and bioassays to arrive at a preliminary classification.[5] Although information on human exposure is not formally integrated into the overall assessment of carcinogenic hazard, these data are valuable for characterizing the timing, duration, and levels of exposure in the population, and in evaluating the quality of the exposure assessment in epidemiological studies. In contrast, mechanistic data may warrant a change from the preliminary classification based on evidence of carcinogenicity in humans and in experimental animals. This is a key step in the evaluation process and is the focus of this chapter.

9.1.2 Recent Examples

Mechanistic studies can provide critical information on the capacity of an agent to produce cancer in mammalian systems, including humans. The evaluation of mechanistic data in IARC evaluations has been a part of the process from the beginning of the Programme. In 1991, IARC convened a Working Group to explore how mechanistic data could be used to identify agents with the potential to cause cancer in humans.[14] Since 1991, it has been possible for an agent to be classified as carcinogenic to humans (group 1) in the absence of sufficient evidence in humans when there is sufficient evidence in experimental animals and "strong evidence in exposed humans that the agent acts through a relevant mechanism of carcinogenicity". The IARC Monographs have used mechanistic data to assign a higher or lower classification based on integration of epidemiological studies and cancer bioassays alone. Mechanistic data[15] have been influential in the overall evaluation of several important agents: an early example was the classification in group 1 of 2,3,7,8-tetrachlorodibenzo-*para*-dioxin (TCDD) based on mechanistic evidence, in volume 69 in 1997. This classification was later affirmed based on sufficient evidence of carcinogenicity in humans from epidemiological studies.[16] In the following sections we review examples of the application of mechanistic and molecular epidemiology studies in recent IARC Monograph carcinogen hazard evaluations.

9.1.2.1 Trichloroethylene

The chlorinated solvent trichloroethylene (TCE) is a ubiquitous environmental pollutant. Worldwide, TCE is commonly found in hazardous waste sites and contaminated water. In the European Union, an estimated 276 000 workers were exposed to TCE in the early 1990s.[3] Although occupational use of TCE has been declining,[4] exposures remain of concern, especially in developing countries, in the workplace and the environment. The general population is exposed through contaminated water and consumer products, including food.[17]

The carcinogenic hazard of TCE was the subject of a 2012 IARC Working Group evaluation (Section 9.1.2.1.3: Carcinogenicity of Trichloroethylene, Tetrachloroethylene, Some Other Chlorinated Solvents, and Their Metabolites). Information on exposures, relevant data from epidemiological studies, bioassays in experimental animals, and toxicity and mechanism of action studies was used to conclude that TCE is carcinogenic to humans (Group 1). Rusyn *et al.*[17] have summarized the key evidence forming the scientific basis for the IARC classification.

9.1.2.1.1 Trichloroethylene: Mechanistic, Epidemiological, and Other Supporting Evidence of Carcinogenic Hazard. TCE has been extensively studied for its carcinogenicity. Occupational case–control studies provide sufficient evidence of an association between TCE exposure and kidney cancer. Statistically significant risks of modest magnitude were reported in two independent meta-analyses of case–control and cohort studies of kidney cancer.[18,19] A meta-analysis also found positive associations between occupational TCE exposure and risks for non-Hodgkin lymphoma and cancer of the liver.[19] However, the IARC Monographs Volume 106 Working Group characterized this epidemiological evidence as limited. Evidence of carcinogenicity for cancers of the lung, cervix and esophagus was inadequate. Mouse liver and rat kidney tumors in both sexes were reported in multiple experimental animal bioassays. Other rodent tumors included mouse lung and rat testis, with less evidence of leukemias and lymphomas in rodents.[17]

There are two main pathways of TCE metabolism, yielding multiple toxic, mutagenic, and carcinogenic metabolites that are likely to contribute to the carcinogenicity of the parent compound (Figures 9.1 and 9.2).

The cytochrome P450 (CYP)-mediated oxidation of TCE predominates quantitatively in all species studied, especially at environmentally relevant concentrations or doses of TCE. The CYP-mediated oxidation pathway metabolites are, for the most part, chemically stable with evidence of systemic

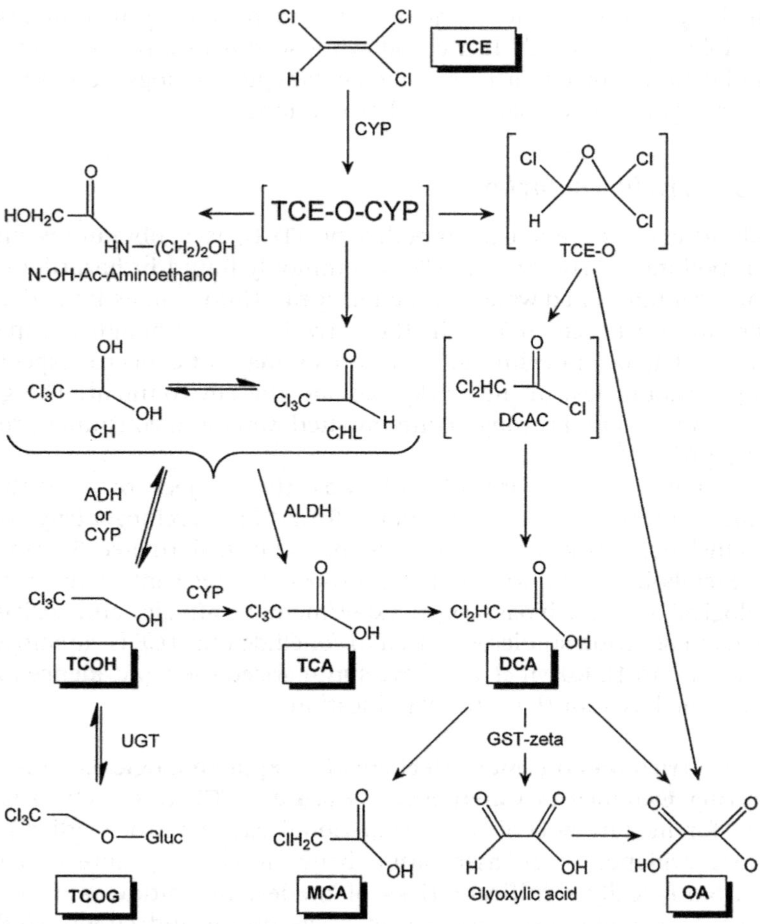

Figure 9.1 Scheme for biotransformation of trichloroethylene (TCE) by oxidative metabolism. TCE undergoes cytochrome P450 (CYP)-dependent oxidation to form either a TCE–CYP intermediate or an epoxide intermediate. Further processing through either non-enzymatic rearrangements or actions of aldehyde dehydrogenase (ALDH), alcohol dehydrogenase (ADH), CYPs, or glutathione (GSH) *S*-transferase zeta (GSTZ) yield a variety of metabolites, including chloral (CHL) and chloral hydrate (CH), dichloroacetate (DCA), trichloroacetate (TCA), trichloroethanol (TCOH) and its glucuronide (TCOG), monochloroacetate (MCA), and oxalate (OA). Names of metabolites that are recovered in urine are shown in boxes and those that are chemically unstable or reactive are shown in brackets. TCE-O: trichloroethylene oxide; DCAC: dichloroacetyl chloride; *N*-OH-acaminoethanol: *N*-hydroxyacetyl aminoethanol; UGT: UDP-glucuronosyltransferase.
Reprinted from Mutation Research/Reviews in Mutation Research, Vol. 762, L. H. Lash, W. A. Chiu, K. Z. Guyton and I. Rusyn, Trichloroethylene biotransformation and its role in mutagenicity, carcinogenicity and target organ toxicity, Pages 22–36, Copyright (2014), with permission from Elsevier.[20]

availability and urinary excretion (Figure 9.1). Other than CHL/CH, these oxidative metabolites are not genotoxic and contribute to carcinogenicity through mechanisms other than direct interaction with DNA. In contrast, metabolites generated by the glutathione (GSH) *S*-transferase (GST) pathway are chemically reactive and have the capacity for direct genotoxicity and mutagenicity (Figure 9.2).

Lash *et al.*[20] highlight recent advances in the understanding of TCE metabolism. The CYP-mediated oxidation pathway metabolites trichloroacetate (TCA), dichloroacetate (DCA), trichloroethanol and its glucuronide (TCOH and TCOG), as well as the GST-pathway metabolite *N*-acetyl-*S*-(1,2-dichlorovinyl)-L-cysteine (NAcDCVC), have been identified in the urine of exposed humans and other species (mostly rats and mice). For the GSH pathway, inter-organ processing occurs. *S*-(1,2-dichlorovinyl)glutathione (DCVG) primarily forms in the liver, and is efficiently excreted into bile or plasma. In the bile, the cysteine conjugate DCVC is formed, and returns to the liver *via* enterohepatic recirculation. In the liver, most DCVC is *N*-acetylated to form the mercapturate NAcDCVC, which is in turn excreted into plasma. Once in plasma, DCVG and NAcDCVC are extracted by the kidneys. Formation of DCVG *in situ* in the kidney also occurs, although to a lesser extent.

Questions remain concerning the extent of metabolism variability across species and individuals, and how the resulting reactive moieties may contribute to TCE-induced mutation and cancer. A recent analysis evaluated TCE metabolism data from genetically diverse mouse strains to inform population-based estimates of toxicokinetic variability.[21] Oxidative and GSH conjugation metabolism data concerning TCE were collected from 16 inbred and one hybrid mouse strains. These data were then used in the calibration and extension of physiologically based pharmacokinetic (PBPK) models. TCE metabolism to oxidative and GSH conjugation metabolites was variable across strains, as evidenced by concentration–time profiles. Interestingly, the extent of variability was less for median metabolic flux through oxidation (five-fold) than that for GSH conjugation (10-fold). Less variability was seen in median predictions for TCA (two-fold) than DCA production (five-fold). Nonetheless, the predicted variability for DCA was exceeded by the uncertainty bounds.

TCE metabolites can lead to target organ mutagenicity, and species and individual sex and tissue differences in catalytic enzyme activity are important determinants of the types and amounts of metabolites formed. Evidence of genotoxicity of TCE and its metabolites is strongest for metabolites formed from the GSH pathway in the kidney.

As noted in the IARC Monograph Volume 106, a recent US National Cancer Institute study on TCE found increased renal cancer risk among individuals carrying at least one intact GSH-*S*-transferase GSTT1 allele (active genotype), but not among individuals with two deleted GSTT1 alleles (null genotype).[22] As discussed, TCE-associated renal genotoxicity occurs predominantly through reductive metabolism involving GST conjugation. Thus the enhanced risk among individuals carrying polymorphisms in genes that are important in the reductive metabolism of TCE supports the biological plausibility of the cancer association. This additional mechanistic insight strengthened the overall hazard classification in Monograph Volume 106 and, importantly, revealed susceptibilities due to gene polymorphisms relevant for identifying potentially vulnerable populations.[23]

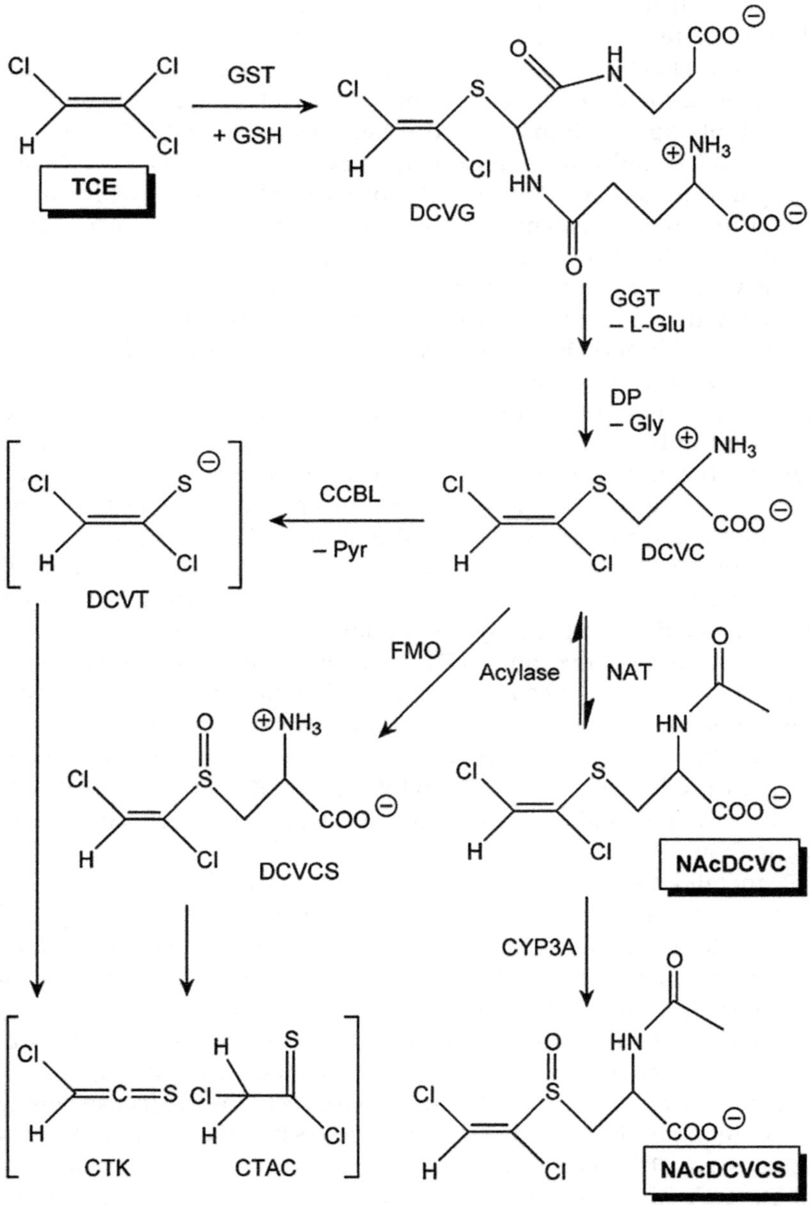

The metabolism and fate of TCE, and the mechanisms by which TCE causes carcinogenicity continue to be the subject of scientific investigation. In the liver, multiple mechanisms may contribute to carcinogenicity, including immune dysregulation, epigenetic alterations, cytotoxicity and secondary oxidative stress, alteration of proliferation and/or apoptosis.[23]

Few studies are informative of the cancer mechanisms operative for other cancers, especially for non-Hodgkin lymphomas in humans. Evidence of inter-individual variability in humans and experimental animals has highlighted the critical role of TCE metabolism in its carcinogenicity. Studies are ongoing to characterize the metabolites and mechanisms that impact TCE carcinogenicity, and the influence of co-exposures to other carcinogens. Additionally, susceptibility to TCE carcinogenicity in exposed individuals is being investigated in epidemiological studies.[23]

Overall, the classification of TCE as carcinogenic to humans (Group 1) was based on sufficient evidence for the carcinogenicity of TCE in humans. Sufficient evidence for carcinogenicity in experimental animals was provided by the induction of liver tumors in more than two independent studies in mice. Additionally, toxicokinetic and mechanistic data support and provide biological plausibility for a causal link between TCE exposure and cancer.[23]

9.1.2.1.2 Human Noncancer Health Effects of Trichloroethylene: Key Findings and Scientific Issues. Aside from its carcinogenicity, TCE also poses a potential human health hazard due to noncancer toxicity. Specifically, target organs of TCE toxicity are the central nervous system, kidney,

Figure 9.2 Scheme for biotransformation of trichloroethylene (TCE) by the glutathione (GSH) conjugation pathway. TCE undergoes conjugation with GSH to yield the GSH *S*-conjugate DCVG. After processing to yield the cysteine *S*-conjugate DCVC, three potential fates are detoxication to yield the mercapturate NAcDCVC or bioactivation by either the cysteine conjugate *b*-lyase to yield 1,2-dichlorovinylthiol (DCVT), which rearranges to yield thioacylating species, or the flavin-containing mono-oxygenase to yield DCVC sulfoxide. The mercapturate can also be deacetylated to regenerate DCVC or it can undergo CYP3A-dependent sulfoxidation. Names of metabolites that are recovered in urine are shown in boxes and those that are chemically unstable or reactive are shown in brackets. CTAC: chlorothionoacetyl chloride; CTK: chlorothioketene; CYP3A: cytochrome P-450 3A; DCVC: *S*-(1,2-dichlorovinyl)-L-cysteine; DCVG: *S*-(1,2-dichlorovinyl)glutathione; DCVCO: DCVC sulfoxide; DP: dipeptidase; FMO: flavin-containing mono-oxygenase; GGT: *g*-glutamyltransferase; GST: GSH *S*-transferase; NAcDCVC: *N*-acetyl-*S*-(1,2-dichlorovinyl)-L-cysteine; NAcDCVCS: NAcDCVC sulfoxide; NAT: *N*-acetyltransferase.
Reprinted from Mutation Research/Reviews in Mutation Research, Vol. 762, L. H. Lash, W. A. Chiu, K. Z. Guyton and I. Rusyn, Trichloroethylene biotransformation and its role in mutagenicity, carcinogenicity and target organ toxicity, Pages 22–36, Copyright (2014), with permission from Elsevier.[20]

liver, immune system, male reproductive system, and the developing embryo/fetus. These conclusions are from a review of key findings and scientific issues by the US Environmental Protection Agency (EPA).[24] Studies on the health effects of TCE continue to report on effects such as kidney carcinogenicity and toxicity.[18,25,26]

9.1.2.1.3 Carcinogenicity of Trichloroethylene, Tetrachloroethylene, Some Other Chlorinated Solvents, and Their Metabolites. As indicated by the discussion above, the biotransformation of TCE is well characterized in humans and animals. The primary pathway involves oxidative metabolism by CYP enzymes and a secondary pathway entails GSH conjugation by GSH *S*-transferase enzymes. Oxidative metabolites include TCA, DCA, and chloral hydrate (CH). The IARC Monograph Volume 106 evaluated the carcinogenicity of several TCE metabolites, as well as of the related chlorinated solvent tetrachloroethylene.[23]

Tetrachloroethylene is one of the most widely used chlorinated solvents, primarily in dry cleaning. It is a widespread environmental contaminant that is found in ambient air, indoor air, soil, drinking water, and groundwater, leading to general population exposures. Absorption following exposure to humans and laboratory animal species is rapid. Distribution to tissues occurs *via* the systemic circulation. Tetrachloroethylene is metabolized, and excreted primarily in breath as unchanged tetrachloroethylene or CO_2, or in urine as metabolites.[23]

Similar to TCE, tetrachloroethylene is also metabolized *via* CYP-mediated oxidation and GSH conjugation. TCA is the predominant metabolite. An updated PBPK model has recently been developed and applied to understand the role of metabolism in the toxicity of tetrachloroethylene.[27] In humans, oxidative metabolism was predicted to be low. In contrast, GSH conjugation metabolism is more uncertain and may be high, low, and/or highly variable. These PBPK model predictions informed the extent and nature of metabolism in different target tissues, and the extrapolation across species and routes of exposure.[28]

The IARC Monograph Working Group concluded that the evidence for tetrachloroethylene carcinogenicity in humans was limited, with positive associations observed for cancer of the bladder.[23] Sufficient evidence of carcinogenicity in experimental animals was provided by rodent neoplasms. These included mouse liver tumors, as well as rat tumors of the hematopoietic system, testes, kidney, and brain. Based on the limited evidence in humans and the sufficient evidence in experimental animals, tetrachloroethylene was classified as probably carcinogenic to humans (Group 2A). The US EPA[29] reached a similar conclusion following the *Guidelines for Carcinogen Risk Assessment*.[30] The characterization of tetrachloroethylene by the US EPA as "likely to be carcinogenic to humans" by all routes of exposure was based on suggestive evidence of carcinogenicity in epidemiological studies and conclusive evidence of increased tumor incidence in exposed rodents.[31–33] Recent analyses of tetrachlorethylene exposure and cancer continue to add support for the carcinogenicity classifications of tetrachloroethylene.[34–39]

In terms of noncancer outcomes, tetrachloroethylene exposure has been associated with neurotoxic outcomes in human studies of controlled exposure, occupational, and residential designs as well as in experimental animal studies. Evidence supporting the neurotoxicity of tetrachloroethylene also continues to accumulate, with recent studies examining effects such as illicit drug use;[40] mental illness;[41] visual effects;[42] visuospatial functioning, learning and memory, motor, attention, and mood;[43] and Parkinson's disease.[44]

CH is a byproduct of water disinfection, and is used as a sedative. Evidence of carcinogenicity in humans was inadequate in the IARC Monograph Volume 106 evaluation. In experimental animals, sufficient evidence of carcinogenicity was provided by an increased incidence of hepatocellular adenomas and carcinomas in several studies of mice. The overall classification of CH was upgraded to group 2A, on the basis of mechanistic evidence. This included evidence of CH genotoxicity in studies in many experimental systems. Importantly, micronucleus induction was observed in infants administered CH as a sedative.[45]

Epidemiological evidence was inadequate for DCA, TCA, and 1,1,1,2-tetrachloroethane, and 1,1,2,2-tetra-chloroethane. However, several chronic bioassays in mice show that DCA, TCA, 1,1,1,2-tetrachloroethane, and 1,1,2,2-tetrachloroethane increased the incidence of hepatocellular tumors. Mechanistic evidence was not strong for any of the agents, and did not impact the overall carcinogenicity classification. Accordingly, these four chlorinated agents were classified by IARC as possibly carcinogenic to humans (Group 2B) on the basis of sufficient evidence for carcinogenicity in animals.

9.2 Predicting Chemical Carcinogenicity Using Mechanistic Data

Experimental methods for predicting chemical carcinogenicity have not been substantially updated in the last 20 years, despite considerable advances in scientific understanding of the public health significance of these exposures. As discussed by Guyton *et al.*,[46] new bioanalytical approaches (*e.g.* transcriptomics, proteomics, and metabolomics) could be applied in human, animal, and *in vitro* studies to better characterize a wider array of hazard traits and to improve prediction of the potential carcinogenicity of chemicals. Responses to single or multiple chemicals may be evident at multiple levels. Indeed, deleterious effects at the genetic level (*e.g.* DNA damage or genome instability), may be accompanied by other changes in epigenetic endpoints (*e.g.*, DNA methylation alterations), or in signaling pathways (*e.g.* apoptosis or proliferation). While current carcinogenicity predictive systems are primarily focused on genetic damage, the other accompanying changes may also afford opportunities to characterize and predict cancer. Several examples are elucidated in this section.

9.2.1 Toxicogenomics Studies to Predict Carcinogenicity

As discussed in previous chapters in this volume, toxicogenomics and statistical classification methods have progressed such that it may be possible to use them in acute or subchronic studies to predict carcinogenicity. Cancer-relevant gene sets have been identified that appear to discriminate compounds that cause cancer from those that do not. The observed changes are multifactorial in nature, indicating that a combination of pathway-associated gene expression profiles could provide insights into underlying mechanisms might have utility in predicting chemical carcinogenesis. Together, these studies demonstrate that toxicogenomics, aided by statistical classification methods and pathway analysis tools, can identify gene expression profiles characteristic of diverse carcinogenic mechanisms.

9.2.2 *In vitro* Testing for Carcinogenicity—Moving Beyond Batteries

The genetic toxicology battery has been widely implemented due to its utility for identifying carcinogens that induce endpoints such as mutations, chromosome damage, and DNA damage (*e.g.*, strand breaks and DNA adducts). Nonetheless, this test battery is not designed to detect carcinogens operating through other mechanisms, including by disrupting epigenetic or signaling pathways. Alternative high-throughput methodologies are being explored to address these deficiencies. For instance, a DNA deletion recombination assay in yeast, which has high genomic sequence homology to humans, was able to discern a variety of carcinogens, including those with no activity in bacterial mutagenesis tests (reviewed by Ku *et al.*[47]).

Another promising possibility concerns detection of pathway-level changes in human cell-based systems.[48] In human TK6 cells, gene expression signatures were able to discriminate between four types of cellular stress relevant to carcinogenesis, including oxidative stress (arsenite and hydrogen peroxide), ionizing radiation, DNA damage (ultraviolet radiation, methyl methanesulfonate, adriamycin, camptothecin, and *cis* platinum(II) diammine dichloride), and other stresses that do not directly target DNA (heat shock, osmotic shock, and 12-*O*-tetradecanoylphorbol 13-acetate).[48] Toxicogenomic analysis can facilitate characterization of such stress agents, including those that interact with but do not covalently bind DNA.[49] A gene expression signature was identified in yeast that could discriminate genotoxic from cytotoxic stress.[50] Such gene signatures of the cellular responses to exposure have the advantage of incorporating information on multiple operant mechanisms. Accordingly, such mechanistic assays could bridge between screening tests and carcinogenesis bioassays by elucidating mechanisms most relevant to human cancer hazard evaluation. In addition to aiding interpretation of results from the current genotoxicity battery, they may provide a basis for novel assays. Finally, the knowledge gained on molecular pathways can also inform biomarker selection in epidemiological studies of occupational or environmental exposures, or in clinical intervention trials.

9.2.3 Using Biomarkers of Cancer Risk in Molecular Epidemiology

Human biomarker studies have potential not only in identifying possible carcinogens, but in characterizing susceptibility factors as well. Chromosome aberrations have particular promise for use in a classification scheme. Chromosome aberrations have been validated as biomarkers of future cancer risk in humans, and several investigations have demonstrated their ability to predict increases in future cancer risk.[51–55] They also have utility as biomarkers of early biological effect, for example in populations with carcinogenic exposures. The predictive ability of micronuclei for future cancer risk has been recently shown in humans,[51] but further confirmatory investigations are needed to affirm this predictive ability of the biomarker.

Microarray studies have shown that no more than 10% of genes screened in peripheral blood lymphocytes or solid tissues (*e.g.* liver) are differentially expressed with carcinogen exposures, including to benzene,[56] dioxin,[57] arsenic,[58–60] welding fumes,[61] or metal fumes.[62] A recent review by Sen *et al.*[63] supported the potential of gene signatures associated with exposure to cigarette smoke to be validated as predictive cancer biomarkers. Similarly, patterns of genes altered in arsenic-induced skin lesions,[58] the aryl hydrocarbon receptor pathway for dioxins and polycyclic aromatic hydrocarbons, the DNA damage response, and aneuploidy-associated genes[64] could also be applied in cancer prediction. Finally, profiling of DNA methylation, histone modifications and micro-RNAs[65] also hold promise, although the genome-wide methodology poses greater technical challenges than toxicogenomics and population-level data have not been developed. In addition to their application in cancer hazard classification, the relationships between dose and response at low levels of exposure could also be explored by applying these methods. Finally, cataloguing an increased number of chemicals according to these endpoints will advance understanding of the mechanisms underlying carcinogenesis.

It is increasingly evident that multiple biological changes occur during the conversion of a normal cell to a transformed cell and ultimately a tumor.[66] Chemicals can impact this complex process in multiple ways. Proof-of-concept studies of toxicogenomic and systems biology approaches have given an initial demonstration of their potential to predict mechanistically diverse cancer hazards.[48,67–72] Further investigation of this potential and, moreover, validation efforts, are warranted. A broad consideration of how chemical carcinogens change biology may help to stimulate new avenues for identifying possible or probable carcinogens.

The IARC Monographs recently updated the assessments of the more than 100 agents classified as Group 1, carcinogenic to humans (IARC Monographs Volume 100). This endeavor underscored the lack of a systematic method for evaluating mechanistic data to support conclusions regarding human hazard from exposure to carcinogens. In subsequent workshops, an international Working Group of experts identified 10 key characteristics, one or more of which are commonly exhibited by known human carcinogens. The 10 characteristics are the abilities of an agent to: (1) act as an electrophile either directly or after metabolic activation; (2) be genotoxic; (3) alter DNA repair or cause genomic instability; (4) induce epigenetic alterations; (5) induce oxidative stress; (6) induce chronic inflammation; (7) be immunosuppressive; (8) modulate receptor-mediated effects; (9) cause immortalization; and (10) alter cell proliferation, cell death, or nutrient supply.[73]

Future efforts could entail development of novel predictive *in vitro* and *in vivo* assays to evaluate the carcinogenic potential of agents. Additionally, biomarkers could be developed to aid study in human populations exposed to carcinogenic compounds. Especially considering the cost and time requirements of animal bioassays, and that few such assays are performed, novel approaches for categorizing chemicals as to their human carcinogenic are merited. Such approaches range from *in vitro* and experimental animal assays to biomarker-based studies in the exposed human population. Moreover, additional characterization of the diverse carcinogenic mechanisms of chemicals can improve understanding of susceptibilities, clarify human dose–response relationships and, accordingly, support scientifically based public health decisions.[74]

9.2.4 Filling Data Gaps: Predicting Chemical Carcinogenicity Using Modeling can Aid in Prioritization Exercises

As reviewed by Rusyn *et al.*,[75] quantitative structure–activity relationship (QSAR) models are widely employed to predict toxicity *in vivo*.[76,77] They are applied in the early evaluation of drug candidates, adding value to candidate selection. They also have application in sustainability evaluations of chemicals in commerce. Such tools also hold promise for applications in prioritizing agents for cancer hazard evaluation. Traditional QSAR models, utilizing numerical descriptors to represent inherent chemical properties, can be applied to many different molecules. Nonetheless, their predictive value is often limited by the availability of data on relevant endpoints. As part of the Tox21 program of the US government, quantitative *in vitro* screening data on thousands of environmental chemicals in a broad array of experimental systems are now accumulating.[78] In addition, toxicogenomics data collected on hundreds of chemicals and made publicly accessible could prove potentially useful for predictive toxicity modeling. Hybrid QSAR-like quantitative models could exploit these molecular bioactivity data, in combination with chemical structural information, to advance carcinogenicity prediction. The benefits include improvements in the accuracy of models, enhanced interpretation of the most predictive features, and expanded applicability domains for wider chemical space coverage.[75]

9.2.5 Addressing Human Variability

Human variability in toxic response is a key factor in the selection and application of dose-extrapolation approaches, but it can also play an important role in identifying cancer hazards. Zeise *et al.*[79] reviewed opportunities to apply data from novel methodologies in addressing human variability. Multiple aspects of biological variability can be informed by data from emerging evidence streams, and these have application in a number of different chemical screening and evaluation approaches.[79,80] In one recent example, a Bayesian PBPK model-based analysis was used to address the complex toxicokinetics of TCE, systematically estimating model parameters and characterizing their uncertainty and variability.[81] Additionally, evidence emerging from rodent models and from *in vitro* screening efforts that have incorporated population-based genetic determinants (reviewed by Rusyn *et al.*[82]), has advanced the understanding of genetic variability in toxicant response. This approach has also aided identification of factors contributing to genetic susceptibility to the toxic phenotypes induced by a diversity of toxicants. Another promising approach is to characterize population-based variability by integrating information on genetics, molecular pathway data, and clinical biomarkers and observations. Complementary to these are genome-wide[83] and exposure-wide[84] approaches for assessing human population variability in toxic response. Predictive models with application in a range of decision-making contexts can be developed and deployed using data from these emerging evidence streams.

The toxic response to chemicals is controlled by many factors. Of critical importance is the extent of chemical metabolism. In some instances, metabolism can enhance chemical toxicity, but metabolism can also lead to chemical detoxification and excretion. The metabolism of chemicals can be influenced by exposure, lifestyle, diet, and other factors that induce or inhibit the activity of key enzymes. Additionally, numerous genetic variants (polymorphisms) that affect the function of metabolic enzymes have been identified. These polymorphisms have been demonstrated to impact the toxic response to therapeutic drugs, as well as to environmental chemicals. Other polymorphisms have been characterized in detoxification and repair genes. All of these polymorphisms may influence the extent of population variability in toxic response, due to differences in internal dose as well as in response and repair mechanisms. As an early example, drug disposition studies indicated a high extent of variability in the reaction of patients to isoniazid (a tuberculosis drug) or debrisoquine (an antihypertensive). The variability was attributed to polymorphisms, in the acetylation pathway for isoniazid (acetylator phenotype) or in CYP for debrisoquine (poor metabolizer phenotype). Subsequently, the genetic basis for pharmacokinetic variability has been explored for a number of additional metabolic enzymes.

A recent series of analyses[80,85–91] explored the impact on population variability of genetic polymorphisms in a range of enzymes. These analyses were carried out on three phase-I metabolism enzymes [CYP2E1, CYP2D6, and paraoxonase-1 (PON1)], four phase-II conjugating enzymes [GSTs, *N*-acetyltransferases (NATs), UDP-glucronyltransferase, and sulfotransferases], and three enzymes involved in metabolic detoxification [epoxide hydrolase, nicotinamide adenine dinucleotide phosphate quinine oxidoreductase 1, and aldehyde dehydrogenase 2 (ALDH-2)]. For the enzymes studied, polymorphisms have been identified that could contribute to considerable population variability in enzyme activity. Key substrates were identified, and the role of the enzyme in chemical activation or detoxification was considered. The approach entailed characterization of the type and number of variants, their effect on enzyme function, and their frequency within the human population. Among them, it was feasible to generate distributional analyses using Monte Carlo simulations, based on the major polymorphisms identified for CYP2D6 and PON1 (phase I), GSTM1/T1/P1 and NAT1/2 (phase II) and ALDH2 (detoxication) (Table 9.1). The distributional analyses could predict and characterize the population distribution to inform inter-individual variability in enzyme activity.

For enzymes that were the subject of full Monte Carlo analysis (CYP2D6, PON1, GSTs, NATs, and ALDH2), the resulting inter-individual variability

Table 9.1 Enzyme distributional analyses developed for genetic polymorphisms.[a]

Enzyme	Type	Major polymorphisms included in distributional analysis	Ref.
CYP2D6	Phase I	7 Different SNPs contribute to poor metabolizer phenotype	Neafsey *et al.*, 2009[90]
PON1	Phase I	1 Coding region SNP alters activity in substrate-dependent manner; 2 regulatory sequence SNPs decrease gene expression	Ginsberg *et al.*, 2009[87]
GST M1/T1/P1	Phase II	Null polymorphisms in M1 and T1; 3 coding region SNPs in P1 alter activity in substrate-dependent manner	Ginsberg *et al.*, 2009[86]
NAT1/NAT2	Phase II	5 Different SNPs in NAT2 contribute to slow acetylator phenotype	Walker *et al.*, 2009[91]
ALDH2	Detoxification	*2: Null activity	Ginsberg *et al.*, 2002[85]

[a]Abbreviations: ALDH: aldehyde dehydrogenase; CYP: cytochrome P450; GST: glutathione *S*-transferase; NAT: *N*-acetyltransferase; PON: paraoxonase; SNP: single nucleotide polymorphism.

in enzyme activity was substantial. In particular, comparison of the 99th percentile to the population median activity indicated that a significant fraction of the population (≥20%) would be more than a half log (3.2-fold) different from the median. The 3.2-fold difference is a useful benchmark because toxicokinetics is commonly assumed, in the absence of other data, to contribute 3.2-fold to the overall 10-fold inter-individual variability uncertainty factor. The distributional analyses revealed instances of >3.2-fold variability. For instance, a high proportion of individuals are estimated to be more than 10-fold different from the median for GSTM1 and T1, due to the common null polymorphisms in these enzymes. For ALDH2, NAT2, and CYP2D6, a smaller (<10%) proportion of the population emerged as distinct subgroups with greater than 10-fold different median activity.

Overall, the characterizations demonstrate the substantial influence of the null phenotype for enzymes such as ALDH2,[85] GSTM1,[86] and GSTT1.[86] Indeed, a large percentage of the population lacks any activity of GSTM1 and T1. In contrast, multimodal distributions were evidenced for PON1, NAT2, and CYP2D6. These multimodal distributions indicate clear separation of phenotypic subgroups of activity. Accordingly, to better capture human variability in internal dose and risk, the resulting enzyme variability distributions could be used as input information when modeling internal dose (using PBPK modeling). In developing such models, consideration of several other factors is warranted, to best elucidate potential impact of enzyme variability in toxicant metabolism and clearance. (For instance, hepatic blood flow can limit the importance of enzyme variance and compensatory clearance pathways can also affect the impact of a specific polymorphism.)[92,93] In addition, because probe drugs were used in developing the distributions for certain enzymes (particularly for CYP2D6, NAT2, and ALDH2), the general applicability to other chemicals requires consideration. As suggested by molecular epidemiological evidence indicating associations with health outcomes, this information may be important for understanding toxicant impacts. Accordingly, the results may best be utilized as inputs to PBPK modeling that can incorporate and integrate the results in the context of these other important factors.

Additionally, these analyses support occupational epidemiology findings concerning the potential impact of specific genotypes. For instance, a relationship between susceptibility to cancer or DNA adducts and the presence of the polymorphism has been demonstrated for GST-T1 + workers exposed to TCE,[22,94] and for GST-M1-null polycyclic aromatic hydrocarbon-exposed coke oven workers.[95] These findings highlight the significant potential for genetic polymorphisms to modify toxicant response, demonstrable in studies where exposure dose and polymorphism are identified in workers.

The analyses can also indicate whether it is appropriate to consider internal dose variability in PBPK modeling, and provides published variability distributions for this purpose. These distributions could be used to extend and update analyses of, for example, the modulation of methylene chloride cancer risk by the GSTT1 polymorphism, as has been shown.[96,97] As another example, the role of polymorphisms in modulating the internal dose to warfarin and parathion has been analyzed.[98,99] As an alternative, pathway-specific adjustment factors that could be could be developed and applied instead of the default 3.2-fold pharmacokinetic uncertainty factor. This is particularly promising for ALDH-2, NAT2 and CYP2D6, for which *in vivo* drug clearance data were utilized.[100] Such an approach could advance screening-level approaches applicable to a particular metabolic pathway, but would lack a more complete characterization of variability and uncertainty for a particular toxicant as is possible in PBPK modeling.

In summary, the population distributions of enzyme activity can be incorporated into predictive PBPK/Monte Carlo assessments. They afford the advantage of a population-level simulation of internal dose. Additionally, specific subgroups can be addressed (*e.g.* those having low ALDH-2, or GSTM1 and T1 double-null genotype). Overall, these analyses can aid in refining estimates of the variability in internal dose to better inform full consideration of subpopulations carrying influential polymorphisms in hazard and risk assessment of chemicals.

9.3 Future Directions in Mechanistic Data Evaluations to Support IARC Monographs

9.3.1 Systematic Approaches to Identifying and Presenting Mechanistic Evidence in Human Health Assessments

As reported by Kushman *et al.*[101] and recently recommended in National Research Council reports to the US EPA,[102] better documentation of how the relevant literature was identified and selected for presentation can improve transparency in chemical hazard assessments. A systematic approach to the search, selection, and presentation of studies was explored in a recent case study concerning the carcinogenic of di(2-ethylhexyl)phthalate (DEHP) in the liver.[101] An extensive literature, comprising >3000 publications, was available. Based on published reviews,[103–109] relevant mechanistic events were identified (Figure 9.3). A PubMed search strategy incorporated terms for DEHP and its metabolites (*e.g.* MEHP), the identified mechanistic events, and the species of interest. Pertinent studies were selected after application of exclusion criteria to the retrieved literature.

Literature trees (Figure 9.4) were used to document the number of identified and selected articles. Evidence tables and accompanying succinct narratives were used to clearly summarize the included studies. Documentation of the literature review process was achieved using the Health and Environmental Research Online (http://hero.epa.gov/) database. Overall, this case study demonstrated the feasibility of review a large and complex literature database. Transparency, reproducibility and subsequent synthesis of information across studies was improved.

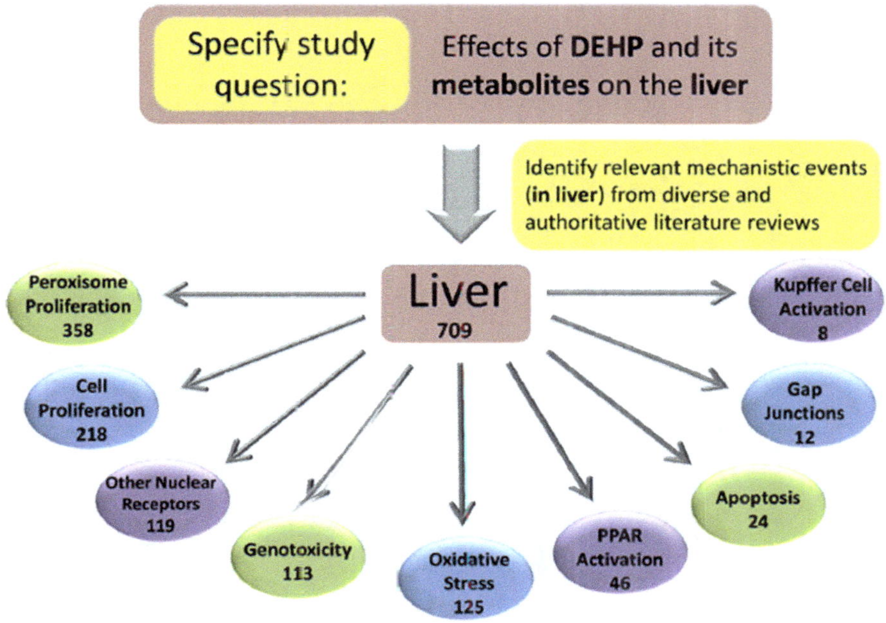

Figure 9.3 Mechanistic events relevant to hepatic carcinogenesis of di(2-ethylhexyl)-phthalate (DEHP), identified from reviews. Numbers indicate the number of articles identified from PubMed searches for DEHP, liver and the identified mechanistic events. PPAR: peroxisome proliferator-activated receptor.
Reprinted from Regulatory Toxicology and Pharmacology, Vol. 67, M. E. Kushman, A. D. Kraft, K. Z. Guyton, W. A. Chiu, S. L. Makris, I. Rusyn, A systematic approach for identifying and presenting mechanistic evidence in human health assessments, Pages 266–277, Copyright (2013), with permission from Elsevier.[101]

9.3.2 High-throughput Data Analysis

A 2014 advisory group to the IARC Monographs endorsed the analysis and appropriate presentation of high-throughput and high-content data streams in Monographs. They highlighted the increasing availability of high-throughput and high-content data streams, including evidence from genomic, transcriptomic, metabolomics, and other such studies. Consistent with this advice, high-throughput screening data generated by the Tox21 and ToxCast research programs of the US government[110,111] were analysed to inform evaluations about the *in vitro* bioactivity of the chemicals included in several recent IARC Monographs.

For volume 110, high-throughput data were evaluated on perfluorooctanoic acid (PFOA). PFOA is a persistent environmental contaminant throughout the world, and has been detected at low levels in the general population. PFOA was classified as possibly carcinogenic to humans

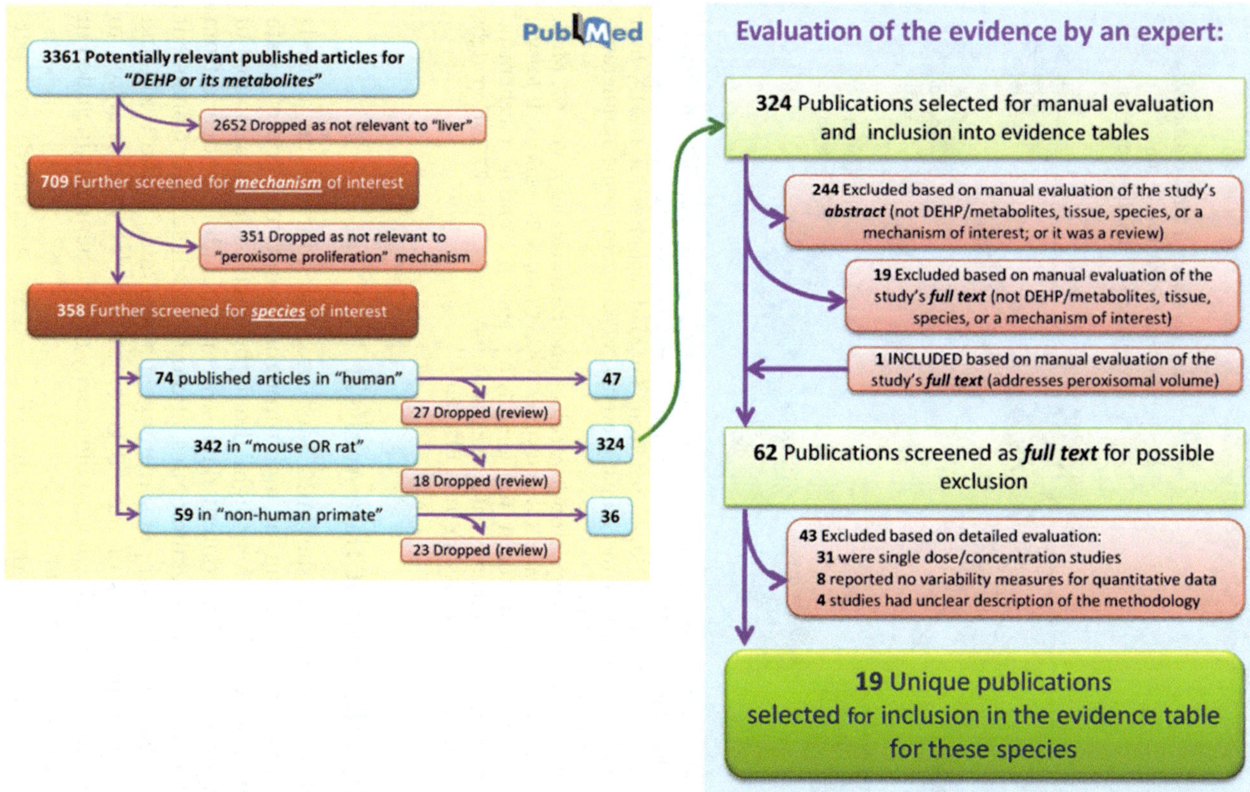

Figure 9.4 Literature tree for the "peroxisome proliferation" mechanistic event in liver. Progression from the initial 3361 identified articles relevant to di(2-ethylhexyl)phthalate (DEHP) or its metabolites to those included in the evidence tables is shown. Reprinted from Regulatory Toxicology and Pharmacology, Vol. 67, M. E. Kushman, A. D. Kraft, K. Z. Guyton, W. A. Chiu, S. L. Makris, I. Rusyn, A systematic approach for identifying and presenting mechanistic evidence in human health assessments, Pages 266–277, Copyright (2013), with permission from Elsevier.[101]

(Group 2B) in consideration of limited evidence in humans for testicular and renal cancer, and limited evidence in experimental animals.[112] Evidence for mechanisms of PFOA-associated carcinogenesis was moderate, which did not change the overall classification. Supporting these conclusion was the Working Group analysis concerning potential receptor-mediated activity of PFOA, based on comparison to other chemicals including rifampicin, phenobarbital, and DEHP and MEHP. The outcome was consistent with observations that multiple nuclear receptors are activated by PFOA *in vivo* in rodents.[113,114] Additionally, unlike the selected comparison compounds, PFOA and its ammonium salt appear to be consistently active in estrogen receptor assays, in keeping with observations on effects on reproductive hormones and tissues.[115–118]

For IARC Monographs Volumes 112 and 113, a range of endpoints from the high-throughput testing database was examined. The volume 112 chemicals diazinon, malathion, and parathion, as well as the oxon metabolites, malaoxon and diazoxon, are among the approximately 1000 chemicals tested across the full ToxCast/Tox21 assay battery as of 3 March 2015. *Z*-Tetrachlorvinphos (CASRN 22248-79-9; a structural isomer of tetrachlorvinphos) and the oxon metabolite of parathion, paraoxon, are among an additional 800 chemicals tested as part of an endocrine profiling effort using a subset of these assays. Glyphosate was not tested in the ToxCast/Tox21 assays. For volume 113, lindane, DDT (*p,p*′-DDT, *o,p*′-DDT, *p,p*′-DDE, *p,p*′-DDD), and 2,4-D (2,4-dichlorophenoxyacetic acid) were among the approximately 1000 chemicals tested across the full ToxCast/Tox21 assay battery as of 27 April 2015. This assay battery includes 342 assays, for which data on 821 assay endpoints are publicly available in the US EPA ToxCast Dashboard (http://www.epa.gov/chemical-research/toxicity-forecasting). Detailed information about the chemicals, assays and associated data analysis procedures is also publicly available from (http://www.epa.gov/chemical-research/toxicity-forecasting). It is of note that while the cell-based assays have a variable degree of metabolic capacity, it is generally limited.

9.4 Conclusions and Future Directions

Identification and incorporation of important, novel scientific findings providing insights into cancer mechanisms continues to be a focus in current IARC Monographs. As recently described by Smith *et al.*,[73] 10 key characteristics of carcinogens have been identified through the Volume 100 review and two subsequent workshops. These characteristics can provide the basis for an objective approach to identifying and organizing results from pertinent mechanistic studies for any carcinogen, including the consideration that certain data may not be available. Overall, these approaches will facilitate comprehensive inclusion of important studies with systematic consideration of the range of available mechanistic data informative of the overall evaluation of carcinogenic hazard. These developments will aid the Monographs to prepare for future evaluations of newly introduced chemicals, including those for which mechanistic data provide the primary evidence of carcinogenicity.

Appendix 9.1: Classification Categories for the Overall Evaluation for the IARC Monographs (IARC 2006)

Group 1: The Agent is Carcinogenic to Humans

This category is used when there is sufficient evidence of carcinogenicity in humans. Exceptionally, an agent may be placed in this category when evidence of carcinogenicity in humans is less than sufficient but there is sufficient evidence of carcinogenicity in experimental animals and strong evidence in exposed humans that the agent acts through a relevant mechanism of carcinogenicity.

Group 2

This category includes agents for which, at one extreme, the degree of evidence of carcinogenicity in humans is almost sufficient, as well as those for which, at the other extreme, there are no human data but for which there is evidence of carcinogenicity in experimental animals. Agents are assigned to either Group 2A (probably carcinogenic to humans) or Group 2B (possibly carcinogenic to humans) on the basis of epidemiological and experimental evidence of carcinogenicity and *mechanistic and other relevant data*. The terms probably carcinogenic and possibly carcinogenic have no quantitative significance and are used simply as descriptors of different levels of evidence of human carcinogenicity, with probably carcinogenic signifying a higher level of evidence than possibly carcinogenic.

Group 2A: The Agent is Probably Carcinogenic to Humans

This category is used when there is limited evidence of carcinogenicity in humans and sufficient evidence of carcinogenicity in experimental animals. In some cases, an agent may be classified in this category when there is inadequate evidence of carcinogenicity in humans and sufficient evidence of carcinogenicity in experimental animals and strong evidence that the carcinogenesis is mediated by a mechanism that also operates in humans. Exceptionally, an agent may be classified in this category solely on the basis of limited evidence of carcinogenicity in humans. An agent may be assigned to this category if it clearly belongs, based on mechanistic considerations, to a class of agents for which one or more members have been classified in Group 1 or Group 2A.

Group 2B: The Agent is Possibly Carcinogenic to Humans

This category is used for agents for which there is limited evidence of carcinogenicity in humans and less than sufficient evidence of carcinogenicity in experimental animals. It may also be used when there is inadequate evidence of carcinogenicity in humans but there is sufficient evidence of carcinogenicity in experimental animals. In some instances, an agent for which there is inadequate evidence of carcinogenicity in humans and less than sufficient evidence of carcinogenicity in experimental animals together with supporting evidence from *mechanistic and other relevant data* may be placed in this group. An agent may be classified in this category solely on the basis of strong evidence from *mechanistic and other relevant data*.

Group 3: The Agent is Not Classifiable as to Its Carcinogenicity to Humans

This category is used most commonly for agents for which the evidence of carcinogenicity is inadequate in humans and inadequate or limited in experimental animals.

Exceptionally, agents for which the evidence of carcinogenicity is inadequate in humans but sufficient in experimental animals may be placed in this category when there is strong evidence that the mechanism of carcinogenicity in experimental animals does not operate in humans.

Agents that do not fall into any other group are also placed in this category.

An evaluation in Group 3 is not a determination of noncarcinogenicity or overall safety. It often means that further research is needed, especially when exposures are widespread or the cancer data are consistent with differing interpretations.

Group 4: The Agent is Probably Not Carcinogenic to Humans

This category is used for agents for which there is evidence suggesting lack of carcinogenicity in humans and in experimental animals. In some instances, agents for which there is inadequate evidence of carcinogenicity in humans but evidence suggesting lack of carcinogenicity in experimental animals, consistently and strongly supported by a broad range of *mechanistic and other relevant data*, may be classified in this group.

References

1. N. E. Pearce, A. Blair, P. Vineis, W. Ahrens, A. Andersen, J. M. Anto, B. K. Armstrong, A. A. Baccarelli, F. A. Beland, A. Berrington, P. A. Bertazzi, L. S. Birnbaum, R. C. Brownson, J. R. Bucher, K. P. Cantor, E. Cardis, J. W. Cherrie, D. C. Christiani, P. Cocco, D. Coggon, P. Comba, P. A. Demers, J. M. Dement, J. Douwes, E. A. Eisen, L. S. Engel, R. A. Fenske, L. E. Fleming, T. Fletcher, E. Fontham, F. Forastiere, R. Frentzel-Beyme, L. Fritschi, M. Gerin, M. Goldberg, P. Grandjean, T. K. Grimsrud, P. Gustavsson, A. Haines, P. Hartge, J. Hansen, M. Hauptmann, D. Heederik, K. Hemminki,

D. Hemon, I. Hertz-Picciotto, J. A. Hoppin, J. Huff, B. Jarvholm, D. Kang, M. R. Karagas, K. Kjaerheim, H. Kjuus, M. Kogevinas, D. Kriebel, P. Kristensen, H. Kromhout, F. Laden, P. Lebailly, G. LeMasters, J. H. Lubin, C. F. Lynch, E. Lynge, A. t. Mannetje, A. J. McMichael, J. R. McLaughlin, L. Marrett, M. Martuzzi, J. A. Merchant, E. Merler, F. Merletti, A. Miller, F. E. Mirer, R. Monson, K. K. Nordby, A. F. Olshan, M. E. Parent, F. P. Perera, M. J. Perry, A. C. Pesatori, R. Pirastu, M. Porta, E. Pukkala, C. Rice, D. B. Richardson, L. Ritter, B. Ritz, C. M. Ronckers, L. Rushton, J. A. Rusiecki, I. Rusyn, J. M. Samet, D. P. Sandler, S. de Sanjose, E. Schernhammer, A. S. Constantini, N. Seixas, C. Shy, J. Siemiatycki, D. T. Silvermann, L. Simonato, A. H. Smith, M. T. Smith, J. J. Spinelli, M. R. Spitz, L. Stallones, L. T. Stayner, K. Steenland, M. Stenzel, B. W. Stewart, P. A. Stewart, E. Symanski, B. Terracini, P. E. Tolbert, H. Vainio, J. Vena, R. Vermeulen, C. G. Victora, E. M. Ward, C. R. Weinberg, D. Weisenburger, C. Wesseling, E. Weiderpass and S. H. Zahm, IARC Monographs: 40 Years of Evaluating Carcinogenic Hazards to Humans, *Environ. Health Perspect.*, **123**, 507–514.

2. R. C. Brownson, J. F. Chriqui and K. A. Stamatakis, Understanding evidence-based public health policy, *Am. J. Public Health*, 2009, **99**, 1576–1583.
3. C. Espina, M. Porta, J. Schuz, I. H. Aguado, R. V. Percival, C. Dora, T. Slevin, J. R. Guzman, T. Meredith, P. J. Landrigan and M. Neira, Environmental and occupational interventions for primary prevention of cancer: a cross-sectorial policy framework, *Environ. Health Perspect.*, 2013, **121**, 420–426.
4. J. M. Samet, Epidemiology and policy: the pump handle meets the new millennium, *Epidemiol. Rev.*, 2000, **22**, 145–154.
5. IARC, Preamble, *IARC Monogr. Eval. Carcinog. Risks Hum.*, 2014, **106**, 7–30, International Agency for Research on Cancer.
6. L. Tomatis, The IARC program on the evaluation of the carcinogenic risk of chemicals to man, *Ann. N. Y. Acad. Sci.*, 1976, **271**, 396–409.
7. IARC, Some inorganic substances, chlorinated hydrocarbons, aromatic amines, *N*-nitroso compounds, and natural products, 1972, **1**, 17–184, International Agency for Research on Cancer, Lyon, France.
8. IARC, Preamble, *IARC Monogr. Eval. Carcinog. Risks Hum.*, 2006.
9. IARC, *Report of the Advisory Group to Recommend Priorities for IARC Monographs during 2010–2014*, 2014.
10. D. Loomis, K. Straif, K. Guyton, Y. Grosse, B. Lauby-Secretan, F. El Ghissassi, *et al.*, Future priorities for the IARC monographs, *Lancet Oncol.*, 2014, **15**, 683–684.
11. C. P. Wild and V. J. Cogliano, A plea on behalf of expert evaluation and the experts involved, *Int. J. Epidemiol.*, 2011, **40**, 253–261.
12. C. P. Wild and K. Straif, Authors' response. Expert working groups—a reliable recipe, *Int. J. Epidemiol.*, 2011, **40**, 1730–1731.
13. IARC, *Preamble, Instructions to Authors*, 2014.

14. H. Vainio, P. Magee, D. McGregor and A. J. McMichael, IARC Scientific Publication No. 116, Lyon, 1992.
15. V. J. Cogliano, R. A. Baan, K. Straif, Y. Grosse, B. Secretan and F. El Ghissassi, Use of mechanistic data in IARC evaluations, *Environ. Mol. Mutagen.*, 2008, **49**, 100–109.
16. IARC, *Chemical Agents and Related Occupations*, International Agency for Research on Cancer, 2012.
17. I. Rusyn, W. A. Chiu, L. H. Lash, H. Kromhout, J. Hansen and K. Z. Guyton, Trichloroethylene: Mechanistic, epidemiologic and other supporting evidence of carcinogenic hazard, *Pharmacol. Ther.*, 2014, **141**, 55–68.
18. S. Karami, Q. Lan, N. Rothman, P. A. Stewart, K. M. Lee, R. Vermeulen and L. E. Moore, Occupational trichloroethylene exposure and kidney cancer risk: a meta-analysis, *Occup. Environ. Med.*, 2012, **69**, 858–867.
19. C. S. Scott and J. Jinot, Trichloroethylene and cancer: systematic and quantitative review of epidemiologic evidence for identifying hazards, *Int. J. Environ. Res. Public Health*, 2011, **8**, 4238–4272.
20. L. H. Lash, W. A. Chiu, K. Z. Guyton and I. Rusyn, Trichloroethylene biotransformation and its role in mutagenicity, carcinogenicity and target organ toxicity, *Mutat. Res., Rev. Mutat. Res.*, 2014, **762**, 22–36.
21. J. A. Bjork, C. Lau, S. C. Chang, J. L. Butenhoff and K. B. Wallace, Perfluorooctane sulfonate-induced changes in fetal rat liver gene expression, *Toxicology*, 2008, **251**, 8–20.
22. L. E. Moore, P. Boffetta, S. Karami, P. Brennan, P. S. Stewart, R. Hung, D. Zaridze, V. Matveev, V. Janout, H. Kollarova, V. Bencko, M. Navratilova, N. Szeszenia-Dabrowska, D. Mates, J. Gromiec, I. Holcatova, M. Merino, S. Chanock, W. H. Chow and N. Rothman, Occupational trichloroethylene exposure and renal carcinoma risk: evidence of genetic susceptibility by reductive metabolism gene variants, *Cancer Res.*, 2010, **70**, 6527–6536.
23. N. Guha, D. Loomis, Y. Grosse, B. Lauby-Secretan, F. El Ghissassi, V. Bouvard, L. Benbrahim-Tallaa, R. Baan, H. Mattock, K. Straif and International Agency for Research on Cancer Monograph Working Group, Carcinogenicity of trichloroethylene, tetrachloroethylene, some other chlorinated solvents, and their metabolites, *Lancet Oncol.*, 2012, **13**, 1192–1193.
24. W. A. Chiu, J. Jinot, C. S. Scott, S. L. Makris, G. S. Cooper, R. C. Dzubow, A. S. Bale, M. V. Evans, K. Z. Guyton, N. Keshava, J. C. Lipscomb, S. Barone, Jr., J. F. Fox, M. R. Gwinn, J. Schaum and J. C. Caldwell, Human health effects of trichloroethylene: key findings and scientific issues, *Environ. Health Perspect.*, 2013, **121**, 303–311.
25. R. Vermeulen, L. Zhang, A. Spierenburg, X. Tang, J. V. Bonventre, B. Reiss, M. Shen, M. T. Smith, C. Qiu, Y. Ge, Z. Ji, J. Xiong, J. He, Z. Hao, S. Liu, Y. Xie, F. Yue, W. Guo, M. Purdue, L. E. Beane Freeman, V. Sabbisetti, L. Li, H. Huang, N. Rothman and Q. Lan, Elevated

urinary levels of kidney injury molecule-1 among Chinese factory workers exposed to trichloroethylene, *Carcinogenesis*, 2012, **33**, 1538–1541.
26. H. D. Hosgood, 3rd, L. Zhang, X. Tang, R. Vermeulen, C. Qiu, M. Shen, M. T. Smith, Y. Ge, Z. Ji, J. Xiong, J. He, B. Reiss, S. Liu, Y. Xie, W. Guo, N. Galvan, L. Li, Z. Hao, N. Rothman, H. Huang and Q. Lan, Decreased Numbers of CD4(+) Naive and Effector Memory T Cells, and CD8(+) Naive T Cells, are Associated with Trichloroethylene Exposure, *Front. Oncol.*, 2011, **1**, 53.
27. W. A. Chiu and G. L. Ginsberg, Development and evaluation of a harmonized physiologically based pharmacokinetic (PBPK) model for perchloroethylene toxicokinetics in mice, rats, and humans, *Toxicol. Appl. Pharmacol.*, 2011, **253**, 203–234.
28. K. Z. Guyton, K. A. Hogan, C. S. Scott, G. S. Cooper, A. S. Bale, L. Kopylev, S. Barone, S. L. Makris, B. Glenn, R. P. Subramaniam, M. R. Gwinn, R. C. Dzubow and W. A. Chiu, Human health effects of tetrachloroethylene: key findings and scientific issues, *Environ. Health Perspect.*, 2014, **122**, 325–334.
29. EPA, *Tetrachloroethylene (Perchloroethylene) (CASRN: 127-18-4)*, http://www.epa.gov/iris/subst/0106.htm.
30. EPA, *Guidelines for Carcinogen Risk Assessment*, 2005.
31. JISA, *Carcinogencity Study of Tetrachloroethylene by Inhalation in Rats and Mice*, Japan Industrial Safety and Health Association, 1993.
32. NCI, *Bioassay of Tetrachloroethylene for Possible Carcinogenicity CAS No. 127-18-4*, 1977.
33. NTP, *Toxocology and Carcinogenesis Studies of Tetrachloroethylene (CAS No. 127-18-4) in F344/N Rats and B6C3F1 Mice (Inhalation Studies)*, National Institute of Environmental Health Sciences.
34. K. Y. Christensen, D. Vizcaya, H. Richardson, J. Lavoue, K. Aronson and J. Siemiatycki, Risk of selected cancers due to occupational exposure to chlorinated solvents in a case-control study in Montreal, *J. Occup. Environ. Med.*, 2013, **55**, 198–208.
35. L. Lipworth, J. S. Sonderman, M. T. Mumma, R. E. Tarone, D. E. Marano, J. D. Boice, Jr. and J. K. McLaughlin, Cancer mortality among aircraft manufacturing workers: an extended follow-up, *J. Occup. Environ. Med.*, 2011, **53**, 992–1007.
36. A. M. Ruder, J. H. Yiin, M. A. Waters, T. Carreon, M. J. Hein, M. A. Butler, G. M. Calvert, K. E. Davis-King, P. A. Schulte, J. S. Mandel, R. F. Morton, D. J. Reding, K. D. Rosenman, P. A. Stewart and G. Brain, Cancer Collaborative Study, The Upper Midwest Health Study: gliomas and occupational exposure to chlorinated solvents, *Occup. Environ. Med.*, 2013, **70**, 73–80.
37. D. Vizcaya, K. Y. Christensen, J. Lavoue and J. Siemiatycki, Risk of lung cancer associated with six types of chlorinated solvents: results from two case-control studies in Montreal, Canada, *Occup. Environ. Med.*, 2013, **70**, 81–85.

38. J. Vlaanderen, K. Straif, E. Pukkala, T. Kauppinen, P. Kyyronen, J. I. Martinsen, K. Kjaerheim, L. Tryggvadottir, J. Hansen, P. Sparen and E. Weiderpass, Occupational exposure to trichloroethylene and perchloroethylene and the risk of lymphoma, liver, and kidney cancer in four Nordic countries, *Occup. Environ. Med.*, 2013, **70**, 393–401.
39. J. Vlaanderen, K. Straif, A. Ruder, A. Blair, J. Hansen, E. Lynge, B. Charbotel, D. Loomis, T. Kauppinen, P. Kyyronen, E. Pukkala, E. Weiderpass and N. Guha, Tetrachloroethylene exposure and bladder cancer risk: a meta-analysis of dry-cleaning-worker studies, *Environ. Health Perspect.*, 2014, **122**, 661–666.
40. A. Aschengrau, J. M. Weinberg, P. A. Janulewicz, M. E. Romano, L. G. Gallagher, M. R. Winter, B. R. Martin, V. M. Vieira, T. F. Webster, R. F. White and D. M. Ozonoff, Affinity for risky behaviors following prenatal and early childhood exposure to tetrachloroethylene (PCE)-contaminated drinking water: a retrospective cohort study, *Environ. Health*, 2011, **10**, 102.
41. A. Aschengrau, J. M. Weinberg, P. A. Janulewicz, M. E. Romano, L. G. Gallagher, M. R. Winter, B. R. Martin, V. M. Vieira, T. F. Webster, R. F. White and D. M. Ozonoff, Occurrence of mental illness following prenatal and early childhood exposure to tetrachloroethylene (PCE)-contaminated drinking water: a retrospective cohort study, *Environ. Health*, 2012, **11**, 2.
42. K. D. Getz, P. A. Janulewicz, S. Rowe, J. M. Weinberg, M. R. Winter, B. R. Martin, V. M. Vieira, R. F. White and A. Aschengrau, Prenatal and early childhood exposure to tetrachloroethylene and adult vision, *Environ. Health Perspect.*, 2012, **120**, 1327–1332.
43. P. A. Janulewicz, R. F. White, B. M. Martin, M. R. Winter, J. M. Weinberg, V. Vieira and A. Aschengrau, Adult neuropsychological performance following prenatal and early postnatal exposure to tetrachloroethylene (PCE)-contaminated drinking water, *Neurotoxicol. Teratol.*, 2012, **34**, 350–359.
44. S. M. Goldman, P. J. Quinlan, G. W. Ross, C. Marras, C. Meng, G. S. Bhudhikanok, K. Comyns, M. Korell, A. R. Chade, M. Kasten, B. Priestley, K. L. Chou, H. H. Fernandez, F. Cambi, J. W. Langston and C. M. Tanner, Solvent exposures and Parkinson disease risk in twins, *Ann. Neurol.*, 2012, **71**, 776–784.
45. M. Ikbal, A. Tastekin, H. Dogan, I. Pirim and R. Ors, The assessment of genotoxic effects in lymphocyte cultures of infants treated with chloral hydrate, *Mutat. Res.*, 2004, **564**, 159–164.
46. K. Z. Guyton, A. D. Kyle, J. Aubrecht, V. J. Cogliano, D. A. Eastmond, M. Jackson, N. Keshava, M. S. Sandy, B. Sonawane, L. Zhang, M. D. Waters and M. T. Smith, Improving prediction of chemical carcinogenicity by considering multiple mechanisms and applying toxicogenomic approaches, *Mutat. Res.*, 2009, **681**, 230–240.
47. W. W. Ku, J. Aubrecht, R. J. Mauthe, R. H. Schiestl and A. J. Fornace, Jr., Genetic toxicity assessment: employing the best science for human

safety evaluation Part VII: Why not start with a single test: a transformational alternative to genotoxicity hazard and risk assessment, *Toxicol. Sci.*, 2007, **99**, 20–25.

48. S. A. Amundson, K. T. Do, L. Vinikoor, C. A. Koch-Paiz, M. L. Bittner, J. M. Trent, P. Meltzer and A. J. Fornace, Jr., Stress-specific signatures: expression profiling of p53 wild-type and -null human cells, *Oncogene*, 2005, **24**, 4572–4579.
49. H. H. Li, J. Aubrecht and A. J. Fornace, Jr., Toxicogenomics: overview and potential applications for the study of non-covalent DNA interacting chemicals, *Mutat. Res.*, 2007, **623**, 98–108.
50. E. Caba, D. A. Dickinson, G. R. Warnes and J. Aubrecht, Differentiating mechanisms of toxicity using global gene expression analysis in *Saccharomyces cerevisiae*, *Mutat. Res.*, 2005, **575**, 34–46.
51. S. Bonassi, A. Znaor, M. Ceppi, C. Lando, W. P. Chang, N. Holland, M. Kirsch-Volders, E. Zeiger, S. Ban, R. Barale, M. P. Bigatti, C. Bolognesi, A. Cebulska-Wasilewska, E. Fabianova, A. Fucic, L. Hagmar, G. Joksic, A. Martelli, L. Migliore, E. Mirkova, M. R. Scarfi, A. Zijno, H. Norppa and M. Fenech, An increased micronucleus frequency in peripheral blood lymphocytes predicts the risk of cancer in humans, *Carcinogenesis*, 2007, **28**, 625–631.
52. L. Hagmar, S. Bonassi, U. Stromberg, A. Brogger, L. E. Knudsen, H. Norppa and C. Reuterwall, Chromosomal aberrations in lymphocytes predict human cancer: a report from the European Study Group on Cytogenetic Biomarkers and Health (ESCH), *Cancer Res.*, 1998, **58**, 4117–4121.
53. A. Fucic, A. Znaor, M. Strnad, O. van der Hel, A. Aleksandrov, S. Miskov, J. Grah, M. Sedlar, A. M. Jazbec, M. Ceppi, R. Vermeulen, P. Boffetta, H. Norppa and S. Bonassi, Chromosome damage and cancer risk in the workplace: the example of cytogenetic surveillance in Croatia, *Toxicol. Lett.*, 2007, **172**, 4–11.
54. S. Bonassi, L. Hagmar, U. Stromberg, A. H. Montagud, H. Tinnerberg, A. Forni, P. Heikkila, S. Wanders, P. Wilhardt, I. L. Hansteen, L. E. Knudsen and H. Norppa, Chromosomal aberrations in lymphocytes predict human cancer independently of exposure to carcinogens. European Study Group on Cytogenetic Biomarkers and Health, *Cancer Res.*, 2000, **60**, 1619–1625.
55. S. Bonassi, A. Znaor, H. Norppa and L. Hagmar, Chromosomal aberrations and risk of cancer in humans: an epidemiologic perspective, *Cytogenet. Genome Res.*, 2004, **104**, 376–382.
56. M. S. Forrest, Q. Lan, A. E. Hubbard, L. Zhang, R. Vermeulen, X. Zhao, G. Li, Y. Y. Wu, M. Shen, S. Yin, S. J. Chanock, N. Rothman and M. T. Smith, Discovery of novel biomarkers by microarray analysis of peripheral blood mononuclear cell gene expression in benzene-exposed workers, *Environ. Health Perspect.*, 2005, **113**, 801–807.
57. C. M. McHale, L. Zhang, A. E. Hubbard, X. Zhao, A. Baccarelli, A. C. Pesatori, M. T. Smith and M. T. Landi, Microarray analysis of gene

expression in peripheral blood mononuclear cells from dioxin-exposed human subjects, *Toxicology*, 2007, **229**, 101–113.

58. M. Argos, M. G. Kibriya, F. Parvez, F. Jasmine, M. Rakibuz-Zaman and H. Ahsan, Gene expression profiles in peripheral lymphocytes by arsenic exposure and skin lesion status in a Bangladeshi population, *Cancer Epidemiol., Biomarkers Prev.*, 2006, **15**, 1367–1375.
59. T. Lu, J. Liu, E. L. LeCluyse, Y. S. Zhou, M. L. Cheng and M. P. Waalkes, Application of cDNA microarray to the study of arsenic-induced liver diseases in the population of Guizhou, China, *Toxicol. Sci.*, 2001, **59**, 185–192.
60. M. M. Wu, H. Y. Chiou, I. C. Ho, C. J. Chen and T. C. Lee, Gene expression of inflammatory molecules in circulating lymphocytes from arsenic-exposed human subjects, *Environ. Health Perspect.*, 2003, **111**, 1429–1438.
61. K. T. Rim, K. K. Park, J. H. Sung, Y. H. Chung, J. H. Han, K. S. Cho, K. J. Kim and I. J. Yu, Gene-expression profiling using suppression-subtractive hybridization and cDNA microarray in rat mononuclear cells in response to welding-fume exposure, *Toxicol. Ind. Health*, 2004, **20**, 77–88.
62. Z. Wang, D. Neuburg, C. Li, L. Su, J. Y. Kim, J. C. Chen and D. C. Christiani, Global gene expression profiling in whole-blood samples from individuals exposed to metal fumes, *Environ. Health Perspect.*, 2005, **113**, 233–241.
63. B. Sen, B. Mahadevan and D. M. DeMarini, Transcriptional responses to complex mixtures: a review, *Mutat. Res.*, 2007, **636**, 144–177.
64. L. D. Wood, D. W. Parsons, S. Jones, J. Lin, T. Sjoblom, R. J. Leary, D. Shen, S. M. Boca, T. Barber, J. Ptak, N. Silliman, S. Szabo, Z. Dezso, V. Ustyanksky, T. Nikolskaya, Y. Nikolsky, R. Karchin, P. A. Wilson, J. S. Kaminker, Z. Zhang, R. Croshaw, J. Willis, D. Dawson, M. Shipitsin, J. K. Willson, S. Sukumar, K. Polyak, B. H. Park, C. L. Pethiyagoda, P. V. Pant, D. G. Ballinger, A. B. Sparks, J. Hartigan, D. R. Smith, E. Suh, N. Papadopoulos, P. Buckhaults, S. D. Markowitz, G. Parmigiani, K. W. Kinzler, V. E. Velculescu and B. Vogelstein, The genomic landscapes of human breast and colorectal cancers, *Science*, 2007, **318**, 1108–1113.
65. S. M. Reamon-Buettner, V. Mutschler and J. Borlak, The next innovation cycle in toxicogenomics: environmental epigenetics, *Mutat. Res.*, 2008, **659**, 158–165.
66. D. Hanahan and R. A. Weinberg, The hallmarks of cancer, *Cell*, 2000, **100**, 57–70.
67. K. Tsujimura, M. Asamoto, S. Suzuki, N. Hokaiwado, K. Ogawa and T. Shirai, Prediction of carcinogenic potential by a toxicogenomic approach using rat hepatoma cells, *Cancer Sci.*, 2006, **97**, 1002–1010.
68. K. Nakayama, Y. Kawano, Y. Kawakami, N. Moriwaki, M. Sekijima, M. Otsuka, Y. Yakabe, H. Miyaura, K. Saito, K. Sumida and T. Shirai, Differences in gene expression profiles in the liver between

carcinogenic and non-carcinogenic isomers of compounds given to rats in a 28-day repeat-dose toxicity study, *Toxicol. Appl. Pharmacol.*, 2006, **217**, 299–307.

69. A. Y. Nie, M. McMillian, J. B. Parker, A. Leone, S. Bryant, L. Yieh, A. Bittner, J. Nelson, A. Carmen, J. Wan and P. G. Lord, Predictive toxicogenomics approaches reveal underlying molecular mechanisms of nongenotoxic carcinogenicity, *Mol. Carcinog.*, 2006, **45**, 914–933.
70. R. S. Thomas, T. M. O'Connell, L. Pluta, R. D. Wolfinger, L. Yang and T. J. Page, A comparison of transcriptomic and metabonomic technologies for identifying biomarkers predictive of two-year rodent cancer bioassays, *Toxicol. Sci.*, 2007, **96**, 40–46.
71. H. Ellinger-Ziegelbauer, B. Stuart, B. Wahle, W. Bomann and H. J. Ahr, Comparison of the expression profiles induced by genotoxic and non-genotoxic carcinogens in rat liver, *Mutat. Res.*, 2005, **575**, 61–84.
72. J. Aubrecht and E. Caba, Gene expression profile analysis: an emerging approach to investigate mechanisms of genotoxicity, *Pharmacogenomics*, 2005, **6**, 419–428.
73. M. T. Smyth, K. Z. Guyton, C. F. Gibbons, J. M. Fritz, C. J. Portier, I. Rusyn, D. M. DeMarini, J. C. Caldwell, R. J. Kavlock, P. Lambert, S. Hecht, J. R. Bucher, B. W. Stewart, R. Baan, V. Cogliano and K. Straif, Key Characteristics of Carcinogens as a Basis for Organizing Data on Mechanisms of Carcinogenesis, *Environ. Health Perspect.*, DOI: 10.1289/ehp.1509912.
74. R. L. Melnick, M. C. Kohn and C. J. Portier, Implications for risk assessment of suggested nongenotoxic mechanisms of chemical carcinogenesis, *Environ. Health Perspect.*, 1996, **104**(Suppl 1), 123–134.
75. I. Rusyn, A. Sedykh, Y. Low, K. Z. Guyton and A. Tropsha, Predictive modeling of chemical hazard by integrating numerical descriptors of chemical structures and short-term toxicity assay data, *Toxicol. Sci.*, 2012, **127**, 1–9.
76. F. Nigsch, N. J. Macaluso, J. B. Mitchell and D. Zmuidinavicius, Computational toxicology: an overview of the sources of data and of modelling methods, *Expert Opin. Drug Metab. Toxicol.*, 2009, **5**, 1–14.
77. L. G. Valerio, Jr., In silico toxicology for the pharmaceutical sciences, *Toxicol. Appl. Pharmacol.*, 2009, **241**, 356–370.
78. F. S. Collins, G. M. Gray and J. R. Bucher, Toxicology. Transforming environmental health protection, *Science*, 2008, **319**, 906–907.
79. L. Zeise, F. Y. Bois, W. A. Chiu, D. Hattis, I. Rusyn and K. Z. Guyton, Addressing human variability in next-generation human health risk assessments of environmental chemicals, *Environ. Health Perspect.*, 2013, **121**, 23–31.
80. G. Ginsberg, S. Smolenski, P. Neafsey, D. Hattis, K. Walker, K. Z. Guyton, D. O. Johns and B. Sonawane, The influence of genetic polymorphisms on population variability in six xenobiotic-metabolizing enzymes, *J. Toxicol. Environ. Health, Part B*, 2009, **12**, 307–333.
81. W. A. Chiu, M. S. Okino and M. V. Evans, Characterizing uncertainty and population variability in the toxicokinetics of trichloroethylene and

metabolites in mice, rats, and humans using an updated database, physiologically based pharmacokinetic (PBPK) model, and Bayesian approach, *Toxicol. Appl. Pharmacol.*, 2009, **241**, 36–60.
82. I. Rusyn, D. M. Gatti, T. Wiltshire, S. R. Kleeberger and D. W. Threadgill, Toxicogenetics: population-based testing of drug and chemical safety in mouse models, *Pharmacogenomics*, 2010, **11**, 1127–1136.
83. C. M. Hutter, J. Chang-Claude, M. L. Slattery, B. M. Pflugeisen, Y. Lin, D. Duggan, H. Nan, M. Lemire, J. Rangrej, J. C. Figueiredo, S. Jiao, T. A. Harrison, Y. Liu, L. S. Chen, D. L. Stelling, G. S. Warnick, M. Hoffmeister, S. Kury, C. S. Fuchs, E. Giovannucci, A. Hazra, P. Kraft, D. J. Hunter, S. Gallinger, B. W. Zanke, H. Brenner, B. Frank, J. Ma, C. M. Ulrich, E. White, P. A. Newcomb, C. Kooperberg, A. Z. LaCroix, R. L. Prentice, R. D. Jackson, R. E. Schoen, S. J. Chanock, S. I. Berndt, R. B. Hayes, B. J. Caan, J. D. Potter, L. Hsu, S. Bezieau, A. T. Chan, T. J. Hudson and U. Peters, Characterization of gene-environment interactions for colorectal cancer susceptibility loci, *Cancer Res.*, 2012, **72**, 2036–2044.
84. H. V. Patel and K. Kalia, Sub-chronic arsenic exposure aggravates nephrotoxicity in experimental diabetic rats, *Indian J. Exp. Biol.*, 2010, **48**, 762–768.
85. G. Ginsberg, S. Smolenski, D. Hattis and B. Sonawane, Population distribution of aldehyde dehydrogenase-2 genetic polymorphism: implications for risk assessment, *Regul. Toxicol. Pharmacol.*, 2002, **36**, 297–309.
86. G. Ginsberg, S. Smolenski, D. Hattis, K. Z. Guyton, D. O. Johns and B. Sonawane, Genetic Polymorphism in Glutathione Transferases (GST): Population distribution of GSTM1, T1, and P1 conjugating activity, *J. Toxicol. Environ. Health, Part B*, 2009, **12**, 389–439.
87. G. Ginsberg, P. Neafsey, D. Hattis, K. Z. Guyton, D. O. Johns and B. Sonawane, Genetic polymorphism in paraoxonase 1 (PON1): Population distribution of PON1 activity, *J. Toxicol. Environ. Health, Part B*, 2009, **12**, 473–507.
88. G. Ginsberg, K. Guyton, D. Johns, J. Schimek, K. Angle and B. Sonawane, Genetic polymorphism in metabolism and host defense enzymes: implications for human health risk assessment, *Crit. Rev. Toxicol.*, 2010, **40**, 575–619.
89. P. Neafsey, G. Ginsberg, D. Hattis, D. O. Johns, K. Z. Guyton and B. Sonawane, Genetic polymorphism in CYP2E1: Population distribution of CYP2E1 activity, *J. Toxicol. Environ. Health, Part B*, 2009, **12**, 362–388.
90. P. Neafsey, G. Ginsberg, D. Hattis and B. Sonawane, Genetic polymorphism in cytochrome P450 2D6 (CYP2D6): Population distribution of CYP2D6 activity, *J. Toxicol. Environ. Health, Part B*, 2009, **12**, 334–361.
91. K. Walker, G. Ginsberg, D. Hattis, D. O. Johns, K. Z. Guyton and B. Sonawane, Genetic polymorphism in *N*-Acetyltransferase (NAT): Population distribution of NAT1 and NAT2 activity, *J. Toxicol. Environ. Health, Part B*, 2009, **12**, 440–472.
92. G. L. Kedderis, Extrapolation of *in vitro* enzyme induction data to humans *in vivo*, *Chem. Biol. Interact.*, 1997, **107**, 109–121.

93. J. C. Lipscomb and G. L. Kedderis, Incorporating human interindividual biotransformation variance in health risk assessment, *Sci. Total Environ.*, 2002, **288**, 13–21.
94. T. Bruning, M. Lammert, M. Kempkes, R. Thier, K. Golka and H. M. Bolt, Influence of polymorphisms of GSTM1 and GSTT1 for risk of renal cell cancer in workers with long-term high occupational exposure to trichloroethene, *Arch. Toxicol.*, 1997, **71**, 596–599.
95. M. Rojas, K. Alexandrov, I. Cascorbi, J. Brockmoller, A. Likhachev, K. Pozharisski, G. Bouvier, G. Auburtin, L. Mayer, A. Kopp-Schneider, I. Roots and H. Bartsch, High benzo[a]pyrene diol-epoxide DNA adduct levels in lung and blood cells from individuals with combined CYP1A1 MspI/Msp-GSTM1*0/*0 genotypes, *Pharmacogenetics*, 1998, **8**, 109–118.
96. H. A. El-Masri, D. A. Bell and C. J. Portier, Effects of glutathione transferase theta polymorphism on the risk estimates of dichloromethane to humans, *Toxicol. Appl. Pharmacol.*, 1999, **158**, 221–230.
97. F. Jonsson and G. Johanson, A Bayesian analysis of the influence of GSTT1 polymorphism on the cancer risk estimate for dichloromethane, *Toxicol. Appl. Pharmacol.*, 2001, **174**, 99–112.
98. L. T. Haber, A. Maier, P. R. Gentry, H. J. Clewell and M. L. Dourson, Genetic polymorphisms in assessing interindividual variability in delivered dose, *Regul. Toxicol. Pharmacol.*, 2002, **35**, 177–197.
99. P. R. Gentry, C. E. Hack, L. Haber, A. Maier and H. J. Clewell, 3rd, An approach for the quantitative consideration of genetic polymorphism data in chemical risk assessment: examples with warfarin and parathion, *Toxicol. Sci.*, 2002, **70**, 120–139.
100. A. G. Renwick and N. R. Lazarus, Human variability and noncancer risk assessment–an analysis of the default uncertainty factor, *Regul. Toxicol. Pharmacol.*, 1998, **27**, 3–20.
101. M. E. Kushman, A. D. Kraft, K. Z. Guyton, W. A. Chiu, S. L. Makris and I. Rusyn, A systematic approach for identifying and presenting mechanistic evidence in human health assessments, *Regul. Toxicol. Pharmacol.*, 2013, **67**, 266–277.
102. *Review of the Environmental Protection Agency's Draft IRIS Assessment of Formaldehyde*, NRC, Washington, DC, 2011.
103. I. Rusyn, J. M. Peters and M. L. Cunningham, Modes of action and species-specific effects of di-(2-ethylhexyl)phthalate in the liver, *Crit. Rev. Toxicol.*, 2006, **36**, 459–479.
104. J. M. Peters, C. Cheung and F. J. Gonzalez, Peroxisome proliferator-activated receptor-alpha and liver cancer: where do we stand?, *J. Mol. Med.*, 2005, **83**, 774–785.
105. J. C. Caldwell, DEHP: genotoxicity and potential carcinogenic mechanisms-a review, *Mutat. Res.*, 2012, **751**, 82–157.
106. I. Rusyn and J. C. Corton, Mechanistic considerations for human relevance of cancer hazard of di(2-ethylhexyl) phthalate, *Mutat. Res.*, 2012, **750**, 141–158.
107. K. Z. Guyton, W. A. Chiu, T. F. Bateson, J. Jinot, C. S. Scott, R. C. Brown and J. C. Caldwell, A reexamination of the PPAR-alpha activation mode

of action as a basis for assessing human cancer risks of environmental contaminants, *Environ. Health Perspect.*, 2009, **117**, 1664–1672.
108. J. E. Klaunig, M. A. Babich, K. P. Baetcke, J. C. Cook, J. C. Corton, R. M. David, J. G. DeLuca, D. Y. Lai, R. H. McKee, J. M. Peters, R. A. Roberts and P. A. Fenner-Crisp, PPARalpha agonist-induced rodent tumors: modes of action and human relevance, *Crit. Rev. Toxicol.*, 2003, **33**, 655–780.
109. R. A. Roberts, P. E. Ganey, C. Ju, L. M. Kamendulis, I. Rusyn and J. E. Klaunig, Role of the Kupffer cell in mediating hepatic toxicity and carcinogenesis, *Toxicol. Sci.*, 2007, **96**, 2–15.
110. R. Kavlock, K. Chandler, K. Houck, S. Hunter, R. Judson, N. Kleinstreuer, T. Knudsen, M. Martin, S. Padilla, D. Reif, A. Richard, D. Rotroff, N. Sipes and D. Dix, Update on EPA's ToxCast program: providing high throughput decision support tools for chemical risk management, *Chem. Res. Toxicol.*, 2012, **25**, 1287–1302.
111. R. R. Tice, C. P. Austin, R. J. Kavlock and J. R. Bucher, Improving the human hazard characterization of chemicals: a Tox21 update, *Environ. Health Perspect.*, 2013, **121**, 756–765.
112. L. Benbrahim-Tallaa, B. Lauby-Secretan, D. Loomis, K. Z. Guyton, Y. Grosse, F. El Ghissassi, V. Bouvard, N. Guha, H. Mattock, K. Straif and International Agency for Research on Cancer Monograph Working Group, Carcinogenicity of perfluorooctanoic acid, tetrafluoroethylene, dichloromethane, 1,2-dichloropropane, and 1,3-propane sultone, *Lancet Oncol.*, 2014, **15**, 924–925.
113. C. R. Elcombe, B. M. Elcombe, J. R. Foster, D. G. Farrar, R. Jung, S. C. Chang, G. L. Kennedy and J. L. Butenhoff, Hepatocellular hypertrophy and cell proliferation in Sprague-Dawley rats following dietary exposure to ammonium perfluorooctanoate occurs through increased activation of the xenosensor nuclear receptors PPARalpha and CAR/PXR, *Arch. Toxicol.*, 2010, **84**, 787–798.
114. M. B. Rosen, B. D. Abbott, D. C. Wolf, J. C. Corton, C. R. Wood, J. E. Schmid, K. P. Das, R. D. Zehr, E. T. Blair and C. Lau, Gene profiling in the livers of wild-type and PPARalpha-null mice exposed to perfluorooctanoic acid, *Toxicol. Pathol.*, 2008, **36**, 592–607.
115. L. B. Biegel, R. C. Liu, M. E. Hurtt and J. C. Cook, Effects of ammonium perfluorooctanoate on Leydig cell function: *in vitro*, *in vivo*, and *ex vivo* studies, *Toxicol. Appl. Pharmacol.*, 1995, **134**, 18–25.
116. J. C. Cook, S. M. Murray, S. R. Frame and M. E. Hurtt, Induction of Leydig cell adenomas by ammonium perfluorooctanoate: a possible endocrine-related mechanism, *Toxicol. Appl. Pharmacol.*, 1992, **113**, 209–217.
117. B. Zhao, Y. Chu, D. O. Hardy, X. K. Li and R. S. Ge, Inhibition of 3beta- and 17beta-hydroxysteroid dehydrogenase activities in rat Leydig cells by perfluorooctane acid, *J. Steroid Biochem. Mol. Biol.*, 2010, **118**, 13–17.
118. Y. Zhao, Y. S. Tan, S. Z. Haslam and C. Yang, Perfluorooctanoic acid effects on steroid hormone and growth factor levels mediate stimulation of peripubertal mammary gland development in C57BL/6 mice, *Toxicol. Sci.*, 2010, **115**, 214–224.

CHAPTER 10

Conazoles and Cancer: A Review

STEPHEN NESNOW

Stephen Nesnow, Consulting, Chapel Hill, NC, USA
Email: snconsult@pobox.com

10.1 Introduction and Background

Conazoles are azole-containing compounds almost solely used as antifungal agents in medicine and agriculture. Azoles by definition are five-membered ring heterocycles containing one nitrogen atom (pyrrole); two nitrogen atoms (imidazole, with nitrogen atoms at ring positions 1,3 and pyrazole with nitrogen atoms at ring positions 1,2); three nitrogen atoms (triazole) with nitrogen atoms at ring positions 1,2,3 or 1,2,4; four nitrogen atoms (tetrazole); and five nitrogen atoms (pentazole). The term conazole was introduced with the development of miconazole [(*R*,*S*)-1-(2-(2,4-dichlorobenzyloxy)-2-(2,4-dichlorophenyl)ethyl)-1*H*-imidazole], synthesized in 1969[1] and marketed as miconazole nitrate (Daktarin) in 1971 by Janssen Pharmaceuticals. It seems probable that this term was derived from the words *con*taining *azole*, hence the name conazole. Almost all conazoles contain the imidazole or triazole moiety. The structures of two representative conazoles, propiconazole, a triazole-containing conazole, and ketoconazole, an imidazole-containing conazole, are shown in Figure 10.1.

The history of conazole development began in 1944 when Woolley first reported that the azole, benzimidazole, possessed antifungal activity.[2] In 1958, chlormidazole was the first azole antifungal to be marketed and used

Issues in Toxicology No. 28
Toxicogenomics in Predictive Carcinogenicity
Edited by Michael D. Waters and Russell S. Thomas

Published by the Royal Society of Chemistry, www.rsc.org

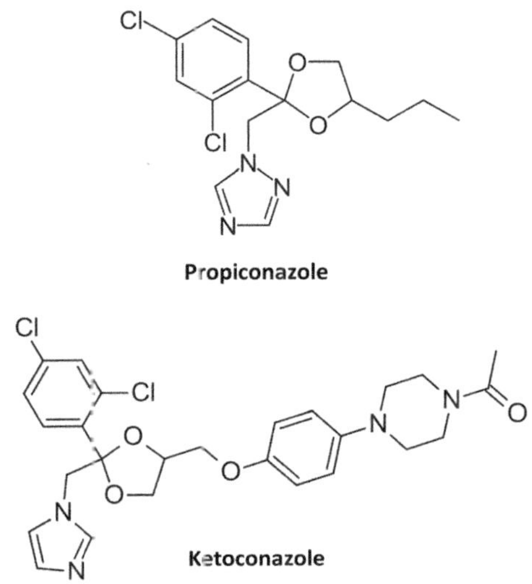

Figure 10.1 The structures of propiconazole and ketoconazole.

as a topical cream. Several years later the imidazole-containing conazoles: clotrimazole, miconazole, econazole and ketoconazole were developed and used clinically.[3] The development of these conazoles was followed in the 1990s by the development of a newer group of conazoles, almost all of which were triazoles: fluconazole and itraconazole, the drugs of choice for candidal and crypyococccal infections (fluconazole) and histoplasmosis and blastomyosis (itraconazole). Newer generation triazole conazoles have been developed recently, such as albaconazole, isvuconazole, ravuconazole, pramiconazole, posaconazole and voriconazole that possess increased activity against resistant and emerging pathogens.[4–6]

The development of azole-containing antifungals for agricultural use paralleled those used as pharmaceuticals with the appearance of one of the simplest compounds, amitrole or 2-amino-1,3,4-triazole and thiabendazole in the 1970s. In the 1980s and 1990s the conazoles epoxiconazole, flusilazole, hexaconazole, myclobutanil, prochloraz, penconazole, propiconazole, triadimefon, triadimenol and tebuconazole to name a few, became used as antifungal agents on seeds, grasses and crops.[6]

10.2 Inhibition of CYP51 by Conazoles

10.2.1 Introduction to CYP51

In order to interpret the adverse health effects of conazoles it is important to review the background of the designed function of conazoles as antifungal agents; that is, the inhibition of ergosterol biosynthesis by the inhibition of a critical enzyme in the biosynthetic pathway, lanosterol-14α-demethylase or CYP51 (cytochrome P450, family 51). It is important to note that conazole antifungal agents do not kill fungi *per se*; that is, they are not fungicides in the strictest sense, as they inhibit the replication of fungi and are thus fungistatic.

The structural and functional properties of the CYP51 superfamily have diversified extensively over the course of evolution.[7] CYP51 is found in 82 organisms including all animals, plants, yeast, fungi, in many protozoa and in some bacteria. CYP51 is the most widely distributed of the known P450 cytochromes.[8] Fungi are eukaryotes and have some biochemical and structural similarities to mammals as they share an evolutionary lineage with animals.[9] Fungi and mammals share several structural features, including the cytoplasmic organelles, mitochondria and membrane-bound nuclei. The nuclei of both species contain chromosomes and the nuclear DNA of both species contains introns and exons,[10] ribosomal RNA (80S), some similar metabolic pathways and sterol-containing membranes.[11] Ergosterol (ergosta-5,7,22-trien-3β-ol) is a sterol found in cell membranes of fungi and protozoa, serving many of the same functions that cholesterol serves in animal cells. These include multiple effects on the physical properties of membranes, such as membrane rigidity, fluidity, phase behavior and membrane thickness.[12–14]

CYP51 is required for sterol biosynthesis in different phyla.[15] CYP51 is found in the both the fungal ergosterol biosynthesis and mammalian cholesterol biosynthesis pathways. The biosyntheses of ergosterol and cholesterol share common pathways which are complex, extensive and require many steps. Briefly, in the first portion of the sequence, two molecules of acetyl-coenzyme A (CoA) combine to form acetoacetyl-CoA. Acetoacetyl-CoA combines with acetyl-CoA to form 3-hydroxy-3-methylglutaryl-CoA (HMG-CoA). HMG-CoA is reduced to mevalonate. Through a number of transformations mevalonate is converted to isopentylpyrophosphate, thus the creation of activated isoprene units that eventually become linked to form the basic sterol backbone. Isopentylpyrophosphate is converted to geranyl pyrophosphate, then to farnesyl pyrophosphate. Farnesylpyrophosphate is converted into squalene, then to squalene-2,3-epoxide which undergoes ring cyclization to create the sterol tetracyclic structure lanosterol. After CYP51 demethylation, the product, 4,4-dimethylcholesta-8(9),14,24-trien-3β-ol [also known as 4,4-dimethyl-5α-cholesta-8,14,24-trien-3ß-ol or 14-demethyl-14-dehydrolanosterol or follicular fluid meiosis-activating sterol (FF-MAS)] undergoes an extensive set of transformation processes (approximately 19) within separate pathways to give ergosterol or cholesterol. An abbreviated description of the biosynthesis of ergosterol, cholesterol and other sterols and the role of CYP51 is presented in Figure 10.2.

The mechanism of CYP51-mediated demethylation of lanosterol involves sequential multiple oxidation processes and occurs in the post-squalene part of the ergosterol pathway. CYP51 catalyzes a three-step oxidation process removing the C14α-methyl group of lanosterol. The first step is oxidation from methyl to hydroxymethyl, then to formyl and then the final oxidation step induces the elimination of formic acid and thus removal of the

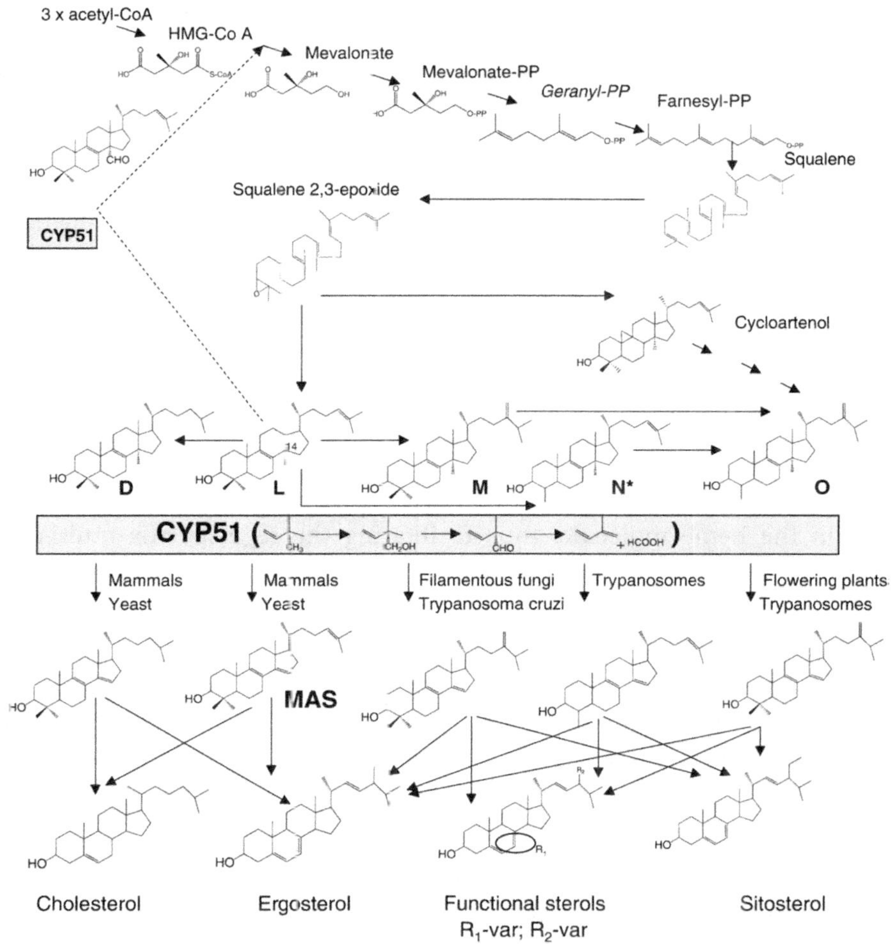

Figure 10.2 An abbreviated description of sterol biosynthesis in various phyla and the role of CYP51. D: 24,25-dihydrolanosterol; L: lanosterol; M: 24-methylenedihydrolanosterol; N: norlanosterol; O: obtusifoliol; R1: alkyl or alkylene group; R2: C6 or C7 double bond; MAS: meiosis-activating sterols. The dashed line shows a major site of regulation of sterol biosynthesis.
Image reprinted from[8] G. I. Lepesheva and M. R. Waterman, Sterol 14α-demethylase cytochrome P450 (CYP51), a P450 in all biological kingdoms. *Biochim. Biophys. Acta,* 2011, **1770**(3), 467–477, Copyright 2007, with permission from Elsevier.

methyl group. This elimination of formic acid creates a double bond within the D-ring of the sterol backbone (Figure 10.2).[14,16,17] Like many CYPs, CYP51 is generally membrane-bound in eukaryotes and is found in the mitochondrial inner membrane and the membrane of the endoplasmic reticulum. In prokaryotes, CYP51 is present in a soluble form.[18]

10.2.2 Mechanism of the Inhibitory Action of CYP51

Based on a three-dimensional molecular model of the fungal form of CYP51 from *Saccharomyces cerevisiae*, the endogenous ligand lanosterol binds to the active site of CYP51 in a specific orientation. Aoyama *et al.* proposed in 1989 that the 3-hydroxy group of lanosterol was essential for orienting lanosterol in the CYP51 active site.[19] This conclusion was based on demethylation studies of lanosterol derivatives where the 3-hydroxy group was oxidized to 3-oxo-lanosterol (3-oxolanosta-8,24-diene), acetylated to 3-acetoxy-lanosterol or replaced with a methylene group (3-methylene-lanost-8-ene).[19] The demethylation rate of the 3-oxo derivative was lower than lanosterol and the 3-acetoxy and methylene derivatives could not be enzymatically demethylated. In addition, it was found that the 8-position double bond of lanosterol plays an important role in the enzyme–substrate interaction of CYP51.[20] It was reported later that significant amino acid interactions within the CYP51 active site position orient the C14-methyl group of the lanosterol ligand to be in close proximity to the oxy-ferryl atom within the heme molecule, thus facilitating the regiospecific multi-step oxidation process.[21]

The molecular basis of inhibition of CYP51 by conazoles is the binding of one of the nitrogens in the azole N-heterocycle within the active site of CYP51. This can be seen in the model of crystal structure of the active site chamber of CYP51 from *Mycobacterium tuberculosis* bound to fluconazole (Figure 10.3). The tetraporphyrin heme is red, fluconazole is light green, and the protein loops are light blue. The heme iron atom is in the center of the tetraporphyrin heme molecule and near the N4 of the triazole ring nitrogen (in dark blue) of one of the two triazole rings of fluconazole. An azole nitrogen binds to the 6th axial position of the heme iron atom. The tetraporphyrin molecule coordinates with the four equatorial positions of the heme iron atom, while the cysteine sulfur binds to the 5th axial position (not shown in the figure) of the heme iron.

One of the characteristics of the conazole–CYP51 complex are the spectral changes that are observed when performing an absorption difference spectrum. This is accomplished essentially by subtracting the ultraviolet (UV)–visible (VIS) spectrum of the enzyme alone from the UV–VIS spectrum of enzyme and inhibitor. This generates a difference spectrum and these have been characterized mainly as type I or type II and mixed types depending on the shape and wavelengths of the maxima and minima of the absorbance spectra. In general, azole antifungals form a 1-to-1 complex with oxidized CYP51 and induce type II difference spectral changes with minima at about 398 nm (390–410 nm) and maxima at about 430 nm (425–430 nm). This spectral change is associated with the interaction of inhibitors that bind directly to the iron atom, replacing the water molecule as the sixth axial ligand.[22] A summary of conazoles that produce type II difference spectra with CYP51 from different species is found in Table 10.1. The conclusion that we can draw from this table is that these conazoles interact with CYP51

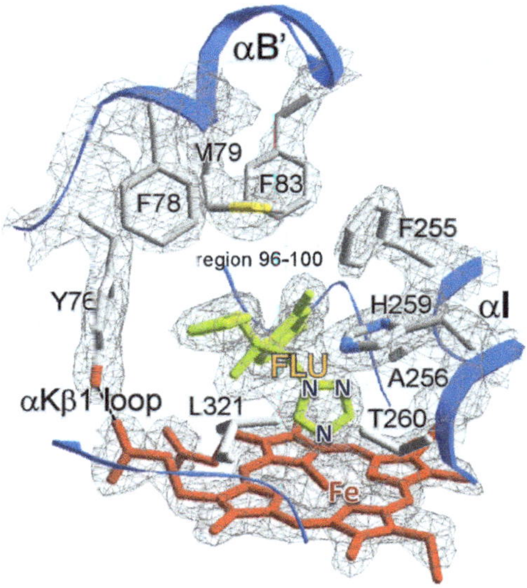

Figure 10.3 Model of the crystal structure of the active site chamber of CYP51 from *Mycobacterium tuberculosis*, bound to fluconazole. The heme is colored red, fluconazole is light green, and the protein loops are blue. Image reprinted from[23] L. M. Podust, T. L. Poulos and M. R. Waterman, Crystal structure of cytochrome P450 14α-sterol demethylase (CYP51) from *Mycobacterium tuberculosis* in complex with azole inhibitors. *Proc. Natl. Acad. Sci. U. S. A.*, 2001, **98**(6), 3068–3073, Copyright (2001), National Academy of Sciences, U. S. A. Reproduced with permission.

from several fungal species and humans in a similar fashion. Conazole potency and selectivity are determined by the type of nitrogen heterocycle and, to a much greater extent, by the chemical structure, three dimensional structure and physiochemical properties of the *N*-1 substituent.[18] Conazoles bind mostly through hydrophobic interactions with residues within the active site and also induce substantial conformational changes in helix and loop regions.[17,23]

It is noted that a second mechanism may be operative for some conazoles. This is a fungicidal mechanism where conazoles induce direct membrane damage. This mechanism is dependent on the growth phase of the fungus and requires high concentrations.[31–33]

10.2.3 Inhibition of Fungal CYP51

An extensive number of studies have been published that describe the inhibitory activities of conazoles in fungal systems, with the majority being the pharmaceutical azoles (see reviews by Fromtling, Maertens and Niwa *et al.*).[3,4,34] An example of the activity of one of the most used conazoles, fluconazole, against five species of fungi, compared to several later generation conazoles, itraconazole and voriconazole is shown in Table 10.2 (using

Table 10.1 Conazoles that induce type II P450 binding spectra with CYP51 from different species.

Azole	Fungal CYP51	Human CYP51	Rodent CYP51	Ref.
Clotrimazole	*C. albicans*[a]	hCYP51[b]	Rat liver	Warrilow *et al.*[24] Strushkevich *et al.*[17] Ronis *et al.*[25]
Cyproconazole		hCYP51[b]		Strushkevich *et al.*[17]
Econazole	*C. albicans*[a]	hCYP51[b]		Warrilow *et al.*[24] Strushkevich *et al.*[17]
Epoxiconazole	*C. albicans*[a]	Δ60HsCYP51[c]		Warrilow *et al.*[26]
Fluconazole	*C. albicans*[a]	Δ60HsCYP51[c] hCYP51[b]		Warrilow *et al.*[24] Warrilow *et al.*[26] Strushkevich *et al.*[17]
Itraconazole	*C. albicans*[a] *S. cerevisae*[d]	Δ60HsCYP51[c]		Warrilow *et al.*[24] Yoshida and Aoyama[27] Warrilow *et al.*[26]
Ketoconazole	*C. albicans*[a,e] *S. cerevisae*[d,f]	hCYP51[b]	Rat liver	Warrilow *et al.*[24] Hitchcock *et al.*[28] Yoshida and Aoyama[27] Venkateswarlu *et al.*[29] Strushkevich *et al.*[17] Ronis *et al.*[25]
Miconazole	*C. albicans*[a]	hCYP51[b]		Warrilow *et al.*[24] Strushkevich *et al.*[17]
Propiconazole	*C. albicans*[a]	Δ60HsCYP51[c] hCYP51[b]	Rat liver	Warrilow *et al.*[26] Strushkevich *et al.*[17] Ronis *et al.*[25]
Tebuconazole		hCYP51[b]		Strushkevich *et al.*[17]
Triadimenol	*S. cerevisae*[d]	hCYP51[b]		Yoshida and Aoyama[30] Strushkevich *et al.*[17]
Voriconazole	*C. albicans*[a]			Warrilow *et al.*[24]

[a]Purified CYP51 protein from *Escherichia coli* expressing a transfected *Candida albicans* CYP51 gene.
[b]Purified human CYP51 protein from *E. coli* expressing a transfected CYP51 cDNA.
[c]Purified Δ60 truncated *Homo sapiens* CYP51 protein from *E. coli* expressing a transfected Δ60 truncated *H. sapiens* CYP51 gene.
[d]Purified CYP51 protein from microsomes from *Saccharomyces cerevisae* cells grown semi-anaerobically.
[e]Purified CYP51 protein from *C. albicans* microsomes.
[f]Membrane fraction from *E. coli* expressing a transfected *S. cerevisae* CYP51 gene.

data taken from Dickinson *et al.*[35]). Of the three *Candida* species evaluated (*C. albicans*, *C. krusei* and *C. glabrata*), *C. albicans* is the most common species and can cause thrush or candidiasis in humans. *Aspergillus fumigatus* is a cause of invasive fungal infection in immunosuppressed individuals, as is *Cryptococcus neoformans*. In Table 10.2, *C. albicans* and *C. krusei* are resistant to fluconazole, while itraconazole and voriconazole are potent inhibitors with minimum inhibitory concentrations at the 70–90 nM levels. In all comparisons the newer generation conazoles are an order of magnitude or more potent than fluconazole.[35]

Table 10.2 Inhibition of a series of fungal species by pharmaceutical conazoles.

	Conazole MIC[a,b] (μM)		
Species	Voriconazole	Itraconazole	Fluconazole
Aspergillus fumigatus	0.26	0.55	>163
Candida albicans	0.09	0.17	3.3
Candida krusei	0.69	0.07	>82
Candida glabrata	0.54	0.27	6.2
Cryptococcus neoformans	1.1	0.55	31

[a]Data recalculated from Dickinson *et al.*[35]
[b]MIC: minimium inhibitory concentration.

10.2.4 Comparison of the Inhibition of Fungal and Human CYP51

Two significant studies compared the inhibitory potencies of a series of conazoles on fungal and mammalian CYP51 (Table 10.3). Trosken *et al.*[36] and Warrilow *et al.*[26] used different human and *C. albicans* CYP51 protein sources, different assay conditions and different mass spectrometric methods to quantitate the demethylated lanosterol metabolite. Trosken *et al.* compared 13 agricultural conazoles and six antifungal conazole drugs. Microsomes obtained from a recombinant human CYP51 (hCYP51) and a baculovirus-expressed *C. albicans* (cCYP51) CYP51 were used in the inhibition assays.[36] For hCYP51, the inhibitory concentration 50% (IC_{50}) values for the 13 agricultural conazoles were all in the lower micromolar scale, with a range of 1.3–36.1 μM. Within the group of agricultural conazoles the activities of selected conazoles were: bitertanol (1.3 μM), epoxiconazole (1.95 μM), tebuconazole (3.61 μM), propiconazole (8.25 μM), triadimefon (9.95 μM), hexaconazole (15.3 μM), penconazole (19.3 μM), cyproconazole (22.8 μM) and myclobutanil (29 μM). For cCYP51 the IC_{50} values for the 13 agricultural conazoles were an order of magnitude lower, with a range of 0.059–0.35 μM. Warrilow *et al.*[26] compared five agricultural conazoles and three antifungal conazole drugs. A *C. albicans* CYP51 gene and a truncated human CYP51 gene were each expressed in *Escherichia coli* and the CYP51 proteins isolated and purified. In general, the IC_{50} data from Warrilow *et al.* were 10-fold higher than the Trosken *et al.* data for the three drugs tested in common. The IC_{50} data on the agricultural conazoles from Warrilow *et al.*: epoxiconazole, prochloraz, propiconazole, tebuconazole and triadimenol, had similarities and differences to those from Trosken *et al.* for both fungal and human CYP51 proteins. The conclusions of these studies are that the conazole antifungal agents are more potent inhibitors of fungal CYP51 compared to the human CYP51 ortholog, but the extent of this specificity varies greatly from study to study and from conazole to conazole, independent of its use (drug or agricultural).

Table 10.3 Inhibition of *Candida albicans* CYP51 and human CYP51 by azoles.[a]

Pharmaceutical azoles	cCYP51[b]	IC_{50} (μM) CaCYP51[c]	*C. albicans* CYP51[d]	hCYP51[e]	Δ60HsCYP51[f]
Pharmaceutical azoles					
Bifonazole	0.3			0.8	
Clotrimazole	0.091			0.85	
Fluconazole	0.051	0.6			>1300
Itraconazole	0.039	0.4	0.0076		70
Ketoconazole	0.064	0.5	0.008	0.43	4.5
Miconazole	0.072			0.057	
Agricultural azoles					
Bitertanol	0.059			1.3	
Cyproconazole	0.1			22.8	
Epoxiconazole	0.22	0.5		1.95	12
Flusilazole	0.085			3.36	
Hexaconazole	0.066			15.6	
Imazalil	0.082			36.1	
Myclobutanil	0.14			29	
Penconazole	0.076			19.3	
Prochloraz	0.098	0.7		5	8
Propiconazole	0.15	0.6		8.25	110
Tebuconazole	0.35	0.9		3.61	1.3
Tetraconazole			0.35		
Triadimefon	0.13			9.95	
Triadimenol	0.33	1.3		37.2	30

[a]IC_{50}: inhibitory concentration 50%.
[b]Inhibition studies used cell membrane preparations from cells expressing *C. albicans* CYP51 and co-expressed with *C. tropicalis* oxidoreductase in a baculovirus system (cCYP51).[36]
[c]Inhibition studies used purified CYP51 protein from *Escherichia coli* expressing a transfected *C. albicans* CYP51 gene (CaCYP51).[26]
[d]Inhibition studies used microsomes from genetically-modified *Saccharomyces cerevisae* expressing a transfected *C. albicans* CYP51 gene (*C. albicans* CYP51).[37,38]
[e]Inhibition studies used human CYP51 supersomes (hCYP51).[36]
[f]Inhibition studies used purified CYP51 protein from *E. coli* expressing a transfected truncated 60HsCYP51gene (Δ60HsCYP51).[26]

10.2.5 Inhibition of Other Mammalian CYPs

By design, conazoles were targeted to inhibit CYP51. However, they also have wide inhibitory activity across many other human CYP classes. Table 10.4 summarizes studies of conazoles inhibiting major classes of CYPs: CYP1A, CYP2A, CYP2B, CYP2C, CYP2D, CYP2E and CYP3A using different substrates and oxidative processes (hydroxylation, de-ethylation and debenzylation). Some important examples in Table 10.4 are noted. Of the conazoles studied, miconazole, sulconazole and tioconazole were potent CYP1A2 inhibitors, clotrimazole and miconazole were potent CYP2A6 inhibitors and itraconazole, miconazole, sulconazole and tioconazole were potent CYP2B6 inhibitors. The most potent CYP2C9, CYP2C19 and CYP2D6 inhibitors were miconazole, sulconazole and tioconazole. Tioconazole was the most potent CYP2E1 inhibitor. Many conazoles were very active CYP3A4 inhibitors while

Table 10.4 Examples of inhibition of common mammalian cytochrome P450 (CYP) by conazoles.[a]

CYP	CYP source	Inhibitor	Substrate	Reaction	IC_{50} (μM)	K_i (μM)	Ref.
CYP1A2	cDNA-expressed human CYP	Clotrimazole	Phenacetin	De-ethylation		7.9	Zhang *et al.*[40]
CYP1A2	Human liver microsomes	Fluconazole	Phenacetin	De-ethylation	>100		Kozakai *et al.*[41]
CYP1A2	Human liver microsomes	Itraconazole	Phenacetin	De-ethylation	>10		Kozakai *et al.*[41]
CYP1A2	Human liver microsomes	Ketoconazole	7-Ethoxyresorufin	De-ethylation	13		Baldwin *et al.*[42]
CYP1A2	Human liver microsomes	Miconazole	7-Ethoxyresorufin	De-ethylation	2.9		Niwa *et al.*[43]
CYP1A2	cDNA-expressed human CYP	Miconazole	Phenacetin	De-ethylation		3.2	Zhang *et al.*[40]
CYP1A2	cDNA-expressed human CYP	Sulconazole	Phenacetin	De-ethylation		0.4	Zhang *et al.*[40]
CYP1A2	cDNA-expressed human CYP	Tioconazole	Phenacetin	De-ethylation		0.4	Zhang *et al.*[40]
CYP2A6	cDNA-expressed human CYP	Clotrimazole	Coumarin	Hydroxylation		0.19	Draper *et al.*[44]
CYP2A6	cDNA-expressed human CYP	Clotrimazole	Coumarin	Hydroxylation		2.7	Zhang *et al.*[40]
CYP2A6	Human liver microsomes	Itraconazole	Coumarin	Hydroxylation	>10		Kozakai *et al.*[41]
CYP2A6	Human liver microsomes	Fluconazole	Coumarin	Hydroxylation	43.5		Kozakai *et al.*[41]
CYP2A6	cDNA-expressed human CYP	Miconazole	Coumarin	Hydroxylation		0.19	Draper *et al.*[44]
CYP2A6	cDNA-expressed human CYP	Miconazole	Coumarin	Hydroxylation		2.2	Zhang *et al.*[40]
CYP2A6	cDNA-expressed human microsomes	Sulconazole	Coumarin	Hydroxylation		2	Zhang *et al.*[40]
CYP2A6	cDNA-expressed human CYP	Tioconazole	Coumarin	Hydroxylation		1.7	Zhang *et al.*[40]
CYP2B6	cDNA-expressed human CYP	Clotrimazole	7-Ethoxy-4-trifluoromethyl coumarin	De-ethylation		12	Zhang *et al.*[40]
CYP2B6	cDNA-expressed human CYP	Clotrimazole	Bupropion	Hydroxylation	0.243		Walsky *et al.*[45]
CYP2B6	Human liver microsomes	Fluconazole	Bupropion	Hydroxylation	>100		Kozakai *et al.*[41]
CYP2B6	Human liver microsomes	Itraconazole	Bupropion	Hydroxylation	0.2		Kozakai *et al.*[41]
CYP2B6	cDNA-expressed human CYP	Itraconazole	Bupropion	Hydroxylation	0.0895		Walsky *et al.*[45]
CYP2B6	cDNA-expressed human CYP	Ketoconazole	Bupropion	Hydroxylation	3.18		Walsky *et al.*[45]
CYP2B6	cDNA-expressed human CYP	Miconazole	7-Ethoxy-4-trifluoromethyl coumarin	De-ethylation		0.050	Zhang *et al.*[40]
CYP2B6	cDNA-expressed human CYP	Sulconazole	7-Ethoxy-4-trifluoromethyl coumarin	De-ethylation		0.04	Zhang *et al.*[40]

Table 10.4 (*Continued*)

CYP	CYP source	Inhibitor	Substrate	Reaction	IC_{50} (μM)	K_i (μM)	Ref.
CYP2B6	cDNA-expressed human CYP	Tioconazole	7-Ethoxy-4-trifluoromethyl coumarin	De-ethylation		0.1	Zhang *et al.*[40]
CYP2C9	cDNA-expressed human CYPs	Clotrimazole	Diclofenac	Hydroxylation		12	Zhang *et al.*[40]
CYP2C9	Human liver microsomes	Fluconazole	Tolbutamide	Hydroxylation	30.3		Niwa *et al.*[46]
CYP2C9	Human liver microsomes	Fluconazole	Diclofenac	Hydroxylation	26.3		Kozakai *et al.*[41]
CYP2C9	Human liver microsomes	Itraconazole	Diclofenac	Hydroxylation	>10		Kozakai *et al.*[41]
CYP2C9	Human liver microsomes	Itraconazole	Tolbutamide	Hydroxylation	>10		Niwa *et al.*[46]
CYP2C9	Human liver microsomes	Ketoconazole	Tolbutamide	Hydroxylation	13		Baldwin *et al.*[42]
CYP2C9	Human liver microsomes	Miconazole	Tolbutamide	Hydroxylation	2		Niwa *et al.*[46]
CYP2C9	cDNA-expressed human CYP	Miconazole	Diclofenac	Hydroxylation		1.2	Zhang *et al.*[40]
CYP2C9	cDNA-expressed human CYP	Sulconazole	Diclofenac	Hydroxylation		0.01	Zhang *et al.*[40]
CYP2C9	cDNA-expressed human CYP	Tioconazole	Diclofenac	Hydroxylation		0.7	Zhang *et al.*[40]
CYP2C9	Human liver microsomes	Voriconazole	Tolbutamide	Hydroxylation	8.4		Niwa *et al.*[46]
CYP2C19	cDNA-expressed human CYP	Clotrimazole	Omeprazole	Hydroxylation		1.4	Zhang *et al.*[40]
CYP2C19	Human liver microsomes	Fluconazole	*S*-Mephenyltoin	Hydroxylation	18.3		Kozakai *et al.*[41]
CYP2C19	Human liver microsomes	Fluconazole	*S*-Mephenytoin	Hydroxylation	12.3		Niwa *et al.*[46]
CYP2C19	Human liver microsomes	Itraconazole	*S*-Mephenyltoin	Hydroxylation	>10		Kozakai *et al.*[41]
CYP2C19	Human liver microsomes	Itraconazole	*S*-Metphenytoin	Hydroxylation	>10		Niwa *et al.*[46]
CYP2C19	Human liver microsomes	Miconazole	*S*-Metphenytoin	Hydroxylation	0.33		Niwa *et al.*[46]
CYP2C19	cDNA-expressed human CYP	Miconazole	Omeprazole	Hydroxylation		0.01	Zhang *et al.*[40]
CYP2C19	cDNA-expressed human CYP	Sulconazole	Omeprazole	Hydroxylation		0.008	Zhang *et al.*[40]
CYP2C19	cDNA-expressed human CYP	Tioconazole	Omeprazole	Hydroxylation		0.04	Zhang *et al.*[40]
CYP2C19	Human liver microsomes	Voriconazole	*S*-Metphenytoin	Hydroxylation	8.7		Niwa *et al.*[46]
CYP2D6	cDNA-expressed human CYP	Clotrimazole	Dextromethorphan	Demethylation		13	Zhang *et al.*[40]
CYP2D6	Human liver microsomes	Fluconazole	Bufuralol	Hydroxylation	>100		Kozakai *et al.*[41]
CYP2D6	Human liver microsomes	Itraconazole	Bufuralol	Hydroxylation	>10		Kozakai *et al.*[41]
CYP2D6	cDNA-expressed human CYP	Miconazole	Dextromethorphan	Demethylation		0.7	Zhang *et al.*[40]
CYP2D6	Human liver microsomes	Miconazole	Debrisoquine	Hydroxylation	6.46		Niwa *et al.*[43]

CYP2D6	cDNA-expressed human CYP	Sulconazole	Dextromethorphan	Demethylation		0.4	Zhang *et al.*[40]
CYP2D6	cDNA-expressed human microsomes	Tioconazole	Dextromethorphan	Demethylation		1	Zhang *et al.*[40]
CYP2E1	cDNA-expressed human CYP	Clotrimazole	Chlorzoxazone	Hydroxylation		67	Zhang *et al.*[40]
CYP2E1	Human liver microsomes	Ketoconazole	Chlorzoxazone	Hydroxylation	90		Baldwin *et al.*[42]
CYP2E1	cDNA-expressed human microsomes	Miconazole	Chlorzoxazone	Hydroxylation		7.3	Zhang *et al.*[40]
CYP2E1	cDNA-expressed human CYP	Sulconazole	Chlorzoxazone	Hydroxylation		3.5	Zhang *et al.*[40]
CYP2E1	cDNA-expressed human microsomes	Tioconazole	Chlorzoxazone	Hydroxylation		0.38	Zhang *et al.*[40]
CYP3A4	cDNA-expressed human CYP	Clotrimazole	Dibenzylfluorescein	Debenzylation	0.005		Stresser *et al.*[47]
CYP3A4	cDNA-expressed human CYP	Clotrimazole	Benzyloxyquinoline	Debenzylation	0.013		Stresser *et al.*[47]
CYP3A4	cDNA-expressed human CYP	Clotrimazole	Benzyloxyresorufin	Debenzylation	0.019		Stresser *et al.*[47]
CYP3A4	cDNA-expressed human CYP	Clotrimazole	Omeprazole	Sulfonation		0.018	Zhang *et al.*[40]
CYP3A4	Human intestinal microsomes	Fluconazole	Midazolam	Hydroxylation		10.4	Gibbs *et al.*[48]
CYP3A4	Human liver microsomes	Fluconazole	Midazolam	Hydroxylation		10.7	Gibbs *et al.*[48]
CYP3A4	Human liver microsomes	Fluconazole	Nifedipine	Oxidation	13.1		Niwa *et al.*[46]
CYP3A4	cDNA-expressed human CYP	Fluconazole	Midazolam	Hydroxylation		7.4	Isoherranen *et al.*[49]
CYP3A4	cDNA-expressed human CYP	Fluconazole	Midazolam	Hydroxylation		9.21	Gibbs *et al.*[48]
CYP3A4	Human liver microsomes	Itraconazole	Diazepam	Hydroxylation	0.19		Tran *et al.*[50]
CYP3A4	cDNA-expressed human CYP	Itraconazole	Dibenzylfluorescein	Debenzylation	0.038		Stresser *et al.*[47]
CYP3A4	cDNA-expressed human CYP	Itraconazole	Benzyloxyquinoline	Debenzylation	0.35		Stresser *et al.*[47]
CYP3A4	Recombinant human microsomes	Itraconazole	Benzyloxyresorufin	Debenzylation	0.176		Stresser *et al.*[47]
CYP3A4	cDNA-expressed human CYP	Itraconazole	Midazolam	Hydroxylation	0.24		Shirasaka *et al.*[51]
CYP3A4	Human liver microsomes	Itraconazole	Nifedipine	Oxidation	0.0326		Niwa *et al.*[46]
CYP3A4	cDNA-expressed human CYP	Ketoconazole	Dibenzylfluorescein	Debenzylation	0.002		Stresser *et al.*[47]
CYP3A4	cDNA-expressed human CYP	Ketoconazole	Benzyloxyquinoline	Debenzylation	0.072		Stresser *et al.*[47]
CYP3A4	cDNA-expressed human CYP	Ketoconazole	Benzyloxyresorufin	Debenzylation	0.034		Stresser *et al.*[47]
CYP3A4	Human intestinal microsomes	Ketoconazole	Midazolam	Hydroxylation		0.017	Gibbs *et al.*[48]
CYP3A4	cDNA-expressed human CYP	Ketoconazole	Midazolam	Hydroxylation	0.040		Shirasaka *et al.*[51]

Table 10.4 (*Continued*)

CYP	CYP source	Inhibitor	Substrate	Reaction	IC_{50} (μM)	K_i (μM)	Ref.
CYP3A4	Human liver microsomes	Ketoconazole	Diazepam	Hydroxylation	0.0059		Tran *et al.*[50]
CYP3A4	Human liver microsomes	Ketoconazole	Midazolam	Hydroxylation		0.0149	Gibbs *et al.*[48]
CYP3A4	cDNA-expressed human CYP	Ketoconazole	Midazolam	Hydroxylation		0.0267	Gibbs *et al.*[48]
CYP3A4	cDNA-expressed human CYP	Ketoconazole	Benzyloxyquinoline	Debenzylation	0.072		Stresser *et al.*[47]
CYP3A4	cDNA-expressed human CYP	Miconazole	Benzyloxyresorufin	Debenzylation	0.175		Stresser *et al.*[47]
CYP3A4	cDNA-expressed human CYP	Miconazole	Benzyloxyquinoline	Debenzylation	0.302		Stresser *et al.*[47]
CYP3A4	cDNA-expressed human CYP	Miconazole	Omeprazole	Sulfonation		0.028	Zhang *et al.*[40]
CYP3A4	Human liver microsomes	Miconazole	Nifedipine	Oxidation	0.0742		Niwa *et al.*[46]
CYP3A4	cDNA-expressed human CYP	Miconazole	Dibenzylfluorescein	Debenzylation	0.058		Stresser *et al.*[47]
CYP3A4	cDNA-expressed human CYP	Sulconazole	Omeprazole	Sulfonation		0.01	Zhang *et al.*[40]
CYP3A4	cDNA-expressed human CYP	Tioconazole	Omeprazole	Sulfonation		0.02	Zhang *et al.*[40]
CYP3A4	Human liver microsomes	Voriconazole	Nifedipine	Oxidation	10.5		Niwa *et al.*[46]
CYP3A5	Human recombinant	Fluconazole	Midazolam	Hydroxylation		53	Isoherranen *et al.*[49]
CYP3A5	cDNA-expressed human CYP	Fluconazole	Midazolam	Hydroxylation		84.6	Gibbs *et al.*[48]
CYP3A5	cDNA-expressed human CYP	Ketoconazole	Midazolam	Hydroxylation		0.109	Gibbs *et al.*[48]
CYP3A5	Human recombinant	Ketoconazole	Midazolam	Hydroxylation	0.200		Shirasaka *et al.*[51]
CYP3A5	cDNA-expressed human CYP	Itraconazole	Midazolam	Hydroxylation	2.3		Shirasaka *et al.*[51]

[a] IC_{50}: inhibitory concentration 50%; K_i: inhibition constant.

ketoconazole was a potent inhibitor of CYP3A5. While the literature has many examples of the inhibitory effects of conazoles on human CYPs, the studies solely use pharmaceutical conazoles. No examples of agricultural conazoles inhibiting the CYPs cited in Table 10.4 were found. Several reviews document examples of clinical interactions of conazoles on oxidative drug metabolism and toxicity.[34,39]

10.3 Induction of CYPs by Conazoles

10.3.1 Introduction to Nuclear Receptors

Nuclear receptors are transcription factors that include nuclear hormone receptors and orphan nuclear receptors. Nuclear hormone receptors have an identified hormonal ligand such as the thyroid hormone, glucocorticoids and sex steroids, while the ligands for the orphan receptors or nuclear X receptors are unknown. Two of the most common nuclear X receptors are the constitutive androstane receptor (CAR) and the pregnane X receptor (PXR). PXR and CAR are predominantly expressed in the liver, although extrahepatic expression of PXR has also been reported.[52,53] CAR and PXR are predominantly localized in the cytoplasm in their inactive states.[53] In general, when bound to a ligand they translocate to the nucleus.[54] These complexes form a heterodimer with the 9-*cis* retinoic acid receptor (retinoid X receptor, RXR) that binds to the xenobiotic response element in the promoter region of responsive genes that elicits expression of gene products.[55] Two mechanisms have been described to explain the activation of CAR: direct ligand binding and an indirect activation mechanism.[53] TCPOBOP (1,4-bis-[2-(3,5-dichloropyridyloxy)]benzene-3,3′,5,5′-tetrachloro-1,4-bis(pyridyloxy)benzene) directly binds to mouse CAR (mCAR) at its ligand-binding domain with interactions of coregulators. A similar mechanism is proposed for the conazole clotrimazole with human CAR (hCAR). The indirect mechanism for CAR nuclear translocation has been reported for phenobarbital where there is no direct ligand binding to CAR. There is evidence that the dephosphorylation of Ser-202 is a required step that regulates the xenobiotic-dependent nuclear translocation of mCAR.[56] After translocation, CAR and PXR recruit coactivators to drive gene transcription.[53]

CAR and PXR regulate the expression of overlapping sets of genes encoding xenobiotic detoxification, elimination and transport functions encompassing phase I, phase II and transport enzymes. Both hPXR and hCAR regulate the gene expression of phase I CYPs, including *CYP2B6*, *CYP2C9*, *CYP2C19*, *CYP3A4* and *CYP3A5*. The expression of the phase II enzyme, *UGT1A1* is also regulated by both hCAR and hPXR. The mouse GST genes such as *Gsta1*, *Gstm1* and *Gstm2* have been shown to be regulated by both mPXR and mCAR. PXR and CAR also regulate the expression of human drug transporters *ABCB1* and *ABCC2*. CAR and PXR also regulate the expression of distinctive sets of genes in metabolism and excretion. PXR regulates the expression of the phase I genes, *CYP7A1* and *CYP4F12*, while CAR has been found to regulate *CYP1A1* and *CYP1A2* expression, a finding generally associated with the aryl hydrocarbon nuclear receptor (Ahr).[57]

10.3.2 Induction of CYPs

A summary of the effects of conazoles on the induction of specific CYPs after *in vivo* treatment is presented in Table 10.5. CYPs were identified by Western immunoblot techniques using CYP-specific antibodies or by using metabolic probes specific for each CYP. These probes, enzymatic activities and specific CYPs wereas follows. CYP1A1: ethoxyresorufin and 7-ethoxyresorufin-*O*-deethylase (EROD); CYP1A2: 7-methoxyresorufin and 7-methoxyresorufin-*O*-demethylase (MROD); CYP2B: 7-pentoxyresorufin and 7-pentoxyresorufin-*O*-depentylase (PROD); and CYP3A2: erythromycin and erythromycin-*N*-demethylase. Studies using mice and rats and routes of administration of oral gavage, feed or intraperitoneal (IP) are presented, with the liver being the predominant organ studied. All of the nine pharmaceutical and agricultural conazoles evaluated induced isoforms of CYP2B and CYP3A in either or both mice and rats. CYP1A1 induction was also reported for clotrimazole (rats), fluconazole (mice and rats), imazalil (mice), myclobutanil (mice and rats), propiconazole (mice and rats) and triadimefon (mice). The magnitude of inducing ability of conazoles varied with substrate, species and conazoles. For example, in mice, PROD, a measure of Cyp2b activity, was the most strongly affected by conazoles with maximal induction at 30 days of treatment of 37.5-, 17- and 14.9-fold over control for propiconazole, myclobutanil and triadimefon, respectively. These high levels of induction for PROD activity continued at 90 days of treatment.[58] In rats treated with the same conazoles, even higher levels of PROD induction were recorded at 30 days of treatment: 125-, 91- and 80.8-fold over control for propiconazole, myclobutanil and triadimefon, respectively.[59] These data indicate that CYP induction by itself could not explain the liver or thyroid tumorigenic responses of these three conazoles in chronic studies in mice and rats (propiconazole: mouse liver carcinogen, no tumors in rats; myclobutanil: no tumors in mice or rats; triadimefon: liver tumors in mice, thyroid tumors in rats).[58,59]

10.4 Hepatic Effects Induced by Conazoles

10.4.1 Increased Liver Weight and Hypertrophy

Hepatic hypertrophy is a commonly observed histological effect in studies of conazoles and is directly associated with increased liver weight. The mean absolute liver weights of male CD-1 mice given diets containing propiconazole at levels of 0, 850 or 2500 ppm for up to 60 days increased with both dose and time to $\sim$3.2 g at 850 ppm and $\sim$4.52 g at 2500 ppm at 60 days.[70] Allen *et al.* found that centrilobular, and in some cases, midzonal, hepatic

Table 10.5 Cytochrome P450s (CYPs) induced by conazoles *in vivo*.

Conazole	CYP(s) induced	Western or enzymatic activity	Species, gender	Strain	Route of administration	Organ	Ref.
Climbazole	CYP2B1 CYP3A2	Aminopyrine-*N*-demethylase Erythromycin-*N*-demethylase	Rat, male, female	Wistar	IP	Liver	Kobayashi *et al.*[60]
Climbazole	CYP2B1	Western	Rat, male	Wistar	IP	Liver	Kobayashi *et al.*[60]
Climbazole	CYP3A2	Western	Rat, male, female	Wistar	IP	Liver	Kobayashi *et al.*[60]
Climbazole	CYP2B1 CYP3A2	Western Erythromycin-*N*-demethylase	Rat, female	Long-Evans	IP	Liver	Kobayashi *et al.*[61]
Clotrimazole	CYP2B1 CYP3A2	Western	Rat, male, female	Wistar	IP	Liver	Kobayashi *et al.*[60]
Clotrimazole	CYP1A1/2 CYP2B1/2 CYP3A	Western	Rat, male, female	Sprague-Dawley	Oral gavage	Liver	Ronis *et al.*[25]
Cyproconazole	CYP2B	Western	Mouse, male	C3H/HeNClrBR	Oral feed	Liver	Peffer *et al.*[62]
Cyproconazole	CYP2A	Coumarin hydroxylase	Mouse, male	C3H/HeNClrBR	Oral feed	Liver	Peffer *et al.*[62]
Fenbuconazole	CYP2B	PROD[a]	Mouse, female	CD-1	Oral feed	Liver	Juberg *et al.*[63]
Fenbuconazole	CYP2B CYP3A	Western	Mouse, NS	NS	NS	Liver	Juberg *et al.*[63]
Fluconazole	CYP1A1 CYP2B CYP3A	Western EROD, PROD	Rat, male	Sprague-Dawley	Oral gavage	Liver	Sun *et al.*[64]
Fluconazole	CYP1A1 CYP1A2 CYP2B	EROD, MROD, PROD	Mouse, male	CD-1	Oral gavage	Liver	Sun *et al.*[64]
Imazalil	CYP1A1 CYP2B	EROD, PROD	Mouse, male	ddY	Oral gavage	Liver	Muto *et al.*[65]

Table 10.5 (*Continued*)

Conazole	CYP(s) induced	Western or enzymatic activity	Species, gender	Strain	Route of administration	Organ	Ref.
Imazalil	CYP2B CYP2C CYP3A	Western	Mouse, male	ddY	Oral gavage	Liver, Small intestine mucosa	Muto *et al.*[65]
Myclobutanil	CYP2B	PROD	Rat, male	Sprague-Dawley	Oral gavage	Liver	Sun *et al.*[66]
Myclobutanil	CYP2B1 CYP3A1	Western	Rat, male	Sprague-Dawley	Oral gavage	Liver	Sun *et al.*[66]
Myclobutanil	CYP2B CYP3A	Western	Mouse, male	CD-1	IP	Liver	Chen *et al.*[67]
Myclobutanil	CYP1A1 CYP1A2 CYP2B	EROD, MROD, PROD	Mouse, male	CD-1	Oral feed	Liver	Allen *et al.*[58]
Myclobutanil	CYP1A1 CYP2B	EROD, PROD	Rat, male	Wistar/Han	Oral feed	Liver	Wolf *et al.*[59]
Propiconazole	CYP1A1/2 CYP2B1/2 CYP3A	Western EROD, MROD, PROD	Rat, male	Sprague-Dawley	Oral gavage	Liver	Ronis *et al.*[25]
Propiconazole	CYP1A	EROD	Rat, male	Sprague-Dawley	IP	Liver	Leslie *et al.*[68]
Propiconazole	CYP1A1 CYP1A2 CYP2B	EROD, MROD, PROD	Rat, male	Sprague-Dawley	Oral gavage	Liver	Sun *et al.*[69]

Propiconazole	CYP1A2 CYP2B	MROD, PROD	Mouse, male	CD-1	Oral gavage	Liver	Sun *et al.*[69]
Propiconazole	CYP2B CYP3A	Western	Mouse, male	CD-1	IP	Liver	Chen *et al.*[67]
Propiconazole	CYP1A1 CYP1A2 CYP2B	EROD, MROD, PROD	Mouse, male	CD-1	Oral feed	Liver	Allen *et al.*[58]
Propiconazole	CYP1A1 CYP1A2 CYP2B	EROD, MROD, PROD	Rat, male	Wistar/Han	Oral feed	Liver	Wolf *et al.*[59]
Triadimefon	CYP2B	PROD	Rat, male	Sprague-Dawley	Oral gavage	Liver	Sun *et al*[66]
Triadimefon	CYP2B1 CYP3A1 CYP3A2	Western	Rat, male	Sprague-Dawley	Oral gavage	Liver	Sun *et al.*[66]
Triadimefon	CYP2B CYP3A	Western	Mouse, male	CD-1	IP	Liver	Chen *et al.*[67]
Triadimefon	CYP1A1 CYP1A2 CYP2B	EROD, MROD, PROD	Mouse, male	CD-1	Oral feed	Liver	Allen *et al.*[58]
Triadimefon	CYP1A1 CYP1A2 CYP2B	EROD, MROD, PROD	Rat, male	Wistar/Han	Oral feed	Liver	Wolf *et al.*[59]

[a]EROD: 7-Ethoxyresorufin-*O*-deethylase; IP: intraperitoneal; MROD: 7-methoxyresorufin-*O*-demethylase; NS: not stated; PROD: 7-pentoxyresorufin-*O*-depentylase.

hypertrophy was increased in male CD-1 mice fed propiconazole (2500 ppm), myclobutanil (2000 ppm) or triadimefon (1800 ppm) after 4, 30 and 90 days of treatment.[58] Concomitantly, liver weights were elevated at each time point. In some cases liver:body weight measures increased significantly, up to 60%.[58] In a companion study using the same conazoles fed to male Wistar/Han rats, Wolf *et al.* found the incidence of hepatic hypertrophy significantly increased by each conazole with non-significant increases in liver:body weights.[59] Female mice fed fenbuconazole (200–1300 ppm) for 1, 4, 52 or 78 weeks had increased hepatic hypertrophy.[63] At 1 and 4 weeks liver:body weights were significantly increased by 70% and 50%, respectively, at the 1300 ppm dose level. Histologically, there was evidence of centrilobular hepatocellular hypertrophy with increased cytoplasmic eosinophilia and panlobular hepatocellular vacuolation and some focal hepatocellular necrosis at 1 week of treatment. At 4 weeks of treatment hepatocellular hypertrophy was present in the centrilobular-to-midzonal regions of the liver, with the hepatocellular vacuolation localized to periportal regions of the liver. When the dosing was terminated in mice dosed for 4 weeks, and these mice were held for an additional 6 weeks, the liver:body weights returned to control values and there were no histological changes. Male CD-1 mice fed three hepatocarcinogenic conazoles were found to have increased liver:body weights after 30 days of treatment with propiconazole, doubling the liver values.[71] Cyproconazole induced an increase in liver:body weight and hepatocyte hypertrophy in male CD-1 C57BL/6J and C3H/HeNClrBR mice.[62] In other studies using routes of administration other than feed, propiconazole significantly increased liver:body weights in male CD-1 mice and Sprague-Dawley rats after gavage administration, and in male mice after IP administration.[67,69] Myclobutanil significantly increased liver:body weights in male Sprague-Dawley rats after gavage administration,[66] and fluconazole significantly increased liver:body weights and centrilobular-to-midzonal hepatic hypertrophy in male CD-1 mice and Sprague-Dawley rats after gavage administration.[64] A group of dietary studies using a series of dietary levels of propiconazole (500, 850, 1450 and 2500 ppm) in male mice over the course of 100 weeks indicated hepatic hypertrophy increased over the entire 100 week period (Table 10.6). Cyproconazole, tebuconazole and fluconazole each induced dose-dependent liver weight increases and hepatocellular hypertrophy when administrated in the diet to male C3H/HeNCrlCrlj mice for 1 week.[72] A longer-term study was conducted with male C3HHeNCrlCrlj and CAR-knockout mice treated with cyproconazole, tebuconazole or fluconazole in the diet for 27 weeks, after initiation by diethylnitrosamine. The results indicated that CAR was the main mediator of liver hypertrophy induced by cyproconazole and fluconazole, but CAR was not the main mediator of liver hypertrophy induced by tebuconazole.[73]

10.4.2 Cell Proliferation

Hepatic cell proliferation induced by conazoles has been reported in a number of studies using different techniques to determine cell proliferation

Table 10.6 Summarized key events associated with tumor induction in male mice fed diets containing propiconazole (submitted studies).[a]

Key event	Dose (ppm)	Weeks 4	8	9	13		14	17	52–53		78	100
MRID		−502[a]	−502[a]	−401[a]	−502[a]	−501[a]	−401[a]	−501[a]	−401[a]	570[a]	−401[a]	−570[a]
↑Liver/body weight	500[b]			10%	18%				13%	12%	10%	6%
	850			33%	34%				29%		20%	
	1450				56%							
	2500				94%					41%		50%
↑Hypertrophy	500[b]	100%	80%[c]	60%[c]	80%			20%	60%	18%[d]	56%	33%[d]
	850	90%	100%	100%	100%			70%	80%		58%	
	1450	100%	100%		100%			100%				
	2500	90%	100%		100%			100%		78%[d]		86%[d]
↓Serum cholesterol	500[b]	NC[f]	NC[f]	−11%	NC[f]	NC[f]	−14%	−12%	−15%	−15%	−24%	−6%
	850	−29%	−9%	−25%	−30%	−10%	−24%	−24%	−34%		−29%	
	1450	−37%	−19%		−39%	−42%		−45%				
	2500	−64%	−50%		−45%	−41%		−44%		−28%		+44%
Single cell necrosis	500[b]	40%	80%	10%	25%			0	0	0	0	0
	850	70%	40%	30%	65%			0	0		0	
	1450	90%	90%		90%			10%				
	2500	90%	90%		100%			60%		0		0
Necrosis	500[b]	10%	10%	30%	25%			10%	0	0	0	0
	850	30%	0	−50%	15%			20%	0		0	
	1450	30%	50%		30%			40%				
	2500	70%	60%		30%			60%		0		0
Tumors[e]	500[b]										10%	37%
	850										24%	
	1450											
	2500											86%

[a]Reproduced from EPA (Table 2).[74] Data are presented as % of control values, unless otherwise stated. Data were derived from Master Record Identification (MRID) numbers 42050**502**, 42050**401**, 42050**501**, 45215**801** and 000129**570**.
[b]Results for lower doses (≤100 ppm) were generally comparable to the vehicle control group.
[c]MRID 45215801: reread of histological slides from MRID 42050502 (8 weeks) and 44381401 (9 weeks).
[d]Listed as "hepatocyte enlargement", presumed to be hepatocyte hypertrophy.
[e]Combined adenomas and carcinomas.
[f]NC, no change.

at different feed levels and different strains of mice and in rats. Female CD-1 mice fed fenbuconazole at its tumorigenic dose (1300 ppm) for 1 week showed a very large increase in cell proliferation (852%) as measured by bromodeoxyuridine (BrdU) labeling. However, at 4 weeks of fenbuconazole treatment, cell proliferation returned to control levels.[63] Propiconazole, epoxiconazole and cyproconazole each significantly increased cell proliferation after 30 days of treatment in the livers of male CD-1 mice fed tumorigenic levels of each conazole (2500, 500 and 200 ppm, respectively), with an approximate doubling of the labeling indices as measured by Ki-67 immunohistochemistry.[71] Propiconazole (2500 ppm) and myclobutanil (2000 ppm) increased cell proliferation in male mice after 4 days of treatment, while myclobutanil and triadimefon (1800 ppm) increased cell proliferation at 30 days of treatment in male CD-1 mice, as measured by proliferating cell nuclear antigen (PCNA) immunohistochemistry.[58] The effects of propiconazole on cell proliferation were studied in wild-type and CAR-null mice (based on a C57BL/6 background). Mice were gavaged with propiconazole (210 mg kg^{-1}) each day for 7 days. Liver:body weight ratios were increased in wild-type mice but not in CAR-null mice. The labeling index was significantly increased in hepatocytes from wild-type but not CAR-null mice exposed to propiconazole.[75] Propiconazole induced cell proliferation in male CD-1 mice for 7 days at its tumorigenic dose (2500 ppm). At 850 ppm it increased cell proliferation for 4 days as measured by BrdU incorporation.[76] Cyproconazole did not induce cell proliferation in male C3H/HeNClrBR, C57BL/6J or CD-1 mice at its tumorigenic dose (200 ppm) for 2, 7 or 14 days, but did increase cell proliferation in male C3H/HeNClrBR and C57BL/6J mice at 7 days and in male C57BL/6J mice at 14 days at a higher dose level (450 ppm) as measured by BrdU incorporation.[62] Cell proliferation was increased in C57BL/6J mice treated with cyproconazole at 200 ppm for 7 days, but not in Car-null mice under the same treatment conditions. At 450 ppm cyproconazole, cell proliferation was increased in wild-type mice at 7 days and to a lesser extent in Car-null mice.[62] Increased cell proliferation was measured in male Wistar/Han rats fed myclobutanil (2000 ppm), propiconazole (2500 ppm) and triadimefon (1800 ppm) for 4 days as measured by PCNA immunohistochemistry. After 30 days of treatment, only myclobutanil produced increased cell proliferation.[59]

10.5 Effects on Serum Cholesterol and Triglyceride Levels

Studies in humans were the first to show the effects of conazoles on cholesterol and lipids. Male prostate cancer patients receiving high-dose ketoconazole therapy had reduced total serum cholesterol values with no effect on serum triglyceride levels. Both measures remained at these levels for 5 months in almost all of the patients. Also found was a reduction in low-density lipoprotein (LDL) cholesterol levels without associated changes in high-density lipoprotein (HDL) cholesterol levels. Concomitantly, serum lanosterol levels increased during ketoconazole treatment.[77] Additional studies in cultured normal human fibroblasts showed that ketoconazole inhibited cholesterol synthesis by blocking the conversion of lanosterol to cholesterol. Clotrimazole exerted similar effects. Confluent cells exposed to ketoconazole had a major reduction in 3-hydroxy-3-methylglutaryl coenzyme A reductase activity. The authors concluded that ketoconazole had several effects on cholesterol synthesis; inhibiting CYP51 and indirectly suppressing total sterol synthesis *via* feedback inhibition by sterol intermediates of 3-hydroxy-3-methylglutaryl coenzyme A reductase activity.[78] Additional studies in cancer and non-cancer patients reported significant lowering of serum cholesterol levels by ketoconazole with no effect on serum triglycerides.[79,80]

Studies of the effects of conazoles on serum cholesterol levels have been reported in both mice and rats. Male CD1 (ICR) BR mice received diets containing propiconazole at levels of 0, 20, 500, 850, 1450 or 2500 ppm for up to 13 weeks. Serum cholesterol levels decreased with dose at both 4 and 13 weeks of treatment with maximal reductions to 36% and 55% of control levels at 4 and 13 weeks, respectively.[70] In another study male and female CD1 (ICR) BR mice received diets containing propiconazole at levels of 0, 20, 500, 850, 1450 or 2500 ppm for males and 0, 20, 500 or 2500 ppm for females, for 17 weeks. Serum cholesterol level decreased to 56% of control for males but was not altered in females, mirroring the tumorigenic responses of male and female mice in 2 year chronic feeding studies.[70] In male CD-1 mice fed conazoles over a 90 day period, myclobutanil (2000 ppm), propiconazole (2500 ppm) and triadimefon (1800 ppm) produced statistically significant (triadimefon at 4 and 30 days of treatment and propiconazole and myclobutanil at 30 and 90 days of treatment) lowering of serum cholesterol levels. Triglyceride levels were generally elevated with statistically significant increases for all three conazoles at 30 days of treatment.[58] A group of studies in male mice used a series of dietary levels of propiconazole (500, 850, 1450 and 2500 ppm) over the course of 100 weeks. Lower serum cholesterol levels were observed at many time periods at 2500 or 850 ppm up to 100 weeks. A greater than 40% reduction in serum cholesterol levels was observed up to 17 weeks of treatment and a 28% reduction in serum cholesterol levels at 52–53 weeks of treatment at 2500 ppm. At 850 ppm, mice had reduced serum cholesterol levels at every time period measured, with a 29% decrease at 78 weeks (Table 10.6). Cyproconazole fed to male CD-1 mice also decreased serum cholesterol levels.[62] Both male and female NMRI mice fed tebuconazole (180 ppm) for 52 weeks had significantly lower serum cholesterol levels.[81] Under the same dosing conditions as used in the Allen *et al.*[58] mouse conazole study, studies in male Wistar/Han rats revealed that myclobutanil, propiconazole and triadimefon did not decrease serum cholesterol levels, but either statistically significantly increased them or had no effect. Triglyceride levels were decreased by triadimefon at 30 and 90 days of treatment.[59] Total serum cholesterol increased in female Crl:CD(BR)SD rats fed uniconazole-P (1000 ppm) for extended time periods up to 104 weeks.[82] Hepatic levels of free and esterified cholesterol in rats fed ketoconazole (0.05%) were not changed, while serum cholesterol levels were slightly decreased (−19%).[83]

10.6 Genotoxicity

Pesticides submitted to the US Environmental Protection Agency (EPA) for registration and use are required to have been evaluated in a battery of tests for genotoxicity. These tests can include a bacterial reverse mutation test, an *in vitro* mammalian cell gene mutation test, an *in vitro* mammalian chromosome aberration test, a mammalian bone marrow chromosomal aberration test, an unscheduled DNA synthesis in mammalian cells in culture test and *in vitro* and *in vivo* tests for sister chromatid exchanges.[84] While this chapter is not a comprehensive review of the numerous findings of these tests for conazoles, some generalizations can be made. The California EPA Office of Environmental Health Hazard Assessment provides a summary document on the genotoxicity of 14 commonly used conazoles (Table 10.7).[85] In this table the results of tests for gene mutation, chromosomal effects and unscheduled DNA synthesis are listed. All 14 conazoles tested were negative in tests for gene mutation in bacteria. Of the 14 conazoles tested, fluconazole and uniconazole were positive for the induction of micronuclei. This test is generally conducted using bone marrow of treated rodents. Propiconazole (or its metabolites) are another example of a conazole with the lack of general DNA reactivity or direct genotoxicity in standardized tests. Propiconazole was investigated *in vitro* for its ability to induce gene mutations in bacteria and mammalian cells, chromosomal aberrations in mammalian cells and unscheduled DNA synthesis in mammalian cells. Propiconazole was also tested *in vivo* in two studies of micronucleus induction for the induction of dominant lethal mutations. All studies gave negative results.[70] A transgenic mouse study using propiconazole at its tumorigenic dose (2500 ppm for 4 days) found an increase in mutation frequency in the liver and a shift in mutation spectra.[86,87] It was also found that propiconazole increased the levels of a specific group of endogenous DNA adducts in the livers of the transgenic mice[88] (see Section 10.10.5). With that exception, the general conclusion to date is that conazoles or their metabolites as a class of agents are neither directly DNA reactive nor directly genotoxic.

10.7 Tumorigenic Effects of Conazoles

The organs susceptible to the tumorigenic effects of 27 conazoles based on available data are found in Table 10.8. Much of these data are not found in the peer-reviewed literature but are from government sponsored reports of reviews of unpublished studies commissioned by industry for submission to government regulators. The majority of these reports used in this review are

Table 10.7 Genotoxicity findings for various triazole antifungal agents.[a,b]

Chemical	Gene mutation		Chromosomal effects		UDS
	Salmonella	Other	Micronucleus	Other	
Cyproconazole (US EPA, 1991; 1992; 2008)				+ (CHO cells)	NA
Difenoconazole (US EPA, 1994)				NA	
Epoxiconazole (EPA 2001b)					
Etaconazole (US EPA, 1998; 2000)			NA	NA	NA
Fenbuconazole (US EPA, 2001a)			NA		
Fluconazole (Fucic *et al.*, 2008)	NA	NA	+ (*in vivo*: mouse)	NA	NA
Flusilazole (IPCS, 1995)					
Hexaconazole (US EPA, 2000)					
Myclobutanil (Ross *et al.*, 2009)			NA		
Propiconazole (Ross *et al.*, 2009)		+ (*in vivo:* mouse)			
Tebuconazole (US EPA, 2010; CDPR, 2003)					
Triadimefon (Ross *et al.*, 2009)		+ (*in vivo:* mouse)			
Triadimenol (US EPA, 1998; CDPR, 2000)					
Uniconazole (US EPA, 1998; 2000)			+ (*in vivo*: mouse)	+ (CHO cells)	

[a]Table reproduced from Office of Environmental Health Hazard Assessment OEHHA.[85] See the original document for references cited in this table.
[b]Abbreviations: UDS: unscheduled DNA synthesis; CHO: Chinese hamster ovary; NA: not assessed.

from the International Programme on Chemical Safety (IPCS) INCHEM website (www.inchem.org). INCHEM reports are a product of several agencies including the IPCS and the Canadian Centre for Occupational Health and Safety. The other major data source used in this table is the European Food Safety Authority, an independent European agency associated with the European Union risk-assessment review program regarding food and feed safety. A picture of species and organ specificity presents itself based on *in vivo* long-term cancer studies of conazoles in mice and rats. Mouse liver is the most common target for conazole's tumorigenic effects. Table 10.8 shows that 17 out of the 27 conazoles (63%) were reported to induce liver tumors in mice and eight conazoles (30%) induced various tumors in rats. In the rat, conazoles mainly targeted the thyroid gland, although the frequency of this occurrence was less than those that induced liver tumor in mice. Amitrole, which is not a true conazole by definition (*i.e.* does not contain an aliphatic and halo-aromatic side chain), induced liver and thyroid tumors in rats and liver tumors in mice in a feeding study, thus being the least

Table 10.8 The tumorigenic effects of conazoles.[a]

Conazole/azole	CAS No.	Azole type	Mice	Rat	Ref.
Amitrol	61-82-5	Triazole	Liver, thyroid tumors	Thyroid, liver and pituitary tumors	NTP[92]
Bitertanol	55179-31-2	Triazole	No tumors	No tumors	INCHEM[93]
Bromuconazole	116255-48-2	Triazole	No tumors	No tumors	OPPTS[94]
Cyproconazole	94361-06-5	Triazole	Liver tumors	No tumors	Syngenta[95] OEHHA[85]
Difenoconazole	119446-68-3	Triazole	Liver tumors	No tumors	FAO[96] OEHHA[85]
Diniconazole	83657-18-5	Triazole		Thyroid tumors	Hosokawa *et al.*[97]
Epoxiconazole	133855-98-8	Triazole	Liver tumors	Adrenal cortex tumors, liver cholangiomas, liver tumors and ovarian tumors	EPA[98] OEHHA[85]
Etaconazole	60207-93-4	Triazole	Liver tumors		OEHHA[85]
Fenbuconazole	114369-43-6	Triazole	Liver tumors	Thyroid tumors	INCHEM[99] OEHHA[85]
Fluquinconazole	136426-54-5	Triazole	Liver tumors	Thyroid and liver tumors	EFSA[100]
Flusilazole	85509-19-9	Triazole	Liver tumors	Bladder and Leydig cell tumors	FAO[96] OEHHA[85]
Hexaconazole	79983-71-4	Triazole	No tumors	Leydig cell tumors	INCHEM[101]
Imazalil	35554-44-0	Imidazole	Liver tumors	No tumors	EPA[102]
Ipconazole	125225-28-7	Triazole	No tumors	No tumors	EFSA[103]
Metconazole	125116-23-6	Triazole	Liver tumors	No tumors	EPA[104]
Myclobutanil	88671-89-0	Triazole	No tumors	No tumors	INCHEM[105]

Paclobutrazol	76738-62-0	Triazole	No tumors	No tumors	INCHEM[106]
Penconazole	66246-88-6	Triazole	No tumors	No tumors	INCHEM[107]
Prochloraz	67747-09-5	Imidazole	Liver tumors	No tumors	INCHEM[90] EPA[108]
Propiconazole	60207-90-1	Triazole	Liver tumors	No tumors	INCHEM[109] OEHHA[85]
Tebuconazole	107534-96-3	Triazole	Liver tumors	No tumors	EFSA[81] EC[110] OEHHA[85]
Tetraconazole	112281-77-3	Triazole	Liver tumors	No tumors	EFSA[89]
Triadimefon	43121-43-3	Triazole	Liver tumors	Thyroid tumors	INCHEM[111] OEHHA[85]
Triadimenol	55219-65-3	Triazole	Liver tumors	No tumors	EPA[112] OEHHA[85]
Triflumizole	68694-11-1	Imidazole	No tumors	No tumors	JMPR[113]
Triticonazole	131983-72-7	Triazole	No tumors	No tumors	ORNL[114,115]
Uniconazole-P	83657-22-1	Triazole	Liver tumors	No tumors	EPA[82] OEHHA[85]

[a]The data and citations in this table are almost all from secondary sources such as government committee reports that reviewed reports from industry sponsored studies. CAS: Chemical Abstracts Service; EC: European Commission; EFSA: European Food Safety Authority; EPA: Environmental Protection Agency; FAO: Food and Agriculture Organization of the United Nations; JMPR: Joint FAO/WHO Meeting on Pesticide Residues; NTP: National Toxicology Program; OEHHA: Office of Environmental Health Hazard Assessment; OPPTS: Office of Prevention, Pesticides and Toxic Substances; ORNL: Oak Ridge National Laboratory.

organ-specific in its tumorigenic effects. Many conazoles only induced liver tumors in mice. Although the prevalent mouse liver tumor type was adenoma, malignant adenocarcinomas were induced by a number of these agents. For example, chronic exposure in the diet to tetraconazole increased both malignant and benign liver tumors.[89] Malignant hepatic adenocarcinomas were also reported for cyproconazole, difenoconazole, etaconazole, epoxiconazole, fenbuconazole, fusilazole, metconazole, prochloraz, propiconazole, tebuconazole and uniconazole.[85,90] In addition to inducing malignant hepatic tumors in mice as a complete carcinogen, propiconazole was also found to be a rat liver tumor promoter in two different studies in rats initiated with diethylnitrosamine and fed propiconazole. One study used glutathione *S*-transferase placental form positive foci[91] and the second used γ-glutamyltranspeptidase-positive foci[70] as markers of liver tumor formation. Triadimefon was also found to be a tumor promoter in the glutathione *S*-transferase placental form positive foci assay system.[91]

10.8 Toxicogenomic Studies in Mice

10.8.1 Genomic Studies Using Liver Samples from Mice Treated with Conazoles

Liver samples from male CD-1 mice fed triadimefon (a mouse liver tumorigen), propiconazole (a mouse liver carcinogen) or myclobutanil (not tumorigenic in mice fed a diet of myclobutanil for 2 years in a continuous oral-dose regimen) for 4, 30 or 90 days[58] were assessed using analytic genomic methods.[116] Differentially expressed genes and pathways were identified using Affymetrix GeneChips. Gene–pathway associations were obtained from the Kyoto Encyclopedia of Genes and Genomes, Biocarta and MetaCore compendia. The pathway profiles of each conazole were different at each time point. In general, the number of altered metabolism (nuclear receptors, CYPs and xenobiotic metabolizing enzymes) cell signaling and growth pathways increased with time and dose and were greatest with propiconazole. All conazoles had effects on nuclear receptors as evidenced by increased expression and enzymatic activities of a series of related CYPs. A subset of altered genes and pathways distinguished the three conazoles from each other. Both triadimefon and propiconazole altered apoptosis, cell cycle, adherens junction, calcium signaling and epidermal growth factor receptor signaling pathways. There were significant transcriptional changes in cholesterol biosynthesis genes in mouse liver in high-dose animals compared to control animals at 4 days. Triadimefon significantly overexpressed acetoacetyl-CoA synthetase (*Aacs*), HMG-CoA synthase 1 (*Hmgcs1*), mevalonate kinase (*Mvk*), mevalonate decarboxylase (*Mvd*), isopentenyl pyrophosphate (*Ipp*) isomerase (*Idi1*), farnesyl phosphate transferase 1 (*Fdft1*), squalene epoxidase (*Sqle*), lanosterol 14α-demethylase (*CYP51*) and 7-dehydrocholesterol reductase (*Dhcr7*). Propiconazole significantly overexpressed squalene epoxidase (*Sqle*), lanosterol 14α-demethylase (*CYP51*) and

7-dehydrocholesterol reductase (*Dhcr7*), while myclobutanil did not significantly overexpress any of these genes. Both triadimefon and propiconazole affected retinoic acid metabolism. Propiconazole had greater effects on genes responding to oxidative stress and on the IGF/P13K/AKt/PTEN/mTor and Wnt-β-catenin pathways. Propiconazole induced *Akr1b7* overexpression at all days of treatment with the highest recorded level of 8.7-fold after 30 days of treatment. Akr1b7, an aldo-keto reductase is associated with the detoxification of lipid peroxidation products, the cleavage of isocaproaldehyde from the cholesterol side chain. Lipid peroxidation products can arise from reactive oxygen species (ROS) generated from high levels and activities of CYP enzymes. Concomitant with *Akr1b7* overexpression induced by propiconazole was the overexpression of *Gadd45a* (2.3-fold) and *Gadd45b* (4.7-fold) after 30 days of treatment. *Gadd45a* and *Gadd45b* overexpression have been associated with DNA damage, apoptosis and general genotoxic stress. This data suggested a possible role of reactive aldehydes and ROS in the mode of action of propiconazole. While triadimefon, propiconazole and myclobutanil had similar effects in mouse liver on hepatomegaly, histology, CYP activities, cell proliferation and serum cholesterol,[58] genomic analyses revealed major differences in their gene expression profiles suggesting different modes of action for each conazole.[116] A genomic analysis was conducted on liver tissues from wild-type and CAR-null mice (based on a C57BL/6 background) gavaged with propiconazole (210 mg kg^{-1}) or triadimefon (165 mg kg^{-1}) each day for 7 days.[75] A series of marker genes for CAR and PXR were examined for changes in gene expression. *Cyp2b10* was increased by propiconazole and triadimefon in wild-type mice, but not significantly altered in CAR-null mice, suggesting it was dependent on CAR. *Birc5, Cyp3a11* and *Gsta2* genes were overexpressed by both propiconazole and triadimefon in both CAR-null and wild-type mice, but with lower expression levels in CAR-null mice. *Gstm3* was over expressed by both conazoles in both CAR-null and wild-type mice, but with higher expression levels in CAR-null mice. *Cyp51* expression levels were overexpressed by both conazoles in both CAR-null and wild-type mice with similar expression levels.

In a study by Nesnow *et al.*,[117] the microarray data from liver samples from male CD-1 mice fed triadimefon and propiconazole for 4 and 30 days[58] were reanalyzed by a second set of genomic methods different to those used by Ward *et al.*[116] These new methods used Rosetta Resolver software to import the raw signal intensities from each scan (.celfiles) and to determine which probe sets in a treatment group differed from the control group. The scans were converted into profiles within this software and the data were normalized using a Rosetta Resolver error model. Statistically filtered sequences were identified using one-way ANOVA with a false-discovery rate (Benjamini–Hochberg test) of $p<0.05$ followed by a *post hoc* test (Tukey–Kramer) for significance. Fold-change for each sequence was computed for each calculated ratio representative of a given sequence. Statistical lists were used in Venn diagrams to find sequences common and unique for each treatment. Using Ingenuity Pathways Analysis (IPA) (Ingenuity Systems;

www.ingenuity.com), these common and unique significant transcripts were evaluated for relevance to IPA canonical pathways and IPA Tox lists. The "compare experiments" function within GeneGo MetaCore (http://portal.genego.com/cgi/index.cgi#Information) was used to create significant networks from differentially expressed genes obtained from Venn analyses. The *p*-values for MetaCore processes were calculated using a hypergeometric distribution. Targeted transcriptomic analyses were conducted at the gene level, examining differentially expressed genes and subsets of differentially expressed genes: cell-cycle genes and transcription factors. Analyses were also conducted on function, pathway and network levels examining IPA Tox lists and canonical pathways, and Gene-Go MetaCore dynamic networks and their central hubs. The hub analysis was critical to understanding the altered cellular functions induced by propiconazole and triadimefon. The brief treatment period (4 days) of propiconazole or triadimefon produced the same hubs that altered the expression of groups of genes controlled by *Ahr*, *Maml1* (mastermind-like 1 (*Drosophila*)), *Hnf4a* (hepatocyte nuclear factor 4-α), *Myc* (myelocytomatosis oncogene1) and *Esr1* (estrogen receptor 1 (α)). After extended (30 days) treatment, both propiconazole and triadimefon produced the same hubs that altered the expression of groups of genes controlled by *Crpb* (C-reactive protein, precursor), *Ern1* (serine/threonine-protein kinase/endoribonuclease IRE1 precursor), *Ppap2a* (lipid phosphate phosphohydrolase 1), *Hnf4a* (hepatocyte nuclear factor 4-α), *Myc* (myelocytomatosis oncogene1) and *Esr1* (estrogen receptor 1 (α)). Triadimefon also altered groups of genes controlled by *Cdk2* (cyclin-dependent kinase 2), *Rb1* (retinoblastoma 1) and *Cdkn1b(p27)* (cyclin-dependent kinase inhibitor1b).

The toxicological and transcriptional responses of three carcinogenic conazoles: cyproconazole, epoxiconazole and propiconazole, were compared using liver tissues from male CD-1 mice fed diets of these conazoles at their tumorigenic dose levels for 30 days.[71] A common set of toxicological effects altered by these conazoles was found: hepatomegaly, hepatocellular hypertrophy, decreased serum cholesterol levels, decreased levels of hepatic all-trans-retinoic acid and increased hepatic cell proliferation. Microarray-based transcriptional analysis revealed 330 significantly altered probe sets common to these conazoles, many of which showed strong dose responses for cytochrome P450, glutathione *S*-transferase and oxidative stress genes. Subsequent data mining of the 330 significantly altered probe sets common to all three conazoles identified a subset of 80 altered genes common to the three conazoles that were associated with cancer. Pathways associated with these genes included xenobiotic metabolism, oxidative stress, cell signaling and cell proliferation. A common transforming growth factor (TGF)α-centric pathway was identified within the 80-gene set. In this pathway TGFα signals to Rras which signals to mitogen-activated protein kinases (MAPKs) to initiate a cascade of signals to Myc/Jun and E2f to initiate hepatocyte proliferation. All of these genes were overexpressed by each conazole. TGFα overexpression in the liver is the persistent induction of hepatocyte proliferation.[118] TGFα is known to signal through Rras to MAPK to initiate a cascade of signals to Myc/Jun and E2f.[119,120] Enhanced E2f activity during hepatocarcinogenesis has been reported in c-myc and c-myc/TGFα transgenic mice, along with the induction of E2f target genes Cdc2 and c-Myc.[121] E2f has also been reported to be involved in murine hepatocyte proliferation *in vitro*.[122]

The concordance of transcriptional and apical benchmark dose levels for five conazoles based on conazole-induced liver effects after 30 days of treatment in mice was examined by Bhat *et al.*[123] These conazoles were cyproconazole, epoxiconazole, myclobutanil, propiconazole and triadimefon. Apical data from Allen *et al.*[58] and Hester *et al.*,[71] genomic data from Nesnow *et al.*[117] and Hester *et al.*[71] and tumor data from various sources were used in these benchmark dose calculations. Transcriptional benchmark dose levels were estimated for a subset dose-responsive genes with large (≥5-fold) increases or decreases in signal intensity at the highest dose. These genes encompassed CAR/RXR activation, P450 metabolism, liver hypertrophy, glutathione depletion, lipopolysaccharide/interleukin-1-mediated inhibition of RXR and nuclear factor erythroid 2-related factor (NRF2)-mediated oxidative stress pathways. The median transcriptional benchmark dose levels estimates were within one-half order of magnitude of the chronic apical benchmark dose levels for hepatocellular tumors. The potency differences in the dose-responsive transcription of certain phase II metabolism, bile acid detoxification and lipid oxidation genes paralleled each conazole's tumorigenic potency. The transcriptional benchmark dose levels corresponded to tumorigenic potency on a mg kg^{-1} day^{-1} basis with cyproconazole > epoxiconazole > propiconazole > triadimefon > myclobutanil (non-tumorigenic).

10.8.2 Proteomic Studies Using Liver Samples from Mice Treated with Conazoles

Soluble proteins from liver samples from mice fed propiconazole (2500 ppm) for 4 days and liver samples from control mice were subjected to two-dimensional gel electrophoresis, mass spectrometry analysis.[124] Sixty-two proteins were found to be altered. Several of these protein changes detected were verified by Western blot analyses. These differentially expressed proteins were mapped to known canonical pathways and Tox lists and subjected to network analysis. The protein network analysis produced interacting subnetworks centered on hepatocyte nuclear factor 4 α (HNF4 α), MYC, proteasome subunit type 4 α and glutathione *S*-transferase. There was significant overlap between altered protein pathways found by proteomic analysis and those found by a previous genomic analysis.[117] Both proteomic and genomic approaches indicated that propiconazole altered the expression of the following pathways in mouse liver: nuclear receptor activation, metabolism of xenobiotics, metabolism of biochemical intermediates, biosynthesis of biochemical intermediates and oxidative stress.

10.8.3 Metabolomic Studies Using Liver Samples from Mice Treated with Conazoles

Endogenous biochemicals from liver samples from mice fed propiconazole (2500 ppm) for 4 days and those from control mice were profiled using liquid chromatography/mass spectrometry and gas chromatography/mass spectrometry methods.[125] Of the 261 biochemicals detected, the levels of 159 of these had been significantly altered by propiconazole treatment compared to control mice. Alterations in the levels of biochemicals were found in the glycogen metabolism, glycolysis, lipolysis, carnitine, the tricarboxylic acid cycle pathways and in pathways involved in the metabolism and clearance of propiconazole: glucuronate, glutathione and cysteine pathways. Propiconazole induced changes in metabolomic responses also known to be altered from toxicological and genomic studies and related to propiconazole-induced mouse liver cancer: oxidative stress and increases in the cholesterol biosynthesis pathway. Furthermore, the metabolomic data showed that propiconazole induced a series of metabolic states in the mouse liver that mimicked those found in tumors or tumor cells, and/or animals or humans bearing tumors. These states were: the increased need for energy [through glycolysis/glutamate metabolism/fatty acid metabolism/tricarboxylic acid (TCA) cycle], the increased levels of a polyamine, spermidine and the increased levels of sarcosine. In the metabolomic studies, propiconazole altered cellular glycogen metabolism, glycolysis, lipolysis, carnitine esterification and metabolism, and the TCA cycle consistent with an overall increase in energy production in the liver. Glycogen metabolism and glycolysis ultimately produce pyruvate, which forms acetyl CoA. Lipolysis of triglycerides produces fatty acids. Carnitine is involved in the transport of activated (CoA) long-chain fatty acids from the cytosol to the mitochondria, where they are subject to β-oxidation.[126] Fatty acids are degraded producing energy in the β-oxidation process and creating acetyl CoA, a substrate in the TCA cycle. However, metabolomic analysis did not detect changes in fatty acid metabolism. These results paralleled the transcriptional analyses of livers from propiconazole treated mice with respect to fatty acid metabolism, glycerolipid metabolism, pyruvate metabolism, linoleic acid metabolism, arachidonic acid and synthesis and degradation of ketone bodies.[117] Overall, these results suggested that propiconazole increased the flux from glycogen, glucose and glutamine through the TCA cycle, and triglycerides to fatty acids. Rapidly dividing cells or malignant tumor cells typically have a very high glycolytic rate compared to normal tissues.[127] Many tumors and cancer cells exhibit the Warburg effect: the production of energy by the high flux of glucose through glycolysis producing high levels of pyruvate and lactate, while normal differentiated cells exhibit a comparatively low rate of glycolysis followed by a high rate of mitochondrial oxidation of pyruvate.[128] Livers from mice exposed to propiconazole also exhibit a high flux of glucose to produce energy without high levels of lactate being produced.[125] These results are consistent with both the increased proliferative effect in mouse liver induced by propiconazole treatment[58,71] and the increased energy requirements needed to metabolize and eliminate propiconazole.

Propiconazole treatment affected the urea cycle and polyamine biosynthesis. Overall, aspartate and asparagine (formed from aspartate) levels were reduced while spermidine levels were concomitantly increased by propiconazole treatment. Transcriptomic analysis of livers from propiconazole treated mice for 4 days indicated that the urea cycle had been altered with arginase (*Arg2*) up-regulated 1.9-fold.[116] Arginase converts to arginine to ornithine and urea. In transcriptomic studies specific genes associated with polyamine biosynthesis were upregulated: ornithine decarboxylase (*Odc*) mRNA levels increased 1.5-fold, while ornithine decarboxylase antizyme 2 mRNA (*Oaz2*) was decreased 2.1-fold.[116] Ornithine decarboxylase antizyme inactivates and targets Odc protein for rapid degradation by the 26S proteasome, and its down-regulation would enhance putrescine and spermidine production.[129,130] The enzyme spermidine synthase converts putrescine to spermidine and 5′-methylthioadenosine. Transcriptomic analysis of livers from propiconazole-treated mice for 4 days indicated that spermidine synthase (*Srm*) and spermine synthase (*Sms*) were both up-regulated.[116] Spermidine synthase (*Srm*) RNA levels were increased 3.3-fold by propiconazole after 4 days of treatment with 2500 ppm propiconazole.[117] Polyamines are involved in tumor promotion and carcinogenesis.[131] In two studies, propiconazole was found to be a tumor promoter in rats, the only species tested.[70,91] Tumors secrete higher levels of polyamines than normal tissues[132] related to increased activities of ornithine decarboxylase and spermidine synthase.[133,134] cMyc is a proto-oncogene and a transcription factor that transactivates *Odc* and *Srm*, and is frequently overexpressed in cancers.[133] Propiconazole treatment increased *myc* levels 1.7-fold.[116,117] Thus, there is a consistent body of information that relates molecular, metabolomic and toxicological data on propiconazole to hepatic polyamine formation and tumor promotion. Sarcosine levels in the livers of propiconazole treated mice were altered to the greatest extent of all endogenous metabolites detected. Proteomic analysis of livers from propiconazole-treated mice revealed that dimethylglycine dehydrogenase was down-regulated.[124] The higher levels of spermidine, 5′-methylthioadenosine and spermidine synthase mRNA suggests that higher levels of *S*-adenosylmethionine were present, and that, combined with the lower protein levels of dimethylglycine dehydrogenase, suggests that propiconazole treatment favors the biosynthesis of sarcosine from glycine. Sarcosine has been recently identified as a potential biomarker for human prostate cancer progression.[135] The relevance of these findings to mouse liver neoplasia is unknown; however, proteomic analysis of human hepatocellular carcinomas identified lower levels of sarcosine dehydrogenase compared to normal liver. Since sarcosine dehydrogenase metabolizes sarcosine to glycine, the lower levels of this enzyme suggest that sarcosine levels could increase in the tumor tissue.[136] Hepatic pseudouridine levels were increased by propiconazole. As pseudouridine is a component of transfer RNA, its presence shows increased RNA turnover since it is not utilized for *de novo* or salvage RNA nucleotide synthesis.[137] Pseudouridine has been shown to be a biochemical marker as urinary nucleoside levels including pseudouridine are higher in cancer patients;[138,139] therefore it is a potential cancer biomarker for propiconazole.

10.9 Toxicogenomic Studies in Rats

10.9.1 Toxicological Studies in Rats

Toxicological studies of three conazoles: propiconazole (2500 ppm), triadimefon (1800 ppm) and myclobutanil (2000 ppm) were conducted in male Wistar/Han rats.[59] These conazoles were selected for their ability or inability to induce thyroid tumor responses in rats: triadimefon (a thyroid tumorigen), propiconazole and myclobutanil (do not induce thyroid tumors). Thyroid and liver tissues were analyzed from rats fed these conazoles (and a control group) for up to 90 days at their tumorigenic and non-tumorigenic doses.[59] Histologically, livers from all high-dose treated rats had centrilobular hepatocyte hypertrophy after 4 days, triadimefon- and propiconazole-treated rats had hepatocyte hypertrophy after 30 days and triadimefon-treated rats had hepatocyte hypertrophy after 90 days. Thyroid follicular cell hypertrophy, increased follicular cell proliferation and colloid depletion were present only after 30 days in rats treated with the high dose of triadimefon. PROD activity in liver tissues increased with a dose-related response at each time point examined (4, 30 and 90 days) with all conazoles. Uridine diphosphoglucuronosyl (UDP) transferase, the enzyme that metabolizes T4, was induced to the same extent after 30 and 90 days for all three conazoles. A dose-dependent decrease in serum T4 was present after 4 days with all three conazoles, but only the high doses of propiconazole and triadimefon produced decreased T4 after 30 days. Serum T3 was decreased after high-dose triadimefon after 4 days and in a dose-dependent manner for all compounds after 30 days. Serum thyroid hormone levels did not differ from control values after 90 days and serum thyroid-stimulating hormone (TSH) was not increased in any exposure group. A unique pattern of responses from triadimefon was not apparent to explain the thyroid tumor formation. The combined histological and biochemical data suggested that these alterations do not result in sufficient thyroid disruption to result in an increase in TSH. These data suggest that an alternate pathway, not requiring persistently elevated TSH, is active in thyroid gland follicular tumor development associated with triadimefon. In another study, RAI rats were dosed with propiconazole at 0, 20, 80, 160 or 320 mg kg^{-1} body weight per day by gavage for 2 weeks. Rats showed increases in liver weight at a dose of 20 mg kg^{-1} body weight per day of propiconazole. At $\geq$80 mg kg^{-1} body weight per day, there were increases in microsomal protein, phospholipid and cytochrome P450 content and activities of ethoxycoumarin deethylase, epoxide hydrolase, UDP-glucuronyltransferase and γ-glutamyl transferase. Glutathione *S*-transferase activity was increased at $\geq$160 mg kg^{-1} body weight per day and a marked proliferation of smooth endoplasmic reticulum was seen in livers of rats treated at 320 mg kg^{-1} body weight per day.[70]

10.9.2 Genomic Studies Using Liver Tissues from Rats Treated with Conazoles

Gene expression patterns in liver samples from rats fed propiconazole, triadimefon and myclobutanil obtained from the Wolf *et al.*[59] study were analyzed by microarray techniques.[140] Differential gene expression was assessed at the probe level using Robust Multichip Average analysis. Principal component analysis by treatment and time showed within-group sample similarity and that the treatment groups were distinct from each other. The number of significantly altered genes varied by treatment, dose and time. The greatest number of altered genes was induced by triadimefon and propiconazole after 90 days of treatment, while myclobutanil had minimal effects at that time point. Pathway-level analyses revealed that after 90 days of treatment the most significant numbers of altered pathways were related to cell signaling, growth and metabolism. Pathway-level analysis for triadimefon and propiconazole resulted in 71 altered pathways common to both chemicals. These pathways controlled cholesterol metabolism, activation of nuclear receptors and N-ras and K-ras signaling. There were 37 pathways uniquely changed by propiconazole. Triadimefon uniquely altered 34 pathways, including a repressed expression of the thyroid-specific genes (*TPO*, *TRH*, *TRHR* and *TSHβ*) observed only with triadimefon treatment. Groups of genes in the nuclear receptors and xenobiotic metabolizing enzyme induction processes including *CYP1A1*, *CYP2B15*, *CYP4A*, *ALDH1A1*, *ALDH1A2*, *ALDH1*, *ADH4* and *CES2* were significantly overexpressed by all conazoles at 90 days. These observations are consistent with nuclear receptor CAR and PXR activation. General cellular stress response genes were consistently overexpressed, including GSTs and glutathione peroxidase (2 and 4), NAD(P)H: quinone oxidoreductase and heme oxygenase 2.

10.9.3 Genomic Studies Using Thyroid Tissues from Rats Treated with Conazoles

Gene expression changes in thyroid tissues from the Wolf *et al.*[59] study with male Wistar/Han rats fed triadimefon (a rat thyroid tumorigen) and myclobutanil (which was not tumorigenic in rats after a 2 year bioassay) for up to 90 days and a control group were analyzed by the gene set expression analyses method.[141] Using this analytical approach, the triadimefon responses were separated from the myclobutanil and control responses. Core genes from these gene sets were mapped to canonical, metabolic and GeneGo processes and these processes compared across group and treatment time. Gene sets from the triadimefon (30 days treatment group) were over-represented in fatty acid metabolism, oxidation and degradation processes [including peroxisome proliferator-activated receptor (PPAR)γ and CYP involvement] and in cell proliferation responses. The core genes in each gene set were compared with genes known to be associated with human thyroid cancer. Genes common in the rat and human cancer data sets were *Acaca*, *Asns*, *Cebpg*, *Crem*, *Ddit3*, *Gja1*, *Grn*, *Jun*, *Junb* and *Vegf*. Thus, a mode of action for the induction of thyroid tumors by triadimefon proposed involving the induction of oxidative response genes and activation of the nuclear receptor, Pparγ, initiating transcription of gene products and signaling to a series of genes involved in cell proliferation. This proposed mode of action based on transcriptional analysis in thyroid tissues is consistent with the toxicological findings of Wolf *et al.*[59] and provides a plausible mechanism for thyroid tumor induction by triadimefon.

10.10 Mode of Action of Propiconazole: Introduction

The complexity of the biochemical and biological effects of propiconazole in mice (as described earlier) projects to its proposed mode of action, which is also complex, consisting of a series of linked key events leading to the induction of mouse liver adenocarcinomas and adenomas (Figure 10.4). This proposed mode of action features propiconazole's multiple activities on CYPs. It is both an inducer of a number of CYPs (*i.e.* gene and protein expression and enzymatic activity) as well as an inhibitor of CYP51 enzymatic activities. These events and the sequalae downstream of these events lead to mouse liver cancer. Based on the results of a series of toxicological, biochemical, molecular, genomic, proteomic and metabolomic studies, a mode of carcinogenic action has been proposed for propiconazole and is described below.[142] This mode of action was created in concert with the conceptual framework in considering the chemical induction of a specific tumor in animals.[143]

10.10.1 Mode of Carcinogenic Action of Propiconazole, a Series of Key Events Leading to Cancer

Propiconazole was negative in a battery of standard tests for genotoxicity including *in vitro* tests for bacterial (*Salmonella typhimurium*), yeast (*S. cerevisiae*) and mammalian cell gene mutation (L5178Y mouse lymphoma cells), cell transformation (BALBc 3T3 mouse embryo fibroblasts), DNA repair (rat hepatocytes and human fibroblasts), chromosomal aberration (human peripheral lymphocytes) and unscheduled DNA synthesis (Tif:RAIF rat hepatocytes). Propiconazole was negative in *in vivo* tests: micronuclei (Chinese hamster bone marrow), sister-chromatid exchange (Chinese hamster bone marrow), host-mediated assay (*S. typhimurium* in mice and L5178Y mouse lymphoma cells in mice), dominant lethal mutations (Tif:MAGf SPF mice), nucleus anomaly test (Chinese hamster bone marrow) and chromosomal aberration (mouse spermatogonia and spermatocytes).[70,109] It is

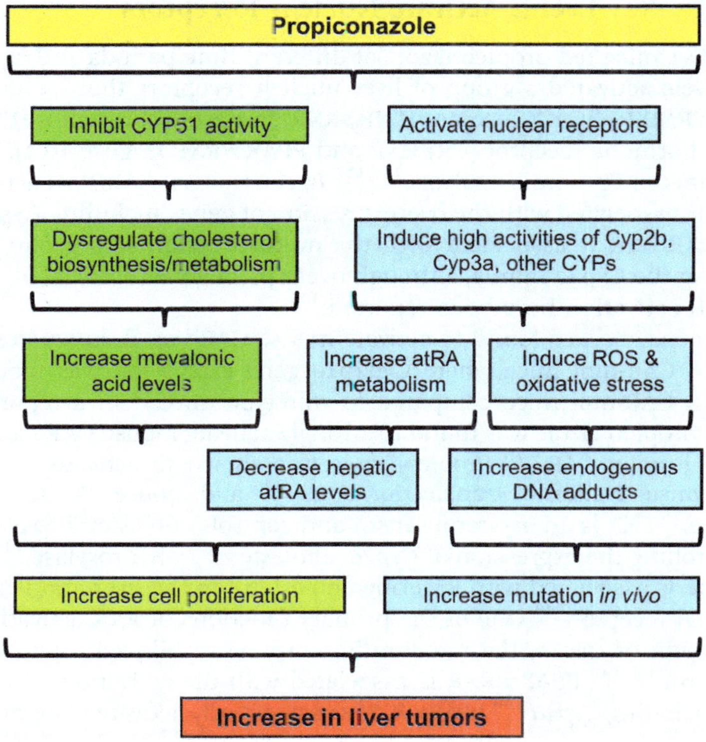

Figure 10.4 The mode of action of propiconazole describings its hepatocarcinogenic activity in mice. Each box represents a key event in the mode of action. CYP: cytochrome P450; atRA: all-*trans*-retinoic acid; ROS: reactive oxygen species.

noted that none of these tests for genotoxicity use liver tissues from propiconazole-treated mice, the tumor target organ. It is well known that many metabolically reactive metabolites of chemical carcinogens are more genotoxic and cytotoxic than the parent carcinogen. The question of the potential involvement of metabolites of propiconazole in the toxicity and genotoxicity of propiconazole was addressed in a study by Chen *et al.*[144] Propiconazole was shown to be metabolized to a series of metabolites.[70] The cytotoxic effects of propiconazole and five of its metabolites were examined in three hepatic cell types: the mouse hepatoma Hepa1c1c7 cell line, the human hepatoma HepG2 cell line and primary cultures of mouse hepatocytes.[144] All five propiconazole metabolites were less cytotoxic in both mouse and human tumor cell lines compared to the propiconazole. While propiconazole was cytotoxic in primary mouse hepatocytes, none of the five propiconazole metabolites exerted cytotoxic activities in these primary cells. It was concluded that these propiconazole metabolites would not contribute to the propiconazole-induced cytotoxicity and potential genotoxicity processes in primary mouse hepatocytes.

10.10.2 Key Event: Activate Nuclear Receptors

Male CD-1 mice fed propiconazole at different time periods and at different dose levels activated a group of liver nuclear receptors that included CAR/RXR, PXR/RXR, liver X receptor (LXR)/RXR, farnesoid X receptor (FXR)/RXR, thyroid hormone receptor (TR)/RXR and PPARα/RXR (Figure 10.4). This was determined by genomic analysis.[116,117] In the mouse, CAR/RXR activation is generally associated with the overexpression of genes including Cyp2b10/20, while PXR/RXR is associated with the overexpression of a group of genes including the Cyp3a family, although overlaps in genes associated with CAR and PXR activation have been reported.[145]

Propiconazole was found to activate mouse CAR based on studies in wild-type and CAR-null mice where Cyp2b10 gene expression was markedly reduced in CAR-null mice compared to wild-type mice.[75] In a reporter assay system, propiconazole was found to strongly activate mouse CAR3 and weakly activate human CAR3.[76] Propiconazole was shown to activate human PXR using transactivation or transfection assays[146] and mouse PXR using knockout mice.[75] LXR is an oxysterol sensor and controls cholesterol levels, in part, by controlling the expression of *Cyp7a*, cholesterol 7α-hydroxylase.[147] Thyroid hormone is associated with lowering cholesterol in the liver and is mediated by the TR receptor.[148] One of the primary functions of FXR activation is the suppression of *Cyp7a*, the rate-limiting enzyme in bile acid synthesis from cholesterol.[147,149] PPARα/RXR is associated with the activation of a group of genes including *Cyp4a*.[150] *PPARα/RXR* was negatively activated (suppressed) by propiconazole treatment in rat and mouse liver.[25,117] Nuclear receptor activation is a key event in the proposed mode of action of propiconazole.

10.10.3 Key Event: CYP Induction

Propiconazole treatment of mice and rats increases absolute and relative liver weights and hepatic CYPs within the CYP1A, CYP2B and CYP3A families, as determined by Western immunoblots and/or enzymatic activities such as alkoxyresorufin dealkylase and 16β-hydroxytestosterone activities. These studies were conducted in various strains of mice and rats using a variety of routes of administration (see Table 10.5).[25,58,67–69] Groups of male MAG mice given gavage doses of propiconazole up to 320 mg kg^{-1} body weight per day for 14 days showed increased hepatic CYP content, hepatic enzymatic activities of EROD, UDP-glucuronyl transferase, epoxide hydrolase and glutathione *S*-transferase activities at the 160 and 320 mg kg^{-1} body weight per day dose levels.[70] Male CD-1 mice were given diets of propiconazole (850 ppm and 2500 ppm) for 14 days. These mice showed elevated activity (at both dose levels) of a series of microsomal hepatic enzymatic enzymes related to the hydroxylation of testosterone and coumarin, as well as increased PROD, EROD and epoxide hydrolase activities. Increased immunoreactivity to Cyp2b and Cyp3a antibodies was also found at both dose levels.[70]

Genomic analyses of liver tissues from male CD-1 mice fed propiconazole at a tumorigenic dose level (2500 ppm) revealed altered expression of two additional families of CYPs, a large up-regulation of *Cyp2c54* and *Cyp2c55* and a down-regulation of *Cyp4a10*.[117] A time– and dose–response study using male CD-1 mice fed a diet of propiconazole indicated both time- and dose-dependent changes in both alkoxyresorufin dealkylase activities and gene expression changes associated with the CYPs.[58,116] At the tumorigenic dose, propiconazole-induced PROD activity (a measure of Cyp2b protein) increased 13.3-fold at 4 days, 37.5-fold at 30 days and 35.9-fold at 90 days of treatment. Concomitant increases in *Cyp2b10/20* gene expression were also observed. CYP induction is a key event in the proposed mode of action of propiconazole (Figure 10.4).

10.10.4 Key Event: Induction of ROS and Oxidative Stress

10.10.4.1 Induction of ROS

ROS production can be induced by specific CYPs after exposure to xenobiotics.[151–153] An explanation for ROS production by CYPs is through the uncoupling of the CYP catalytic cycle.[154–156] A series of CYPs are known to be susceptible to uncoupling of their catalytic cycle, thus generating ROS and oxidative stress: CYP1A1, CYP1A2, CYP1B1, CYP2B1, CYP3A2 and CYP2E1.[151–154] Aroclor-1254-induced rat liver microsomes (known to contain CYP1A1 and CYP1A2) produce ROS in the presence of nicotinamide adenine dinucleotide phosphate (NADPH) and 3,3′,4,4′-tetrachlorobiphenyl. Recombinant human CYP1A1 and CYP1A2 microsomes in the presence of NADPH also produced ROS.[156] Overexpression of CYP2E1 has been linked to ROS production, lipid peroxidation and oxidative stress.[157] Purified CYP2B1 proteins in a reconstituted system in the presence of NADPH produced lipid peroxidation and hydrogen peroxide and superoxide radical.[158] This was confirmed *in vivo* in mice and rats as phenobarbital treatment increased F(2)-isoprostane levels (products of free radical peroxidation of arachidonate) in livers and urine.[159,160] Rats treated with dexamethasone responded with 50-fold increases in hepatic CYP3A proteins. When these rats were pulsed with salicylic acid, a 17.2-fold increase in plasma 2,3-dihydroxybenzoic acid (a salicylic acid metabolite and a measure of ROS formation) was detected.[161] A highly reactive ROS form, the hydroxyl radical, was detected in hepatic microsomes from phenobarbital-treated rats (containing CYP2B and CYP3A proteins) in electron spin resonance studies using a spin trap agent. The formation of the hydroxyl radical was dependent on NADPH and was inhibited by the CYP3A inhibitor ketoconazole,[162] and the flavoprotein CYP reductase inhibitor, diphenylene iodonium chloride.[163] In the same study the purified rat liver CYPs CYP1A2, CYP2B1, CYP2C11, CYP3E1, CYP3A2 and CYP4A2 all produced the hydroxyl radical to differing extents, with CYP2B1 and CYP3A2 giving the highest levels of hydroxyl radical formation.[164]

The source of the ROS induced by propiconazole was identified using traditional *in vitro* and *in vivo* biochemical approaches, and using both AML12 immortalized mouse hepatocytes in culture and liver tissues from mice.[165] ROS were induced in AML12 cells by propiconazole as measured by fluorescence detection and their formation was ameliorated by *N*-acetylcysteine. Propiconazole induced the levels of GSTα proteins and increased the levels of thiobarbituric acid reactive substances (TBARS) in AML12 cells. The TBARS levels were decreased by diphenylene iodonium chloride revealing the role of CYPs in ROS generation. It has been previously reported that Cyp2b and Cyp3a proteins were induced in mouse liver by propiconazole and that Cyp2b and Cyp3a proteins undergo uncoupling of their CYP catalytic cycle, releasing ROS. Therefore, salicylic acid hydroxylation was used as a probe for ROS formation using microsomes from male CD-1 mice treated with propiconazole (2500 ppm) for 4 days. These studies showed that 2,3-dihydroxybenzoic acid levels were decreased by ketoconazole, melatonin and diphenylene iodonium chloride. These data suggested that specific CYP proteins whose levels were increased by propiconazole could explain the propiconazole-induced ROS formation and oxidative stress through uncoupling of the CYP catalytic cycle.

10.10.4.2 Induction of Oxidative Stress

Propiconazole induced oxidative stress in mouse liver in studies conducted in liver tissues from male CD-1 mice fed propiconazole for up to 90 days. This conclusion was a result of data from genomic, proteomic, metabolomic and biochemical studies using liver tissues from mice fed propiconazole at a tumorigenic dose level (2500 ppm). Genomic analyses indicated that propiconazole induced oxidative stress.[116] Genes associated with oxidative stress that were overexpressed were *Akr1b7*, *Gstt3* and *Gstm3.* Akr1b7 protein is an aldo-keto reductase that is responsible for detoxifying isocaproaldehyde generated by the conversion of cholesterol to pregnenolone, and detoxifying 4-hydroxynonenal, a lipid peroxidation byproduct.[166,167] Gst theta3 and Gstmu3 are glutathione *S*-transferases and are phase II conjugating enzymes. Genomic analyses combined with proteomic studies identified canonical pathways associated with oxidative stress that were altered by propiconazole treatment in mouse liver: oxidative stress, Nrf2-mediated oxidative stress response, glutathione metabolism and cysteine metabolism.[117,124] Metabolomic analysis of homogenates of livers from propiconazole-treated mice revealed suppressed levels of reduced glutathione and cysteine consistent with ROS formation.[125] Biochemical investigations of indicators of ROS and oxidative stress in livers from mice fed propiconazole showed increases in the formation of protein carbonyls and oxidized proteins, decreases in the reduced/oxidized glutathione ratio, increases in glutathione *S*-transferase activities,[168] increases in hepatic malondialdehyde levels and increases in Gstα protein levels. However, no effects on hepatic catalase or superoxide dismutase activities were observed.[165] ROS and oxidative stress induction are components of a key event in the proposed mode of action of propiconazole.

10.10.5 Key Event: Increase in Endogenous DNA Adduct Levels and Mutations

The spectra of endogenous DNA adducts in the livers of Big Blue® mice fed a control diet were compared to those fed a diet of propiconazole at a tumorigenic dose level (2500 ppm) for 4 days using ^{32}P-postlabelling analyses. The same DNA adducts were observed in both control mice and in mice who were fed the propiconazole diet. Quantitatively, a subgroup of 13 adducts with the highest chromatographic mobility were present at significantly higher levels in the livers of mice treated with propiconazole when compared to their concurrent controls.[88] The structures and identities of these propiconazole-generated adducts are presently unknown. A substantial number of endogenous DNA adducts or modified bases have been characterized in human tissues and in tissues of exposed experimental animals. Many of these endogenous DNA adducts are formed by the direct reactions of ROS with nucleic acids (*e.g.* 8-oxodeoxyguanosine) and from the products generated by these ROS forms (*e.g.* malondialdehyde) that react with nucleic acids to form DNA adducts.[169,170] Increase in endogenous DNA adducts is a key event in the proposed mode of action of propiconazole (Figure 10.4).

The Big Blue® mice transgenic mutation assay was used to evaluate the mutagenic activity of propiconazole. Propiconazole induced mutations in the livers of Big Blue® mice after 4 days of dietary administration at its tumorigenic dose level (2500 ppm) compared to Big Blue® mice fed a control diet. Propiconazole induced a 1.97-fold increase in mutant frequency in the liver.[87] The mutants recovered from the livers of propiconazole-fed mice were compared to those mice receiving the control diet using DNA sequencing. DNA sequencing revealed that the propiconazole-induced mutations did not represent a general clonal expansion of background mutations, and supported the hypothesis that they arose from the accumulation of endogenous DNA adducts formed by increased levels of propiconazole-induced ROS within the liver.[86] Increase in mutations *in vivo* is a key event in the proposed mode of action of propiconazole (Figure 10.4).

Additional mutation and DNA adduct studies using Big Blue® mice fed diets of triadimefon (1800 ppm), a conazole that was hepatotumorigenic and myclobutanil (2000 ppm), a non-tumorigenic conazole, were conducted.[86–88] Triadimefon was found to be mutagenic while myclobutanil was not in the Big Blue® mouse assay. As found with propiconazole, the group of 13 endogenous DNA adducts with the highest chromatographic mobility were, as a group, present at significantly higher levels in the livers of mice fed diets of triadimefon compared to the concurrent controls, whereas this same group of DNA adducts in the myclobutanil-fed mice were not different from controls. The levels of DNA adducts from this same group of endogenous adducts were significantly correlated with mutant frequency across all treatment groups, as were total endogenous DNA adduct levels. This correlation of endogenous DNA adducts levels with the *in vivo* mutant frequencies for three conazoles add strength to the conclusion that propiconazole is an *in vivo* mutagen in mouse liver.[88]

10.10.6 Key Event: Increased atRA Metabolism

The first report that the retinol/retinoic acid metabolism pathway was overexpressed by propiconazole came from genomic analyses.[116] This genomic finding was confirmed using biochemical and molecular approaches.[67] All-*trans*-retinoic acid (atRA) is metabolized in a two-step process. First, atRA is metabolized to 4-hydroxy-atRA and second, 4-hydroxy-atRA is metabolized to 4-oxo-atRA.[171,172] Liver microsomes from male CD-1 mice injected by a single IP injection of propiconazole per day (210 mg kg^{-1} per day) for 4 consecutive days showed increased activity in both oxidation steps increasing the overall metabolism of atRA by 2.3-fold.[67] Using a combination of both chemical and immuno-inhibition approaches, it was shown that Cyp26a1, Cyp2b and Cyp3a, but not Cyp1a1 proteins were involved in atRA metabolism. Confirmatory quantitative reverse transcription polymerase chain reaction studies showed that *Cyp1a2*, *Cyp2b10/20*, *Cyp2c65*, *Cyp3a11* and *Cyp26a1* genes were significantly overexpressed in the livers of propiconazole-treated mice and Western immunoblot analyses indicated propiconazole-induced increases in Cyp2b and Cyp3a proteins.[67] Increased atRA metabolism is a key event in the proposed mode of action of propiconazole (Figure 10.4).

10.10.7 Key Event: Decreased Hepatic Levels of atRA

As the metabolism of atRA was increased by propiconazole, there was a concomitant decrease in mouse hepatic tissue levels of atRA. Male CD-1 mice injected by a single IP injection of propiconazole per day (210 mg kg^{-1} per day) for 4 consecutive days showed decreased hepatic atRA tissue levels of treatment to 55% of control values.[67] The hepatic levels of atRA in male CD-1 mice fed propiconazole (2500 ppm) for 30 days were 54% of the control values.[71] Decreased hepatic levels of atRA are a key event in the proposed mechanism of action of propiconazole (Figure 10.4).

10.10.8 Key Event: Inhibit CYP51 Activity

Propiconazole inhibits human CYP51 with an IC_{50} of 8.25 μM.[36] Hepatic tissue levels of propiconazole in mice have not been reported, nor have physiologically based pharmacokinetic models been developed to estimate these levels. The level of hepatic CYP51 activity in these propiconazole-treated mice has also not been determined. It is noted that *CYP51* gene expression was increased ~2-fold by propiconazole in both CAR-null and wild-type mice suggesting that CAR was not transcriptionally regulating this gene.[75] Inhibition of CYP51 activity is a key event in the proposed mechanism of action of propiconazole (Figure 10.4).

10.10.9 Key Event: Dysregulation of Cholesterol Biosynthesis and Metabolism

Propiconazole's ability to decrease serum cholesterol levels in mice is described in Section 10.5. In a genomic study using livers from mice fed propiconazole, several elements within the cholesterol biosynthesis pathway (see Figure 10.2) were found to be overexpressed.[116] Specifically, propiconazole treatment increased the expression of genes in the cholesterol biosynthesis pathway including squalene epoxidase (*Sqle*), lanosterol-14α-demethylase (*CYP51*) and 7-dehydrocholesterol reductase (*Dhcr7*). Propiconazole overexpressed *Cyp7a* (cholesterol-7α-hydroxylase) within the cholesterol metabolism pathway. Propiconazole also overexpressed *Akr1d1* (aldo-keto reductase family 1, member D1) and *Aldh1a7* (aldehyde dehydrogenase family 1, subfamily A7) within the bile acid biosynthesis pathway.[116,117] A metabolomics study using liver tissues from mice fed diets of propiconazole at tumorigenic and non-tumorigenic levels indicated no changes in hepatic cholesterol levels; however, increased levels of both 7α-hydroxycholesterol and 7β-hydroxycholesterol levels were reported as well as increased levels of taurochenodeoxycholate and chenodeoxycholate, bile acids formed from 7α-hydroxycholesterol.[125] Proteomic studies using the same mouse liver tissues revealed increases in Akr1d1 and Aldh1a1, both proteins involved in bile acid biosynthesis.[124] Overall, these genomic, proteomic and metabolomic data strongly suggested an increased flux in cholesterol biosynthesis induced by propiconazole, presumably a result of a negative feedback loop due to inhibition of CYP51 by propiconazole. Dysregulation of cholesterol biosynthesis and metabolism is a key event in the proposed mode of action of propiconazole (Figure 10.4).

10.10.10 Key Event: Increase in Cell Proliferation by Decreased Levels of atRA and by a Mevalonic Acid/Cholesterol Biosynthesis Feedback Mechanism

Hepatic cell proliferation was determined by immunohistochemical staining using a PCNA antibody, and by counting 900–1000 labeled cells per mouse in mice fed dietary levels of propiconazole for 4, 30 and 90 days.[58] Using this technique, the tumorigenic dose level of propiconazole (2500 ppm) increased hepatic cell proliferation 8.9-fold over control mice after 4 days of feeding. Longer treatment times did not produce increases in cell proliferation.[58] A subsequent study was conducted where the slides from the Allen *et al.*[58] study were reanalyzed using immunohistochemistry using a Ki-67 antibody and by counting ~8000 and 5200 cells per mouse in the control and treated groups, respectively. In that study, statistically significant 2-fold increases in cell proliferation were observed at the tumorigenic dose level in mice fed propiconazole for 30 days.[71] In another cell proliferation study, male CD-1 mice were fed 850 and 2500 ppm propiconazole for up to 60 days. Using BrdU incorporation to measure cell proliferation, proliferation was significantly increased at both dose levels at day 1 to day 4, and at the

2500 ppm dose level to day 7. By day 14, cell proliferation returned to control levels.[70,76] This result differed from that reported by Hester *et al.*[71] where cell proliferation in male CD-1 mice fed 2500 ppm propiconazole for 30 days showed a 2-fold increase. Increase in cell proliferation is a key event in the proposed mode of action of propiconazole (Figure 10.4).

10.10.10.1 Key Event: Decreased Levels of atRA Leads to Increased Cell Proliferation

One mechanism that can explain the increase in cell proliferation is the depression of hepatic atRA. Propiconazole depressed the levels of hepatic atRA in mice after 4 and 30 days of treatment (Section 10.10.7). Concurrently, propiconazole treatment of mice increased hepatic labeling indices at both 4 and 30 days of treatment.[58,71] Retinoids control cell proliferation in epithelial cells. For example, the growth of MCF-7 breast cancer cells was inhibited by atRA that was associated with inhibition and accumulation of cells in G1 phase of the cell cycle due to the inhibition of cdk2 mRNA and protein production, and by a decrease in cdk2 activity.[173] Retinoblastoma protein phosphorylation was also partially reduced and the level of retinoblastoma protein was decreased.[173] The converse has also been reported. Ethanol-induced liver cancer in rats was associated with increased CYP2E1 induction, increased cell proliferation and decreased retinoic acid levels.[174,175] The cell proliferation responses were reversed with dietary atRA supplementation.[176] These data indicated that decreased atRA levels provided a tumor-promoting environment.[175] The effects of atRA on cell proliferation were studied in AML12 cells, a non-transformed immortalized mouse hepatocyte cell line.[177] At various time points, control and treated cells were counted to determine the effect of atRA on AML12 cell proliferation. At 5 days of atRA treatment, there was a 56% decrease in the number of atRA-treated cells when compared to control cells. This effect was reversible, as an increase in cell number was observed 2 days after removal of atRA from the treatment media. Cell cycle analysis by flow cytometry using propidium iodide stained AML12 cells showed a significant reduction in S phase cells, and a significant increase in G1 phase cells at 48 hours of atRA treatment when compared to control cells, indicating a block at the G1/S checkpoint. Western immunoblot analysis of control and treated whole-cell lysates revealed a decrease in phosphorylated (Serine807/Serine811) retinoblastoma protein. Taken together, these results indicated that proliferation in AML12 hepatocytes can be reversibly modified by atRA by regulating these cells at the G1/S checkpoint with alterations in the levels of phosphorylated retinoblastoma protein.[177] These results on the effects of altering atRA levels on cell proliferation in mouse hepatocytes are in general agreement with the effects of propiconazole on the depression of hepatic atRA levels and the increase in hepatic cell proliferation in mice. Increase in hepatic cell proliferation by decreased levels of atRA metabolism is a key event in the proposed mode of action of propiconazole (Figure 10.4).

10.10.10.2 *Key Event: Increase Mevalonic Acid Levels Leads to Increased Cell Proliferation*

Another mechanism that can explain an increase in cell proliferation is by a mevalonic acid/cholesterol biosynthesis feedback mechanism. Genomic, proteomic and metabolomic studies using the livers of treated mice revealed that propiconazole increased the expression of genes and proteins within the cholesterol biosynthesis, cholesterol metabolism and bile acid biosynthesis pathways and increased the levels of hepatic cholesterol metabolites and bile acids; however, hepatic cholesterol levels were unchanged.[116,117,124,125] It was concluded that propiconazole increased the flux of the cholesterol and bile acid pathways even though propiconazole is a CYP51 inhibitor and should decrease the hepatic levels of cholesterol and its downstream metabolites. This conundrum was explained by examining the rate-limiting steps in cholesterol biosynthesis. The rate-limiting step in cholesterol biosynthesis is 3-hydroxy-3-methyl-glutaryl (HMG)-CoA reductase and not CYP51.[178,179] This would allow lower CYP51 activities (due to propiconazole inhibition, see Table 10.3) to still produce increasing amounts of cholesterol metabolites and bile acids. HMG-CoA reductase, a key enzyme in the cholesterol biosynthesis, catalyzes the rate limiting conversion of HMG-CoA to mevalonate.[178,179] Mevalonate was found to increase cell proliferation *in vitro* using cancer cell lines and to increase the growth of human cancer cell derived tumors in mice.[180] Conversely, HMG-CoA reductase inhibitors decreased cell growth in a number of cell lines.[178] Mevalonate is metabolically converted to farnesyl pyrophosphate and geranylgeranylpyrophosphate which are substrates for farnesyl and geranylgeranyl transferases that prenylate proteins.[179] Protein prenylation is a lipid post-translational modification that generally directs the membrane localization, protein–protein interactions and activation of proteins such as Ras GTPases and nuclear lamins. Farnesylation of Ras facilitates plasma membrane localization where Ras cycles through inactive/active forms with the exchange of GDP/GTP. There, Ras initiates a signaling cascade involving Raf and Erk initiating the MAPK pathway cascade.[181] Activation of this pathway ultimately leads to transcriptional activation of genes that are crucial for the S phase checkpoint of the cell cycle and cell proliferation. The effects of propiconazole on this process were studied using the immortalized non-tumorigenic AML12 hepatocyte cell line to determine if propiconazole increases cell proliferation by dysregulation of Ras farnesylation as a result of changes in the cholesterol biosynthesis pathway.[182] Mevalonic acid increased cell proliferation in AML12 cells, which was reduced by farnesyl transferase inhibitors or an HMG-CoA reductase inhibitor. Similarly, cell proliferation in AML12 cells was increased by propiconazole, which was reduced by a farnesyl transferase inhibitor or an HMG-CoA reductase inhibitor. Molecular analysis of treated cells confirmed that both propiconazole and mevalonic acid increased the translocation of Ras from the membrane to the nucleus, and increased the activation of Erk1/2. Erk kinase activity is important in regulation of the c-Myc transcription factor that controls transcriptional activation of a number of important cell cycle genes, including cyclin D1 and Rb protein.[183] These data indicated that propiconazole increases cell proliferation by increasing the levels of cholesterol biosynthesis intermediates presumably through a negative feedback mechanism within the pathway, a result of CYP51 inhibition.[182] This feedback mechanism increases Erk1/2 signaling through mevalonate-mediated Ras activation. These results provide an explanation for the observed effects of propiconazole on hepatic cholesterol pathways, and on the increased hepatic cell proliferation induced by propiconazole in mice. Increase in mevalonic acid levels resulting in increased cell proliferation is a key event in the proposed mode of action of propiconazole (Figure 10.4).

10.10.11 Key Event: Hepatocellular Tumors

Propiconazole has been evaluated as a carcinogen in mice in two dietary studies: a 79 week study and a 104 week study, each with a different set of upper dose levels. Groups of 80 Crl:CD1 (ICR) BR male mice were given diets containing propiconazole at a concentration of 0, 100, 500 or 850 ppm for 18 months. Fifty animals per group were used for evaluation of oncogenic potential. At 79 weeks mice fed 850 ppm had statistically significant increases in hepatocellular carcinomas and combined hepatocellular adenomas and carcinomas.[70]

Groups of 52 male and 52 female CD-1 mice received diets containing propiconazole at concentrations of 0, 100, 500 or 2500 ppm for 104 weeks. The gross examination revealed a higher incidence of liver masses and/or enlarged livers in males and females receiving propiconazole at a dietary concentration of 2500 ppm. The gross findings were confirmed by microscopic examination. A statistically significant increase in the incidence of malignant and benign liver cell tumors (adenocarcinomas and adenomas) was seen in male mice at the 2500 ppm propiconazole feed level.[70]

Two ancillary tumor promotion studies in rats are germane to the discussion of the carcinogenic activity of propiconazole. Groups of male and female Tif:RAIf rats were given *N*-diethylnitrosamine (15 mg kg^{-1} body weight) or saline by IP administration. After weaning on day 22, rats were fed diets containing propiconazole (2000 ppm) or a control diet. Subgroups of rats of each sex were sacrificed after 14, 28 or 56 days of treatment. Livers were analyzed for the presence of γ-glutamyl transpeptidase-positive foci. Rats initiated with *N*-diethylnitrosamine and fed propiconazole had significant increases in liver weights and increases in the numbers of γ-glutamyl transpeptidase-positive foci with the increases being more marked with time of treatment.[70] In a related study, male F344 rats were given a single IP dose of *N*-diethylnitrosamine (200 mg kg^{-1} body weight) and starting 2 weeks later, were treated with propiconazole (2500 ppm) in the diet for 6 weeks.[91] All rats were subjected to a two-thirds partial hepatectomy at week 3. Carcinogenic potential was scored by comparing the numbers and areas of induced glutathione *S*-transferase placental form (GST-P)-positive foci in the livers. Propiconazole significantly increased both measures of GST-P foci.[91] Each study concluded that propiconazole had tumor promoting activities. An increase in hepatocellular tumors is a key event in the proposed mode of action of propiconazole (Figure 10.4).

10.11 Mode of Carcinogenic Action of Propiconazole: Discussion

Chemically-induced liver carcinogenesis is a multi-stage process. One stage involves alterations to DNA leading to a mutagenic event (initiation). The DNA alteration may be a result of the direct interaction of the chemical (or its metabolically activated reactive form) with DNA or by an indirect pathway such as chemical induction of ROS that can modify DNA, or by intercalation of the chemical with DNA. A second stage involves increased cell proliferation (promotion) which increases clonal expansion and inhibits apoptosis of initiated cells. The third stage is progression that can involve karyotypic instability, activation of protooncogenes and loss of functional tumor suppressor genes.[184,185] There are a myriad of altered cellular events and/or pathways within each stage that can contribute to the mechanism of action of a chemical. These altered events and/or pathways can be a result of the pleiotropic effects of the chemical, such as altering intermediary metabolism, inducing oxidative stress, activation of key nuclear receptors and alterations in cell signaling pathways, to name some examples.

The advent of genomics and related technologies including proteomics and metabolomics are having dramatic effects on insights into understanding the mechanisms of action of chemicals for various toxicological endpoints.[186–191] In the past, conventional rodent toxicity studies generally identified apical endpoints such as clinical signs or pathological states. Targeted mechanistic studies using both *in vivo* and *in vitro* approaches had been used to develop potential mechanisms of action on a single pathway by pathway basis. Many toxicologically induced disease states (*e.g.* cancer, reproductive or neurological disease) are generally the results of complex processes that can result from multiple pathways. The power of omic approaches is the ability, at one time, to uncover a collection of pathways and networks of genes, proteins and metabolites that are altered by chemical treatment in the target or related tissues. This allows the experimentalist with knowledge of toxicology to visualize a more complete picture of the breadth of changes and begin *de novo* to formulate mechanisms of action for the agent. These mechanisms can be validated by further experimental studies targeted at each step or key event in the process. In essence, omics provide a discovery-driven rather than hypothesis-driven approach towards developing mechanisms of action that can be integrated with toxicological alterations/apical endpoints into a more complete description of the inherent mechanisms.

The mode of action developed for propiconazole combines toxicological and omic data. Genomic, proteomic and metabolomic data were obtained from livers of mice treated with propiconazole under the conditions used in the chronic bioassay that produced hepatocellular cancer. These data were analyzed and interpreted within a framework of pathways that could lead to tumor development and tumor progression. In many cases the toxicogenomic data findings and interpretations were further confirmed by additional molecular, biochemical or biological studies.

The mode of action for propiconazole is based on its ability to be an inducer of a group of CYPs and an inhibitor of CYP51. Propiconazole activates nuclear receptors that lead to the induction of a group of CYPs. The induction of these CYPs creates two pathways. One pathway leads to the increased metabolism of atRA that lowers hepatic levels of atRA resulting in an increase in cell proliferation. The second pathway resulting from CYP induction is the release of ROS and the increase in oxidative stress. The ROS increase the levels of a group of endogenous DNA adducts that result in an increase in mutations in the liver. The inhibition of CYP51 leads to a dysregulation of cholesterol biosynthesis and metabolism. This results in an increase in mevalonic acid levels that stimulate cell proliferation through Ras farnesylation leading to the MAPK pathway cascade. All of these events lead to an increased incidence of hepatocellular adenocarcinomas and adenomas. Additional supporting data on propiconazole-induced hepatic alterations related to cancer induction came from the dose–response metabolomic study where propiconazole increased the need for energy (through glycolysis/glutamate metabolism/fatty acid metabolism/TCA cycle), increased the levels of a polyamine, spermidine, and increased the levels of sarcosine. These three metabolic alterations are found in tumors or tumor cells, and/or animals or humans bearing tumors.

One of the key events in the proposed mode of action of propiconazole is the induction of mutations *in vivo*. Data that provided this conclusion came a from Big Blue® mice transgenic mutation assay where propiconazole induced a 1.97-fold increase in mutant frequency in the liver after 4 days of dietary administration at its tumorigenic dose level (2500 ppm) compared to Big Blue® mice fed a control diet.[87] Moreover, DNA sequencing of the mutants revealed that the propiconazole-induced mutations did not represent general clonal expansion of background mutations.[86] Furthermore, the mutation data were supported by the endogenous DNA adduct data formed by increased levels of propiconazole-induced ROS within the liver.[88] Finally, a comparative Big Blue® mice transgenic mutation assay analysis of three conazole fungicides, two tumorigens (propiconazole and triadimefon) and one non-tumorigen (myclobutanil) indicated a concordance of mutation and endogenous DNA adduct results, thus strengthening the conclusion that propiconazole was a mutagen *in vivo*.[88] The methods, analyses and conclusions from the Big Blue® mice transgenic mutation assay studies were re-evaluated in a commentary by Shane *et al.*,[192] who reported that a re-analysis of the mutation spectra using a conventional method indicated no treatment-related differences in the spectra. They concluded that propiconazole did not act as a mutagen *in vivo*, a conclusion that conflicted with the conclusions of Ross and Leavitt[86] and Ross *et al.*,[88] who reported increases in mutation frequency with supporting evidence for increases in endogenous DNA adducts. The results and conclusions of the Shane *et al.*[192] commentary were questioned in a response by Ross and Leavitt.[193] In this response it was pointed out that the commentary contained factual errors and mischaracterizations of the Ross and Leavitt[86] data that undermined the commenter's conclusions regarding the mutagenic activity of propiconazole.

Propiconazole only induces hepatocellular carcinoma in male mice, not in female mice or male or female rats.[70] The sex preference is an interesting finding as it mirrors hepatocellular cancer in humans: the incidence of hepatocellular carcinoma is higher in men than in women.[194] The sex preference in liver carcinogenesis in mice is due to the multiple effects of sex hormones on liver tumor induction, and studies demonstrate that androgens promote liver tumors, whereas ovarian hormones inhibit their development.[195] Male rats treated with propiconazole demonstrated a number of the same responses as reported in mice: increased liver:body weight ratio, increased hepatic hypertrophy,[59,69] increased mRNA expression of CYP2B1, CY3A1 and CYP3A2 indicative of CAR and PXR activation[69] and increased PROD activity.[59,69] Hepatic cell proliferation was increased by propiconazole in rats after a brief treatment time,[59] while mice responded with increased cell proliferation after both brief and extended treatment times.[58,71,74,182] The lack of tumorigenic activity of propiconazole in rats could be related to their lack of alteration of cholesterol biosynthesis and metabolism. In mice, propiconazole treatment over a 90 day period (2500 ppm) decreased serum cholesterol levels, while virtually no effect or a slight increase in serum cholesterol levels was reported in rats treated under the same conditions.[58,59] This suggest that the complex dynamic of dysregulation of cholesterol biosynthesis and metabolism leading to increased hepatic cell proliferation observed in mice may not occur in rats, which could explain their resistance to the carcinogenic effects of propiconazole.

10.12 Mode of Carcinogenic Action: Human Relevance

10.12.1 Species Comparison Across Key Events

The relevance of the mode of action of propiconazole-induced hepatocarcinogenesis in mice to humans can be approached by comparing some of the key events in the proposed mode of action found in mice to those that could occur in exposed humans. It has been shown that propiconazole induces DNA-damaging ROS in mice *via* CAR and PXR activation leading to increased CYP activities. Many of these same events have been reported using pharmaceutical conazoles in human hepatic systems. Propiconazole was found to activate both human and mouse PXR.[75,146] Activation of the

PXR nuclear receptor is known to lead to the induction of CYP3A4.[196] CYP3A4 is the most abundant CYP in human liver, comprising about 70% of the total CYP present in the liver.[197] Clotrimazole, an azole-containing fungicide similar to propiconazole, was found to activate human PXR in transfected human hepatoblastoma cells and to induce CYP3A4 protein levels in primary cultured human hepatocytes.[198] Increased CYP3A4 expression in human hepatoma cells was found to increase the levels of ROS.[199] Therefore, it is reasonable to propose that propiconazole could induce CYP3A4 in exposed humans, which could lead to the induction of ROS. Propiconazole reduces hepatic levels of atRA in mice that can lead to an increase in cell proliferation. The decreases in hepatic levels of atRA were a result of increased atRA metabolism due to increased CYP activities induced by propiconazole. Chemical and immuno-inhibition metabolism studies revealed that Cyp2b and Cyp3a proteins (CYPs induced by propiconazole) were involved in atRA metabolism in the mouse.[67] Similar decreases in hepatic atRA levels in mice by other agents that activate PXR transcribed CYPs have been reported.[200] In humans, increased urinary excretion of atRA metabolites and decreased atRA plasma levels in leukemia patients treated with atRA has been, in part, ascribed to an induction of accelerated catabolism by a CYP enzyme system.[201,202] Therefore, propiconazole, which is known to activate human PXR, could increase CYP3A4 protein levels in exposed humans and reduce atRA levels leading to increased cell proliferation.

The dysregulation of cholesterol biosynthesis and metabolism leading to increased mevalonic acid levels and increased cell proliferation reported in mice exposed to propiconazole has not been studied in humans. However, studies in humans showed ketoconazole therapy in male prostate cancer patients reduced total serum cholesterol levels and concomitantly increased serum lanosterol levels.[77] There are ancillary studies in humans that point to the relationship of statin use and the lower rates of cancer in several organs, including liver.[203] Statins are a class of drugs used to lower cholesterol levels by inhibiting the enzyme HMG-CoA reductase. HMG-CoA reductase catalyzes the reduction of 3-hydroxy-3-methylglutaryl-CoA to mevalonic acid, a known enhancer of cell proliferation. In two human hepatocellular tumor cell lines, simvastatin reduced cell growth and impaired cell cycle progression.[204] Therefore, mevalonic acid levels in humans would seem to be a determinant of tumor outcome, and agents like propiconazole that increase mevalonic acid levels could increase human liver cancer.

10.12.2 Comparison of Toxicologic and Genomic Studies Comparing Mouse Liver Responses to Conazoles and Phenobarbital

The similarity of a group of characteristic apical responses (non-genotoxicity, CAR activation, induction of CYP2B and CYP3A proteins, increase in cell proliferation and liver tumors) that are common to many conazoles and phenobarbital has prompted the proposal that mouse tumorigenic conazoles have the same mode of carcinogenic action as phenobarbital. This presumed association is important in risk assessment as extensive epidemiological studies of patients who have received long-term phenobarbital therapy show no increase in the incidence of hepatic cancer, even though their phenobarbital plasma levels were close to those found in susceptible rodents.[205] For that reason, chemicals that are mouse liver tumorigens and produce phenobarbital-associated responses in mouse liver are considered to be "phenobarbital-like" and their tumorigenic responses have been deemed not relevant to humans.[206] The induction of cell proliferation by propiconazole and phenobarbital was compared in male CD-1 given diets containing propiconazole at a concentration of 0, 850 or 2500 ppm or with phenobarbital (850 ppm).[70] Hepatocyte cell proliferation was measured by BrdU-immunohistochemistry image analysis over the course of 60 days. Hepatocellular hypertrophy was observed in mice treated with propiconazole that progressed in severity with time of treatment. Mice treated with phenobarbital showed a progression of hepatocellular hypertrophy somewhat similar to that observed in the propiconazole at 2500 ppm group. Dewhurst and Dellarco[70] state:

> The hypertrophy caused by propiconazole affected all lobular compartments of the liver, but was predominantly centrilobular, while phenobarbital almost exclusively affected centrilobular or midzonal hepatocytes. Hepatocellular necrosis was present in all groups, showing a dose related increase in incidence and severity in the groups treated with propiconazole. Panlobular hepatocyte vacuolation (minimal) was observed in some animals from all treatment groups at the beginning of the study. Centrilobular vacuolation was observed towards the end of the treatment with a high incidence (almost all animals) and severity (minimal to marked) from day 7 to day 60 in animals treated with propiconazole at 2500 ppm, but only sporadically in animals treated with propiconazole or phenobarbital at 850 ppm. Mitotic activity and incorporation of BrdU was increased markedly in most treated animals during day 1 to day 4 of treatment, returning to control values by day 14. The increased incorporation of BrdU was primarily in the centrilobular/midzonal hepatocytes. The magnitude of this response was lower with propiconazole at 2500 ppm than with phenobarbital at 850 ppm, contrary to the results for liver weight, [*author*: At 4 days mice fed propiconazole at 2500 ppm and mice fed phenobarbital at 850 ppm had approximately the same mean absolute liver weights], indicating that a short mitogenic response is not the sole mechanism underlying the hepatocellular proliferative effects of propiconazole. This study showed propiconazole to be a mouse liver mitogen with a degree of similarity to phenobarbital, but of a lower potency in terms of BrdU incorporation.

This cell proliferation study was also reported by Currie *et al.*[76] The histology findings regarding differences in hypertrophy induced by propiconazole and phenobarbital were mirrored in a study comparing three other conazoles and phenobarbital.[72] Male C3HHeNCrlCrlj and CAR-knockout mice were fed three dose levels of cyproconazole, tebuconazole, fluconazole and phenobarbital for 1 week. The authors concluded that cyproconazole and fluconazole mainly induced CAR-mediated liver hypertrophy, while tebuconazole slightly involved CAR-mediated hypertrophy and that all three triazoles had non-CAR-mediated liver hypertrophy pathways. They further concluded that the hypertrophy induced by these triazoles differed from that of phenobarbital.

A comparison of propiconazole and phenobarbital was conducted on the induction of microsomal enzymes in male Crl:CD1(ICR)BR mice.[70] Mice received diets containing propiconazole (0, 850 or 2500 ppm) or phenobarbital (850 ppm) for 14 days (Table 10.9). A series of biochemical parameters were studied: protein content of microsomal and cytosolic fractions, microsomal cytochrome P450, microsomal EROD, PROD, microsomal coumarin 7-hydroxylase activity; regio- and stereoselective microsomal hydroxylation of testosterone; microsomal lauric acid 11- and 12-hydroxylation, microsomal UDP-glucuronosyltransferase activity; cystolic glutathione *S*-transferase activity; and microsomal epoxide hydrolase activity. The authors concluded:

> The pattern of enzyme induction produced by propiconazole was similar, but not identical, to that produced by phenobarbital. The main differences were in the induction of testosterone oxidation to androstenedione and of testosterone 2α-hydroxylation that were induced by propiconazole but not phenobarbital ... Within the range of microsomal enzyme parameters measured, the overall response to propiconazole is more consistent with a phenobarbital-type mechanism than peroxisome proliferation or aromatic hydrocarbon mechanism. However, mechanisms other than one similar to that used by phenobarbital cannot be conclusively excluded.

The majority of responses induced by propiconazole increased in magnitude from 850 ppm to 2500 ppm propiconazole, suggesting similar of mechanisms of enzyme induction at both doses known to induce mouse liver tumors.

A genomic study was conducted to compare conazoles and phenobarbital using intensive detailed genomic evaluations to examine this presumed association that was based on apical responses.[117] The microarray data from liver samples from male CD-1 mice fed triadimefon (1800 ppm) and propiconazole (2500 ppm) for 4 and 30 days[58] from an earlier study were analyzed with genomic data from male CD-1 mice administered diets containing phenobarbital (850 ppm) for 4 and 30 days. The dose levels used were selected because they produced mouse liver tumors in 2 year studies. Targeted

Table 10.9 Induction of microsomal enzymes in male mice given diets containing propiconazole or phenobarbital for 14 days (% of control values).[a]

	Dietary concentration (ppm)		
	Phenobarbital	Propiconazole	
Parameter	850	850	2500
Absolute liver weight	163	140	199
Microsomal cytochrome P450 content	239	300	389
Microsomal pentoxyresorufin *O*-depentylase	3534	3024	5524
Microsomal coumarin 7-hydroxylase	480	534	2384
Microsomal testosterone 2β-hydroxylation	466	298	531
Microsomal testosterone 6α-hydroxylation	779	502	715
Microsomal testosterone 6β-hydroxylation	500	366	524
Microsomal testosterone 15β-hydroxylation	526	316	977
Microsomal testosterone 16β-hydroxylation	5.7[b]	5.2[b]	6.4[b]
Microsomal testosterone oxidation to androstenedione	179	589[c]	652[c]
Microsomal testosterone 2α-hydroxylation	0[b]	2.7[b,c]	2.7[b,c]
Total microsomal testosterone oxidation	356	440	555
Microsomal epoxide hydrolase[d]	179	172	321
Immunoblot goat anti-rat CYP2B1 (Cyp2b) Band 1	2743	2608	3049
Immunoblot goat anti-rat CYP2B1 (Cyp2b) Band 2	581	810	579
Immunoblot goat anti-rat CYP2B1 (Cyp2b) Band 3	351	439	617
Immunoblot mAb p6 (Cyp3a)	577	658	1068
Microsomal ethoxyresorufin *O*-deethylase[e]	232	219	388
Microsomal testosterone 16α-hydroxylation	194	214	262
Microsomal lauric acid 11-hydroxylase	271	267	305
Microsomal lauric acid 12-hydroxylase[d]	163	153	161
Microsomal UDP-glucuronosyltransferase	156	156	139
Cystosolic glutathione *S*-transferase	187	158	184
Microsomal testosterone 7α-hydroxylation	229	175	125
Microsomal protein content	97	109	113
Cytosolic protein content	92	97	94
Immunoblot mAb d15 (Cyp1a)[e]	86	124	89
Immunoblot mAb clo4 (Cyp4a)[d]	129	140	157

[a]Reproduced Table 24 from Dewhurst and Dellarco.[70] Table data are taken from the report of Beilstein (1998) (see Dewhurst and Dellarco for complete reference).
[b]Absolute values, control value below limit of detection.
[c]Notable difference between phenobarbital and propiconazole.
[d]Indicative of peroxisomal proliferation.
[e]Indicative of aromatic hydrocarbon response.

transcriptomic analyses were conducted at the gene level examining differentially expressed genes and subsets containing cell cycle genes, and transcription factors. Analyses were also conducted on function, pathway and network levels examining IPA Tox lists and canonical pathways, and Gene-Go MetaCore dynamic networks and their central hubs. Genes expressed by phenobarbital or the two conazoles were also compared with those genes associated with human hepatocellular cancer. The Tox list and canonical pathway analyses revealed marked differences and similarities between

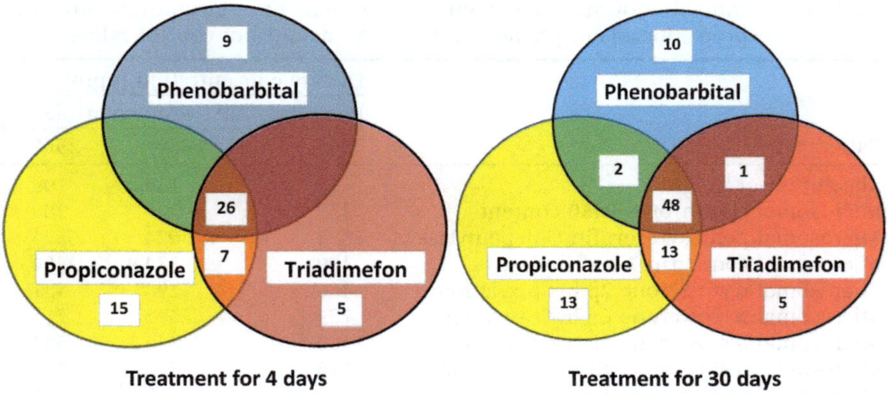

Figure 10.5 Venn diagrams showing commonalities and differences in the numbers of altered pathways between propiconazole, triadimefon and phenobarbital based on genomic analyses of livers from mice treated for 4 and 30 days with these mouse liver carcinogens.

phenobarbital, propiconazole and triadimefon. After 4 days of treatment, of the 62 altered lists and pathways, 26 (42%) were common to phenobarbital and the two conazoles, 27 (44%) were altered only by conazoles and nine (15%) were altered only by phenobarbital. Comparing only propiconazole with phenobarbital at 4 days indicates 22 lists or pathways solely altered by propiconazole and nine solely altered by phenobarbital (Figure 10.5). After 30 days of treatment, of the 92 altered lists and pathways, 51 (55%) were common to phenobarbital and one of both conazoles, 31 (34%) were altered only by conazoles and 10 (11%) were altered only by phenobarbital. At 30 days there were 26 lists or pathways solely altered by propiconazole and 10 solely altered by phenobarbital (Figure 10.5). Cell cycle analyses within phases of the cell cycle revealed more commonalities between propiconazole and triadimefon and more differences between the conazoles and phenobarbital. Transcription factor analyses indicated that eight transcription factors out of 25 were common to phenobarbital and the two conazoles, 10 transcription factors were altered by one or both conazoles and six transcription factors were altered only by phenobarbital. On a global basis, principal component analyses of differential expressed genes (*i.e.* each gene normalized to its respective control) at 30 days clearly depict marked and significant differences between each agent. Most importantly, the conazoles and phenobarbital show dramatic differences in the altered expression of a group of genes known to be involved in human liver cancer. It was concluded that, whereas all three agents induce mouse hepatocarcinogenesis, they each possess different modes of action, possibly with some common key events. This is based on the proposition that the transcriptional profiles of tissues exposed to toxic chemicals inherently contain their mechanisms of toxicity. Therefore, the chemically induced alterations of the molecular processes that are involved in the transition of normal hepatocytes to neoplasia are inherently different between these conazoles and phenobarbital. Therefore, the mechanisms or mode of action of these conazoles are unlike phenobarbital and their tumorigenic effects are potentially relevant to informing human health assessment. This conclusion is consistent with those of Dewhurst and Dellarco[70] based on their analysis of the cell proliferation data comparing propiconazole and phenobarbital. A re-analysis of the genomic data published in Nesnow *et al.*[117] was conducted by Currie *et al.*[76] They examined experimental design differences between the *in vivo* studies with propiconazole and phenobarbital and how they may impact the comparisons of toxicogenomics results.[76] They also demonstrated that different pathway analysis tools can give different conclusions on specific mechanistic pathways when using the same data sets. They concluded that despite these differences, the results across three different analyses (Nesnow *et al.*[117] and Currie *et al.*[76]) showed a striking degree of similarity observed for propiconazole and phenobarbital when the expression data are viewed as major signaling pathways and cell processes affected. Currie *et al.*[76] also presented the results of cell-based splice variant CAR3 reporter assays (human and engineered mouse CAR) using a COS-1 kidney cell line from an African green monkey. Propiconazole was a strong mouse CAR3 activator and showed a markedly reduced response with human CAR3. They concluded that with regard to the events which are hallmarks of CAR-induced effects that are key events in the mode of action of mouse liver carcinogenesis with phenobarbital, propiconazole-induced tumors can be viewed as being promoted by a similar phenobarbital-like CAR-dependent mode of action. This conclusion differs from that of Nesnow *et al.*[117] who concluded that although phenobarbital and the conazoles propiconazole and triadimefon induce mouse liver tumors and exhibit some similar toxicological responses, their transcriptional profiles are significantly different and thus their mechanisms of tumorigenic action are likely to differ.

The role of nuclear receptor signaling in the control of cholesterol homeostasis is complex involving a series of nuclear receptors: LXR, FXR, PXR, PPAR and CAR. These receptors coordinate the regulation of transcriptional programs that control the storage, transport and catabolism of sterols including cholesterol and their metabolites.[207] As described earlier, the dysregulation of cholesterol biosynthesis plays an important role in the mode of action of propiconazole's hepatocarcinogenic activity. Propiconazole has been shown to activate each of these receptors (Section 10.10.2). This dysregulation is evidenced *in vivo* as decreased serum cholesterol levels in mice fed diets of propiconazole and is persistent for extended time periods (Section 10.5). This is not the case in mice or humans treated with phenobarbital. In several studies in mice fed phenobarbital, serum cholesterol levels were unchanged at 7 days, while mixed results were found at shorter times of treatment. Serum cholesterol levels in three strains of mice fed phenobarbital (850 ppm) for 7 days were compared. Serum cholesterol levels in male CD-1, C57BL/6J or C3H/HeNClrBR mice fed phenobarbital (850 ppm) for 7 days were unchanged compared to control mice.[62] In the same report, a shorter term dosing study (3 days) was conducted in male

C3H/HeNClrBR mice and related Car-null mice fed phenobarbital (850 ppm). Serum cholesterol levels were slightly reduced in both wild-type and Car-null mice compared to their respective controls to about the same extent (15–19%).[62] This suggested that at 3 days phenobarbital alterations of cholesterol levels were Car-independent. In another study male and female CD-1 mice were administered dietary phenobarbital at 0, 0.15, 1.5, 15, 75 or 150 mg kg^{-1} per day for 2 or 7 days. No statistically significant changes in serum cholesterol were observed at either 2 or 7 days at any dose level.[208] In several human studies patients on phenobarbital exhibited no changes or increases in serum cholesterol levels. Patients receiving phenobarbital for at least 6 months showed no statistically significant differences in serum cholesterol (total cholesterol or HDL-cholesterol) and triglyceride levels, while LDL-cholesterol levels increased 17% as measured in 25 epileptic patients receiving chronic antiepileptic (phenobarbital) monotherapy and in age- and sex-matched controls.[209] In another study, serum lipids, lipoproteins and apolipoproteins A and B were measured in 101 epileptic patients receiving chronic treatment with antiepileptic drugs and in 75 age- and sex-matched control subjects. In relation to controls, subjects treated with phenobarbital showed significantly higher values of total cholesterol, HDL-cholesterol, apolipoprotein A and apolipoprotein B levels.[210]

Overall, based on the metabolic, toxicological, histological and toxicogenomic data, and the differential effects on cholesterol biosynthesis, transport, storage and metabolism, the modes of action described for propiconazole and phenobarbital differ. Moreover, the modes of action described for propiconazole and phenobarbital differ as the mode of action for propiconazole contains both CAR-dependent and CAR-independent key events required for cancer induction. This differs from the mode of action for phenobarbital which is solely CAR-dependent, as all key events emanate from, and are specific to CAR activation.[211]

10.13 Conclusions

The development of conazoles as inhibitors of fungal CYP51 for pharmaceutical applications and their subsequent use on seeds, grasses and crops has engendered a body of research on their potential adverse health effects in humans. The mode of action for propiconazole's induction of mouse liver tumors was based on toxicological approaches and incorporated omic and related technologies. In addition, most of the key events first identified through toxicogenomic analyses were subsequently confirmed by additional biochemical or molecular studies using liver tissues from propiconazole-treated mice. Conazoles exert multiple toxicological effects (hepatotoxicity and reproductive toxicity) in mammalian systems in addition to many being tumorigenic. Conazoles have the ability to induce CYPs and to inhibit many CYPs, including those involved in steroid biosynthesis and metabolism such as aromatase (CYP19).[36,212] Inhibition of aromatase has been associated in part with reproductive toxicity.[213] Conazoles also can affect ecological systems based on their ability to induce CYPs or inhibit aromatase.[214,215] The complexity and pleiotropic effects of conazoles offer challenges to fully understanding their ability to alter critical pathways that lead to toxicity.

The use of omic information as a discovery tool, coupled with its integration with toxicological, molecular, biochemical and related findings is a powerful approach to understanding the process of chemically induced disease states. The use of global molecular profiling technologies is evolving as applied to the understanding of toxicity mechanisms. Moreover, the technologies themselves are evolving with the development of new high-throughput assays and high-throughput sequencing technologies and these will play a larger role in the future in furthering mechanistic understanding of chemically-induced toxicity. Finally, omics technology has great potential to characterize and possibly predict human toxicity. It allows for the molecular understanding of the behavior of cells, tissues, organs and fluids obtained from humans in response to toxicant or environmental stress. These technologies combined with targeted molecular, biochemical and biological measurements also have the potential to be incorporated into the development of predictive toxicology tools that could provide further insight into mechanisms of human toxicity as they can encompass multiple scales of complexity, from molecular to cellular and tissue to organismal levels.

Acknowledgements

I would like to thank Marie Nesnow for her assistance in preparing this manuscript.

References

1. E. F. Godefroi, J. Heeres, J. Van Cutsem and P. A. Janssen, The preparation and antimycotic properties of derivatives of 1-phenethylimidazole, *J. Med. Chem.*, 1969, **12**, 784–791.
2. D. W. Woolley, Some biological effects produced by benzimidazole and their reversal by purines, *J. Biol. Chem.*, 1944, **152**, 225–232.
3. R. A. Fromtling, Overview of medically important antifungal azole derivatives, *Clin. Microbiol. Rev.*, 1988, **1**, 187–217.
4. J. A. Maertens, History of the development of azole derivatives, *Clin. Microbiol. Infect.*, 2004, **10**(Suppl. 1), 1–10.
5. J. Heeres, L. Meerpoel and P. L. Conazoles, *Molecules*, 2010, **15**, 4129–4188.
6. J. E. Parker, A. G. Warrilow, C. L. Price, J. G. Mullins, D. E. Kelly and S. L. Kelly, Resistance to antifungals that target CYP51, *J. Chem. Biol.*, 2014, **7**, 143–161.
7. R. Becher and S. G. Wirsel, Fungal cytochrome P450 sterol 14alpha-demethylase (CYP51) and azole resistance in plant and human pathogens, *Appl. Microbiol. Biotechnol.*, 2012, **95**, 825–840.

8. G. I. Lepesheva and M. R. Waterman, Structural basis for conservation in the CYP51 family, *Biochim. Biophys. Acta*, 2011, **1814**, 88–93.
9. K. Shalchian-Tabrizi, M. A. Minge, M. Espelund, R. Orr, T. Ruden, K. S. Jakobsen and T. Cavalier-Smith, Multigene phylogeny of choanozoa and the origin of animals, *PLoS One*, 2008, **3**, e2098.
10. M. Spingola, L. Grate, D. Haussler and M. Ares Jr., Genome-wide bioinformatic and molecular analysis of introns in *Saccharomyces cerevisiae*, *RNA*, 1999, **5**, 221–234.
11. A. Muto, M. Hattori and M. Kanehisa, Analysis of common substructures of metabolic compounds within the different organism groups, *Genome Inf.*, 2007, **18**, 299–307.
12. L. W. Parks and W. M. Casey, Physiological implications of sterol biosynthesis in yeast, *Annu. Rev. Microbiol.*, 1995, **49**, 95–116.
13. E. L. Crockett, Cholesterol Function in Plasma Membranes from Ectotherms: Membrane-Specific Roles in Adaptation to Temperature, *Am. Zool.*, 1998, **38**, 291–304.
14. G. I. Lepesheva and M. R. Waterman, Sterol 14alpha-demethylase cytochrome P450 (CYP51), a P450 in all biological kingdoms, *Biochim. Biophys. Acta*, 2007, **1770**, 467–477.
15. G. I. Lepesheva and M. R. Waterman, CYP51–the omnipotent P450, *Mol. Cell. Endocrinol.*, 2004, **215**, 165–170.
16. Y. Aoyama, Y. Yoshida, Y. Sonoda and Y. Sato, Deformylation of 32-oxo-24,25-dihydrolanosterol by the purified cytochrome P-45014DM (lanosterol 14 alpha-demethylase) from yeast evidence confirming the intermediate step of lanosterol 14 alpha-demethylation, *J. Biol. Chem.*, 1989, **264**, 18502–18505.
17. N. Strushkevich, S. A. Usanov and H. W. Park, Structural basis of human CYP51 inhibition by antifungal azoles, *J. Mol. Biol.*, 2010, **397**, 1067–1078.
18. H. Vanden Bossche and P. Marichal, Mode of action of anti-Candida drugs: focus on terconazole and other ergosterol biosynthesis inhibitors, *Am. J. Obstet. Gynecol.*, 1991, **165**, 1193–1139.
19. Y. Aoyama, Y. Yoshida, Y. Sonoda and Y. Sato, The 3-hydroxy group of lanosterol is essential for orienting the substrate site of cytochrome P-450(14DM) (lanosterol 14 alpha-demethylase), *Biochim. Biophys. Acta*, 1989, **1006**, 209–213.
20. Y. Aoyama, Y. Yoshida, Y. Sonoda and Y. Sato, Role of the 8-double bond of lanosterol in the enzyme-substrate interaction of cytochrome P-450(14DM) (lanosterol 14 alpha-demethylase), *Biochim. Biophys. Acta*, 1989, **1001**, 196–200.
21. D. F. Lewis, A. Wiseman and M. H. Tarbit, Molecular modelling of lanosterol 14 alpha-demethylase (CYP51) from *Saccharomyces cerevisiae via* homology with CYP102, a unique bacterial cytochrome P450 isoform: quantitative structure-activity relationships (QSARs) within two related series of antifungal azole derivatives, *J. Enzyme Inhib.*, 1999, **14**, 175–192.

22. T. L. Poulos and A. J. Howard, Crystal structures of metyrapone- and phenylimidazole-inhibited complexes of cytochrome P-450cam, *Biochemistry*, 1987, **26**, 8165–8174.
23. L. M. Podust, T. L. Poulos and M. R. Waterman, Crystal structure of cytochrome P450 14alpha-sterol demethylase (CYP51) from *Mycobacterium tuberculosis* in complex with azole inhibitors, *Proc. Natl. Acad. Sci. U. S. A.*, 2001, **98**, 3068–3073.
24. A. G. Warrilow, C. M. Martel, J. E. Parker, N. Melo, D. C. Lamb, W. D. Nes, D. E. Kelly and S. L. Kelly, Azole binding properties of *Candida albicans* sterol 14-alpha demethylase (CaCYP51), *Antimicrob. Agents Chemother.*, 2010, **54**, 4235–4245.
25. M. J. Ronis, M. Ingelman-Sundberg and T. M. Badger, Induction, suppression and inhibition of multiple hepatic cytochrome P450 isozymes in the male rat and bobwhite quail (*Colinus virginianus*) by ergosterol biosynthesis inhibiting fungicides (EBIFs), *Biochem. Pharmacol.*, 1994, **48**, 1953–1965.
26. A. G. Warrilow, J. E. Parker, D. E. Kelly and S. L. Kelly, Azole affinity of sterol 14alpha-demethylase (CYP51) enzymes from *Candida albicans* and *Homo sapiens*, *Antimicrob. Agents Chemother.*, 2013, **57**, 1352–1360.
27. Y. Yoshida and Y. Aoyama, Interaction of azole antifungal agents with cytochrome P-45014DM purified from *Saccharomyces cerevisiae* microsomes, *Biochem. Pharmacol.*, 1987, **36**, 229–235.
28. C. A. Hitchcock, K. Dickinson, S. B. Brown, E. G. Evans and D. J. Adams, Interaction of azole antifungal antibiotics with cytochrome P-450-dependent 14 alpha-sterol demethylase purified from *Candida albicans*, *Biochem. J.*, 1990, **266**, 475–480.
29. K. Venkateswarlu, D. E. Kelly and S. L. Kelly, Characterization of *Saccharomyces cerevisiae* CYP51 and a CYP51 fusion protein with NADPH cytochrome P-450 oxidoreductase expressed in *Escherichia coli*, *Antimicrob. Agents Chemother.*, 1997, **41**, 776–780.
30. Y. Yoshida and Y. Aoyama, Stereoselective interaction of an azole antifungal agent with its target, lanosterol 14 alpha-demethylase (cytochrome P-45014DM): a model study with stereoisomers of triadimenol and purified cytochrome P-45014DM from yeast, *Chirality*, 1990, **2**, 10–15.
31. N. D. Lees, M. C. Broughton, D. Sanglard and M. Bard, Azole susceptibility and hyphal formation in a cytochrome P-450-deficient mutant of *Candida albicans*, *Antimicrob. Agents Chemother.*, 1990, **34**, 831–836.
32. I. J. Sud and D. S. Feingold, Heterogeneity of action of mechanisms among antimycotic imidazoles, *Antimicrob. Agents Chemother.*, 1981, **20**, 71–74.
33. I. J. Sud and D. S. Feingold, Mechanisms of action of the antimycotic imidazoles, *J. Invest. Dermatol.*, 1981, **76**, 438–441.
34. T. Niwa, Y. Imagawa and H. Yamazaki, Drug interactions between nine antifungal agents and drugs metabolized by human cytochromes P450, *Curr. Drug Metab.*, 2014, **15**, 651–679.

35. R. P. Dickinson, A. S. Bell, A. Hitchcock, S. Narayanaswami, S. J. Ray, K. Richardson and P. F. Troke, Novel antifungal 2-aryl-I-(IH-1,2,4-triazol-1-YL)butan-2-ol derivatives with high activity against *Aspergillus Fumigatus*, *Bioorg. Med. Chem. Lett.*, 1996, **6**, 2031–2036.
36. E. R. Trosken, M. Adamska, M. Arand, J. A. Zarn, C. Patten, W. Volkel and W. K. Lutz, Comparison of lanosterol-14 alpha-demethylase (CYP51) of human and *Candida albicans* for inhibition by different antifungal azoles, *Toxicology*, 2006, **228**, 24–32.
37. D. C. Lamb, D. E. Kelly, B. C. Baldwin, F. Gozzo, P. Boscott, W. G. Richards and S. L. Kelly, Differential inhibition of *Candida albicans* CYP51 with azole antifungal stereoisomers, *FEMS Microbiol. Lett.*, 1997, **149**, 25–30.
38. D. C. Lamb, D. E. Kelly, B. C. Baldwin and S. L. Kelly, Differential inhibition of human CYP3A4 and *Candida albicans* CYP51 with azole antifungal agents, *Chem. Biol. Interact.*, 2000, **125**, 165–175.
39. K. Venkatakrishnan, L. L. von Moltke and D. J. Greenblatt, Effects of the antifungal agents on oxidative drug metabolism: clinical relevance, *Clin. Pharmacokinet.*, 2000, **38**, 111–180.
40. W. Zhang, Y. Ramamoorthy, T. Kilicarslan, H. Nolte, R. F. Tyndale and E. M. Sellers, Inhibition of cytochromes P450 by antifungal imidazole derivatives, *Drug Metab. Dispos.*, 2002, **30**, 314–318.
41. K. Kozakai, Y. Yamada, M. Oshikata, T. Kawase, E. Suzuki, Y. Haramaki and H. Taniguchi, Cocktail-substrate approach-based high-throughput assay for evaluation of direct and time-dependent inhibition of multiple cytochrome P450 isoforms, *Drug Metab. Pharmacokinet.*, 2014, **29**, 198–207.
42. S. J. Baldwin, J. C. Bloomer, G. J. Smith, A. D. Ayrton, S. E. Clarke and R. J. Chenery, Ketoconazole and sulphaphenazole as the respective selective inhibitors of P4503A and 2C9, *Xenobiotica*, 1995, **25**, 261–270.
43. T. Niwa, S. Inoue-Yamamoto, T. Shiraga and A. Takagi, Effect of antifungal drugs on cytochrome P450 (CYP) 1A2, CYP2D6, and CYP2E1 activities in human liver microsomes, *Biol. Pharm. Bull.*, 2005, **28**, 1813–1816.
44. A. J. Draper, A. Madan and A. Parkinson, Inhibition of coumarin 7-hydroxylase activity in human liver microsomes, *Arch. Biochem. Biophys.*, 1997, **341**, 47–61.
45. R. L. Walsky, A. V. Astuccio and R. S. Obach, Evaluation of 227 drugs for in vitro inhibition of cytochrome P450 2B6, *J. Clin. Pharmacol.*, 2006, **46**, 1426–1438.
46. T. Niwa, T. Shiraga and A. Takagi, Effect of antifungal drugs on cytochrome P450 (CYP) 2C9, CYP2C19, and CYP3A4 activities in human liver microsomes, *Biol. Pharm. Bull.*, 2005, **28**, 1805–1808.
47. D. M. Stresser, A. P. Blanchard, S. D. Turner, J. C. Erve, A. A. Dandeneau, V. P. Miller and C. L. Crespi, Substrate-dependent modulation of CYP3A4 catalytic activity: analysis of 27 test compounds with four fluorometric substrates, *Drug Metab. Dispos.*, 2000, **28**, 1440–1448.

48. M. A. Gibbs, K. E. Thummel, D. D. Shen and K. L. Kunze, Inhibition of cytochrome P-450 3A (CYP3A) in human intestinal and liver microsomes: comparison of Ki values and impact of CYP3A5 expression, *Drug Metab. Dispos.*, 1999, **27**, 180–187.
49. N. Isoherranen, S. R. Ludington, R. C. Givens, J. K. Lamba, S. N. Pusek, E. C. Dees, D. K. Blough, K. Iwanaga, R. L. Hawke, E. G. Schuetz, P. B. Watkins, K. E. Thummel and M. F. Paine, The influence of CYP3A5 expression on the extent of hepatic CYP3A inhibition is substrate-dependent: an *in vitro-in vivo* evaluation, *Drug Metab. Dispos.*, 2008, **36**, 146–154.
50. T. H. Tran, L. L. Von Moltke, K. Venkatakrishnan, B. W. Granda, M. A. Gibbs, R. S. Obach, J. S. Harmatz and D. J. Greenblatt, Microsomal protein concentration modifies the apparent inhibitory potency of CYP3A inhibitors, *Drug Metab. Dispos.*, 2002, **30**, 1441–1445.
51. Y. Shirasaka, S. Y. Chang, M. F. Grubb, C. C. Peng, K. E. Thummel, N. Isoherranen and A. D. Rodrigues, Effect of CYP3A5 expression on the inhibition of CYP3A-catalyzed drug metabolism: impact on modeling CYP3A-mediated drug-drug interactions, *Drug Metab. Dispos.*, 2013, **41**, 1566–1574.
52. S. A. Kliewer, J. T. Moore, L. Wade, J. L. Staudinger, M. A. Watson, S. A. Jones, D. D. McKee, B. B. Oliver, T. M. Willson, R. H. Zetterstrom, T. Perlmann and J. M. Lehmann, An orphan nuclear receptor activated by pregnanes defines a novel steroid signaling pathway, *Cell*, 1998, **92**, 73–82.
53. H. Li and H. Wang, Activation of xenobiotic receptors: driving into the nucleus, *Expert Opin. Drug Metab. Toxicol.*, 2010, **6**, 409–426.
54. T. Kawamoto, T. Sueyoshi, I. Zelko, R. Moore, K. Washburn and M. Negishi, Phenobarbital-responsive nuclear translocation of the receptor CAR in induction of the CYP2B gene, *Mol. Cell. Biol.*, 1999, **19**, 6318–6322.
55. S. A. Kliewer, B. Goodwin and T. M. Willson, The nuclear pregnane X receptor: a key regulator of xenobiotic metabolism, *Endocr. Rev.*, 2002, **23**, 687–702.
56. F. Hosseinpour, R. Moore, M. Negishi and T. Sueyoshi, Serine 202 regulates the nuclear translocation of constitutive active/androstane receptor, *Mol. Pharmacol.*, 2006, **69**, 1095–1102.
57. Y. M. Wang, S. S. Ong, S. C. Chai and T. Chen, Role of CAR and PXR in xenobiotic sensing and metabolism, *Expert Opin. Drug Metab. Toxicol.*, 2012, **8**, 803–817.
58. J. W. Allen, D. C. Wolf, M. H. George, S. D. Hester, G. Sun, S. F. Thai, D. A. Delker, T. Moore, C. Jones, G. Nelson, B. C. Roop, S. Leavitt, E. Winkfield, W. O. Ward and S. Nesnow, Toxicity profiles in mice treated with hepatotumorigenic and non-hepatotumorigenic triazole conazole fungicides: Propiconazole, triadimefon, and myclobutanil, *Toxicol. Pathol.*, 2006, **34**, 853–862.
59. D. C. Wolf, J. W. Allen, M. H. George, S. D. Hester, G. Sun, T. Moore, S. F. Thai, D. Delker, E. Winkfield, S. Leavitt, G. Nelson, B. C. Roop,

C. Jones, J. Thibodeaux and S. Nesnow, Toxicity profiles in rats treated with tumorigenic and nontumorigenic triazole conazole fungicides: Propiconazole, triadimefon, and myclobutanil, *Toxicol. Pathol.*, 2006, **34**, 895–902.

60. Y. Kobayashi, M. Suzuki, N. Ohshiro, T. Sunagawa, T. Sasaki, S. Tokuyama, T. Yamamoto and T. Yoshida, Climbazole is a new potent inducer of rat hepatic cytochrome P450, *J. Toxicol. Sci.*, 2001, **26**, 141–150.
61. Y. Kobayashi, M. Suzuki, N. Ohshiro, T. Sunagawa, T. Sasaki, T. Oguro, S. Tokuyama, T. Yamamoto and T. Yoshida, Induction and inhibition of cytochrome P450 and drug-metabolizing enzymes by climbazole, *Biol. Pharm. Bull.*, 2002, **25**, 53–57.
62. R. C. Peffer, J. G. Moggs, T. Pastoor, R. A. Currie, J. Wright, G. Milburn, F. Waechter and I. Rusyn, Mouse liver effects of cyproconazole, a triazole fungicide: role of the constitutive androstane receptor, *Toxicol. Sci.*, 2007, **99**, 315–325.
63. D. R. Juberg, D. R. Mudra, G. A. Hazelton and A. Parkinson, The effect of fenbuconazole on cell proliferation and enzyme induction in the liver of female CD1 mice, *Toxicol. Appl. Pharmacol.*, 2006, **214**, 178–187.
64. G. Sun, S. F. Thai, G. R. Lambert, D. C. Wolf, D. B. Tully, A. K. Goetz, M. H. George, R. D. Grindstaff, D. J. Dix and S. Nesnow, Fluconazole-induced hepatic cytochrome P450 gene expression and enzymatic activities in rats and mice, *Toxicol. Lett.*, 2006, **164**, 44–53.
65. N. Muto, H. Hirai, T. Tanaka, N. Itoh and K. Tanaka, Induction and inhibition of cytochrome P450 isoforms by imazalil, a food contaminant, in mouse small intestine and liver, *Xenobiotica*, 1997, **27**, 1215–1223.
66. G. Sun, R. D. Grindstaff, S. F. Thai, G. R. Lambert, D. B. Tully, D. J. Dix and S. Nesnow, Induction of cytochrome P450 enzymes in rat liver by two conazoles, myclobutanil and triadimefon, *Xenobiotica*, 2007, **37**, 180–193.
67. P. J. Chen, W. T. Padgett, T. Moore, W. Winnik, G. R. Lambert, S. F. Thai, S. D. Hester and S. Nesnow, Three conazoles increase hepatic microsomal retinoic acid metabolism and decrease mouse hepatic retinoic acid levels *in vivo*, *Toxicol. Appl. Pharmacol.*, 2009, **234**, 143–155.
68. C. Leslie, G. F. Reidy and N. H. Stacey, The effects of propiconazole on hepatic xenobiotic biotransformation in the rat, *Biochem. Pharmacol.*, 1988, **37**, 4177–4181.
69. G. Sun, S. F. Thai, D. B. Tully, G. R. Lambert, A. K. Goetz, D. C. Wolf, D. J. Dix and S. Nesnow, Propiconazole-induced cytochrome P450 gene expression and enzymatic activities in rat and mouse liver, *Toxicol. Lett.*, 2005, **155**, 277–287.
70. I. Dewhurst and V. Dellarco, *International Programme on Chemical Safety Evaluations 2004, Part 2 Propiconazole*, 2004.
71. S. Hester, T. Moore, W. T. Padgett, L. Murphy, C. E. Wood and S. Nesnow, The hepatocarcinogenic conazoles: cyproconazole,

epoxiconazole, and propiconazole induce a common set of toxicological and transcriptional responses, *Toxicol. Sci.*, 2012, **127**, 54–65.
72. K. Tamura, K. Inoue, M. Takahashi, S. Matsuo, K. Irie, Y. Kodama, S. Ozawa, A. Nishikawa and M. Yoshida, Dose-response involvement of constitutive androstane receptor in mouse liver hypertrophy induced by triazole fungicides, *Toxicol. Lett.*, 2013, **221**, 47–56.
73. K. Tamura, K. Inoue, M. Takahashi, S. Matsuo, K. Irie, Y. Kodama, T. Gamo, S. Ozawa and M. Yoshida, Involvement of constitutive androstane receptor in liver hypertrophy and liver tumor development induced by triazole fungicides, *Food Chem. Toxicol.*, 2015, **78**, 86–95.
74. EPA, *Integrated Approaches to Testing and Assessment Strategy: Use of New Computational and Molecular Tools. US Environmental Protection Agency, Office of Pesticide Programs*, Washington, DC, 2011.
75. K. Oshida, N. Vasani, C. Jones, T. Moore, S. Hester, S. Nesnow, S. Auerbach, D. R. Geter, L. M. Aleksunes, R. S. Thomas, D. Applegate, C. D. Klaassen and J. C. Corton, Identification of chemical modulators of the constitutive activated receptor (CAR) in a gene expression compendium, *Nucl. Recept. Signaling*, 2015, **13**, e002.
76. R. A. Currie, R. C. Peffer, A. K. Goetz, C. J. Omiecinski and J. I. Goodman, Phenobarbital and propiconazole toxicogenomic profiles in mice show major similarities consistent with the key role that constitutive androstane receptor (CAR) activation plays in their mode of action, *Toxicology*, 2014, **321**, 80–88.
77. F. B. Kraemer and A. Pont, Inhibition of cholesterol synthesis by ketoconazole, *Am. J. Med.*, 1986, **80**, 616–622.
78. F. B. Kraemer and S. D. Spilman, Effects of ketoconazole on cholesterol synthesis, *J. Pharmacol. Exp. Ther.*, 1986, **238**, 905–911.
79. H. Gylling, H. Vanhanen and T. A. Miettinen, Effects of ketoconazole on cholesterol precursors and low density lipoprotein kinetics in hypercholesterolemia, *J. Lipid Res.*, 1993, **34**, 59–67.
80. T. A. Miettinen, Cholesterol metabolism during ketoconazole treatment in man, *J. Lipid Res.*, 1988, **29**, 43–51.
81. EFSA, Conclusion on the peer review of the pesticide risk assessment of the active substance tebuconazole, *EFSA J.*, 2014, **12**, 98.
82. EPA, *Uniconazole-P Human Health Risk Assessment for Proposed Uses on Fruiting Vegetables (except Cucurbits), Crop Group 8*, Report EPA-HQ-OPP-2007-1199-0004, US EPA OPPTS, Washington, DC, 2008.
83. T. E. Strandberg, R. S. Tilvis and T. A. Miettinen, Effects of ketoconazole on cholesterol synthesis and precursor concentrations in the rat liver, *Lipids*, 1987, **22**, 1020–1024.
84. EPA, Pesticide Registration Data Requirements, www2.epa.gov/pesticide-registration/data-requirements, 2015.
85. OEHHA, *Triazole Antifungal Agents*, Office of Environmental Health Hazard Assessment, CAL EPA 2011.
86. J. A. Ross and S. A. Leavitt, Analysis of the mutations induced by conazole fungicides *in vivo*, *Mutagenesis*, 2010, **25**, 231–234.

87. J. A. Ross, T. Moore and S. A. Leavitt, *In vivo* mutagenicity of conazole fungicides correlates with tumorigenicity, *Mutagenesis*, 2009, **24**, 149–152.
88. J. A. Ross, S. A. Leavitt, J. E. Schmid and G. B. Nelson, Quantitative changes in endogenous DNA adducts correlate with conazole in vivo mutagenicity and tumorigenicity, *Mutagenesis*, 2012, **27**, 541–549.
89. EFSA, *Conclusion on the peer review of the pesticide risk assessment of the active substance tetraconazole*, 2008.
90. INCHEM, *Monograph 641 Prochloraz (Pesticide residues in food: 1983 evaluations)*, 1983.
91. R. Hasegawa and N. Ito, Liver medium-term bioassay in rats for screening of carcinogens and modifying factors in hepatocarcinogenesis, *Food Chem. Toxicol.*, 1992, **30**, 979–992.
92. NTP, *Amitrole in 13th Report on Carcinogens (RoC)*, NTP, Washington, DC, 2014.
93. INCHEM, *Monograph 946 Bitertanol (JMPR Evaluations Part II Toxicological)*, 1998.
94. OPPTS, *Bromuconazole – Report of the Hazard Identification Assessment Review Committee*, US EPA OPPTS, Washington, DC, 2002.
95. Syngenta, *Response to OEHHA Notice Regarding Prioritization: Chemicals for Consultation by The Carcinogen Identification Committee*, Syngenta Crop Protection, Greensboro, NC, 2011.
96. FAO, *Pesticide residues in food – 2007 Toxicological evaluations*, Geneva, 2007.
97. S. Hosokawa, J. Nakamura, S. Ito, M. Murakami, M. Ineyama, K. Yoshioka, T. Yamada, T. Seki, M. Matsuo and H. Yamada, Hormonal disregulation mechanism in the rat thyroid tumor induced by diniconazole, *J. Toxicol. Sci.*, 1993, **18**, 57–67.
98. EPA, *Epoxiconazole Pesticide Tolerance*, US EPA OPPTS, Washington, DC, 2006.
99. INCHEM, *Monograph 930 Fenbuconazole: Part II Toxicological and Environmental.*, 1997.
100. EFSA, Conclusion on the peer review of the pesticide risk assessment of the active substance fluquinconazole, *EFSA J.*, 2011, **9**, 112.
101. INCHEM, *Monograph 810 Hexaconazole*, INCHEM, Geneva, 1990.
102. EPA, *R.E.D. Facts: IMAZALIL*, Washington, DC, 2005.
103. EFSA, Conclusion on the peer review of the pesticide risk assessment of the active substance ipconazole, *EFSA J.*, 2013, **11**, 76.
104. EPA, *Pesticide Fact Sheet: Metconazole*, EPA OPPTS, Washington, DC, 2006.
105. INCHEM, *Monograph 847. Myclobutanil (Pesticide residues in food: 1992 evaluations Part II Toxicology)* 1992.
106. INCHEM, *Monograph 778. Paclobutrazol (Pesticide residues in food: 1988 evaluations: Part II Toxicology)* 1988.
107. INCHEM, *Monograph 848 Penconazole (Pesticide residues in food; 1992 evaluations Part II Toxicology)*, INCHEM, 1992.

108. EPA, *IRIS Prochloraz (CASRN 67747-09-5)*, Washington, DC, 1989.
109. INCHEM, *Monograph 768. Propiconazole Part II Toxicology Pesticide residues in food: 1987 Evaluations.*, 1987.
110. EC, *Draft Assessment Report Tebuconazole Volume 3. Annex B.6: Toxicology and metabolism*, EU Review Programme, 2006.
111. INCHEM, *Monograph 566. Triadimefon (Pesticide residues in food:1981 evaluations)*, 1981.
112. EPA, *Triadimenol: HED Chapter of the Tolerance Reassessment Eligibility Decision (TRED) Document (Revised)*, Washington, DC, 2006.
113. JMPR, Triflumizole, *Joint FAO/WHO Meeting on Pesticide Residues*, 2013.
114. ORNL, *Data Evaluation Report RPA 400727 Triticonazole Chronic Toxicity/Oncogenicity Feeding Mouse*, Oak Ridge National Laboratory, Oak Ridge, TN, 2000.
115. ORNL, *Data Evaluation Report RPA 400727 Triticonazole Chronic Toxicity/Oncogenicity Oral study – Rat*, Oak Ridge National Laboratory, Oak Ridge, TN, 2000.
116. W. O. Ward, D. A. Delker, S. D. Hester, S. F. Thai, D. C. Wolf, J. W. Allen and S. Nesnow, Transcriptional profiles in liver from mice treated with hepatotumorigenic and nonhepatotumorigenic triazole conazole fungicides: Propiconazole, triadimefon, and myclobutanil, *Toxicol. Pathol.*, 2006, **34**, 863–878.
117. S. Nesnow, W. Ward, T. Moore, H. Ren and S. D. Hester, Discrimination of tumorigenic triazole conazoles from phenobarbital by transcriptional analyses of mouse liver gene expression, *Toxicol. Sci.*, 2009, **110**, 68–83.
118. H. Murakami, N. D. Sanderson, P. Nagy, P. A. Marino, G. Merlino and S. S. Thorgeirsson, Transgenic mouse model for synergistic effects of nuclear oncogenes and growth factors in tumorigenesis: interaction of c-myc and transforming growth factor alpha in hepatic oncogenesis, *Cancer Res.*, 1993, **53**, 1719–1723.
119. Z. Chen, T. B. Gibson, F. Robinson, L. Silvestro, G. Pearson, B. Xu, A. Wright, C. Vanderbilt and M. H. Cobb, MAP kinases, *Chem. Rev.*, 2001, **101**, 2449–2476.
120. S. H. Yang, A. D. Sharrocks and A. J. Whitmarsh, Transcriptional regulation by the MAP kinase signaling cascades, *Gene*, 2003, **320**, 3–21.
121. E. Santoni-Rugiu, M. R. Jensen and S. S. Thorgeirsson, Disruption of the pRb/E2F pathway and inhibition of apoptosis are major oncogenic events in liver constitutively expressing c-myc and transforming growth factor alpha, *Cancer Res.*, 1998, **58**, 123–134.
122. S. Zellmer, W. Schmidt-Heck, P. Godoy, H. Weng, C. Meyer, T. Lehmann, T. Sparna, W. Schormann, S. Hammad, C. Kreutz, J. Timmer, F. von Weizsacker, P. A. Thurmann, I. Merfort, R. Guthke, S. Dooley, J. G. Hengstler and R. Gebhardt, Transcription factors ETF, E2F, and SP-1 are involved in cytokine-independent proliferation of murine hepatocytes, *Hepatology*, 2010, **52**, 2127–2136.

123. V. S. Bhat, S. D. Hester, S. Nesnow and D. A. Eastmond, Concordance of transcriptional and apical benchmark dose levels for conazole-induced liver effects in mice, *Toxicol. Sci.*, 2013, **136**, 205–215.
124. P. A. Ortiz, M. E. Bruno, T. Moore, S. Nesnow, W. Winnik and Y. Ge, Proteomic analysis of propiconazole responses in mouse liver: comparison of genomic and proteomic profiles, *J. Proteome Res.*, 2010, **9**, 1268–1278.
125. S. Nesnow, W. T. Padgett and T. Moore, Propiconazole induces alterations in the hepatic metabolome of mice: relevance to propiconazole-induced hepatocarcinogenesis, *Toxicol. Sci.*, 2011, **120**, 297–309.
126. F. M. Vaz and R. J. Wanders, Carnitine biosynthesis in mammals, *Biochem. J.*, 2002, **361**, 417–429.
127. R. J. Gillies, I. Robey and R. A. Gatenby, Causes and consequences of increased glucose metabolism of cancers, *J. Nucl. Med.*, 2008, **49**(Suppl. 2), 24S–42S.
128. M. G. Vander Heiden, L. C. Cantley and C. B. Thompson, Understanding the Warburg effect: the metabolic requirements of cell proliferation, *Science*, 2009, **324**, 1029–1033.
129. Y. Murakami, S. Matsufuji, T. Kameji, S. Hayashi, K. Igarashi, T. Tamura, K. Tanaka and A. Ichihara, Ornithine decarboxylase is degraded by the 26S proteasome without ubiquitination, *Nature*, 1992, **360**, 597–599.
130. Y. Murakami, K. Tanaka, S. Matsufuji, Y. Miyazaki and S. Hayashi, Antizyme, a protein induced by polyamines, accelerates the degradation of ornithine decarboxylase in Chinese-hamster ovary-cell extracts, *Biochem. J.*, 1992, **283**(Pt 3), 661–664.
131. H. C. Pitot, Adventures in hepatocarcinogenesis, *Annu. Rev. Pathol.*, 2007, **2**, 1–29.
132. T. Thomas and T. J. Thomas, Polyamine metabolism and cancer, *J. Cell. Mol. Med.*, 2003, **7**, 113–126.
133. T. P. Forshell, S. Rimpi and J. A. Nilsson, Chemoprevention of B-cell lymphomas by inhibition of the Myc target spermidine synthase, *Cancer Prev. Res.*, 2010, **3**, 140–147.
134. E. W. Gerner, Cancer chemoprevention locks onto a new polyamine metabolic target, *Cancer Prev. Res.*, 2010, **3**, 125–127.
135. A. Sreekumar, L. M. Poisson, T. M. Rajendiran, A. P. Khan, Q. Cao, J. Yu, B. Laxman, R. Mehra, R. J. Lonigro, Y. Li, M. K. Nyati, A. Ahsan, S. Kalyana-Sundaram, B. Han, X. Cao, J. Byun, G. S. Omenn, D. Ghosh, S. Pennathur, D. C. Alexander, A. Berger, J. R. Shuster, J. T. Wei, S. Varambally, C. Beecher and A. M. Chinnaiyan, Metabolomic profiles delineate potential role for sarcosine in prostate cancer progression, *Nature*, 2009, **457**, 910–914.
136. S. O. Lim, S. J. Park, W. Kim, S. G. Park, H. J. Kim, Y. I. Kim, T. S. Sohn, J. H. Noh and G. Jung, Proteome analysis of hepatocellular carcinoma, *Biochem. Biophys. Res. Commun.*, 2002, **291**, 1031–1037.

137. E. Borek, B. S. Baliga, C. W. Gehrke, C. W. Kuo, S. Belman, W. Troll and T. P. Waalkes, High turnover rate of transfer RNA in tumor tissue, *Cancer Res.*, 1977, **37**, 3362–3366.
138. A. Seidel, S. Brunner, P. Seidel, G. I. Fritz and O. Herbarth, Modified nucleosides: an accurate tumour marker for clinical diagnosis of cancer, early detection and therapy control, *Br. J. Cancer*, 2006, **94**, 1726–1733.
139. S. Tamura, H. Fujioka, T. Nakano, T. Hada and K. Higashino, Serum pseudouridine as a biochemical marker in small cell lung cancer, *Cancer Res.*, 1987, **47**, 6138–6141.
140. S. D. Hester, D. C. Wolf, S. Nesnow and S. F. Thai, Transcriptional profiles in liver from rats treated with tumorigenic and non-tumorigenic triazole conazole fungicides: Propiconazole, triadimefon, and myclobutanil, *Toxicol. Pathol.*, 2006, **34**, 879–894.
141. S. D. Hester and S. Nesnow, Transcriptional responses in thyroid tissues from rats treated with a tumorigenic and a non-tumorigenic triazole conazole fungicide, *Toxicol. Appl. Pharmacol.*, 2008, **227**, 357–369.
142. S. Nesnow, Integration of toxicological approaches with "omic" and related technologies to elucidate mechanisms of carcinogenic action: propiconazole, an example, *Cancer Lett.*, 2013, **334**, 20–27.
143. C. Sonich-Mullin, R. Fielder, J. Wiltse, K. Baetcke, J. Dempsey, P. Fenner-Crisp, D. Grant, M. Hartley, A. Knaap, D. Kroese, I. Mangelsdorf, E. Meek, J. M. Rice and M. Younes, IPCS conceptual framework for evaluating a mode of action for chemical carcinogenesis, *Regul. Toxicol. Pharmacol.*, 2001, **34**, 146–152.
144. P.-J. Chen, T. Moore and S. Nesnow, Cytotoxic effects of propiconazole and its metabolites in mouse and human hepatoma cells and primary mouse hepatocytes, *Toxicol. In Vitro*, 2008, **22**, 1476–1483.
145. J. M. Maglich, C. M. Stoltz, B. Goodwin, D. Hawkins-Brown, J. T. Moore and S. A. Kliewer, Nuclear pregnane x receptor and constitutive androstane receptor regulate overlapping but distinct sets of genes involved in xenobiotic detoxification, *Mol. Pharmacol.*, 2002, **62**, 638–646.
146. G. Lemaire, W. Mnif, J. M. Pascussi, A. Pillon, F. Rabenoelina, H. Fenet, E. Gomez, C. Casellas, J. C. Nicolas, V. Cavailles, M. J. Duchesne and P. Balaguer, Identification of new human pregnane X receptor ligands among pesticides using a stable reporter cell system, *Toxicol. Sci.*, 2006, **91**, 501–509.
147. M. Hafner, T. Rezen and D. Rozman, Regulation of hepatic cytochromes p450 by lipids and cholesterol, *Curr. Drug Metab.*, 2011, **12**, 173–185.
148. L. Johansson, M. Rudling, T. S. Scanlan, T. Lundasen, P. Webb, J. Baxter, B. Angelin and P. Parini, Selective thyroid receptor modulation by GC-1 reduces serum lipids and stimulates steps of reverse cholesterol transport in euthyroid mice, *Proc. Natl. Acad. Sci. U. S. A.*, 2005, **102**, 10297–10302.

149. A. del Castillo-Olivares and G. Gil, Role of FXR and FTF in bile acid-mediated suppression of cholesterol 7alpha-hydroxylase transcription, *Nucleic Acids Res.*, 2000, **28**, 3587–3593.
150. E. F. Johnson, M. H. Hsu, U. Savas and K. J. Griffin, Regulation of P450 4A expression by peroxisome proliferator activated receptors, *Toxicology*, 2002, **181–182**, 203–206.
151. M. Ingelman-Sundberg and A. L. Hagbjork, On the significance of the cytochrome P-450-dependent hydroxyl radical-mediated oxygenation mechanism, *Xenobiotica*, 1982, **12**, 673–686.
152. S. C. Serron, N. Dwivedi and W. L. Backes, Ethylbenzene induces microsomal oxygen free radical generation: antibody-directed characterization of the responsible cytochrome P450 enzymes, *Toxicol. Appl. Pharmacol.*, 2000, **164**, 305–311.
153. P. J. Vuchetich, D. Bagchi, M. Bagchi, E. A. Hassoun, L. Tang and S. J. Stohs, Naphthalene-induced oxidative stress in rats and the protective effects of vitamin E succinate, *Free Radical Biol. Med.*, 1996, **21**, 577–590.
154. R. M. Green, N. J. Hodges, J. K. Chipman, M. R. O'Donovan and M. Graham, Reactive oxygen species from the uncoupling of human cytochrome P450 1B1 may contribute to the carcinogenicity of dioxin-like polychlorinated biphenyls, *Mutagenesis*, 2008, **23**, 457–463.
155. J. J. Schlezinger, W. D. Struntz, J. V. Goldstone and J. J. Stegeman, Uncoupling of cytochrome P450 1A and stimulation of reactive oxygen species production by co-planar polychlorinated biphenyl congeners, *Aquat. Toxicol.*, 2006, **77**, 422–432.
156. J. J. Schlezinger, R. D. White and J. J. Stegeman, Oxidative inactivation of cytochrome P-450 1A (CYP1A) stimulated by 3,3′,4,4′-tetrachlorobiphenyl: production of reactive oxygen by vertebrate CYP1As, *Mol. Pharmacol.*, 1999, **56**, 588–597.
157. A. A. Caro and A. I. Cederbaum, Oxidative stress, toxicology, and pharmacology of CYP2E1, *Annu. Rev. Pharmacol. Toxicol.*, 2004, **44**, 27–42.
158. G. Ekstrom and M. Ingelman-Sundberg, Mechanisms of lipid peroxidation dependent upon cytochrome P-450 LM2, *Eur. J. Biochem.*, 1986, **158**, 195–201.
159. M. Dostalek, J. D. Brooks, K. D. Hardy, G. L. Milne, M. M. Moore, S. Sharma, J. D. Morrow and F. P. Guengerich, *In vivo* oxidative damage in rats is associated with barbiturate response but not other cytochrome P450 inducers, *Mol. Pharmacol.*, 2007, **72**, 1419–1424.
160. M. Dostalek, K. D. Hardy, G. L. Milne, J. D. Morrow, C. Chen, F. J. Gonzalez, J. Gu, X. Ding, D. A. Johnson, J. A. Johnson, M. V. Martin and F. P. Guengerich, Development of oxidative stress by cytochrome P450 induction in rodents is selective for barbiturates and related to loss of pyridine nucleotide-dependent protective systems, *J. Biol. Chem.*, 2008, **283**, 17147–17157.
161. M. Strolin-Benedetti, G. Brogin, M. Bani, F. Oesch and J. G. Hengstler, Association of cytochrome P450 induction with oxidative stress *in vivo*

as evidenced by 3-hydroxylation of salicylate, *Xenobiotica*, 1999, **29**, 1171–1180.
162. Y. Sai, R. Dai, T. J. Yang, K. W. Krausz, F. J. Gonzalez, H. V. Gelboin and M. Shou, Assessment of specificity of eight chemical inhibitors using cDNA-expressed cytochromes P450, *Xenobiotica*, 2000, **30**, 327–343.
163. Y. Ishihara, D. Shiba and N. Shimamoto, Enhancement of DMNQ-induced hepatocyte toxicity by cytochrome P450 inhibition, *Toxicol. Appl. Pharmacol.*, 2006, **214**, 109–117.
164. S. Imaoka, M. Osada, Y. Minamiyama, T. Yukimura, S. Toyokuni, S. Takemura, T. Hiroi and Y. Funae, Role of phenobarbital-inducible cytochrome P450s as a source of active oxygen species in DNA-oxidation, *Cancer Lett.*, 2004, **203**, 117–125.
165. S. Nesnow, R. D. Grindstaff, G. Lambert, W. T. Padgett, M. Bruno, Y. Ge, P. J. Chen, C. E. Wood and L. Murphy, Propiconazole increases reactive oxygen species levels in mouse hepatic cells in culture and in mouse liver by a cytochrome P450 enzyme mediated process, *Chem. Biol. Interact.*, 2011, **194**, 79–89.
166. P. Kotokorpi, C. Gardmo, C. S. Nystrom and A. Mode, Activation of the glucocorticoid receptor or liver X receptors interferes with growth hormone-induced akr1b7 gene expression in rat hepatocytes, *Endocrinology*, 2004, **145**, 5704–5713.
167. A. M. Lefrancois-Martinez, C. Tournaire, A. Martinez, M. Berger, S. Daoudal, D. Tritsch, G. Veyssiere and C. Jean, Product of side-chain cleavage of cholesterol, isocaproaldehyde, is an endogenous specific substrate of mouse vas deferens protein, an aldose reductase-like protein in adrenocortical cells, *J. Biol. Chem.*, 1999, **274**, 32875–32880.
168. M. Bruno, T. Moore, S. Nesnow and Y. Ge, Protein Carbonyl Formation in Response to Propiconazole-Induced Oxidative Stress, *J. Proteome Res.*, 2009, **8**, 2070–2078.
169. R. De Bont and N. van Larebeke, Endogenous DNA damage in humans: a review of quantitative data, *Mutagenesis*, 2004, **19**, 169–185.
170. J. A. Swenberg, K. Lu, B. C. Moeller, L. Gao, P. B. Upton, J. Nakamura and T. B. Starr, Endogenous versus exogenous DNA adducts: their role in carcinogenesis, epidemiology, and risk assessment, *Toxicol. Sci.*, 2011, **120**(Suppl. 1), S130–145.
171. J. Marill, T. Cresteil, M. Lanotte and G. G. Chabot, Identification of human cytochrome P450s involved in the formation of all-*trans*-retinoic acid principal metabolites, *Mol. Pharmacol.*, 2000, **58**, 1341–1348.
172. J. Marill, N. Idres, C. C. Capron, E. Nguyen and G. G. Chabot, Retinoic acid metabolism and mechanism of action: a review, *Curr. Drug Metab.*, 2003, **4**, 1–10.
173. C. Teixeira and M. A. Pratt, CDK2 is a target for retinoic acid-mediated growth inhibition in MCF-7 human breast cancer cells, *Mol. Endocrinol.*, 1997, **11**, 1191–1202.

174. C. Liu, J. Chung, H. K. Seitz, R. M. Russell and X. D. Wang, Chlormethiazole treatment prevents reduced hepatic vitamin A levels in ethanol-fed rats, *Alcohol.: Clin. Exp. Res.*, 2002, **26**, 1703–1709.
175. X. D. Wang, Retinoids and alcohol-related carcinogenesis, *J. Nutr.*, 2003, **133**, 287S–290S.
176. J. Chung, P. R. Chavez, R. M. Russell and X. D. Wang, Retinoic acid inhibits hepatic Jun N-terminal kinase-dependent signaling pathway in ethanol-fed rats, *Oncogene*, 2002, **21**, 1539–1547.
177. L. Murphy, T. Moore and S. Nesnow, Reversible effect of all-trans-retinoic acid on AML12 hepatocyte proliferation and cell cycle progression [abstract], presented in part at the 101st Annual Meeting of the American Association of Cancer Research, Washington, DC, April 21, 2010, 2010.
178. M. S. Brown and J. L. Goldstein, Multivalent feedback regulation of HMG CoA reductase, a control mechanism coordinating isoprenoid synthesis and cell growth, *J. Lipid Res.*, 1980, **21**, 505–517.
179. J. L. Goldstein and M. S. Brown, Regulation of the mevalonate pathway, *Nature*, 1990, **343**, 425–430.
180. R. E. Duncan, A. El-Sohemy and M. C. Archer, Mevalonate promotes the growth of tumors derived from human cancer cells in vivo and stimulates proliferation in vitro with enhanced cyclin-dependent kinase-2 activity, *J. Biol. Chem.*, 2004, **279**, 33079–33084.
181. K. Pruitt and C. J. Der, Ras and Rho regulation of the cell cycle and oncogenesis, *Cancer Lett.*, 2001, **171**, 1–10.
182. L. A. Murphy, T. Moore and S. Nesnow, Propiconazole-enhanced hepatic cell proliferation is associated with dysregulation of the cholesterol biosynthesis pathway leading to activation of Erk1/2 through Ras farnesylation, *Toxicol. Appl. Pharmacol.*, 2012, **260**, 146–154.
183. M. K. Mateyak, A. J. Obaya and J. M. Sedivy, c-Myc regulates cyclin D-Cdk4 and -Cdk6 activity but affects cell cycle progression at multiple independent points, *Mol. Cell. Biol.*, 1999, **19**, 4672–4683.
184. J. E. Klaunig and L. M. Kamendulis, The role of oxidative stress in carcinogenesis, *Annu. Rev. Pharmacol. Toxicol.*, 2004, **44**, 239–267.
185. H. C. Pitot, Y. P. Dragan, J. Teeguarden, S. Hsia and H. Campbell, Quantitation of multistage carcinogenesis in rat liver, *Toxicol. Pathol.*, 1996, **24**, 119–128.
186. S. L. Burgess-Herbert and S. Y. Euling, Use of comparative genomics approaches to characterize interspecies differences in response to environmental chemicals: Challenges, opportunities, and research needs, *Toxicol. Appl. Pharmacol.*, 2013, **271**, 372–385.
187. S. Y. Euling, L. D. White, A. S. Kim, B. Sen, V. S. Wilson, C. Keshava, N. Keshava, S. Hester, A. M. Ovacik, M. G. Ierapetritou, I. P. Androulakis and K. W. Gaido, Use of genomic data in risk assessment case study: II. Evaluation of the dibutyl phthalate toxicogenomic data set, *Toxicol. Appl. Pharmacol.*, 2013, **271**, 349–362.
188. S. Shindo, S. Numazawa and T. Yoshida, A physiological role of AMP-activated protein kinase in phenobarbital-mediated constitutive

androstane receptor activation and CYP2B induction, *Biochem. J.*, 2007, **401**, 735–741.
189. M. D. Waters, M. Jackson and I. Lea, Characterizing and predicting carcinogenicity and mode of action using conventional and toxicogenomics methods, *Mutat. Res.*, 2010, **705**, 184–200.
190. V. S. Wilson, N. Keshava, S. Hester, D. Segal, W. Chiu, C. M. Thompson and S. Y. Euling, Utilizing toxicogenomic data to understand chemical mechanism of action in risk assessment, *Toxicol. Appl. Pharmacol.*, 2013, **271**, 299–308.
191. S. M. Ihlaseh, K. A. Bailey, S. D. Hester, C. Jones, H. Ren, A. P. Cardoso, M. L. Oliveira, D. C. Wolf and J. L. de Camargo, Transcriptional profile of diuron-induced toxicity on the urinary bladder of male Wistar rats to inform mode of action, *Toxicol. Sci.*, 2011, **122**, 330–338.
192. B. S. Shane, E. Zeiger, W. W. Piegorsch, E. D. Booth, J. I. Goodman and R. C. Peffer, Re-evaluation of the Big Blue(R) mouse assay of propiconazole suggests lack of mutagenicity, *Environ. Mol. Mutagen.*, 2012, **53**, 1–9.
193. J. A. Ross and S. A. Leavitt, Response to commentary: Re-evaluation of the Big Blue(R) mouse assay of propiconazole suggests lack of mutagenicity, *Environ. Mol. Mutagen.*, 2012, **53**, 574–577.
194. D. M. Parkin, F. Bray, J. Ferlay and P. Pisani, Global cancer statistics, 2002, *CA Cancer J. Clin.*, 2005, **55**, 74–108.
195. S. E. Peychal, A. Bilger, H. C. Pitot and N. R. Drinkwater, Predominant modifier of extreme liver cancer susceptibility in C57BR/cdJ female mice localized to 6 Mb on chromosome 17, *Carcinogenesis*, 2009, **30**, 879–885.
196. T. E. Akiyama and F. J. Gonzalez, Regulation of P450 genes by liver-enriched transcription factors and nuclear receptors, *Biochim. Biophys. Acta*, 2003, **1619**, 223–234.
197. F. P. Guengerich, Human cytochrome P450 enzymes, in *Cytochrome P450: Structure, Mechanisms, and Biochemistry*, ed. P. R. Ortiz de Montellano, Plenum, New York, 1995, pp. 473–575.
198. S. R. Faucette, H. Wang, G. A. Hamilton, S. L. Jolley, D. Gilbert, C. Lindley, B. Yan, M. Negishi and E. L. LeCluyse, Regulation of CYP2B6 in primary human hepatocytes by prototypical inducers, *Drug Metab. Dispos.*, 2004, **32**, 348–358.
199. R. C. Zangar, N. Bollinger, T. J. Weber, R. M. Tan, L. M. Markillie and N. J. Karin, Reactive oxygen species alter autocrine and paracrine signaling, *Free Radical Biol. Med.*, 2011, **51**, 2041–2047.
200. T. Wang, X. Ma, K. W. Krausz, J. R. Idle and F. J. Gonzalez, Role of pregnane X receptor in control of all-trans retinoic acid (ATRA) metabolism and its potential contribution to ATRA resistance, *J. Pharmacol. Exp. Ther.*, 2008, **324**, 674–684.
201. J. Muindi, S. R. Frankel, W. H. Miller Jr., A. Jakubowski, D. A. Scheinberg, C. W. Young, E. Dmitrovsky and R. P. Warrell Jr., Continuous treatment with all-trans retinoic acid causes a progressive reduction in plasma drug concentrations: implications for relapse and

retinoid "resistance" in patients with acute promyelocytic leukemia, *Blood*, 1992, **79**, 299–303.
202. J. R. Muindi, S. R. Frankel, C. Huselton, F. DeGrazia, W. A. Garland, C. W. Young and R. P. Warrell Jr., Clinical pharmacology of oral all-*trans* retinoic acid in patients with acute promyelocytic leukemia, *Cancer Res.*, 1992, **52**, 2138–2142.
203. M. Osmak, Statins and cancer: current and future prospects, *Cancer Lett.*, 2012, **324**, 1–12.
204. B. Relja, F. Meder, K. Wilhelm, D. Henrich, I. Marzi and M. Lehnert, Simvastatin inhibits cell growth and induces apoptosis and G0/G1 cell cycle arrest in hepatic cancer cells, *Int. J. Mol. Med.*, 2010, **26**, 735–741.
205. IARC, Phenobarbital and its sodium salt, in *IARC Monographs, Some Thyrotropic Agents*, IARC Press, Lyon, France, 2001, vol. 79, pp. 161–288.
206. M. P. Holsapple, H. C. Pitot, S. M. Cohen, A. R. Boobis, J. E. Klaunig, T. Pastoor, V. L. Dellarco and Y. P. Dragan, Mode of action in relevance of rodent liver tumors to human cancer risk, *Toxicol. Sci.*, 2006, **89**, 51–56.
207. D. S. Ory, Nuclear receptor signaling in the control of cholesterol homeostasis: have the orphans found a home?, *Circ. Res.*, 2004, **95**, 660–670.
208. D. R. Geter, V. S. Bhat, B. B. Gollapudi, R. Sura and S. D. Hester, Dose-response modeling of early molecular and cellular key events in the CAR-mediated hepatocarcinogenesis pathway, *Toxicol. Sci.*, 2014, **138**, 425–445.
209. T. Nikolaos, G. Stylianos, N. Chryssoula, P. Irini, M. Christos, T. Dimitrios, P. Konstantinos and T. Antonis, The effect of long-term antiepileptic treatment on serum cholesterol (TC, HDL, LDL) and triglyceride levels in adult epileptic patients on monotherapy, *Med. Sci. Monit*, 2004, **10**, MT50–52.
210. E. P. Calandre, B. Sinues Porta and D. Garcia de la Calzada, The effect of chronic phenytoin treatment on serum lipid profile in adult epileptic patients, *Epilepsia*, 1992, **33**, 154–157.
211. C. R. Elcombe, R. C. Peffer, D. C. Wolf, J. Bailey, R. Bars, D. Bell, R. C. Cattley, S. S. Ferguson, D. Geter, A. Goetz, J. I. Goodman, S. Hester, A. Jacobs, C. J. Omiecinski, R. Schoeny, W. Xie and B. G. Lake, Mode of action and human relevance analysis for nuclear receptor-mediated liver toxicity: A case study with phenobarbital as a model constitutive androstane receptor (CAR) activator, *Crit. Rev. Toxicol.*, 2014, **44**, 64–82.
212. A. M. Vinggaard, C. Hnida, V. Breinholt and J. C. Larsen, Screening of selected pesticides for inhibition of CYP19 aromatase activity *in vitro*, *Toxicol. In Vitro*, 2000, **14**, 227–234.
213. J. C. Rockett, M. G. Narotsky, K. E. Thompson, I. Thillainadarajah, C. R. Blystone, A. K. Goetz, H. Ren, D. S. Best, R. N. Murrell, H. P. Nichols, J. E. Schmid, D. C. Wolf and D. J. Dix, Effect of conazole

fungicides on reproductive development in the female rat, *Reprod. Toxicol.*, 2006, **22**, 647–658.

214. J. E. Chambers, H. Greim, R. J. Kendall, H. Segner, R. M. Sharpe and G. Van Der Kraak, Human and ecological risk assessment of a crop protection chemical: a case study with the azole fungicide epoxiconazole, *Crit. Rev. Toxicol.*, 2014, **44**, 176–210.
215. C. H. Lin, P. H. Chou and P. J. Chen, Two azole fungicides (carcinogenic triadimefon and non-carcinogenic myclobutanil) exhibit different hepatic cytochrome P450 activities in medaka fish, *J. Hazard. Mater.*, 2014, **277**, 150–158.

CHAPTER 11

Application of Transcriptomics in Exposed Human Populations: Benzene as an Example

CLIONA M. McHALE,* MARTYN T. SMITH AND LUOPING ZHANG

Division of Environmental Health Sciences, Genes and Environment Laboratory, School of Public Health, University of California, Berkeley, California, USA
*Email: cmchale@berkeley.edu

11.1 Application of Toxicogenomics in Occupational Benzene Exposure

Benzene is a major industrial chemical and is also an extensive environmental contaminant due to its presence in traffic exhaust and cigarette smoke.[1,2] Benzene induces myelodysplastic syndrome (MDS) and acute myeloid leukemia (AML),[3] and probably causes non-Hodgkin lymphoma[4] and related neoplasms including acute lymphocytic leukemia (ALL), chronic lymphocytic leukemia and multiple myeloma,[5] as well as childhood ALL and AML.[6–8] It has long been known to induce bone marrow toxicity and decrease the numbers of circulating blood cells,[8–10] resulting in pancytopenia and aplastic anemia at high exposures.[11,12] Benzene is also associated with

Issues in Toxicology No. 28
Toxicogenomics in Predictive Carcinogenicity
Edited by Michael D. Waters and Russell S. Thomas

Published by the Royal Society of Chemistry, www.rsc.org

adverse non-cancer effects on the reproductive, immune, nervous, endocrine, cardiovascular, and respiratory systems.[13]

Despite being studied extensively for many years, the underlying mechanisms of benzene toxicity are unclear and it appears that the current permissible exposure limit of 1 ppm (8 h time-weighted average, TWA) set by the Occupational Safety and Health Administration,[14] is inadequate for worker protection. In recent years, toxicogenomic approaches, particularly transcriptomics, in benzene-exposed human populations have identified gene expression biomarkers of exposure/early effect and expanded our understanding of potential mechanisms of toxicity at a range of exposures including those below 1 ppm. Previously, we reviewed the application of toxicogenomics and transcriptomics in molecular epidemiology studies, including our studies of benzene exposure.[9,15–17] Here, we discuss our recent

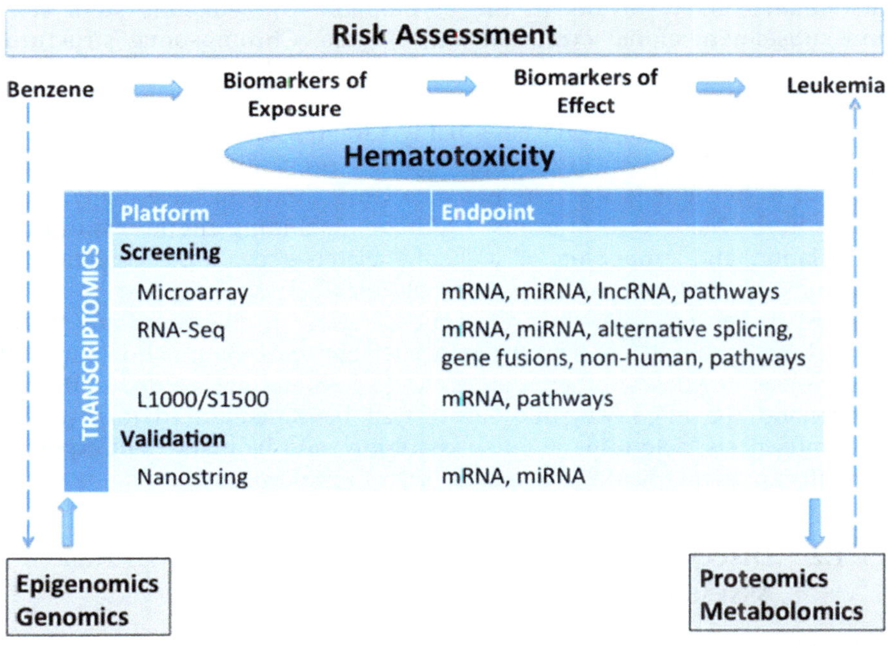

Figure 11.1 Transcriptomics can reveal biomarkers of benzene-induced exposure and effect and toxicity pathways relevant to risk assessment, even at low exposure levels. Altered levels of expression of a variety of transcripts, *via* epigenomics or other means of transcriptional regulation, provides a readout of cellular response to the environment, intermediate to phenotypic effects such as proteomics and metabolomics. Transcriptomic platforms, used or proposed for use in our studies, and the transcriptional endpoints detectable by them, are listed. Hematotoxicity is a phenotypic anchor of benzene exposure and counts of individual blood cell types can be used to adjust the data, as can other sources of confounding, such as experimental batch effect, age, gender, smoking, *etc.*

findings from benzene transcriptomic studies, their relevance to risk assessment and future directions. Our approach is summarized in Figure 11.1. This figure shows how our current and future transcriptomic approaches can identify transcriptional biomarkers of exposure and early-effect at low benzene exposure levels, and thus inform risk assessment.

11.1.1 Mechanisms and Biomarkers of Benzene Toxicity

Benzene, *via* its multiple metabolites, probably induces leukemia by targeting critical genes and pathways in hematopoietic stem cells (HSC) or bone marrow niche stromal cells, leading to altered proliferation and differentiation or apoptosis (see review[9]). Gene dysregulation, mediated by genotoxic effects on genes or chromosomes, chromosomal aneuploidy and rearrangement, or epigenetics, in conjunction with oxidative stress, aryl hydrocarbon receptor dysregulation and reduced immunosurveillance, likely results in the development of leukemic stem cells and subsequent clonal evolution to leukemia. Chromosomal structural and numerical alterations have been detected in lymphocytes of exposed workers in multiple studies.[18] We reported the induction of aneuploidy in mature blood cells using targeted and chromosome-wide approaches[19–21] and in circulating myeloid progenitor cells of workers exposed to benzene.[22] Benzene has also been shown to have many other effects in the blood cells of exposed humans, including alterations in DNA methylation and expression of leukemia-related genes.[9] Dysregulation of cytokines, chemokines, or adhesion molecules or responsiveness to them, induced by benzene or by inherent genetic variation, may contribute to aberrant hematopoiesis.[23,24] Benzene-associated chronic inflammation may lead to the onset of MDS or progression to AML.[23] Studies are needed to further understand the underlying mechanisms of benzene-induced toxicity and leukemogenesis, determinants of susceptibility, and biomarkers of exposure and effect particularly at low doses, in order to minimize risk.

11.1.2 Discerning Low-dose Effects is a Challenge in Risk Assessment

The current permissible occupational exposure limit, set by the Occupational Safety and Health Administration, is 1 ppm (8 h TWA).[25] The American Conference of Governmental Industrial Hygienists' threshold limit value is 0.5 ppm[26] and the National Institute of Occupational Safety and Health's recommended exposure limit is 0.1 ppm (TWA).[27] In the US, occupational exposure levels are typically below 1 ppm.[14] Gas-station attendants are exposed to levels around 0.05 ppm, while environmental exposure levels in US cities are an order of magnitude lower, at around 0.005 ppm. There is an interest in understanding health effects at lower occupational exposure levels (0.05–1 ppm), in order to refine appropriate exposure limits.

Various effects of exposure to benzene at levels below 1 ppm have been reported in human exposure studies. Risk of developing MDS,[28,29] AML, multiple myeloma, and chronic lymphocytic leukaemia (CLL),[30] was increased. Circulating blood cells were reduced in some studies including our own,[31–34] although not in others.[35–37] In our cross-sectional molecular epidemiological study of benzene exposure in factory workers in China, white blood cell counts were decreased in workers exposed to less than 1 ppm benzene compared with controls, and a highly significant dose–response relationship was present,[31] with no apparent threshold within the occupational exposure range (0.2–75 ppm benzene).[38] Increased blood mitochondrial DNA copy number, an indicator of oxidative stress-induced mitochondrial DNA damage and dysfunction, and altered global and gene-specific DNA methylation was reported in various Italian occupational groups with low levels of exposure to benzene.[39] (Previously, we reported that benzene levels >1 ppm increased mitochondrial DNA copy number.[40]) Genotoxic effects in the form of micronuclei, but not chromosome aberrations, were induced in Italian fuel-tank drivers.[41] Reduced expression of the co-stimulatory molecules CD80 and CD86 in monocytes, and increased levels of interleukin-8, suggestive of compromised adaptive immunity and immunosurveillance, were recently reported in gas-station attendants in Brazil.[42] Complementing these targeted endpoint studies, toxicogenomic approaches in low-exposed human populations can further enhance our understanding of the mechanisms of benzene toxicity and identify additional biomarkers of exposure, early effect, and susceptibility useful for risk assessment.

11.1.3 Toxicogenomic Studies

Toxicogenomics involves the genome-wide study of susceptibility genes (genomics), gene expression (transcriptomics), protein expression (proteomics), metabolite profiling (metabolomics), and epigenetic modifications (epigenomics), in exposed cells or tissues. Application to exposed human populations can lead to an understanding of gene–environment interactions, and the discovery of biomarkers of exposure, early effect, and susceptibility. Previously, we discussed how such findings inform risk assessment by elucidating mechanisms of action underlying exposure-related disease, identifying susceptible individuals and detecting response at low doses.[9,16] Integrative analysis of multiple omic datasets for an individual chemical in a systems biology approach[17] can support risk characterization, and aggregate data for multiple chemicals can contribute to hazard identification. As previously reviewed, we have applied a toxicogenomics approach to a molecular epidemiology study of factory workers occupationally exposed to a range of benzene levels in China, including transcriptomics [messenger (m)RNA and micro (mi)RNA] using microarrays, serum proteomics using mass spectrometry, and epigenomics using DNA methylation microarrays.[9,16,17,43] In addition, we have applied functional genomic screening in yeast, followed by validation in human cell lines and exposed populations, to identify and validate susceptibility genes associated with benzene exposure.[15] These ongoing toxicogenomic efforts have increased our understanding of the mechanisms of leukemogenesis and identified potential biomarkers. In particular, analysis of the transcriptome of benzene-exposed workers has revealed new insights and has the potential to advance our understanding even further as outlined in the following sections.

11.2 Transcriptomic Studies of Occupational Benzene Exposure

11.2.1 The Complex Human Transcriptome and Its Analysis

The human transcriptome is dynamic, allowing rapid response to a changing environment. It captures the effects of all gene expression regulatory mechanisms such as DNA and histone modifications, feedback mechanisms, and genetic background. At a single time point, the transcriptome reflects the current status of the tissue being analyzed including responses to chemical exposures, such as adverse outcome pathways and adaptive changes. The human transcriptome is complex, comprising a large number of coding transcripts (mRNA) and an even larger number of non-coding transcripts, including small non-coding (snc)RNA (>200 bp) and long non-coding (lnc)RNA (>200 bp).[43–45] Non-coding (nc)RNAs regulate genes through transcriptional and post-transcriptional gene regulation *via cis-* and *trans*-acting mechanisms, chromatin modification, control of transcription factor binding, and regulation of alternative splicing.[45] sncRNAs comprise many subtypes, including miRNA, and regulate gene expression by targeting mRNAs and other RNAs and epigenetic effects.[46] One novel type of lncRNA, circular (circ)RNA, has recently been extensively studied for its role in gene regulation and disease, particularly cancer.[47,48] Abundant and diverse, circRNAs act as miRNA sponges by competing for binding to miRNAs and they also regulate gene expression through additional mechanisms. Dysregulation of both lncRNA and sncRNA is involved in cancer and other diseases,[48–54] including leukemia,[55–59] and their expression is altered by stress and environmental exposures.[60–68] Additional diversity in the mammalian transcriptome comes from alternative splicing and RNA editing, which affect both coding and non-coding genes and extend functionality, and effects of inherent genetic variation on gene expression. In addition, the transcriptome contains information on genetic mutation, gene fusions and non-human transcripts. All aspects of the transcriptome can potentially be dysregulated on the exposure–disease continuum.

The main methods for analyzing the transcriptome are the long-established probe-based microarray and the more recent and evolving RNA-sequencing (RNA-Seq) platforms. We have discussed the technologies on which they are based and their advantages and disadvantages previously.[43] Broadly, microarrays are affordable and the technology, quality control, and bioinformatics are well established. Limitations include the ability to detect only known targets, the semi-quantitative nature of the measurements, a limited dynamic range, and lack of sensitivity to detect low-abundance transcripts. RNA-Seq is expensive and the quality control and bioinformatics are evolving. It is useful for discovery of novel aspects of the transcriptome. With the desire to analyze small amounts of biological samples in a high-throughput, cost-effective manner, with low technical variability, novel methods such as the hybridization-based multiplexed assay used by the nCounter platform from NanoString and the reductive L1000 approach, are gaining traction. In the following sections, we discuss our application of microarray, RNA-Seq, and nCounter to analyze the peripheral blood mononuclear cell (PBMC) transcriptome of human benzene exposure, and the future potential of the L1000/S1500 approach. Characteristics of these transcriptomic platforms, including target transcripts, applications, throughput, cost, quality control, method, and data complexity, are detailed in Table 11.1.

11.2.2 Microarrays

Microarrays have been applied to identify biomarkers of exposure and early effect, mechanism of action, and to improve risk assessment.[16,69,70] We have performed several studies of benzene-associated gene expression by microarray.[71–74]

11.2.2.1 Genes and Pathways Identified by Benzene Microarrays

In our earliest microarray study, we analyzed PBMC gene expression in six highly exposed shoe-factory workers (mean benzene ± SD 47.3 ± 24.3 ppm) and six matched non-occupationally exposed controls from our molecular epidemiology study, using Affymetrix microarrays.[73] Using robust multiarray analysis, paired *t*-statistics, and bootstrapping to control for a 5% family-wise error rate, we identified 29 differentially expressed genes. We confirmed altered expression of four genes, CXCL16, ZNF331, JUN, and PF4, using reverse transcriptase polymerase chain reaction in 28 subjects and showed that it was not caused by differences in PBMC cell composition. In a later study, we compared genes identified in eight highly exposed and eight matched control subjects on two microarray platforms (Affymetrix and Illumina) and found a 50% concordance in differentially expressed genes with a fold-change ≥1.3 and a role for genes involved in apoptosis and lipid metabolism. CXCL16, ZNF331, JUN, and PF4 were among the top 100 genes identified by both platforms.[71] Our most recent microarray study was much larger, involving 125 subjects exposed to a range of benzene levels, with 59 exposed to <1 ppm.[74] We randomized the samples across experimental variables, included multiple technical replicates, incorporated precise, individual measurements of exposure, and analyzed the data using a mixed effects model, with the aim of removing sources of biological and

Table 11.1 Characteristics of transcriptomic platforms used or proposed in occupational benzene exposure studies.[a]

Approach	Endpoints	Applications	Throughput	Cost	Quality control guidelines	Method complexity	Data analysis complexity
Microarray	mRNA, miRNA, lncRNA, pathways	Screen known human genes, pathways, networks	Medium	Medium	MAQC	Medium	Medium
RNA-seq	mRNA, miRNA, pathways, alternative splicing, RNA editing, gene fusions, non-human	Screen known and novel transcripts, pathways, networks; human and viral	Low	High	SEQC	High	High
L1000/S1500	mRNA, pathways	Screen AOPs	High	Low	n/a	Low	Low
NanoString	mRNA, miRNA	Validation	Medium	Medium	n/a	Low	Low

[a]AOPs: adverse outcome pathways; MAQC: MicroArray Quality Control consortium; SEQC: Sequence Quality Control consortium.

experimental variability (nuisance variability). In agreement with others,[75,76] we found variation due to factors such as RNA extraction, labeling, hybridization, and chip assignment, which required statistical adjustment. We identified AML and immune response as significantly impacted pathways. We also identified a 16-gene expression signature associated with all levels of benzene exposure (Figure 11.2), with roles in immune response, inflammatory response, cell adhesion, cell–matrix adhesion, and blood coagulation.[74] Later, we showed that differential expression of the majority of genes identified was independent of PBMC cell composition.[77] The major strengths of our study were its size, study design, range and assessment of exposure levels, and minimization of nuisance variability. A Flemish population study examined global genes expression in relation to internal markers of exposure for several environmental chemicals including *t,t*-muconic acid (*t,t*-MA), a benzene metabolite and an indicator or biomarker

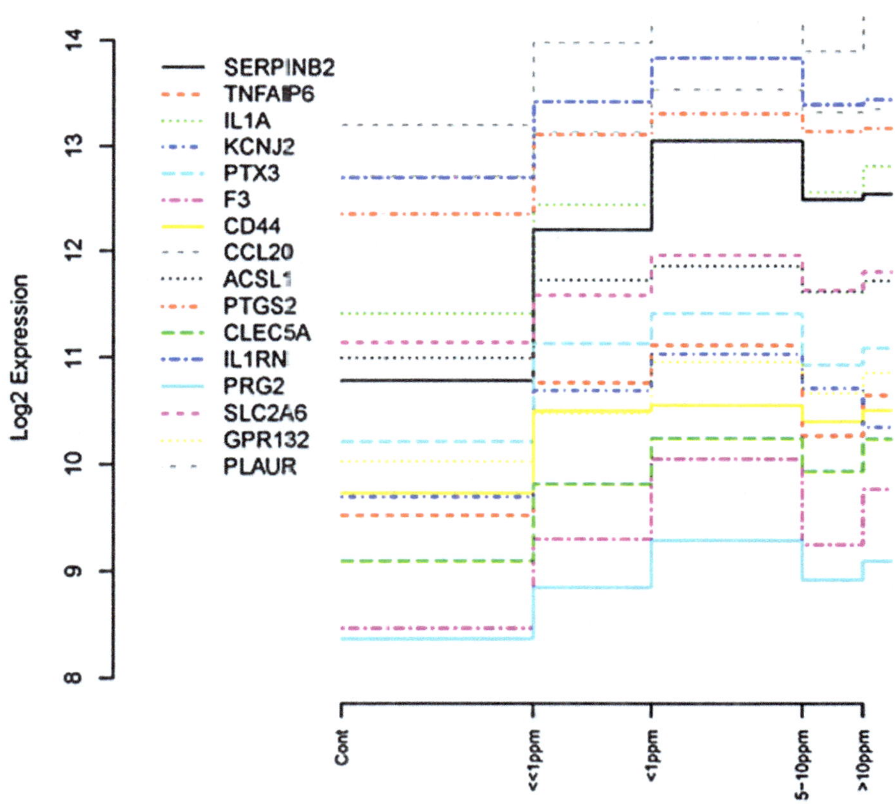

Figure 11.2 Expression of potential biomarkers of benzene exposure plotted across exposure categories. The mean log 2 expression value for 16 potential biomarker genes, whose expression was up- or down-regulated ≥1.5-fold at three or four doses, are plotted for each exposure category. A distinct dose–response curve is common to all of these genes, with a peak in response at the <1 ppm category.
Reproduced from *Environ. Health Perspect.*[74]

of exposure to benzene, in 40 adults.[78] The authors found broadly similar effects on gene expression in males and females, with more pronounced effects in males. Differences in effect on gene expression by gender have been noted in studies of other exposures.[79,80]

A few other groups have analyzed gene expression by microarray in benzene-exposed populations. Gao *et al.* compared profiles from four chronic benzene poisoning patients occupationally exposed to benzene, three benzene-exposed workers, and three unexposed controls and identified altered expression of genes involved in immune response, inflammatory response, chemotaxis, defense response, anti-apoptosis, and signal transduction.[81] From the same data, the group identified differentially expressed lncRNAs with involvement in immune response, hematopoiesis, B-cell receptor signaling pathway, and chronic myeloid leukemia.[82] In the benzene-poisoned subjects, they later identified up-regulated and down-regulated miRNAs.[83] Integrative analysis of miRNA and mRNA data revealed pathways and key target genes. However, in each of these studies, the sample size was very small.

11.2.2.2 Dose–Response Analysis of Benzene-associated Pathways

For transcriptomics data, as for traditional toxicological data, a dose–response relationship increases the weight-of-evidence that the observable effects are related to exposure. In our gene expression dataset from 125 benzene-exposed subjects and controls, we observed dose-dependent effects on expression of genes and biochemical pathways, and a supra-linear response in the expression of a 16-gene signature (Figure 11.2).[74] Later, we applied non-parametric approaches to statistically model the dose–response of AML pathway gene expression using continuous exposure data from all subjects.[77] We estimated air benzene exposure levels resulting from environmental exposure in the control subjects from unmetabolized urinary benzene levels.[77] We also adjusted for proportions of PBMC cell subtypes. We found that AML pathway genes exhibited similar supra-linear responses, similar to the 16-gene signature, and responses at benzene levels as low as 100 ppb in air. The relevance of these findings to risk assessment and regulation are currently unknown. Studies with a large number of subjects exposed to benzene at levels between 0.01 and 1 ppm are needed to further validate these responses.

As with traditional toxicological data, toxicogenomic data can be used to generate relevant points of departure (PoD). Traditionally, the lowest observed adverse-effect level, the lowest dose corresponding to a significant effect on a specific gene or pathway, and the no observed adverse-effect level or no observed transcriptional effect level, the dose below that at which no significant changes are observed, have been applied.[84] Increasingly, benchmark dose (BMD) or benchmark concentration (BMC) modeling, which use all of the experimental data to describe the overall exposure– or dose–response relationship and derive reference doses,[85] are being applied to assess risk. For several carcinogens, it has been shown that BMDs for the most sensitive pathways or gene ontologies in short-term exposure studies were highly correlated with BMDs for standard cancer and non-cancer endpoints in mice exposed for 2 years.[86–88] In a study of the effects of benzo[*a*]pyrene in drinking water in human cell cultures and mice, similar PoDs were generated from biologically relevant pathway data and traditional data.[89,90] From our gene expression data, for a given benzene exposure, we were able to identify the expected log fold-change in the expression of a gene of interest, relative to the mean expression at low exposures, adjusting for confounding.[77] In future analyses, we will use different approaches to identify PoDs, such as BMD or BMC modeling.

11.2.3 RNA-Seq

RNA-Seq, a sequencing based technology, allows the entire transcriptome to be analyzed quantitatively in a high-throughput manner.[91] We and others have reviewed the various technologies, bioinformatics and quality control considerations of RNA-Seq.[43,92] RNA-Seq has a much larger dynamic range than microarrays and the ability to detect genes with low expression intensity[93–96] and cancer-related transcripts.[95,97,98] It can also assess diverse and novel transcript types.[99–101] For example, Beane *et al.* identified differentially expressed ncRNAs, which may play important roles in lung carcinogenesis.[95] In cancer cells, RNA-Seq has revealed fusion transcripts, alternative splicing, allelic imbalance, single nucleotide variations, INDELs (insertions and deletions), exon-skipping events, and RNA editing,[45,102–106] as well as the presence of non-human sequences.[107]

Despite its significant advantages over microarrays, RNA-Seq suffers from limitations,[102,108–110] including experimental bias from multi-step library preparation procedures and transcript length heterogeneity.[111,112] Efforts to standardize approaches and develop best practices are ongoing, including the Sequence Quality Control consortium (http://www.fda.gov/ScienceResearch/BioinformaticsTools/MicroarrayQualityControlProject/), an effort led by the MicroArray Quality Control consortium of the US Food and Drug Administration to compare RNA-Seq platforms, data analysis approaches and data across sequencing laboratories and with microarray data.[113–118] RNA-Seq, particularly the ultra-deep sequencing required to identify certain characteristics,[119] remains expensive relative to microarrays. Despite current challenges and limitations, RNA-Seq has been successfully employed to analyze the transcriptome of multiple diseases[94,102,120–122] and a few exposures[95–97] in people.

11.2.3.1 *Evaluation of RNA-Seq in Benzene Pilot Study*

We evaluated the applicability of RNA-Seq to the analysis of the PBMC transcriptome of benzene exposure in a pilot study, in which we analyzed using RNA-Seq the mRNA transcriptomes of 10 workers highly exposed (>5 ppm) to the leukemogen benzene and 10 unexposed control study subjects matched by age, sex, and smoking status,[72] that had been previously analyzed using microarray.[74] Comparison of the RNA-Seq and microarray data for the 10 exposed *vs* 10 control subjects suggested that RNA-Seq was better able to detect low-intensity gene expression, potentially extending the number of genes identified. Indeed, RNA-Seq revealed differential expression in many more genes ($n = 146$) than did microarray ($n = 1$). Only 89 of the 146 genes had corresponding probes on the Illumina microarray platform; thus RNA-Seq identified additional genes not detectable by microarray. Four pathways, mitogen-activated protein kinase signaling, Fc epsilon signaling, adipokine signaling, and insulin signaling were over-represented in the benzene-related genes identified by both technologies. In addition, type II diabetes was among the top 10 pathways identified by RNA-Seq. In a recent review of the non-cancer health effects of benzene, a potential for an involvement in endocrine disruption, glucose homeostasis and diabetes was discussed.[13] Recent research from the same group showed that daily exposure of rats to benzene caused toxicity to pancreatic islet cells and increased plasma levels of insulin.[123] Although this is a small pilot study, it exemplifies the power of RNA-Seq as a complementary technology to identify additional genes and pathways.

11.2.3.2 Detection of Alternative Splice Variants

In our RNA-Seq pilot data, we identified alternative splicing as a potential mechanism of benzene toxicity.[72] Splicing diversity is involved in hematopoietic function and differentiation[124] and defects in the spliceosomal machinery and alternative splicing are known mechanisms of leukemogenesis[125–127] and MDS.[128] Microarray analysis revealed that almost one-third of expressed transcripts are differentially spliced in AML patients compared to normal donors, including in oncogenes and tumor suppressor genes.[129] RNA-Seq revealed distinct patterns of alternative splicing events associated with mutation in the splicing regulator U2AF1,[130] that could distinguish myeloid malignancies with the mutation from those without,[131] and mutated lung cancer from mutated AML.[132]

In a mouse model of MDS, telomere dysfunction-induced DNA damage has been shown to be associated with altered expression of mRNA splicing/processing genes and altered myeloid progenitor differentiation.[128] In a study of the effects of air pollution on telomeres, telomere length was shorter in traffic officers than in office workers, and decreased with increasing level of personal exposure to benzene and toluene in all subjects combined.[133] However, we reported that mean telomere length was significantly elevated in workers exposed to very high levels of benzene compared with controls.[134] RNA splicing and its underlying mechanisms require further investigation in human benzene exposure studies.

11.2.3.3 Detection of Gene Fusions

Chromosomal translocations and fusion genes are a hallmark of cancers, including leukemias.[135–138] Analysis of the transcriptome by RNA-Seq is a sensitive method to detect fusion transcripts including novel transcripts that may not be detectable by classical/conventional and modern cytogenetic methods.[139–142] A database of gene fusions in cancer based on analysis of published RNA-Seq datasets from 15 kinds of human cancer, including AML, has been created, called FusionCancer (http://donglab.ecnu.edu.cn/databases/FusionCancer/).[143]

There are multiple bioinformatics methods available to detect fusion transcripts from RNA-Seq data[144] from both paired-end reads (*e.g.* ChimeranScan, defuse,[145] and SOAPfusion[146]) and single-end reads (*e.g.* TopHat-Fusion,[147] FusionMap,[148] and FusionFinder[149]). All methods map a single-end read or one paired-end read to each end of two fused genes at the fusion junction site (split reads). Paired-end reads additionally map each read to each of the fused genes independently (bridging reads). Application of fusion finding tools is often limited by many factors, including a high false-detection rate or low sensitivity.[144,150,151] A comparison of six platforms (FusionHunter, FusionMap, FusionFinder, MapSplice, deFuse, and TopHat-fusion) revealed differences in performance and variable dependency of the tools on read length, quality score, and on the number of reads necessary to identify fusions.[150] FusionMap was the least error-prone. Most of these methods are based on short-read data. A sensitive fusion detection method called JAFFA (https://github.com/Oshlack/JAFFA/wiki) was developed that outperforms other methods with reads of 100 bp or greater and works with single- and paired-read data.[152]

These bioinformatics approaches to analyze leukemia transcriptomes have detected novel fusion genes in leukemia patients and have provided prognostic or mechanistic insight. Analysis of T-cell-ALL patient transcriptomes using deFuse revealed gene fusion transcripts, several of which were novel and rare.[106] Cytogenetically normal (CN)-AML occurs in 20% of childhood AML cases and is characterized by heterogeneity in clinical outcome. Analysis of transcriptome sequencing data from seven cases of pediatric CN-AML using deFuse and Chimerascan revealed a novel recurrent CBFA2T3-GLIS2 fusion predictive of poorer outcome.[153] As expression of CBFA2T3-GLIS2 fusion in mice is not sufficient for leukemogenesis, a follow-up study was conducted, and identified a new fusion transcript, DHH-RHEBL1, in the CBFA2T3-GLIS2-positive patients.[154] The DHH-RHEBL1 fusion was subsequently found to be present in 40% of CBFA2T3-GLIS2-positive patients in a validation cohort ($n=55$). Furthermore, CBFA2T3-GLIS2-positive patients with the DHH-RHEBL1 fusion exhibited a specific gene expression signature and had a lower 8 year overall survival rate than that of CBFA2T3-GLIS2 patients lacking this fusion. Analysis of transcriptomes from adult CN-AML cases by FusionSeq and deFuse revealed seven fusion transcripts, one (CIITA-DEXI) of which was found in 48% of

CN-AML cases.[155] Transcriptome sequencing of blast cells from a pediatric AML patient with del(5q), a rare type of childhood leukemia with poor prognosis,[156] identified two fusion transcripts, RUNX1-USP42 and PRDM16-SKI.[157] These fusion transcripts were detected using concordant results from Chimerascan, deFuse, and FusionMap.

We and others have reported increased frequencies of leukemia-related chromosomal translocations and fusion genes in benzene-exposed workers.[19,158,159] As such, fusion genes represent a useful *meet in the middle* biomarker of early effects of benzene on the causal pathway to disease. Preliminary analysis of our benzene pilot RNA-Seq data, along with data for a human reference cell line RNA, by TopHat-Fusion, revealed a greater number of fusion transcripts in cell-line RNA compared to PBMC RNA (unpublished data). However, our ability to detect chromosomal fusions events with high confidence in RNA from PBMCs was limited, possibly by the depth of sequencing. We are conducting additional analyses.

11.2.3.4 Detection of Non-human RNA

According to the International Agency for Research on Cancer, a fifth of cancer cases worldwide are caused by infection, with most caused by viruses.[160,161] The human papilloma virus, Epstein–Barr virus (EBV), hepatitis B virus, and hepatitis C virus are well known,[162] but additional undiscovered human tumor viruses are likely to contribute.[163] As RNA-Seq captures data on both the human and non-human transcriptomes present in a biological sample, computational methods have been developed to characterize the non-human transcripts. In a preliminary proof-of-principle analysis, we applied the PathSeq protocol[107] to identify non-human transcripts in the small portion of our pilot RNA-Seq data[72] that did not map to the human genome. Input reads were filtered to remove low-quality, duplicate, and repetitive sequences and then subjected to subtraction by sequential alignment to human genome and transcriptome reference sequences using MAQ software, Mega BLAST and BLASTN. Candidate microbial reads that remained were then aligned to viral, bacterial and fungal reference databases by BLASTN and BLASTX and assembled *de novo*. We found variable numbers and species of bacteria and viruses across samples, many of which do not map to currently known species (unpublished data).

Other computational pipelines based on similar subtractive approaches have been described.[164–168] RNA CoMPASS (http://rnacompass.sourceforge.net/) is a comprehensive RNA-Seq analysis pipeline for the simultaneous analysis of transcriptomes and metatranscriptomes from diverse biological specimens.[169] Using existing tools and parallel computing technology, it enables the analysis of large datasets through a web-based graphical user interface. RNA CoMPASS identified EBV in 22 B-cell lines generated by the infection of naive B-cells with EBV and in samples from Burkitt's lymphoma that arose in part through EBV infection. An algorithmic method called VirusSeq was developed that accurately detects known viruses and their integration sites with high sensitivity and specificity.[170] It was validated using data from 256 human cancer samples from The Cancer Genome Atlas. Interrogation of RNA-Seq data from 3775 malignant neoplasms in The Cancer Genome Atlas database using VirusSeq identified viral transcripts and viral integration sites in a variety of cancer types.[171] No DNA virus transcripts were detected in certain cancers, including AML.

11.2.4 NanoString

The nCounter platform from NanoString Technologies (www.nanostring.com/products/) hybridizes fluorescent barcodes directly to specific nucleic acid sequences, allowing for the measurement of expression of up to 800 targets in one sample, without a requirement for reverse transcription or amplification.[172,173] This relatively new technology has been applied in various clinical and research studies[174–177] and is comparable to other technologies in the analysis of gene expression.[176,178] Recently, a comprehensive evaluation of the nCounter platform in the analysis of mRNA expression found it to be superior to other techniques in platform robustness, technical variation, sensitivity, and limit of detection across various clinical sample types, amounts, and qualities.[176,177] Comparison of microarray and nCounter data in human leukocyte subsets revealed a high degree of correlation, especially for high-variance genes, and expression-dependent accuracy bias and precision in microarray data.[178] In our benzene study, preliminary comparison of the log intensities of 30 differentially expressed genes associated with benzene exposure identified by microarray, with those generated using nCounter technology, showed a high correlation for most genes (unpublished data). nCounter code sets are also available for lncRNA and miRNA. In the Beijing Truck Driver Air Pollution Study, including 60 truck drivers and 60 office workers, an association between short-term air pollution (elemental carbon) and expression of human blood miRNAs and latent viral miRNAs that may influence immune responses and other biological pathways was reported using the nCounter platform.[179]

11.2.5 L1000 and S1500 Platforms

Pathways and networks comprising proteins encoded by coordinately expressed or co-regulated genes drive the physiological status of cells and tissues. In 2006, transcriptomic data was used to derive a "connectivity map", describing connections between biological perturbations in four human cell lines and treatment with "perturbagens", mostly drugs.[180] Further bioinformatic analyses led to the discovery of a subset of ~1000 "landmark" genes that served as key representative genes for each network and could be used to impute whole transcriptome information. These genes were used to develop a low-cost, high-throughput platform based on Luminex bead-based technology, known as L1000.[181] Through application in studies in the National Institutes of Health Common Fund's Library of Integrated Network-based Cellular Signatures (LINCS) program, the L1000 has generated a "library" of 1 million gene expression signatures characterizing the response of different cell types to varied perturbagens (www.lincscloud.org/l1000/). A browser called LINCS Canvas Browser has been developed to facilitate querying, browsing, and visualizing currently available LINCS L1000 data (www.maayanlab.net/LINCS/LCB).[182]

In phase III of the multi-agency Tox21 initiative to revolutionize toxicity testing,[183] a "1500 genes" high-throughput transcriptomics project was proposed, seeking to develop a platform, known as S1500, that will extend the capability of L1000, include toxicity- and disease-related genes, and optimize extrapolatability and pathway coverage.[184,185] BioSpyder (http://biospyder.com) has developed a high-throughput platform based on targeted sequencing using novel TempO-Seq™ technology. TempO-Seq can analyze expression of hundreds to thousands of genes simultaneously from limited sample material or single cells without pre-amplification. Up to 6144 samples can be pooled in each sequencing run due to sample barcoding and sequencing of short fragments of each gene, making it very cost-effective. The assay is highly accurate and sensitive and data analysis is simple. The goals of S1500 are to identify mechanisms of adverse effects from perturbagen exposures, and generate data pertinent to hazard identification and human health risk assessment.

The L1000 and S1500 platforms are attractive alternatives to microarrays and RNA-Seq for the analysis of transcriptomic responses to a large number of chemicals in exposed human populations. However, as the majority of L1000 data have been generated in cell *lines in vitro*, it is unclear how comparable they will be to the PBMC transcriptomes of chronic exposure to low levels of environmental chemicals.

11.2.6 Transcriptomic Platform Choice and Study Design Considerations

In the previous sections, we have discussed our application of microarrays, RNA-Seq and NanoString technology to the PBMC transcriptome of human benzene exposure. The overall goals of our study were to identify biomarkers of exposure, refine understanding of low-dose effects, and identify pathways and mechanisms. Through careful experimental design of a large microarray study to minimize technical variability, we were able to achieve these goals. The nCounter platform from NanoString is a useful tool for validation and screening targeted gene sets as it has reduced technical variability and some of the advantages of sequencing data, being count based and having a greater dynamic range. It can confirm microarray findings in a high-throughput manner and refine dose–response effects. Our pilot evaluation of RNA-Seq identified additional potential biomarkers among low-expressed genes compared to microarray and alternative splicing as a plausible potential mechanism of benzene-induced leukemogenesis. However, given the cost and ongoing challenges associated with standardizing RNA-Seq technology and data analysis, approaches such as microarray and L1500 remain attractive options, especially for screening large number of samples and chemicals. Some microarray platforms (*e.g.* GeneChip® Human Exon ST Array from Affymetrix) enable the detection of gene expression and alternative splicing. As RNA-Seq discovery-based studies refine our understanding of the make-up and diverse aspects of the transcriptome, these features, *e.g.* lncRNA and circRNA, will become available for analysis in high-throughput, cost-effective technologies. For now, more information regarding these less well-characterized aspects of the transcriptome in human toxicogenomics studies may not necessarily be better.

Regarding study design, as discussed in our previous review,[43] it is important to have sufficient power to detect large number of features in the trancriptome,[45] and sufficient depth in the case of RNA-Seq.[186,187] Technical variability and in-study replication are critical aspects of study design that influence power. In our large microarray study, we found variation due to experimental factors such as RNA extraction, labeling, hybridization, and chip assignment.[74] Statistical adjustment by analysis with a mixed-effects model minimized potential confounding and experimental variability. In a recent study examining factors governing the power to detect miRNA expression changes in human studies where fold changes are small, we again showed that experimental design to reduce and control for technical variability is critical, and furthermore that repeated measures can increase the power to detect changes in miRNA expression in omic studies.[188] We also found that miRNAs differed in the degree to which their measurement was affected by sources of variability, including temporality. Leidinger *et al.* reported that blood cell type specific miRNomes were very consistent between individuals and over time.[189]

Another critical aspect of study design is the optimal cell type to analyze. In our benzene studies, blood (PBMC) is the tissue of choice because of its accessibility and its relevance to hematotoxicity. Many studies have revealed biomarkers of exposure and mechanisms of toxicity in whole blood or peripheral blood leukocytes in populations exposed to multiple exposures and associated with multiple diseases.[9] The peripheral blood transcriptome also partially reflects expression in other tissue types in humans[190–192] and thus may reveal information regarding health effects of benzene beyond the hematopoietic system.[13] One limitation of blood and PBMC is that they comprise mixed cell populations in proportions that vary by individual. Gene expression changes may reflect altered blood cell composition as occurs in hematotoxicity associated with benzene.[31] Statistical approaches can adjust for blood cell counts in transcriptome analysis.[77] If blood count data are unavailable, a new method called CIBERSORT can infer composition of complex tissues from gene expression data.[193] For leukemogens like benzene, the relevant target may be the hematopoietic stem cell or myeloid progenitor cells.[9] Although bone marrow is not readily accessible, cell sorting of circulating myeloid progenitor cells, followed by RNA-Seq or targeted validation may facilitate an understanding of effects at this level. Single-cell expression profiling was applied to elucidate early normal hematopoiesis and AML development.[194,195] Further studies could be conducted in models of the bone marrow niche to understand effects on other relevant cell types such as HSC and mesenchymal stem cells.[196]

Factors related to sample preparation and storage can affect data quality and have been discussed previously and include time to processing, anticoagulant choice,[197] and stabilization and isolation techniques.[198–205] In our study, we standardized blood collection time, minimized time to processing, and used EDTA as the anticoagulant of choice.

11.3 Future Directions and Translation

One of the goals of human transcriptomic studies is to identify blood-based biomarker signatures of exposure, effect, or early disease that can be used to predict exposure or disease. In studies of radiation exposure, development of gene expression profiles in mice alone did not accurately predict human effects; instead, a gene expression classifier generated from analysis of human *ex vivo* irradiated blood and human irradiation-treated patients was able to discriminate exposed and non-exposed individuals and accurately predict radiation dose levels in all human samples tested and led to the development of a rapid and high-throughput assay.[206] In a large prospective study of gene expression profiles of B-cell lymphoma, the EnviroGenomarker Consortium identified transcriptomic signatures of CLL, predictive of disease even in cases diagnosed more than 10 years after blood collection.[207] In our study of benzene exposure, we have identified pathways and a 16-gene signature of benzene exposure with altered expression even at levels below 1 ppm,[74] in line with our hematotoxicity findings.[31] Our finding from bioinformatic analysis of these effects on a continuous exposure scale further suggests the importance of validating these effects in future studies of populations exposed to levels between 0.1 and 1 ppm. In addition, we will apply PoD analysis to our transcriptomic and other toxicity endpoint data to increase the applicability of the data for risk assessment.[84]

Our human benzene transcriptomic data have revealed effects of exposure that may represent mechanisms of leukemogenesis. The potential mechanisms of leukemia induction that we have identified in our benzene transcriptomic studies can be further investigated through mechanistic approaches in relevant cell types, such as HSC or myeloid progenitor cells *in vitro*, or in bone marrow niche models. We have previously conducted validation studies of functional genomic findings in yeast, using RNA interference,[15] and are exploring the application of CRISPR-based[208] approaches.

Our benzene transcriptomic analyses generated important data that may be useful for risk assessment. Integrative omic analysis, including epigenomics and genomics, can provide richer pathway and network information, identify more robust biomarkers, and elucidate mechanisms of toxicity and disease, thus aiding risk characterization.[209–213] As gene regulation is so complex, integrative omic analysis can reveal the means by which biological endpoints are altered and may allow targeted modulation and prevention or intervention. We previously described our systems biology approach to analyzing various toxicogenomic datasets generated from our occupational study of benzene exposure.[17]

Comparison of omic data across chemicals from publicly available toxicogenomic repositories is useful for hazard identification. As there are few known leukemia-causing agents, it is of interest to compare mechanisms of toxicity among known and suspected exposures.[1] Based on hypothesized mechanisms of action, we observed similarities and differences in the effects of hydroquinone, a metabolite of benzene, and etoposide and melphalan, two chemotherapeutic agents that induce therapy-related MDS and AML,[214–216] on global methylation,[217] endoreduplication,[218] and centrosome amplification. On an omic level, we used pathway analysis of Comparative Toxicogenomics Database data to discriminate chemical leukemogens from non-leukemogens.[219] Similar approaches can be applied to characterize the exposure–disease continuum.[210] Transcriptomic and other omic effects are well characterized in leukemia.[56,212] The availability of more toxicogenomic datasets and more endpoints such as alternative splicing in exposure and disease from well-designed studies will strengthen these efforts once bioinformatics challenges are overcome.

A more holistic approach to understanding the causes of leukemia in the context of the exposure–disease continuum, including all environmental and lifestyle contributions, is embodied by the exposome approach. This recognizes that people are exposed to multiple chemicals and stresses throughout their lives from both external and external sources that influence their disease risk.[220] Gene expression and methylation of individual genes can be impacted in a time-dependent manner[221] and may be reversible or irreversible,[222] with implications for diseases of later life, or those with a long latency such as AML. Exposomics extends the single exposure association (and single snapshot in time) approach to measurement of the complete internal chemical environment of an individual using omic technologies and to the relation of static and dynamic data to preclinical and clinical perturbations and disease onset in prospective studies.[223–225] We have proposed an exposomic approach to identify key biomarkers and exposures associated with *de novo* AML, for which only about ~20% of environmental causes are known.[226]

11.4 Conclusion

Through analysis of the PBMC transcriptome of benzene exposure by microarray, we have identified gene and pathway biomarkers of exposure and early effect, at a range of levels, including those below 1 ppm. We are applying dose–response and PoD analyses to refine a gene expression signature of exposure and we will use high-throughput, cost-effective technologies such as NanoString's nCounter and L1500/S1500 to validate it in larger populations exposed to levels below 1ppm in future studies, increasing applicability in risk assessment. Complementary analysis by RNA-Seq in a pilot study has revealed alternative splicing as a potential mechanism of benzene-induced leukemogenesis and has the potential to identify fusion transcripts and non-human sequences, although cost and technical and bioinformatics challenges are limitations. Application of other omic approaches and integrative omic analyses will further identify biomarkers of exposure and early effect and clarify mechanisms of benzene toxicity.

Acknowledgements

Our studies on benzene described here were supported by National Institutes of Health grants (R01ES06721 and P42ES004705 to M.T.S.) and by Environmental Protection Agency contract number EP-11-001398. We are grateful to Nicholas Akers for his analysis of gene fusions and non-human transcripts in the RNA-Seq pilot data.

References

1. D. A. Eastmond, N. Keshava and B. Sonawane, Lymphohematopoietic cancers induced by chemicals and other agents and their implications for risk evaluation: An overview, *Mutat. Res., Rev. Mutat. Res.*, 2014, **761**, 40–64.
2. ATSDR, *Toxicological Profile For Benzene*, http://www.atsdr.cdc.gov/ToxProfiles/tp.asp?id=40&tid=14, U.S. Department Of Health And Human Services. Agency for Toxic Substances and Disease Registry, Atlanta, GA, 2007.
3. R. B. Hayes, Y. Songnian, M. Dosemeci and M. Linet, Benzene and lymphohematopoietic malignancies in humans, *Am. J. Ind. Med.*, 2001, **40**, 117–126.
4. C. Steinmaus, A. H. Smith, R. M. Jones and M. T. Smith, Meta-analysis of benzene exposure and non-Hodgkin lymphoma: biases could mask an important association, *Occup. Environ. Med.*, 2008, **65**, 371–378.
5. IARC, *A Review of Human Carcinogens – Part F: Chemical Agents and Related Occupations*, International Agency for Research on Cancer, Lyon, France, 2012.
6. T. Filippini, J. E. Heck, C. Malagoli, C. Del Giovane and M. Vinceti, A review and meta-analysis of outdoor air pollution and risk of childhood leukemia, *J. Environ. Sci. Health, Part C: Environ. Carcinog. Ecotoxicol. Rev.*, 2015, **33**, 36–66.
7. J. E. Heck, A. S. Park, J. Qiu, M. Cockburn and B. Ritz, Risk of leukemia in relation to exposure to ambient air toxics in pregnancy and early childhood, *Int. J. Hyg. Environ. Health*, 2014, **217**, 662–668.
8. Y. Zhou, S. Zhang, Z. Li, J. Zhu, Y. Bi, Y. Bai and H. Wang, Maternal benzene exposure during pregnancy and risk of childhood acute

lymphoblastic leukemia: a meta-analysis of epidemiologic studies, *PLoS One*, 2014, **9**, e110466.
9. C. M. McHale, L. Zhang and M. T. Smith, Current understanding of the mechanism of benzene-induced leukemia in humans: implications for risk assessment, *Carcinogenesis*, 2012, **33**, 240–252.
10. R. Snyder, Overview of the toxicology of benzene, *J. Toxicol. Environ. Health, Part A*, 2000, **61**, 339–346.
11. B. D. Goldstein, Benzene toxicity: a critical evaluation: hematotoxicity in humans, *J. Toxicol. Environ. Health Suppl.*, 1977, **2**, 69–105.
12. R. Snyder, Leukemia and benzene, *Int. J. Environ. Res. Public Health*, 2012, **9**, 2875–2893.
13. H. Bahadar, S. Mostafalou and M. Abdollahi, Current understandings and perspectives on non-cancer health effects of benzene: a global concern, *Toxicol. Appl. Pharmacol.*, 2014, **276**, 83–94.
14. C. P. Weisel, Benzene exposure: an overview of monitoring methods and their findings, *Chem. Biol. Interact.*, 2010, **184**, 58–66.
15. C. M. McHale, M. T. Smith and L. Zhang, Application of toxicogenomic profiling to evaluate effects of benzene and formaldehyde: from yeast to human, *Ann. N. Y. Acad. Sci.*, 2014, **1310**, 74–83.
16. C. M. McHale, L. Zhang, A. E. Hubbard and M. T. Smith, Toxicogenomic profiling of chemically exposed humans in risk assessment, *Mutat. Res.*, 2010, **705**, 172–183.
17. L. Zhang, C. M. McHale, N. Rothman, G. Li, Z. Ji, R. Vermeulen, A. E. Hubbard, X. Ren, M. Shen, S. M. Rappaport, M. North, C. F. Skibola, S. Yin, C. Vulpe, S. J. Chanock, M. T. Smith and Q. Lan, Systems biology of human benzene exposure, *Chem. Biol. Interact.*, 2010, **184**, 86–93.
18. L. Zhang, D. A. Eastmond and M. T. Smith, The nature of chromosomal aberrations detected in humans exposed to benzene, *Crit. Rev. Toxicol.*, 2002, **32**, 1–42.
19. M. T. Smith, L. Zhang, Y. Wang, R. B. Hayes, G. Li, J. Wiemels, M. Dosemeci, N. Titenko-Holland, L. Xi, P. Kolachana, S. Yin and N. Rothman, Increased translocations and aneusomy in chromosomes 8 and 21 among workers exposed to benzene, *Cancer Res.*, 1998, **58**, 2176–2181.
20. L. Zhang, Q. Lan, W. Guo, A. E. Hubbard, G. Li, S. M. Rappaport, C. M. McHale, M. Shen, Z. Ji, R. Vermeulen, S. Yin, N. Rothman and M. T. Smith, Chromosome-wide aneuploidy study (CWAS) in workers exposed to an established leukemogen, benzene, *Carcinogenesis*, 2011, **32**, 605–612.
21. L. Zhang, N. Rothman, Y. Wang, R. B. Hayes, G. Li, M. Dosemeci, S. Yin, P. Kolachana, N. Titenko-Holland and M. T. Smith, Increased aneusomy and long arm deletion of chromosomes 5 and 7 in the lymphocytes of Chinese workers exposed to benzene, *Carcinogenesis*, 1998, **19**, 1955–1961.

22. L. Zhang, Q. Lan, Z. Ji, G. Li, M. Shen, R. Vermeulen, W. Guo, A. E. Hubbard, C. M. McHale, S. M. Rappaport, R. B. Hayes, M. S. Linet, S. Yin, M. T. Smith and N. Rothman, Leukemia-related chromosomal loss detected in hematopoietic progenitor cells of benzene-exposed workers, *Leukemia*, 2012, **26**, 2494–2498.
23. P. L. Minciullo, M. Navarra, G. Calapai and S. Gangemi, Cytokine network involvement in subjects exposed to benzene, *J. Immunol. Res.*, 2014, **2014**, 937987.
24. Q. Lan, L. Zhang, M. Shen, M. T. Smith, G. Li, R. Vermeulen, S. M. Rappaport, M. S. Forrest, R. B. Hayes, M. Linet, M. Dosemeci, B. P. Alter, R. S. Weinberg, S. Yin, M. Yeager, R. Welch, S. Waidyanatha, S. Kim, S. Chanock and N. Rothman, Polymorphisms in cytokine and cellular adhesion molecule genes and susceptibility to hematotoxicity among workers exposed to benzene, *Cancer Res.*, 2005, **65**, 9574–9581.
25. OSHA, www.osha.gov/dts/chemicalsampling/data/CH_220100.html, Accessed 10-16-20-15, 2015.
26. ACGIH, *Benzene. Threshold limit values for chemical substances and physical agents and biological exposure indices*, American Conference of Governemental Industrial Hygienists, 2004.
27. NIOSH, *Benzene. NIOSH pocket guide to chemical hazards*, www.cdc.gov/niosh/npg/npgd0049.html, Accessed 10-16-2015.
28. A. R. Schnatter, D. C. Glass, G. Tang, R. D. Irons and L. Rushton, Myelodysplastic syndrome and benzene exposure among petroleum workers: an international pooled analysis, *JNCI, J. Natl. Cancer Inst.*, 2012, **104**, 1724–1737.
29. D. C. Glass, C. N. Gray, D. J. Jolley, C. Gibbons, M. R. Sim, L. Fritschi, G. G. Adams, J. A. Bisby and R. Manuell, Leukemia risk associated with low-level benzene exposure, *Epidemiology*, 2003, **14**, 569–577.
30. J. S. Stenehjem, K. Kjaerheim, M. Bratveit, S. O. Samuelsen, F. Barone-Adesi, N. Rothman, Q. Lan and T. K. Grimsrud, U. K. I. B. C. W. Group, Benzene exposure and risk of lymphohaematopoietic cancers in 25 000 offshore oil industry workers, *Br. J. Cancer*, 2015, **112**, 1603–1612.
31. Q. Lan, L. Zhang, G. Li, R. Vermeulen, R. S. Weinberg, M. Dosemeci, S. M. Rappaport, M. Shen, B. P. Alter, Y. Wu, W. Kopp, S. Waidyanatha, C. Rabkin, W. Guo, S. Chanock, R. B. Hayes, M. Linet, S. Kim, S. Yin, N. Rothman and M. T. Smith, Hematotoxicity in workers exposed to low levels of benzene, *Science*, 2004, **306**, 1774–1776.
32. Q. Qu, R. Shore, G. Li, X. Jin, L. C. Chen, B. Cohen, A. A. Melikian, D. Eastmond, S. M. Rappaport, S. Yin, H. Li, S. Waidyanatha, Y. Li, R. Mu, X. Zhang and K. Li, Hematological changes among Chinese workers with a broad range of benzene exposures, *Am. J. Ind. Med.*, 2002, **42**, 275–285.
33. D. H. Koh, H. K. Jeon, S. G. Lee and H. W. Ryu, The relationship between low-level benzene exposure and blood cell counts in Korean workers, *Occup. Environ. Med.*, 2015, **72**, 421–427.

34. S. A. Khuder, M. C. Youngdale, M. S. Bisesi and E. A. Schaub, Assessment of complete blood count variations among workers exposed to low levels of benzene, *J. Occup. Environ. Med.*, 1999, **41**, 821–826.
35. S. P. Tsai, E. E. Fox, J. D. Ransdell, J. K. Wendt, L. C. Waddell and R. P. Donnelly, A hematology surveillance study of petrochemical workers exposed to benzene, *Regul. Toxicol. Pharmacol.*, 2004, **40**, 67–73.
36. G. M. Swaen, L. van Amelsvoort, J. J. Twisk, E. Verstraeten, R. Slootweg, J. J. Collins and C. J. Burns, Low level occupational benzene exposure and hematological parameters, *Chem. Biol. Interact.*, 2010, **184**, 94–100.
37. J. J. Collins, B. K. Ireland, P. A. Easterday, R. S. Nair and J. Braun, Evaluation of lymphopenia among workers with low-level benzene exposure and the utility of routine data collection, *J. Occup. Environ. Med.*, 1997, **39**, 232–237.
38. Q. Lan, R. Vermeulen, L. Zhang, G. Li, P. S. Rosenberg, B. P. Alter, M. Shen, S. M. Rappaport, R. S. Weinberg, S. Chanock, S. Waidyanatha, C. Rabkin, R. B. Hayes, M. Linet, S. Kim, S. Yin, N. Rothman and M. T. Smith, Benzene Exposure and Hematotoxicity: Response, *Science*, 2006, **312**, 998.
39. M. Carugno, A. C. Pesatori, L. Dioni, M. Hoxha, V. Bollati, B. Albetti, H. M. Byun, M. Bonzini, S. Fustinoni, P. Cocco, G. Satta, M. Zucca, D. F. Merlo, M. Cipolla, P. A. Bertazzi and A. Baccarelli, Increased mitochondrial DNA copy number in occupations associated with low-dose benzene exposure, *Environ. Health Perspect.*, 2012, **120**, 210–215.
40. M. Shen, L. Zhang, M. R. Bonner, C. S. Liu, G. Li, R. Vermeulen, M. Dosemeci, S. Yin and Q. Lan, Association between mitochondrial DNA copy number, blood cell counts, and occupational benzene exposure, *Environ. Mol. Mutagen.*, 2008, **49**, 453–457.
41. P. Lovreglio, F. Maffei, M. Carrieri, M. N. D'Errico, I. Drago, P. Hrelia, G. B. Bartolucci and L. Soleo, Evaluation of chromosome aberration and micronucleus frequencies in blood lymphocytes of workers exposed to low concentrations of benzene, *Mutat. Res., Genet. Toxicol. Environ. Mutagen.*, 2014, **770**, 55–60.
42. A. M. Moro, N. Brucker, M. F. Charao, E. Sauer, F. Freitas, J. Durgante, G. Bubols, S. Campanharo, R. Linden, A. P. Souza, C. Bonorino, R. Moresco, D. Pilger, A. Gioda, S. Farsky, A. Duschl and S. C. Garcia, Early hematological and immunological alterations in gasoline station attendants exposed to benzene, *Environ. Res.*, 2015, **137**, 349–356.
43. C. M. McHale, L. Zhang, R. Thomas and M. T. Smith, Analysis of the transcriptome in molecular epidemiology studies, *Environ. Mol. Mutagen.*, 2013, **54**, 500–517.
44. M. Pertea and S. L. Salzberg, Between a chicken and a grape: estimating the number of human genes, *Genome Biol.*, 2010, **11**, 206.
45. M. Pertea, The Human Transcriptome: An Unfinished Story, *Genes*, 2012, **3**, 344–360.

46. E. S. Martens-Uzunova, M. Olvedy and G. Jenster, Beyond microRNA - novel RNAs derived from small non-coding RNA and their implication in cancer, *Cancer Lett.*, 2013, **340**, 201–211.
47. S. Qu, X. Yang, X. Li, J. Wang, Y. Gao, R. Shang, W. Sun, K. Dou and H. Li, Circular RNA: A new star of noncoding RNAs, *Cancer Lett.*, 2015, **365**, 141–148.
48. J. E. Wilusz, Long noncoding RNAs: Re-writing dogmas of RNA processing and stability, *Biochim. Biophys. Acta*, 2016, **1859**, 128–138.
49. R. J. Taft, K. C. Pang, T. R. Mercer, M. Dinger and J. S. Mattick, Non-coding RNAs: regulators of disease, *J. Pathol.*, 2010, **220**, 126–139.
50. M. Esteller, Non-coding RNAs in human disease, *Nat. Rev. Genet.*, 2011, **12**, 861–874.
51. V. A. Moran, R. J. Perera and A. M. Khalil, Emerging functional and mechanistic paradigms of mammalian long non-coding RNAs, *Nucleic Acids Res.*, 2012, **40**, 6391–6400.
52. J. R. Prensner, M. K. Iyer, O. A. Balbin, S. M. Dhanasekaran, Q. Cao, J. C. Brenner, B. Laxman, I. A. Asangani, C. S. Grasso, H. D. Kominsky, X. Cao, X. Jing, X. Wang, J. Siddiqui, J. T. Wei, D. Robinson, H. K. Iyer, N. Palanisamy, C. A. Maher and A. M. Chinnaiyan, Transcriptome sequencing across a prostate cancer cohort identifies PCAT-1, an unannotated lincRNA implicated in disease progression, *Nat. Biotechnol.*, 2011, **29**, 742–749.
53. P. T. Law, H. Qin, A. K. Ching, K. P. Lai, N. N. Co, M. He, R. W. Lung, A. W. Chan, T. F. Chan and N. Wong, Deep sequencing of small RNA transcriptome reveals novel non-coding RNAs in hepatocellular carcinoma, *J. Hepatol.*, 2013, **58**, 1165–1173.
54. M. Rederstorff and A. Huttenhofer, Small non-coding RNAs in disease development and host-pathogen interactions, *Curr. Opin. Mol. Ther.*, 2010, **12**, 684–694.
55. J. E. Gordon, J. J. Wong and J. E. Rasko, MicroRNAs in myeloid malignancies, *Br. J. Haematol.*, 2013, **162**, 162–176.
56. M. Khalaj, M. Tavakkoli, A. W. Stranahan and C. Y. Park, Pathogenic microRNA's in myeloid malignancies, *Front. Genet.*, 2014, **5**, 361.
57. F. Pagano, E. De Marinis, F. Grignani and C. Nervi, Epigenetic role of miRNAs in normal and leukemic hematopoiesis, *Epigenomics*, 2013, **5**, 539–552.
58. C. P. de Souza Melo, C. B. Campos, J. de Oliveira Rodrigues, J. C. Aguirre-Neto, A. Atalla, M. A. Pianovski, E. K. Carbone, L. B. Lares, H. Moraes-Souza, S. Octacilio-Silva, F. S. Pais, A. C. de Souza Ferreira and J. G. Assumpcao, Long non-coding RNAs: biomarkers for acute leukaemia subtypes, *Br. J. Haematol.*, 2015, DOI: 10.1111/bjh.13588.
59. M. Morlando, M. Ballarino and A. Fatica, Long Non-Coding RNAs: New Players in Hematopoiesis and Leukemia, *Front. Med.*, 2015, **2**, 23.
60. C. Jolly and S. C. Lakhotia, Human sat III and Drosophila hsr omega transcripts: a common paradigm for regulation of nuclear RNA processing in stressed cells, *Nucleic Acids Res.*, 2006, **34**, 5508–5514.

61. M. Huarte, M. Guttman, D. Feldser, M. Garber, M. J. Koziol, D. Kenzelmann-Broz, A. M. Khalil, O. Zuk, I. Amit, M. Rabani, L. D. Attardi, A. Regev, E. S. Lander, T. Jacks and J. L. Rinn, A large intergenic noncoding RNA induced by p53 mediates global gene repression in the p53 response, *Cell*, 2010, **142**, 409–419.
62. E. Sonkoly, Z. Bata-Csorgo, A. Pivarcsi, H. Polyanka, A. Kenderessy-Szabo, G. Molnar, K. Szentpali, L. Bari, K. Megyeri, Y. Mandi, A. Dobozy, L. Kemeny and M. Szell, Identification and characterization of a novel, psoriasis susceptibility-related noncoding RNA gene, PRINS, *J. Biol. Chem.*, 2005, **280**, 24159–24167.
63. J. M. Silva, D. S. Perez, J. R. Pritchett, M. L. Halling, H. Tang and D. I. Smith, Identification of long stress-induced non-coding transcripts that have altered expression in cancer, *Genomics*, 2010, **95**, 355–362.
64. X. Wu and Y. Song, Preferential regulation of miRNA targets by environmental chemicals in the human genome, *BMC Genomics*, 2011, **12**, 244.
65. L. Smirnova, A. Sittka and A. Luch, On the role of low-dose effects and epigenetics in toxicology, *EXS*, 2012, **101**, 499–550.
66. S. Choudhuri, Small noncoding RNAs: biogenesis, function, and emerging significance in toxicology, *J. Biochem. Mol. Toxicol.*, 2010, **24**, 195–216.
67. A. K. Marrone, F. A. Beland and I. P. Pogribny, Noncoding RNA response to xenobiotic exposure: an indicator of toxicity and carcinogenicity, *Expert Opin. Drug. Metab. Toxicol.*, 2014, **10**, 1409–1422.
68. K. Vrijens, V. Bollati and T. S. Nawrot, MicroRNAs as potential signatures of environmental exposure or effect: a systematic review, *Environ. Health Perspect.*, 2015, **123**, 399–411.
69. R. A. Currie, Toxicogenomics: the challenges and opportunities to identify biomarkers, signatures and thresholds to support mode-of-action, *Mutat. Res.*, 2012, **746**, 97–103.
70. Y. Cui and R. S. Paules, Use of transcriptomics in understanding mechanisms of drug-induced toxicity, *Pharmacogenomics*, 2010, **11**, 573–585.
71. C. M. McHale, L. Zhang, Q. Lan, G. Li, A. E. Hubbard, M. S. Forrest, R. Vermeulen, J. Chen, M. Shen, S. M. Rappaport, S. Yin, M. T. Smith and N. Rothman, Changes in the peripheral blood transcriptome associated with occupational benzene exposure identified by cross-comparison on two microarray platforms, *Genomics*, 2009, **93**, 343–349.
72. R. Thomas, C. M. McHale, Q. Lan, A. E. Hubbard, L. Zhang, R. Vermeulen, G. Li, S. M. Rappaport, S. Yin, N. Rothman and M. T. Smith, Global gene expression response of a population exposed to benzene: a pilot study exploring the use of RNA-sequencing technology, *Environ. Mol. Mutagen.*, 2013, **54**, 566–573.
73. M. S. Forrest, Q. Lan, A. E. Hubbard, L. Zhang, R. Vermeulen, X. Zhao, G. Li, Y. Y. Wu, M. Shen, S. Yin, S. J. Chanock, N. Rothman and

M. T. Smith, Discovery of novel biomarkers by microarray analysis of peripheral blood mononuclear cell gene expression in benzene-exposed workers, *Environ. Health Perspect.*, 2005, **113**, 801–807.

74. C. M. McHale, L. Zhang, Q. Lan, R. Vermeulen, G. Li, A. E. Hubbard, K. E. Porter, R. Thomas, C. J. Portier, M. Shen, S. M. Rappaport, S. Yin, M. T. Smith and N. Rothman, Global gene expression profiling of a population exposed to a range of benzene levels, *Environ. Health Perspect.*, 2011, **119**, 628–634.
75. C. Schurmann, K. Heim, A. Schillert, S. Blankenberg, M. Carstensen, M. Dorr, K. Endlich, S. B. Felix, C. Gieger, H. Grallert, C. Herder, W. Hoffmann, G. Homuth, T. Illig, J. Kruppa, T. Meitinger, C. Muller, M. Nauck, A. Peters, R. Rettig, M. Roden, K. Strauch, U. Volker, H. Volzke, S. Wahl, H. Wallaschofski, P. S. Wild, T. Zeller, A. Teumer, H. Prokisch and A. Ziegler, Analyzing illumina gene expression microarray data from different tissues: methodological aspects of data analysis in the metaxpress consortium, *PLoS One*, 2012, **7**, e50938.
76. R. R. Kitchen, V. S. Sabine, A. A. Simen, J. M. Dixon, J. M. Bartlett and A. H. Sims, Relative impact of key sources of systematic noise in Affymetrix and Illumina gene-expression microarray experiments, *BMC Genomics*, 2011, **12**, 589.
77. R. Thomas, A. E. Hubbard, C. M. McHale, L. Zhang, S. M. Rappaport, Q. Lan, N. Rothman, R. Vermeulen, K. Z. Guyton, J. Jinot, B. R. Sonawane and M. T. Smith, Characterization of changes in gene expression and biochemical pathways at low levels of benzene exposure, *PLoS One*, 2014, **9**, e91828.
78. S. De Coster, D. M. van Leeuwen, D. G. Jennen, G. Koppen, E. Den Hond, V. Nelen, G. Schoeters, W. Baeyens, J. H. van Delft, J. C. Kleinjans and N. van Larebeke, Gender-specific transcriptomic response to environmental exposure in Flemish adults, *Environ. Mol. Mutagen.*, 2013, **54**, 574–588.
79. K. Hochstenbach, D. M. van Leeuwen, H. Gmuender, R. W. Gottschalk, M. Lovik, B. Granum, U. Nygaard, E. Namork, M. Kirsch-Volders, I. Decordier, K. Vande Loock, H. Besselink, M. Tornqvist, H. von Stedingk, P. Rydberg, J. C. Kleinjans, H. van Loveren and J. H. van Delft, Global gene expression analysis in cord blood reveals gender-specific differences in response to carcinogenic exposure *in utero*, *Cancer Epidemiol., Biomarkers Prev.*, 2012, **21**, 1756–1767.
80. A. Munoz, Y. Chervona, M. Hall, T. Kluz, M. V. Gamble and M. Costa, Sex-specific patterns and deregulation of endocrine pathways in the gene expression profiles of Bangladeshi adults exposed to arsenic contaminated drinking water, *Toxicol. Appl. Pharmacol.*, 2015, **284**, 330–338.
81. A. Gao, J. Yang, G. Yang, P. Niu and L. Tian, Differential gene expression profiling analysis in workers occupationally exposed to benzene, *Sci. Total Environ.*, 2014, **472**, 872–879.

82. W. Bai, J. Yang, G. Yang, P. Niu, L. Tian and A. Gao, Long non-coding RNA NR_045623 and NR_028291 involved in benzene hematotoxicity in occupationally benzene-exposed workers, *Exp. Mol. Pathol.*, 2014, **96**, 354–360.
83. W. Bai, Y. Chen, J. Yang, P. Niu, L. Tian and A. Gao, Aberrant miRNA profiles associated with chronic benzene poisoning, *Exp. Mol. Pathol.*, 2014, **96**, 426–430.
84. J. A. Bourdon-Lacombe, I. D. Moffat, M. Deveau, M. Husain, S. Auerbach, D. Krewski, R. S. Thomas, P. R. Bushel, A. Williams and C. L. Yauk, Technical guide for applications of gene expression profiling in human health risk assessment of environmental chemicals, *Regul. Toxicol. Pharmacol.*, 2015, **72**, 292–309.
85. S. Sand, C. J. Portier and D. Krewski, A signal-to-noise crossover dose as the point of departure for health risk assessment, *Environ. Health Perspect.*, 2011, **119**, 1766–1774.
86. R. S. Thomas, H. J. Clewell, 3rd, B. C. Allen, S. C. Wesselkamper, N. C. Wang, J. C. Lambert, J. K. Hess-Wilson, Q. J. Zhao and M. E. Andersen, Application of transcriptional benchmark dose values in quantitative cancer and noncancer risk assessment, *Toxicol. Sci.*, 2011, **120**, 194–205.
87. R. S. Thomas, H. J. Clewell, 3rd, B. C. Allen, L. Yang, E. Healy and M. E. Andersen, Integrating pathway-based transcriptomic data into quantitative chemical risk assessment: a five chemical case study, *Mutat. Res.*, 2012, **746**, 135–143.
88. R. S. Thomas, S. C. Wesselkamper, N. C. Wang, Q. J. Zhao, D. D. Petersen, J. C. Lambert, I. Cote, L. Yang, E. Healy, M. B. Black, H. J. Clewell, 3rd, B. C. Allen and M. E. Andersen, Temporal concordance between apical and transcriptional points of departure for chemical risk assessment, *Toxicol. Sci.*, 2013, **134**, 180–194.
89. N. L. Chepelev, I. D. Moffat, S. Labib, J. Bourdon-Lacombe, B. Kuo, J. K. Buick, F. Lemieux, A. I. Malik, S. Halappanavar, A. Williams and C. L. Yauk, Integrating toxicogenomics into human health risk assessment: lessons learned from the benzo[a]pyrene case study, *Crit. Rev. Toxicol.*, 2015, **45**, 44–52.
90. I. Moffat, N. L. Chepelev, S. Labib, J. Bourdon-Lacombe, B. Kuo, J. K. Buick, F. Lemieux, A. Williams, S. Halappanavar, A. I. Malik, M. Luijten, J. Aubrecht, D. R. Hyduke, A. J. Fornace, Jr., C. D. Swartz, L. Recio and C. L. Yauk, Comparison of toxicogenomics and traditional approaches to inform mode of action and points of departure in human health risk assessment of benzo[a]pyrene in drinking water, *Crit. Rev. Toxicol.*, 2015, **45**, 1–43.
91. Z. Wang, M. Gerstein and M. Snyder, RNA-Seq: a revolutionary tool for transcriptomics, *Nat. Rev. Genet.*, 2009, **10**, 57–63.
92. B. Ning, Z. Su, N. Mei, H. Hong, H. Deng, L. Shi, J. C. Fuscoe and W. H. Tolleson, Toxicogenomics and cancer susceptibility: advances

with next-generation sequencing, *J. Environ. Sci. Health, Part C: Environ. Carcinog. Ecotoxicol. Rev.*, 2014, **32**, 121–158.
93. S. Liu, L. Lin, P. Jiang, D. Wang and Y. Xing, A comparison of RNA-Seq and high-density exon array for detecting differential gene expression between closely related species, *Nucleic Acids Res.*, 2011, **39**, 578–588.
94. N. Raghavachari, J. Barb, Y. Yang, P. Liu, K. Woodhouse, D. Levy, C. J. O'Donnell, P. J. Munson and G. J. Kato, A systematic comparison and evaluation of high density exon arrays and RNA-seq technology used to unravel the peripheral blood transcriptome of sickle cell disease, *BMC Med. Genomics*, 2012, **5**, 28.
95. J. Beane, J. Vick, F. Schembri, C. Anderlind, A. Gower, J. Campbell, L. Luo, X. H. Zhang, J. Xiao, Y. O. Alekseyev, S. Wang, S. Levy, P. P. Massion, M. Lenburg and A. Spira, Characterizing the impact of smoking and lung cancer on the airway transcriptome using RNA-Seq, *Cancer Prev. Res.*, 2011, **4**, 803–817.
96. N. R. Hackett, M. W. Butler, R. Shaykhiev, J. Salit, L. Omberg, J. L. Rodriguez-Flores, J. G. Mezey, Y. Strulovici-Barel, G. Wang, L. Didon and R. G. Crystal, RNA-Seq quantification of the human small airway epithelium transcriptome, *BMC Genomics*, 2012, **13**, 82.
97. Z. Su, Z. Li, T. Chen, Q. Z. Li, H. Fang, D. Ding, W. Ge, B. Ning, H. Hong, R. G. Perkins, W. Tong and L. Shi, Comparing next-generation sequencing and microarray technologies in a toxicological study of the effects of aristolochic acid on rat kidneys, *Chem. Res. Toxicol.*, 2011, **24**, 1486–1493.
98. J. van Delft, S. Gaj, M. Lienhard, M. W. Albrecht, A. Kirpiy, K. Brauers, S. Claessen, D. Lizarraga, H. Lehrach, R. Herwig and J. Kleinjans, RNA-Seq provides new insights in the transcriptome responses induced by the carcinogen benzo[a]pyrene, *Toxicol. Sci.*, 2012, **130**, 427–439.
99. X. Zhou, L. Ren, Q. Meng, Y. Li, Y. Yu and J. Yu, The next-generation sequencing technology and application, *Protein Cell*, 2010, **1**, 520–536.
100. D. MacLean, J. D. Jones and D. J. Studholme, Application of 'next-generation' sequencing technologies to microbial genetics, *Nat. Rev. Microbiol.*, 2009, **7**, 287–296.
101. S. R. Atkinson, S. Marguerat and J. Bahler, Exploring long non-coding RNAs through sequencing, *Semin. Cell Dev. Biol.*, 2012, **23**, 200–205.
102. V. Costa, M. Aprile, R. Esposito and A. Ciccodicola, RNA-Seq and human complex diseases: recent accomplishments and future perspectives, *Eur. J. Hum. Genet.*, 2013, **21**, 134–142.
103. K. O. Mutz, A. Heilkenbrinker, M. Lonne, J. G. Walter and F. Stahl, Transcriptome analysis using next-generation sequencing, *Curr. Opin. Biotechnol.*, 2013, **24**, 22–30.
104. I. Chepelev, Detection of RNA editing events in human cells using high-throughput sequencing, *Methods Mol. Biol.*, 2012, **815**, 91–102.
105. R. D. Morin, M. D. O'Connor, M. Griffith, F. Kuchenbauer, A. Delaney, A. L. Prabhu, Y. Zhao, H. McDonald, T. Zeng, M. Hirst, C. J. Eaves and M. A. Marra, Application of massively parallel sequencing to microRNA

profiling and discovery in human embryonic stem cells, *Genome Res.*, 2008, **18**, 610–621.

106. Z. K. Atak, V. Gianfelici, G. Hulselmans, K. De Keersmaecker, A. G. Devasia, E. Geerdens, N. Mentens, S. Chiaretti, K. Durinck, A. Uyttebroeck, P. Vandenberghe, I. Wlodarska, J. Cloos, R. Foa, F. Speleman, J. Cools and S. Aerts, Comprehensive analysis of transcriptome variation uncovers known and novel driver events in T-cell acute lymphoblastic leukemia, *PLoS Genet.*, 2013, **9**, e1003997.
107. A. D. Kostic, A. I. Ojesina, C. S. Pedamallu, J. Jung, R. G. Verhaak, G. Getz and M. Meyerson, PathSeq: software to identify or discover microbes by deep sequencing of human tissue, *Nat. Biotechnol.*, 2011, **29**, 393–396.
108. K. D. Hansen, R. A. Irizarry and Z. Wu, Removing technical variability in RNA-seq data using conditional quantile normalization, *Biostatistics*, 2012, **13**, 204–216.
109. Z. Fang and X. Cui, Design and validation issues in RNA-seq experiments, *Briefings Bioinf.*, 2011, **12**, 280–287.
110. S. Schwartz, R. Oren and G. Ast, Detection and removal of biases in the analysis of next-generation sequencing reads, *PLoS One*, 2011, **6**, e16685.
111. K. D. Hansen, S. E. Brenner and S. Dudoit, Biases in Illumina transcriptome sequencing caused by random hexamer priming, *Nucleic Acids Res.*, 2010, **38**, e131.
112. J. H. Bullard, E. Purdom, K. D. Hansen and S. Dudoit, Evaluation of statistical methods for normalization and differential expression in mRNA-Seq experiments, *BMC Bioinf.*, 2010, **11**, 94.
113. SEQC Maqc-III Consortium, A comprehensive assessment of RNA-seq accuracy, reproducibility and information content by the Sequencing Quality Control Consortium, *Nat. Biotechnol.*, 2014, **32**, 903–914.
114. S. Li, P. P. Labaj, P. Zumbo, P. Sykacek, W. Shi, L. Shi, J. Phan, P. Y. Wu, M. Wang, C. Wang, D. Thierry-Mieg, J. Thierry-Mieg, D. P. Kreil and C. E. Mason, Detecting and correcting systematic variation in large-scale RNA sequencing data, *Nat. Biotechnol.*, 2014, **32**, 888–895.
115. S. Li, S. W. Tighe, C. M. Nicolet, D. Grove, S. Levy, W. Farmerie, A. Viale, C. Wright, P. A. Schweitzer, Y. Gao, D. Kim, J. Boland, B. Hicks, R. Kim, S. Chhangawala, N. Jafari, N. Raghavachari, J. Gandara, N. Garcia-Reyero, C. Hendrickson, D. Roberson, J. Rosenfeld, T. Smith, J. G. Underwood, M. Wang, P. Zumbo, D. A. Baldwin, G. S. Grills and C. E. Mason, Multi-platform assessment of transcriptome profiling using RNA-seq in the ABRF next-generation sequencing study, *Nat. Biotechnol.*, 2014, **32**, 915–925.
116. D. Risso, J. Ngai, T. P. Speed and S. Dudoit, Normalization of RNA-seq data using factor analysis of control genes or samples, *Nat. Biotechnol.*, 2014, **32**, 896–902.
117. C. Wang, B. Gong, P. R. Bushel, J. Thierry-Mieg, D. Thierry-Mieg, J. Xu, H. Fang, H. Hong, J. Shen, Z. Su, J. Meehan, X. Li, L. Yang, H. Li,

P. P. Labaj, D. P. Kreil, D. Megherbi, S. Gaj, F. Caiment, J. van Delft, J. Kleinjans, A. Scherer, V. Devanarayan, J. Wang, Y. Yang, H. R. Qian, L. J. Lancashire, M. Bessarabova, Y. Nikolsky, C. Furlanello, M. Chierici, D. Albanese, G. Jurman, S. Riccadonna, M. Filosi, R. Visintainer, K. K. Zhang, J. Li, J. H. Hsieh, D. L. Svoboda, J. C. Fuscoe, Y. Deng, L. Shi, R. S. Paules, S. S. Auerbach and W. Tong, The concordance between RNA-seq and microarray data depends on chemical treatment and transcript abundance, *Nat. Biotechnol.*, 2014, **32**, 926–932.

118. Q. Lan, M. T. Smith, X. Tang, W. Guo, R. Vermeulen, Z. Ji, W. Hu, A. E. Hubbard, M. Shen, C. M. McHale, C. Qiu, S. Liu, B. Reiss, L. Beane-Freeman, A. Blair, Y. Ge, J. Xiong, L. Li, S. M. Rappaport, H. Huang, N. Rothman and L. Zhang, Chromosome-wide aneuploidy study of cultured circulating myeloid progenitor cells from workers occupationally exposed to formaldehyde, *Carcinogenesis*, 2015, **36**, 160–167.

119. K. T. Lin, Y. J. Shann, G. Y. Chau, C. N. Hsu and C. Y. Huang, Identification of latent biomarkers in hepatocellular carcinoma by ultra-deep whole-transcriptome sequencing, *Oncogene*, 2014, **33**, 4786–4794.

120. J. D. Mills, T. Nalpathamkalam, H. I. Jacobs, C. Janitz, D. Merico, P. Hu and M. Janitz, RNA-Seq analysis of the parietal cortex in Alzheimer's disease reveals alternatively spliced isoforms related to lipid metabolism, *Neurosci. Lett.*, 2013, **536**, 90–95.

121. S. J. Jakhesara, P. G. Koringa, V. D. Bhatt, T. M. Shah, S. Vangipuram, S. Shah and C. G. Joshi, RNA-Seq reveals differentially expressed isoforms and novel splice variants in buccal mucosal cancer, *Gene*, 2013, **516**, 24–32.

122. T. Kavanagh, J. D. Mills, W. S. Kim, G. M. Halliday and M. Janitz, Pathway Analysis of the Human Brain Transcriptome in Disease, *J. Mol. Neurosci.*, 2013, **51**, 28–36.

123. H. Bahadar, F. Maqbool, S. Mostafalou, M. Baeeri, M. Rahimifard, M. Navaei-Nigjeh and M. Abdollahi, Assessment of benzene induced oxidative impairment in rat isolated pancreatic islets and effect on insulin secretion, *Environ. Toxicol. Pharmacol.*, 2015, **39**, 1161–1169.

124. S. Tondeur, C. Pangault, T. Le Carrour, Y. Lannay, R. Benmahdi, A. Cubizolle, S. Assou, V. Pantesco, B. Klein, S. Hamamah, J. F. Schved, T. Fest and J. De Vos, Expression map of the human exome in CD34+ cells and blood cells: increased alternative splicing in cell motility and immune response genes, *PLoS One*, 2010, **5**, e8990.

125. H. Makishima, V. Visconte, H. Sakaguchi, A. M. Jankowska, S. Abu Kar, A. Jerez, B. Przychodzen, M. Bupathi, K. Guinta, M. G. Afable, M. A. Sekeres, R. A. Padgett, R. V. Tiu and J. P. Maciejewski, Mutations in the spliceosome machinery, a novel and ubiquitous pathway in leukemogenesis, *Blood*, 2012, **119**, 3203–3210.

126. A. Dolnik, J. C. Engelmann, M. Scharfenberger-Schmeer, J. Mauch, S. Kelkenberg-Schade, B. Haldemann, T. Fries, J. Kronke, M. W. Kuhn, P. Paschka, S. Kayser, S. Wolf, V. I. Gaidzik, R. F. Schlenk, F. G. Rucker, H. Dohner, C. Lottaz, K. Dohner and L. Bullinger, Commonly altered

genomic regions in acute myeloid leukemia are enriched for somatic mutations involved in chromatin remodeling and splicing, *Blood*, 2012, **120**, e83–e92.

127. J. P. Maciejewski and R. A. Padgett, Defects in spliceosomal machinery: a new pathway of leukaemogenesis, *Br. J. Haematol.*, 2012, **158**, 165–173.
128. S. Colla, D. S. Ong, Y. Ogoti, M. Marchesini, N. A. Mistry, K. Clise-Dwyer, S. A. Ang, P. Storti, A. Viale, N. Giuliani, K. Ruisaard, I. Ganan Gomez, C. A. Bristow, M. Estecio, D. C. Weksberg, Y. W. Ho, B. Hu, G. Genovese, P. Pettazzoni, A. S. Multani, S. Jiang, S. Hua, M. C. Ryan, A. Carugo, L. Nezi, Y. Wei, H. Yang, M. D'Anca, L. Zhang, S. Gaddis, T. Gong, J. W. Horner, T. P. Heffernan, P. Jones, L. J. Cooper, H. Liang, H. Kantarjian, Y. A. Wang, L. Chin, C. Bueso-Ramos, G. Garcia-Manero and R. A. DePinho, Telomere dysfunction drives aberrant hematopoietic differentiation and myelodysplastic syndrome, *Cancer Cell*, 2015, **27**, 644–657.
129. S. Adamia, B. Haibe-Kains, P. M. Pilarski, M. Bar-Natan, S. Pevzner, H. Avet-Loiseau, L. Lode, S. Verselis, E. A. Fox, J. Burke, I. Galinsky, I. Dagogo-Jack, M. Wadleigh, D. P. Steensma, G. Motyckova, D. J. Deangelo, J. Quackenbush, R. Stone and J. D. Griffin, A genome-wide aberrant RNA splicing in patients with acute myeloid leukemia identifies novel potential disease markers and therapeutic targets, *Clin. Cancer Res.*, 2014, **20**, 1135–1145.
130. T. Wu and X. D. Fu, Genomic functions of U2AF in constitutive and regulated splicing, *RNA Biol.*, 2015, **12**, 479–485.
131. B. Przychodzen, A. Jerez, K. Guinta, M. A. Sekeres, R. Padgett, J. P. Maciejewski and H. Makishima, Patterns of missplicing due to somatic U2AF1 mutations in myeloid neoplasms, *Blood*, 2013, **122**, 999–1006.
132. A. N. Brooks, P. S. Choi, L. de Waal, T. Sharifnia, M. Imielinski, G. Saksena, C. S. Pedamallu, A. Sivachenko, M. Rosenberg, J. Chmielecki, M. S. Lawrence, D. S. DeLuca, G. Getz and M. Meyerson, A pan-cancer analysis of transcriptome changes associated with somatic mutations in U2AF1 reveals commonly altered splicing events, *PLoS One*, 2014, **9**, e87361.
133. M. Hoxha, L. Dioni, M. Bonzini, A. C. Pesatori, S. Fustinoni, D. Cavallo, M. Carugno, B. Albetti, B. Marinelli, J. Schwartz, P. A. Bertazzi and A. Baccarelli, Association between leukocyte telomere shortening and exposure to traffic pollution: a cross-sectional study on traffic officers and indoor office workers, *Environ. Health*, 2009, **8**, 41.
134. B. A. Bassig, L. Zhang, R. M. Cawthon, M. T. Smith, S. Yin, G. Li, W. Hu, M. Shen, S. Rappaport, F. Barone-Adesi, N. Rothman, R. Vermeulen and Q. Lan, Alterations in leukocyte telomere length in workers occupationally exposed to benzene, *Environ. Mol. Mutagen.*, 2014, **55**, 673–678.
135. M. De Braekeleer, E. De Braekeleer and N. Douet-Guilbert, Geographic/ethnic variability of chromosomal and molecular abnormalities in leukemia, *Expert Rev. Anticancer Ther.*, 2015, **15**, 1093–1102.

136. F. Mertens, B. Johansson, T. Fioretos and F. Mitelman, The emerging complexity of gene fusions in cancer, *Nat. Rev. Cancer*, 2015, **15**, 371–381.
137. J. H. Wang, Mechanisms and impacts of chromosomal translocations in cancers, *Front. Med.*, 2012, **6**, 263–274.
138. J. Zheng, Oncogenic chromosomal translocations and human cancer (review), *Oncol. Rep.*, 2013, **30**, 2011–2019.
139. L. Fernandez-Cuesta, R. Sun, R. Menon, J. George, S. Lorenz, L. A. Meza-Zepeda, M. Peifer, D. Plenker, J. M. Heuckmann, F. Leenders, T. Zander, I. Dahmen, M. Koker, J. Schottle, R. T. Ullrich, J. Altmuller, C. Becker, P. Nurnberg, H. Seidel, D. Bohm, F. Goke, S. Ansen, P. A. Russell, G. M. Wright, Z. Wainer, B. Solomon, I. Petersen, J. H. Clement, J. Sanger, O. T. Brustugun, A. Helland, S. Solberg, M. Lund-Iversen, R. Buettner, J. Wolf, E. Brambilla, M. Vingron, S. Perner, S. A. Haas and R. K. Thomas, Identification of novel fusion genes in lung cancer using breakpoint assembly of transcriptome sequencing data, *Genome Biol.*, 2015, **16**, 7.
140. J. Kim, S. Kim, S. Ko, Y. H. In, H. G. Moon, S. K. Ahn, M. K. Kim, M. Lee, J. H. Hwang, Y. S. Ju, J. I. Kim, D. Y. Noh, S. Kim, J. H. Park, H. Rhee, S. Kim and W. Han, Recurrent fusion transcripts detected by whole-transcriptome sequencing of 120 primary breast cancer samples, *Genes, Chromosomes Cancer*, 2015, **54**, 681–691.
141. C. A. Maher, C. Kumar-Sinha, X. Cao, S. Kalyana-Sundaram, B. Han, X. Jing, L. Sam, T. Barrette, N. Palanisamy and A. M. Chinnaiyan, Transcriptome sequencing to detect gene fusions in cancer, *Nature*, 2009, **458**, 97–101.
142. Q. Zhao, O. L. Caballero, S. Levy, B. J. Stevenson, C. Iseli, S. J. de Souza, P. A. Galante, D. Busam, M. A. Leversha, K. Chadalavada, Y. H. Rogers, J. C. Venter, A. J. Simpson and R. L. Strausberg, Transcriptome-guided characterization of genomic rearrangements in a breast cancer cell line, *Proc. Natl. Acad. Sci. U. S. A.*, 2009, **106**, 1886–1891.
143. Y. Wang, N. Wu, J. Liu, Z. Wu and D. Dong, FusionCancer: a database of cancer fusion genes derived from RNA-seq data, *Diagn. Pathol.*, 2015, **10**, 131.
144. Q. Wang, J. Xia, P. Jia, W. Pao and Z. Zhao, Application of next generation sequencing to human gene fusion detection: computational tools, features and perspectives, *Briefings Bioinf.*, 2013, **14**, 506–519.
145. A. McPherson, F. Hormozdiari, A. Zayed, R. Giuliany, G. Ha, M. G. Sun, M. Griffith, A. Heravi Moussavi, J. Senz, N. Melnyk, M. Pacheco, M. A. Marra, M. Hirst, T. O. Nielsen, S. C. Sahinalp, D. Huntsman and S. P. Shah, deFuse: an algorithm for gene fusion discovery in tumor RNA-Seq data, *PLoS Comput. Biol.*, 2011, **7**, e1001138.
146. J. Wu, W. Zhang, S. Huang, Z. He, Y. Cheng, J. Wang, T. W. Lam, Z. Peng and S. M. Yiu, SOAPfusion: a robust and effective computational fusion discovery tool for RNA-seq reads, *Bioinformatics*, 2013, **29**, 2971–2978.

147. D. Kim and S. L. Salzberg, TopHat-Fusion: an algorithm for discovery of novel fusion transcripts, *Genome Biol.*, 2011, **12**, R72.
148. H. Ge, K. Liu, T. Juan, F. Fang, M. Newman and W. Hoeck, FusionMap: detecting fusion genes from next-generation sequencing data at base-pair resolution, *Bioinformatics*, 2011, **27**, 1922–1928.
149. R. W. Francis, K. Thompson-Wicking, K. W. Carter, D. Anderson, U. R. Kees and A. H. Beesley, FusionFinder: a software tool to identify expressed gene fusion candidates from RNA-Seq data, *PLoS One*, 2012, **7**, e39987.
150. M. Carrara, M. Beccuti, F. Cavallo, S. Donatelli, F. Lazzarato, F. Cordero and R. A. Calogero, State of art fusion-finder algorithms are suitable to detect transcription-induced chimeras in normal tissues?, *BMC Bioinf.*, 2013, **14**(Suppl 7), S2.
151. M. Carrara, M. Beccuti, F. Lazzarato, F. Cavallo, F. Cordero, S. Donatelli and R. A. Calogero, State-of-the-art fusion-finder algorithms sensitivity and specificity, *Biomed. Res. Int.*, 2013, **2013**, 340620.
152. N. M. Davidson, I. J. Majewski and A. Oshlack, JAFFA: High sensitivity transcriptome-focused fusion gene detection, *Genome Med.*, 2015, **7**, 43.
153. R. Masetti, M. Pigazzi, M. Togni, A. Astolfi, V. Indio, E. Manara, R. Casadio, A. Pession, G. Basso and F. Locatelli, CBFA2T3-GLIS2 fusion transcript is a novel common feature in pediatric, cytogenetically normal AML, not restricted to FAB M7 subtype, *Blood*, 2013, **121**, 3469–3472.
154. R. Masetti, M. Togni, A. Astolfi, M. Pigazzi, E. Manara, V. Indio, C. Rizzari, S. Rutella, G. Basso, A. Pession and F. Locatelli, DHH-RHEBL1 fusion transcript: a novel recurrent feature in the new landscape of pediatric CBFA2T3-GLIS2-positive acute myeloid leukemia, *Oncotarget*, 2013, **4**, 1712–1720.
155. H. Wen, Y. Li, S. N. Malek, Y. C. Kim, J. Xu, P. Chen, F. Xiao, X. Huang, X. Zhou, Z. Xuan, S. Mankala, G. Hou, J. D. Rowley, M. Q. Zhang and S. M. Wang, New fusion transcripts identified in normal karyotype acute myeloid leukemia, *PLoS One*, 2012, **7**, e51203.
156. H. Van den Berghe, K. Vermaelen, C. Mecucci, D. Barbieri and G. Tricot, The 5q-anomaly, *Cancer Genet. Cytogenet.*, 1985, **17**, 189–255.
157. R. Masetti, M. Togni, A. Astolfi, M. Pigazzi, V. Indio, B. Rivalta, E. Manara, S. Rutella, G. Basso, A. Pession and F. Locatelli, Whole transcriptome sequencing of a paediatric case of *de novo* acute myeloid leukaemia with del(5q) reveals RUNX1-USP42 and PRDM16-SKI fusion transcripts, *Br. J. Haematol.*, 2014, **166**, 449–452.
158. C. M. McHale, Q. Lan, C. Corso, G. Li, L. Zhang, R. Vermeulen, J. D. Curry, M. Shen, R. Turakulov, R. Higuchi, S. Germer, S. Yin, N. Rothman and M. T. Smith, Chromosome translocations in workers exposed to benzene, *J. Natl. Cancer Inst. Monogr.*, 2008, 74–77.
159. L. Zhang, N. Rothman, G. Li, W. Guo, W. Yang, A. E. Hubbard, R. B. Hayes, S. Yin, W. Lu and M. T. Smith, Aberrations in chromosomes associated with lymphoma and therapy-related leukemia in benzene-exposed workers, *Environ. Mol. Mutagen.*, 2007, **48**, 467–474.

160. V. Bouvard, R. Baan, K. Straif, Y. Grosse, B. Secretan, F. El Ghissassi, L. Benbrahim-Tallaa, N. Guha, C. Freeman, L. Galichet and V. Cogliano, W. H. O. International Agency for Research on Cancer Monograph Working Group. A review of human carcinogens–Part B: biological agents, *Lancet Oncol.*, 2009, **10**, 321–322.
161. D. M. Parkin, The global health burden of infection-associated cancers in the year 2002, *Int. J. Cancer*, 2006, **118**, 3030–3044.
162. R. Sarid and S. J. Gao, Viruses and human cancer: from detection to causality, *Cancer Lett.*, 2011, **305**, 218–227.
163. R. T. Javier and J. S. Butel, The history of tumor virology, *Cancer Res.*, 2008, **68**, 7693–7706.
164. H. Feng, J. L. Taylor, P. V. Benos, R. Newton, K. Waddell, S. B. Lucas, Y. Chang and P. S. Moore, Human transcriptome subtraction by using short sequence tags to search for tumor viruses in conjunctival carcinoma, *J. Virol.*, 2007, **81**, 11332–11340.
165. R. A. Moore, R. L. Warren, J. D. Freeman, J. A. Gustavsen, C. Chenard, J. M. Friedman, C. A. Suttle, Y. Zhao and R. A. Holt, The sensitivity of massively parallel sequencing for detecting candidate infectious agents associated with human tissue, *PLoS One*, 2011, **6**, e19838.
166. R. Naeem, M. Rashid and A. Pain, READSCAN: a fast and scalable pathogen discovery program with accurate genome relative abundance estimation, *Bioinformatics*, 2013, **29**, 391–392.
167. G. Weber, J. Shendure, D. M. Tanenbaum, G. M. Church and M. Meyerson, Identification of foreign gene sequences by transcript filtering against the human genome, *Nat. Genet.*, 2002, **30**, 141–142.
168. Y. Xu, N. Stange-Thomann, G. Weber, R. Bo, S. Dodge, R. G. David, K. Foley, J. Beheshti, N. L. Harris, B. Birren, E. S. Lander and M. Meyerson, Pathogen discovery from human tissue by sequence-based computational subtraction, *Genomics*, 2003, **81**, 329–335.
169. G. Xu, M. J. Strong, M. R. Lacey, C. Baribault, E. K. Flemington and C. M. Taylor, RNA CoMPASS: a dual approach for pathogen and host transcriptome analysis of RNA-seq datasets, *PLoS One*, 2014, **9**, e89445.
170. Y. Chen, H. Yao, E. J. Thompson, N. M. Tannir, J. N. Weinstein and X. Su, VirusSeq: software to identify viruses and their integration sites using next-generation sequencing of human cancer tissue, *Bioinformatics*, 2013, **29**, 266–267.
171. J. D. Khoury, N. M. Tannir, M. D. Williams, Y. Chen, H. Yao, J. Zhang, E. J. Thompson, T. Network, F. Meric-Bernstam, L. J. Medeiros, J. N. Weinstein and X. Su, Landscape of DNA virus associations across human malignant cancers: analysis of 3,775 cases using RNA-Seq, *J. Virol.*, 2013, **87**, 8916–8926.
172. G. K. Geiss, R. E. Bumgarner, B. Birditt, T. Dahl, N. Dowidar, D. L. Dunaway, H. P. Fell, S. Ferree, R. D. George, T. Grogan, J. J. James, M. Maysuria, J. D. Mitton, P. Oliveri, J. L. Osborn, T. Peng, A. L. Ratcliffe, P. J. Webster, E. H. Davidson, L. Hood and K. Dimitrov,

Direct multiplexed measurement of gene expression with color-coded probe pairs, *Nat. Biotechnol.*, 2008, **26**, 317–325.

173. P. P. Reis, L. Waldron, R. S. Goswami, W. Xu, Y. Xuan, B. Perez-Ordonez, P. Gullane, J. Irish, I. Jurisica and S. Kamel-Reid, mRNA transcript quantification in archival samples using multiplexed, color-coded probes, *BMC Biotechnol.*, 2011, **11**, 46.
174. J. M. Balko, R. S. Cook, D. B. Vaught, M. G. Kuba, T. W. Miller, N. E. Bhola, M. E. Sanders, N. M. Granja-Ingram, J. J. Smith, I. M. Meszoely, J. Salter, M. Dowsett, K. Stemke-Hale, A. M. Gonzalez-Angulo, G. B. Mills, J. A. Pinto, H. L. Gomez and C. L. Arteaga, Profiling of residual breast cancers after neoadjuvant chemotherapy identifies DUSP4 deficiency as a mechanism of drug resistance, *Nat. Med.*, 2012, **18**, 1052–1059.
175. T. P. Stricker, A. Morales La Madrid, A. Chlenski, L. Guerrero, H. R. Salwen, Y. Gosiengfiao, E. J. Perlman, W. Furman, A. Bahrami, J. M. Shohet, P. E. Zage, M. J. Hicks, H. Shimada, R. Suganuma, J. R. Park, S. So, W. B. London, P. Pytel, K. H. Maclean and S. L. Cohn, Validation of a prognostic multi-gene signature in high-risk neuroblastoma using the high throughput digital NanoString nCounter system, *Mol. Oncol.*, 2014, **8**, 669–678.
176. M. H. Veldman-Jones, R. Brant, C. Rooney, C. Geh, H. Emery, C. G. Harbron, M. Wappett, A. Sharpe, M. Dymond, J. C. Barrett, E. A. Harrington and G. Marshall, Evaluating Robustness and Sensitivity of the NanoString Technologies nCounter Platform to Enable Multiplexed Gene Expression Analysis of Clinical Samples, *Cancer Res.*, 2015, **75**, 2587–2593.
177. M. H. Veldman-Jones, Z. Lai, M. Wappett, C. G. Harbron, J. C. Barrett, E. A. Harrington and K. S. Thress, Reproducible, Quantitative, and Flexible Molecular Subtyping of Clinical DLBCL Samples Using the NanoString nCounter System, *Clin. Cancer Res.*, 2015, **21**, 2367–2378.
178. A. C. Richard, P. A. Lyons, J. E. Peters, D. Biasci, S. M. Flint, J. C. Lee, E. F. McKinney, R. M. Siegel and K. G. Smith, Comparison of gene expression microarray data with count-based RNA measurements informs microarray interpretation, *BMC Genomics*, 2014, **15**, 649.
179. L. Hou, J. Barupal, W. Zhang, Y. Zheng, L. Liu, X. Zhang, C. Dou, J. P. McCracken, A. Diaz, V. Motta, M. Sanchez-Guerra, K. R. Wolf, P. A. Bertazzi, J. D. Schwartz, S. Wang and A. A. Baccarelli, Particulate Air Pollution Exposure and Expression of Viral and Human MicroRNAs in Blood: The Beijing Truck Driver Air Pollution Study, *Environ. Health Perspect.*, 2015, DOI: 10.1289/ehp.1408519.
180. J. Lamb, E. D. Crawford, D. Peck, J. W. Modell, I. C. Blat, M. J. Wrobel, J. Lerner, J. P. Brunet, A. Subramanian, K. N. Ross, M. Reich, H. Hieronymus, G. Wei, S. A. Armstrong, S. J. Haggarty, P. A. Clemons, R. Wei, S. A. Carr, E. S. Lander and T. R. Golub, The Connectivity Map: using gene-expression signatures to connect small molecules, genes, and disease, *Science*, 2006, **313**, 1929–1935.

181. D. Peck, E. D. Crawford, K. N. Ross, K. Stegmaier, T. R. Golub and J. Lamb, A method for high-throughput gene expression signature analysis, *Genome Biol.*, 2006, **7**, R61.
182. Q. Duan, C. Flynn, M. Niepel, M. Hafner, J. L. Muhlich, N. F. Fernandez, A. D. Rouillard, C. M. Tan, E. Y. Chen, T. R. Golub, P. K. Sorger, A. Subramanian and A. Ma'ayan, LINCS Canvas Browser: interactive web app to query, browse and interrogate LINCS L1000 gene expression signatures, *Nucleic Acids Res.*, 2014, **42**, W449–W460.
183. National Research Council, *Toxicity testing in the 21st century: A vision and a strategy*, Washington, DC, 2007.
184. Department of Health and Human Services, List of Environmentally Responsive Human Genes Selected for Use In Screening Large Numbers of Substances Using Toxicogenomic Approaches, *Fed. Regist.*, 2015, **80**, 20237.
185. Department of Health and Human Services, Request for Information: The National Institute of Environmental Health Sciences/National Toxicology Program Requests the Nomination and Prioritization of Environmentally Responsive Genes for Use in Screening Large Numbers of Substances Using Toxicogenomic Technologies, *Fed. Regist.*, 2013, **78**, 45542.
186. R. Liu, X. Ma, L. Xu, D. Wang, X. Jiang, W. Zhu, B. Cui, G. Ning, D. Lin and S. Wang, Differential microRNA expression in peripheral blood mononuclear cells from Graves' disease patients, *J. Clin. Endocrinol. Metab.*, 2012, **97**, E968–E972.
187. J. C. Marioni, C. E. Mason, S. M. Mane, M. Stephens and Y. Gilad, RNA-seq: an assessment of technical reproducibility and comparison with gene expression arrays, *Genome Res.*, 2008, **18**, 1509–1517.
188. S. I. Daniels, F. C. Sille, A. Goldbaum, B. Yee, E. F. Key, L. Zhang, M. T. Smith and R. Thomas, Improving power to detect changes in blood miRNA expression by accounting for sources of variability in experimental designs, *Cancer Epidemiol., Biomarkers Prev.*, 2014, **23**, 2658–2666.
189. P. Leidinger, C. Backes, B. Meder, E. Meese and A. Keller, The human miRNA repertoire of different blood compounds, *BMC Genomics*, 2014, **15**, 474.
190. C. C. Liew, J. Ma, H. C. Tang, R. Zheng and A. A. Dempsey, The peripheral blood transcriptome dynamically reflects system wide biology: a potential diagnostic tool, *J. Lab. Clin. Med.*, 2006, **147**, 126–132.
191. I. S. Kohane and V. I. Valtchinov, Quantifying the white blood cell transcriptome as an accessible window to the multiorgan transcriptome, *Bioinformatics*, 2012, **28**, 538–545.
192. M. Rotunno, N. Hu, H. Su, C. Wang, A. M. Goldstein, A. W. Bergen, D. Consonni, A. C. Pesatori, P. A. Bertazzi, S. Wacholder, J. Shih, N. E. Caporaso, P. R. Taylor and M. T. Landi, A gene expression signature from peripheral whole blood for stage I lung adenocarcinoma, *Cancer Prev. Res.*, 2011, **4**, 1599–1608.

193. A. M. Newman, C. L. Liu, M. R. Green, A. J. Gentles, W. Feng, Y. Xu, C. D. Hoang, M. Diehn and A. A. Alizadeh, Robust enumeration of cell subsets from tissue expression profiles, *Nat. Methods*, 2015, **12**, 453–457.
194. G. Guo, S. Luc, E. Marco, T. W. Lin, C. Peng, M. A. Kerenyi, S. Beyaz, W. Kim, J. Xu, P. P. Das, T. Neff, K. Zou, G. C. Yuan and S. H. Orkin, Mapping cellular hierarchy by single-cell analysis of the cell surface repertoire, *Cell Stem Cell*, 2013, **13**, 492–505.
195. V. Moignard, I. C. Macaulay, G. Swiers, F. Buettner, J. Schutte, F. J. Calero-Nieto, S. Kinston, A. Joshi, R. Hannah, F. J. Theis, S. E. Jacobsen, M. F. de Bruijn and B. Gottgens, Characterization of transcriptional networks in blood stem and progenitor cells using high-throughput single-cell gene expression analysis, *Nat. Cell Biol.*, 2013, **15**, 363–372.
196. H. Greim, D. A. Kaden, R. A. Larson, C. M. Palermo, J. M. Rice, D. Ross and R. Snyder, The bone marrow niche, stem cells, and leukemia: impact of drugs, chemicals, and the environment, *Ann. N. Y. Acad. Sci.*, 2014, **1310**, 7–31.
197. D. G. Hebels, M. H. van Herwijnen, K. J. Brauers, T. M. de Kok, G. Chalkiadaki, S. A. Kyrtopoulos and J. C. Kleinjans, Elimination of heparin interference during microarray processing of fresh and biobank-archived blood samples, *Environ. Mol. Mutagen.*, 2014, **55**, 482–491.
198. D. G. Hebels, P. Georgiadis, H. C. Keun, T. J. Athersuch, P. Vineis, R. Vermeulen, L. Portengen, I. A. Bergdahl, G. Hallmans, D. Palli, B. Bendinelli, V. Krogh, R. Tumino, C. Sacerdote, S. Panico, J. C. Kleinjans, T. M. de Kok, M. T. Smith and S. A. Kyrtopoulos, Performance in Omics Analyses of Blood Samples in Long-Term Storage: Opportunities for the Exploitation of Existing Biobanks in Environmental Health Research, *Environ. Health Perspect.*, 2013, **121**, 480–487.
199. M. G. Barnes, A. A. Grom, T. A. Griffin, R. A. Colbert and S. D. Thompson, Gene Expression Profiles from Peripheral Blood Mononuclear Cells Are Sensitive to Short Processing Delays, *Biopreserv. Biobanking*, 2010, **8**, 153–162.
200. J. L. Min, A. Barrett, T. Watts, F. H. Pettersson, H. E. Lockstone, C. M. Lindgren, J. M. Taylor, M. Allen, K. T. Zondervan and M. I. McCarthy, Variability of gene expression profiles in human blood and lymphoblastoid cell lines, *BMC Genomics*, 2010, **11**, 96.
201. S. Debey-Pascher, A. Hofmann, F. Kreusch, G. Schuler, B. Schuler-Thurner, J. L. Schultze and A. Staratschek-Jox, RNA-stabilized whole blood samples but not peripheral blood mononuclear cells can be stored for prolonged time periods prior to transcriptome analysis, *J. Mol. Diagn.*, 2011, **13**, 452–460.
202. D. G. Weber, S. Casjens, P. Rozynek, M. Lehnert, S. Zilch-Schoneweis, O. Bryk, D. Taeger, M. Gomolka, M. Kreuzer, H. Otten, B. Pesch,

G. Johnen and T. Bruning, Assessment of mRNA and microRNA Stabilization in Peripheral Human Blood for Multicenter Studies and Biobanks, *Biomarker Insights*, 2010, **5**, 95–102.

203. T. Zander, A. Hofmann, A. Staratschek-Jox, S. Classen, S. Debey-Pascher, D. Maisel, S. Ansen, M. Hahn, M. Beyer, R. K. Thomas, B. Gathof, C. Mauch, K. S. Delank, W. Engel-Riedel, H. E. Wichmann, E. Stoelben, J. L. Schultze and J. Wolf, Blood-based gene expression signatures in non-small cell lung cancer, *Clin. Cancer Res.*, 2011, **17**, 3360–3367.
204. S. Debey, U. Schoenbeck, M. Hellmich, B. S. Gathof, R. Pillai, T. Zander and J. L. Schultze, Comparison of different isolation techniques prior gene expression profiling of blood derived cells: impact on physiological responses, on overall expression and the role of different cell types, *Pharmacogenomics J.*, 2004, **4**, 193–207.
205. S. Debey, T. Zander, B. Brors, A. Popov, R. Eils and J. L. Schultze, A highly standardized, robust, and cost-effective method for genome-wide transcriptome analysis of peripheral blood applicable to large-scale clinical trials, *Genomics*, 2006, **87**, 653–664.
206. J. Lucas, H. K. Dressman, S. Suchindran, M. Nakamura, N. J. Chao, H. Himburg, K. Minor, G. Phillips, J. Ross, M. Abedi, R. Terbrueggen and J. P. Chute, A translatable predictor of human radiation exposure, *PLoS One*, 2014, **9**, e107897.
207. M. Chadeau-Hyam, R. C. Vermeulen, D. G. Hebels, R. Castagne, G. Campanella, L. Portengen, R. S. Kelly, I. A. Bergdahl, B. Melin, G. Hallmans, D. Palli, V. Krogh, R. Tumino, C. Sacerdote, S. Panico, T. M. de Kok, M. T. Smith, J. C. Kleinjans, P. Vineis and S. A. Kyrtopoulos, c. EnviroGenoMarkers project, Prediagnostic transcriptomic markers of Chronic lymphocytic leukemia reveal perturbations 10 years before diagnosis, *Ann. Oncol.*, 2014, **25**, 1065–1072.
208. H. Shen, C. M. McHale, M. T. Smith and L. Zhang, Functional genomic screening approaches in mechanistic toxicology and potential future applications of CRISPR-Cas9, *Mutat. Res., Rev, Mutat. Res.*, 2015, **764**, 31–42.
209. K. S. Enfield, L. A. Pikor, V. D. Martinez and W. L. Lam, Mechanistic Roles of Noncoding RNAs in Lung Cancer Biology and Their Clinical Implications, *Genet. Res. Int.*, 2012, **2012**, 737416.
210. J. M. Gohlke, R. Thomas, Y. Zhang, M. C. Rosenstein, A. P. Davis, C. Murphy, K. G. Becker, C. J. Mattingly and C. J. Portier, Genetic and environmental pathways to complex diseases, *BMC Syst. Biol.*, 2009, **3**, 46.
211. C. Nardini, J. Dent and P. Tieri, Editorial: Multi-omic data integration, *Front. Cell Dev. Biol.*, 2015, **3**, 46.
212. V. Shivarov and L. Bullinger, Expression profiling of leukemia patients: key lessons and future directions, *Exp. Hematol.*, 2014, **42**, 651–660.
213. S. G. van Breda, L. C. Wilms, S. Gaj, D. G. Jennen, J. J. Briede, J. C. Kleinjans and T. M. de Kok, The exposome concept in a human nutrigenomics study: evaluating the impact of exposure to a complex

mixture of phytochemicals using transcriptomics signatures, *Mutagenesis*, 2015.
214. L. Zhang and S. A. Wang, A focused review of hematopoietic neoplasms occurring in the therapy-related setting, *Int. J. Clin. Exp. Pathol.*, 2014, **7**, 3512–3523.
215. IARC, Melphalan, *IARC Monogr. Eval. Carcinog. Risks Hum.*, 2012, **100A**, 107–117.
216. IARC, Etoposide, *IARC Monogr. Eval. Carcinog. Risks Hum.*, 2000, **76**, 177–257.
217. Z. Ji, L. Zhang, V. Peng, X. Ren, C. M. McHale and M. T. Smith, A comparison of the cytogenetic alterations and global DNA hypomethylation induced by the benzene metabolite, hydroquinone, with those induced by melphalan and etoposide, *Leukemia*, 2010, **24**, 986–991.
218. Z. Ji, L. Zhang, W. Guo, C. M. McHale and M. T. Smith, The benzene metabolite, hydroquinone and etoposide both induce endoreduplication in human lymphoblastoid TK6 cells, *Mutagenesis*, 2009, **24**, 367–372.
219. R. Thomas, J. Phuong, C. M. McHale and L. Zhang, Using bioinformatic approaches to identify pathways targeted by human leukemogens, *Int. J. Environ. Res. Public Health*, 2012, **9**, 2479–2503.
220. C. P. Wild, Complementing the genome with an "exposome": the outstanding challenge of environmental exposure measurement in molecular epidemiology, *Cancer Epidemiol., Biomarkers Prev.*, 2005, **14**, 1847–1850.
221. E. R. Mardis, Sequencing the AML genome, transcriptome, and epigenome, *Semin. Hematol.*, 2014, **51**, 250–258.
222. J. Beane, P. Sebastiani, G. Liu, J. S. Brody, M. E. Lenburg and A. Spira, Reversible and permanent effects of tobacco smoke exposure on airway epithelial gene expression, *Genome Biol.*, 2007, **8**, R201.
223. S. M. Rappaport, Discovering environmental causes of disease, *J. Epidemiol. Community Health*, 2012, **66**, 99–102.
224. S. M. Rappaport and M. T. Smith, Epidemiology. Environment and disease risks, *Science*, 2010, **330**, 460–461.
225. C. P. Wild, A. Scalbert and Z. Herceg, Measuring the exposome: a powerful basis for evaluating environmental exposures and cancer risk, *Environ. Mol. Mutagen.*, 2013, **54**, 480–499.
226. M. T. Smith, L. Zhang, C. M. McHale, C. F. Skibola and S. M. Rappaport, Benzene, the exposome and future investigations of leukemia etiology, *Chem. Biol. Interact.*, 2011, **192**, 155–159.

CHAPTER 12

Toxicogenomics Case Study: Furan

A. FRANCINA WEBSTER,[a,b] IAIN B. LAMBERT[a] AND CAROLE L. YAUK*[b]

[a] Department of Biology, Carleton University, 1125 Colonel By Drive, Ottawa, ON, Canada; [b] Environmental Health Science and Research Bureau, Health Canada, Tunney's Pasture, 50 Colombine Driveway, Ottawa, ON, Canada
*Email: carole.yauk@canada.ca

12.1 Introduction

A major goal in human health risk assessment is the identification and management of chemicals that may cause cancer in human populations. The current gold standard for assessing chemical carcinogenicity is the 2 year rodent cancer bioassay,[1] which is animal-, time-, and resource-intensive. The continuing development of genomics technologies and other high-throughput methods has the potential to revolutionize chemical cancer risk assessment by providing relevant data more quickly, using fewer animals, and at a lower cost than the 2 year cancer bioassay. The 2 year cancer bioassay produces a detailed inventory of chemical-dependent lesions; in contrast, toxicogenomics analyzes chemical-dependent changes to global gene expression and provides comprehensive mechanistic data that cannot be obtained using standard tests. In this chapter we show how quantitative, predictive, and mechanistic toxicogenomics have been applied to the evaluation of the hepatocarcinogen furan. We also discuss studies that demonstrate how toxicogenomics may be used to retrospectively analyze

Issues in Toxicology No. 28
Toxicogenomics in Predictive Carcinogenicity
Edited by Michael D. Waters and Russell S. Thomas

Published by the Royal Society of Chemistry, www.rsc.org

formalin-fixed and paraffin embedded (FFPE) samples that collectively represent a rich repository of conventional chemical toxicity data. Finally, we discuss how toxicogenomics can be incorporated into new approaches, such as tiered high-throughput chemical testing and the development of adverse outcome pathways (AOPs) that will be prominent components of the next generation of chemical risk assessment.

12.1.1 Chemical Testing and Toxicogenomics

Since the inception of the 2 year rodent cancer bioassay by the National Toxicology Program (NTP; http://ntp.niehs.nih.gov/) in the 1970s, only ~1500 chemicals have been assessed. Each chemical evaluation costs US$2–4 million and uses 800–1000 rodents.[2] However, international regulatory agencies are now facing the unprecedented challenge of evaluating tens of thousands of "legacy" chemicals. The US Environmental Protection Agency (EPA) and Europe's Registration, Evaluation, Authorisation and Restriction of Chemicals (EU-REACH) regulation list 75 000 and 140 000 chemicals, respectively, that require evaluation. In Canada, the Canadian Environmental Protection Act (CEPA, 1999) requires that chemicals on Canada's Domestic Substances List (DSL) be assessed for potential human health risks. There are more than 23 000 chemicals on the DSL. These chemicals have been divided into phases, based on existing exposure and hazard information. The regulators have completed phase 1 and 2 assessments, and upcoming priorities include completing evaluations for approximately 1700 of these chemicals (phase 3) by 2020. Chemicals are prioritized based on risk of exposure, risk of persistence or bioaccumulation, and estimated toxicity to humans and non-human species (CEPA Section 73, 1999). However, one of the challenges associated with this goal is that the majority of these chemicals are data-poor (*e.g.* have no conventional toxicological testing data associated with them). It is clear that a more efficient system to assess the potentially harmful substances that are present in the environment is necessary in order to address the massive backlog of yet-untested compounds, as well as the large number of novel compounds that enter the market and environment each year.

The toxicogenomics approach assumes that prior to the occurrence of visible phenotypic changes, such as tumors, important changes on a molecular scale must occur. Furthermore, it assumes that similar cancers share some common underlying biology at an early stage in the disease and that this biology is distinct from a generalized cellular toxic- or stress-response mechanism.[3] As reviewed in the introductory chapter of this volume, there are three main toxicogenomics approaches: quantitative, predictive, and mechanistic (Figure 12.1).

The dose response of transcriptomic perturbations (*i.e.* quantitative toxicogenomics) can be modeled to quantify the benchmark dose (BMD), or the dose at which important biological processes depart from background levels.[4] Predictive molecular signatures are derived from controlled

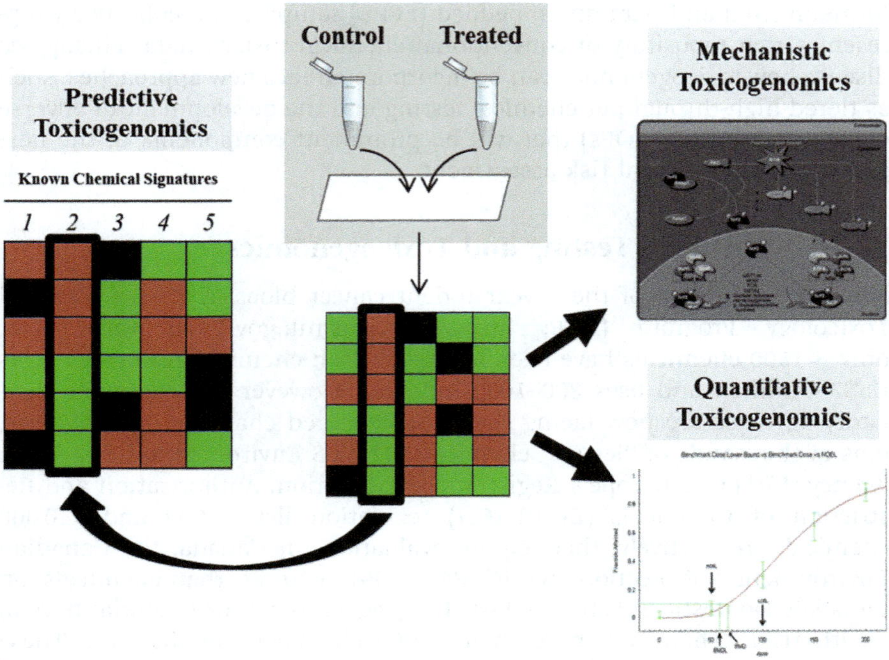

Figure 12.1 Three approaches for analyzing toxicogenomics data. In predictive toxicogenomics novel gene expression signatures are compared to known chemical signatures in order to predict the chemical's potential hazard or mode of action (MoA). In mechanistic toxicogenomics, bioinformatics software are used to identify perturbed pathways and processes in order to understand the compound's MoA. In quantitative toxicogenomics, the dose–response relationship of gene expression data is modeled in order to pinpoint doses at which important biological processes depart from background levels.

experiments on well-characterized chemicals that are established to operate through specific modes of action (MoA), or cause specific phenotypic effects. Meta-analyses can also be applied to explore the correlation between the gene expression profiles derived from a new test chemical to existing repositories of expression profiles from well-characterized chemicals, and to predict the chemical's hazards based on similarities.[2] Finally, mechanistic data are important for understanding and exploring the specific molecular changes that lead to a chemical's toxicity, which can be used to establish human health relevance.[5] Mechanistic assessment involves the application of bioinformatics tools to identify the genes and molecular pathways/functions that are responding to the treatment to more precisely define the molecular mechanisms. All three approaches are usually applied, to varying degrees, in toxicogenomics experiments.

Within the field of cancer toxicology, predictive and mechanistic toxicogenomics are especially important for identifying non-genotoxic and indirectly genotoxic carcinogens. A battery of short-term genotoxicity assays has been developed that is highly effective in the identification of genotoxic carcinogens, which account for >70% of carcinogens.[2] However, non-genotoxic and indirectly genotoxic carcinogens give negative or equivocal results in these tests, which is problematic because the results from these assays are used to prioritize compounds for further carcinogenicity testing. Thus, it has been proposed that short-term toxicogenomic analyses will be useful for identifying a chemical's ability to perturb molecular pathways that are associated with non-genotoxic MoAs in cancer in order to prioritize chemical testing in the 2 year cancer bioassay. Eventually, with the development of robust predictive signatures, the hope is that, together with other short-term testing strategies, toxicogenomics will significantly reduce the need for the 2 year cancer bioassay.[6] This is an appealing prospect as this methodology promises to greatly increase efficiencies, and reduce resources and rodents used. Given the regulatory agenda, it is imperative that a more timely strategy for the identification of rodent (and probable human) carcinogens is developed.

12.1.2 Reducing Animal use in Toxicity Testing

Since toxicity testing cannot be conducted in humans, animals are used as surrogates to measure *in vivo* toxicity. However, the use of animal models comes with its own set of concerns. Toward the reduction of animal testing, the 3Rs—reduction, refinement, and replacement—were originally proposed over half a century ago by Russell and Burch.[7] *Reduction* refers to reducing the number of animals used in toxicity testing, *refinement* refers to the refining of protocols to reduce suffering of experimental animals, and *replacement* refers to the development of *in vitro* and *in silico* assays that can replace *in vivo* testing. While the concept of the 3Rs is widely accepted, animal testing remains a massive part of chemical toxicity testing.[8] The benefits of, and barriers to, adopting alternative testing strategies (defined here as strategies that are mindful of the 3Rs, and, more broadly, alternatives to the current testing paradigm that improve upon efficiency of data generation and accuracy) are discussed below.

There are a number of compelling reasons for the adoption of alternative testing strategies. The most obvious is the ethical problem of enlisting hundreds or thousands of animals in assays whose endpoints are typically expected to be highly advanced disease states, or death. Public concern regarding such studies has led to increased pressure on regulatory agencies to legislate limits on animal testing. For example, the European Union has passed legislation that bans the use of animal testing on cosmetic products sold in the European Union (Regulation (EC) No 1223/2009). Animal testing is also a problem from a financial point of view because the costs associated with housing and caring for experimental animals are enormous. Budgeting constraints of businesses or governmental agencies that perform animal testing limit the amount of testing that can be accomplished, and, therefore, the number of chemicals that can be realistically evaluated within a reasonable timeframe. Finally, from a biological perspective, important questions regarding the predictivity of rodent disease outcomes to human disease outcomes have been raised. Importantly, inbred populations of experimental animals do not adequately represent the genetic or environmental variation present across human populations. Furthermore, there are important inter-species differences in biology that call into question the human relevance of results obtained in rodents. For example, phenobarbital, which is an anti-seizure medication that is routinely prescribed to treat epilepsy in human populations, causes tumors in rodents through a mechanism that is not considered relevant in humans.[5] It has been suggested that human relevancy can be improved by performing *in vitro* assays using human cells, or by measuring global gene expression in (small groups of) exposed animals and establishing human relevance using chemical MoA. Overall, there is a compelling array of ethical, regulatory, financial, and biological reasons to reduce, refine, and replace animal testing.

Despite the appeal of moving away from animal testing, there are many barriers to the elimination of animal use in chemical toxicity testing. First, the results produced in animal-based assays are embedded in today's standard approaches in risk assessment; therefore, a transition away from animal testing would require a massive paradigm change and will take a significant amount of time. Second, animal-based assays have a historical precedent for use, and even in instances where animal tests aren't explicitly required they are often perceived as "less risky" than non-standard approaches. Third, it is not always clear how to interpret *in vitro* data and extrapolate human health significance. Indeed, *in vitro* assays have their own inherent set of limitations. Fourth, data derived from new technologies, particularly genomics technologies, are complex and can be difficult to interpret. Finally, it takes time to establish trust in new assays, which is vital to adopting novel forms of data in a regulatory context. There are several national and international consortia whose purpose is to build this trust through the validation of alternative methods (including the Interagency Coordinating Committee on the Validation of Alternative Methods, International Cooperation on Alternative Testing Methods, the Japanese Center for the Validation of Alternative Methods, European Union Reference Laboratory for Alternatives to Animal Testing, and the National Toxicology Program's Interagency Center for the Evaluation of Alternative Toxicological Methods). However, these validation schemes can be slow and expensive, and do not necessarily lead to assay implementation following their completion. Taken together, barriers to transitioning away from animal testing are often bureaucratic or technical; however, there remain some significant and legitimate hurdles with respect to *in vitro* to *in vivo* extrapolation (IVIVE).

Programs like the US EPA's ToxCast (www.epa.gov/ncct/toxcast/) have shown some success in linking *in vitro* data to *in vivo* outcomes by screening large chemical libraries using a very large collection of *in vitro* assays. ToxCast uses a suite of more than 700 *in vitro* assays which, to date, have been used to screen more than 2000 chemicals. Using this immense dataset as a training set, it was shown that machine learning could be used to identify putative hepatotoxicants.[9] Computer models like these are quickly becoming important tools in IVIVE and animal-free chemical risk assessment. Another study, which focused on ToxCast primary human cell line assays, screened 641 environmental chemicals and 135 failed pharmaceuticals. Chemical targets and molecular pathways that were indicative of chemical effect were identified and used to group chemicals by MoA.[10] ToxCast chemical screening is typically used for chemical prioritization for targeted testing,[11,12] as opposed to being used directly for regulatory decision-making. In addition to the more than 700 *in vitro* assays, ToxCast also uses some model organism assays; for example, zebrafish are used to test chemical effect on development.[13] These model organisms can be cultured in petri dishes and address some of the limitations (described below) of *in vitro* assays.

There are various limitations of *in vitro* assays that may reduce their ability to recapitulate *in vivo* conditions. In particular, the cell lines used in *in vitro* assays are typically immortalized tumor-derived cells that are genetically abnormal and might not respond to a toxicant in the same way that a normal cell would. In addition, cultured cells are not always metabolically competent. Metabolic activation of compounds for most cell models must be achieved prior to addition of the compound to the cells using S9 (a liver homogenate containing metabolic enzymes). Furthermore, *in vitro* models are typically monocultures, and thus interactions between cell types that would be seen *in vivo* (*e.g.* between hepatocytes and Kupffer cells in the liver) are not captured. As a result, toxicities requiring complex interactions between different cell types to fully express their functions (*e.g.* inflammatory systems) are not as measurable *in vitro*. Finally, *in vitro* systems cannot predict the physiological distribution of metabolite(s) following biotransformation, which means that the affected tissues remain unknown. Issues surrounding IVIVE are being actively debated and were recently the topic of a special issue of the journal *Toxicology* (volume 332, June 2015), summarized by Yoon *et al.*[14]

While *in vivo* toxicogenomics does employ animals, it improves upon the 2 year rodent bioassay (with respect to the 3Rs) because it queries a larger number of endpoints using a smaller number of animals that are not taken to a full-blown disease state. However, a barrier to the use of toxicogenomics data is the complexity and magnitude of the data that are produced; their interpretation requires advanced expertise in bioinformatics. Therefore, regulatory agencies seeking to employ toxicogenomics in risk assessment must ensure that they hire staff with training in this area, or invest in their training. Furthermore, during this transitional period, it is critical that the regulatory toxicology community be open to the possibility that new assays that measure different endpoints may be equally predictive of the same adverse outcome. For example, although an omics assay will never provide site-specific tumor data, it may be as useful in predicting carcinogenicity as a cancer bioassay; in contrast, a cancer bioassay cannot reveal molecular MoA.

Overall, it is clear that *in vivo* toxicogenomics can provide data that bridge the gap between high-throughput *in vitro* assays and chronic *in vivo* carcinogenicity studies.

The 3Rs provide different ways to approach the reduction of animal use in toxicity testing. Programs like ToxCast and SEURAT-1 (in Europe; www.seurat-1.eu/) aim to *replace* animal testing with high-throughput *in vitro* screens and computational tools. *In vitro* toxicogenomics is another strategy to *replace* the use of animals; whereas *in vivo* toxicogenomics aims to *reduce* the use of animals and *refine* methods such that suffering is reduced because animals are necropsied at sub-chronic time points. Toxicogenomic analysis of archival tissues (discussed later) has the potential to *replace* (or massively *reduce*) the number of animals required for production of predictive molecular signatures of carcinogenesis. Ultimately, strategies that support the 3Rs are important for the ethical treatment of animals and are currently a focus of many international activities.

12.2 Toxicogenomics Case Study: Furan

Furan is a rodent hepatocarcinogen, and like many chemical toxicants, the adverse effects of furan exposure are mediated by its interaction with cytochrome P450 (CYP) enzymes in the liver. Below, we briefly review liver physiology and consider some of the toxicological responses of the liver following exposure to chemicals such as furan. Furan has been well characterized using standard toxicity and carcinogenicity tests and our approach has been predicated on the assumption that toxicogenomics would build on the existing apical data, and provide valuable insight into the molecular basis of the furan MoA. Indeed, furan represents an exceptional case study showing how toxicogenomics can be used for hazard identification, to inform MoA, to explore human relevance, and to evaluate the potential use of dose–response data from toxicogenomics experiments in quantitative human health risk assessment.

12.2.1 Liver Physiology and Hepatocarcinogenesis

The liver is organized into functional units called lobules. Lobules are primarily composed of hepatocytes that are organized around the liver's blood supplies. Zone 1 of the lobule is the area that is closest to the hepatic artery (which provides oxygenated blood) and the portal vein (which carries nutrient/chemical rich blood from the gastrointestinal tract to the liver) and zone 3 is the furthest from the blood supply. Blood from the hepatic artery and portal vein mixes, travels through the lobule in sinusoidal capillaries, and eventually drains into the central vein. Bile travels in the opposite direction (from zone 3 to 1) using channels called canaliculi, which converge on the canals of Hering that drain into the bile duct.

The three main roles of the liver are biosynthesis, metabolism (detoxification), and bile secretion. Biosynthesis (of serum proteins, bile, and cholesterol) and metabolism (of endogenous chemicals, and xenobiotics) are carried out by hepatocytes, which form the largest cell population in the liver (~60%). Importantly, zone 3 hepatocytes have high levels of cytochrome P450 enzymes, which are a large family of mono-oxygenases involved in chemical detoxification. Following mono-oxidation by P450, conjugation occurs, which facilitates chemical excretion in the bile or urine. Liver function, architecture, and homeostasis are reviewed by Stanger.[15]

Intermediates produced by P450s are often highly reactive electrophiles and may be more harmful to the cell than their parent compound, since they can react with cellular nucleophiles, such as proteins, lipids, and nucleic acids, leading to cellular damage that, in extreme cases, can result in cell death. The liver has the capacity to regenerate itself; however, the concomitant increase in cellular proliferation can result in tumorigenesis. Liver cells are normally quiescent, with only 1–2% turnover under normal circumstances. However, following surgical resection or chemical injury, liver cells rapidly divide to replace damaged tissue. The liver has two modes of regeneration: (1) following surgical resection, the liver regenerates *via* cellular hypertrophy and division of existing cells; (2) following chemically dependent damage, regeneration is thought to occur *via* proliferation of a population of facultative stem cells that is located at the canals of Hering (in zone 1 where canaliculi join and drain into the main bile duct). Facultative stem cells are functional, differentiated cells that will dedifferentiate in response to tissue damage, thereby becoming a population of progenitor cells that can redifferentiate to replace multiple lost cell types. The bile duct is lined with biliary epithelial cells (BECs) and it is thought that, upon toxicant-induced injury, they dedifferentiate into "oval" cells, which can then redifferentiate into hepatocytes or BECs in order to regenerate damaged tissue; this process is known as "ductal expansion".[15,16] Ductal expansion is an important process to consider when studying hepatocarcinogens. An important mode of hepatocarcinogenesis following chemically dependent liver damage is dysregulated regenerative proliferation leading to the development of hepatocellular adenomas (HCA) and hepatocellular carcinomas (HCC). The cell-of-origin of these cancers is typically assumed to be a hepatocyte; however, it is important to also consider that oval cells might represent a reservoir of cancer progenitor cells since they are multipotent and primed for division.[17]

Worldwide, liver cancer is the fifth and seventh most common and the second and sixth most deadly cancer in men and women, respectively.[18] Risk factors for the development of liver cancer include viral hepatitis, alcoholic liver disease, non-alcoholic steatohepatitis (fatty liver disease), and chemical exposure (including cigarette smoke and food contaminants). In particular, exposure to aflatoxin B1 (a metabolite of some fungi of the *Aspergillus* genus) has been strongly implicated in the development of HCC. Human exposure to aflatoxin B1 typically occurs *via* contaminated maize and groundnuts in areas with tropical or sub-tropical climates (sub-Saharan Africa, Eastern Asia, and South America). Aflatoxin B1 is metabolized in the liver to a reactive epoxide that binds to DNA to produce guanine adducts (at position N7). These are highly mutagenic and carcinogenic adducts and, as a result, aflatoxin B1 is a major human health concern.[19,20] More than 80% of primary liver cancers occur in conjunction with cirrhosis, fibrosis, oxidative stress, and/or chronic inflammation.[21] During chemical exposure, chronic oxidative stress is an important contributor to genomic instability, which is necessary for malignant transformation. Furthermore, during necrotic cell death, the cell ruptures and releases its contents into the extracellular space. Released cellular components include damage-associated molecular patterns that trigger an inflammatory response, which further creates a pro-tumorigenic environment. In fact, both mutagenesis and tumor-promoting inflammation were recently recognized among four additional hallmarks of cancer (together with deregulating cellular energetics and avoiding immune destruction).[22] Taken together, the liver's high contact with xenobiotics and its regenerative potential predispose it to carcinogenesis.

The original six hallmarks of cancer are: sustaining proliferative signaling, evading growth suppressors, activating invasion and metastasis, enabling replicative immortality, inducing angiogenesis, and resisting cell death.[23] Cancer development is a multistep process during which mutations are acquired and signaling is altered in order to achieve these hallmarks. Sustaining proliferative signaling is especially relevant here because chronic toxicant exposure typically leads to cytotoxicity, which activates regenerative proliferation in the liver. When liver regeneration occurs on a background of genomic instability, increased rates of cellular proliferation increase the likelihood that driver mutations will become fixed leading to cellular transformation and clonal expansion of a malignant cell. Not surprisingly, a large number of mutations have been found in HCC tumors.[21] Changes in cell physiology can be detected by measuring changes in gene expression. For example, upregulation of c-Jun (part of the dimeric AP-1 transcription factor) was recently identified as an important initiator of HCC in mice.[24] Furthermore, chronic activation of the NRF2 oxidative stress response pathway has also been reported in HCC. It is thought that malignant cells hijack this normally protective pathway to gain a proliferative advantage.[25]

12.2.2 Test Article: Furan

Furan (CAS No. 110-00-9) is a liver toxicant and rodent hepatocarcinogen; it is classified as possibly carcinogenic to humans (group 2B) by the International Agency for Research on Cancer.[26] It is used in the synthesis of nylon, lacquers, insecticides, pharmaceuticals, and stabilizers. Furan was first reported in foods more than 30 years ago.[27] It is formed during heat-treatment of food, probably through thermal decomposition of carbohydrates, and is found in coffee, and canned and jarred foods.[28–32] It is also produced during combustion and is therefore found in engine exhaust fumes, wood smoke, and tobacco smoke.[26] Levels of furan in foods vary considerably. Furan levels tend to be highest in canned foods that contain meat (up to 175 $\mu g\,kg^{-1}$) and in coffee (up to 100 $\mu g\,kg^{-1}$). Of particular concern are high levels of furan in baby food (10–100 $\mu g\,kg^{-1}$), because store-bought baby food is often a very large proportion (up to 100%) of the infant's diet. However, on a population level, brewed coffee is the largest contributor of furan exposure. Because coffee is one of the most widely used beverages around the world, its health effects have been extensively studied. Coffee is a complex mixture of many substances, including antioxidants, that have been shown to have positive health effects. The presence of beneficial chemicals tend to offset any adverse effects of toxic chemicals within the mixture. Thus, coffee has generally been reported to reduce the risk of liver pathologies in high-risk subjects.[33–35] Typical dietary exposure ranges to furan in the USA are 0.3–1 $\mu g\,kg^{-1}$ bodyweight per day. A margin of exposure (MoE) is the ratio of the toxicity effect level to the exposure dose. An MoE ratio of 10 000 would indicate that the toxicity effect level is 10 000 times higher than the expected exposure level, which indicates very low risk of toxicity; whereas a lower MoE, for example 100, would indicate a relatively high risk of toxicity. A typical MoE threshold for a well characterized chemical (like furan) is 1000. MoE calculations for furan-dependent HCC produced values of 750 and 4300 for infants and adults, respectively, which suggest that furan is an important dietary risk factor, especially in infants.[36] Taken together, furan is a rodent hepatocarcinogen and food contaminant that is a potential human health risk.

The National Toxicology Program's (NTP) 2 year cancer bioassay showed that furan induces HCC and HCA in a dose-dependent manner in B6C3F1 mice. F344 rats developed cancer at lower doses than mice and, in addition to HCC and HCA, the rats developed cholangiocarcinoma and mononuclear cell leukemia.[37] Due to equivocal results in the standard battery of genotoxicity tests, it seemed unlikely that furan was producing cancer *via* a directly genotoxic MoA. Additionally, the NTP's 13-week study showed dose-dependent increases of histopathological markers for cytotoxicity, necrosis, and cellular proliferation in B6C3F1 mice and F344 rats (both genders). In their 2 year study, there was evidence of chronic inflammation, hepatic fibrosis, hyperplasia, degeneration, and necrosis in the liver. Therefore, the MoA proposed for furan was chronic cytotoxicity and inflammation followed by dysregulated regenerative proliferation, which is thought to be the most common MoA for spontaneous HCC.[38] This MoA also occurs following exposure to carbon tetrachloride[39] and chloroform,[40,41] and is supported by follow-up studies on furan.[42,43]

Extremely high rates of spontaneous liver cancer in control male mice (26/50 and 7/50 in control males and females, respectively) in the NTP cancer bioassay led Moser *et al.*[42] to conduct their follow-up cancer bioassay using female mice. In addition, Moser *et al.* used a lower dose range (0, 0.5, 1, 2, 4 or 8 $mg\,kg^{-1}$ per day furan) than was used in the NTP cancer bioassay (0, 8, or 15 $mg\,kg^{-1}$ per day), because both NTP doses were carcinogenic (Table 12.1).

Mice were exposed to furan for 3 weeks or 2 years. Consistent with furan's proposed MoA, significant increases in hepatic cytotoxicity (beginning at

Table 12.1 The incidence of furan-dependent liver cancer in female B6C3F1 mice. Benchmark dose lower confidence level (BMDL) values are noted for the Moser *et al.*, study.[42,a]

	Dose of furan mg kg^{-1} per day						
	0	0.5	1	2	4	8	15
NTP, 1993[37]							
HCA[c]	5/50 (10%)					**31/50 (62%)**[b]	**48/50 (96%)**[b]
HCC[c]	2/50 (4%)					**7/50 (14%)**[b]	**27/50 (54%)**[b]
HCC or HCA[c]	7/50 (14%)					**34/50 (68%)**[b]	**50/50 (100%)**[b]
Moser *et al.*, 2009[42]							**BMDL**
HCA[c]	3/36 (8%)	4/72 (6%)	4/53 (8%)	4/41 (10%)	**11/36 (31%)**[b]	**25/39 (64%)**[b]	**2.6 (1.6)**[b]
HCC[c]	0/36 (0%)	4/72 (6%)	2/53 (4%)	1/41 (2%)	2/36 (6%)	**11/39 (28%)**[b]	**5.1 (4.2)**[b]
HCC or HCA[c]	3/36 (8%)	8/72 (11%)	6/53 (11%)	5/41 (12%)	**12/36 (33%)**[b]	**29/39 (74%)**[b]	**2.3 (1.3)**[b]

[a]Shaded areas represent doses not used in the study.
[b]Bold text: significant over control ($p<0.05$).
[c]HCA: hepatocellular adenoma; HCC: hepatocellular carcinoma.

1 mg kg^{-1} per day; measured by serum alanine aminotransferase) and hepatocyte proliferation (at 8 mg kg^{-1} per day; measured by bromodeoxyuridine incorporation) were observed after 3 weeks. Tumorigenesis was observed at 4 and 8 mg kg^{-1} per day after 2 years. The ability of furan to induce sufficient cytotoxicity to promote regenerative proliferation is related to the chemical's metabolism. Experiments in F344 rats using radio-labeled furan have demonstrated that 80% of ingested furan is eliminated in the urine or expired within 24 hours; the remainder [14C] was associated with liver proteins.[44] Furan does not react with DNA *in vivo.*[45] While Burka *et al.* detected 10 metabolites of furan in the urine,[44] the most important metabolite of furan is *cis*-2-butene-1,4-dial (BDA). The furan heterocycle is opened by oxidization to form BDA by cytochrome P450 2E1 (CYP2E1), which is a mono-oxygenase.[46] The role of CYP2E1 in chemical toxicity is well documented and is reviewed by Gonzalez.[47] *In vitro* experiments have demonstrated that metabolic activation to BDA is necessary for furan-induced cytotoxicity,[48] and inhibition of CYP2E1 is sufficient to prevent cytotoxicity in female B6C3F1 mice and in mouse, rat or human microsomes.[43,49] The latter indicates that CYP2E1 is solely responsible for the xenobiotic metabolism of furan. Further, the mechanism for furan oxidation to BDA by CYP2E1 is similar in mouse, rat, and human hepatocytes, which suggests that the furan-dependent adverse effects seen in rodents are applicable to humans.[49,50]

The CYP2E1 catalytic cycle is prone to producing reactive oxygen species (ROS) due to dissociation of superoxide radical from complex I.[47,51] CYP2E1 is unique among P450 enzymes because it is substrate-stabilized; therefore, chronic exposures will have the effect of chronically increasing the oxidative status of the cell. Indeed, CYP2E1-overexpressing HepG2-E47 cells have elevated glutathione (GSH), NRF2 [messenger (m)RNA and protein], and NRF2 target gene expression, compared to standard HepG2 cells that do not express CYP2E1, which is demonstrative of the elevated oxidative status of these cells.[52] GSH plays important roles both as an antioxidant molecule and a conjugant in xenobiotic metabolism. BDA has been shown to bind cellular GSH and reduce it to 30% of control levels in F344 rat hepatocytes, which reduces the amount of GSH available for the reduction of ROS, and is sufficient to produce cytolethality.[53] In addition to the resultant cytotoxicity, evidence of oxidative stress-induced genomic damage in response to furan includes 8-oxo-dG adducts[54] and oxidized purine and pyrimidine bases[55] in rat liver. Therefore, it seems clear that cellular damage due to oxidative stress is an important part of furan's MoA.

Genomics data provide valuable insight into toxicant MoA, building on apical data and enabling more informed assessment of the relevance of the molecular perturbations observed in the rodent studies to human risk. Gene expression analysis of furan exposed rodent livers provides valuable insight into the molecular underpinnings of the furan MoA. In addition, furan is extremely well characterized using standard toxicity tools. Thus, furan provides an exceptional case study to examine how toxicogenomics can be used for hazard identification, to inform MoA, to explore human relevance, and to evaluate the potential use of dose–response data from toxicogenomics experiments in quantitative human health risk assessment.

12.2.3 Quantitative Toxicogenomics

Quantitative toxicogenomics is an extremely useful tool because it can be completed quickly and it provides BMD values that indicate the dose at which the transcriptome becomes abnormal. It has been proposed that the BMD values can be used to identify the compound's point of departure (PoD), which can then be used to calculate the MoE and set acceptable exposure limits. This approach does not provide any indication as to what adverse health outcome the compound might produce, nor does it provide any mechanistic insight. We were interested to know: (1) if transcriptomic BMD values are similar to cancer BMD values; (2) the effect of the genomics platform used (*e.g.* microarrays *versus* RNA-seq) on BMD values; and (3) how to best choose a single transcriptional PoD.[56] Based on Moser *et al.*,[42] the furan-dependent cancer BMD (lower confidence limit BMD; BMDL)s in mice are 2.6 (1.6) mg kg^{-1} per day (for HCA) and 5.1 (4.2) mg kg^{-1} per day (for HCC). We found that furan-dependent changes in gene expression produced BMD values that are consistent with cancer BMD values regardless of genomics platform used. For example: (1) the modes of the pathway BMD(L)-mean values, which represent the dose at which the majority of the transcriptional response occurs, were 3.2 (2.3), 3.7 (2.5), and 4.4 (2.9) mg kg^{-1} per day for RNA-seq, microarray, and quantitative polymerase chain reaction (qPCR), respectively; (2) the BMD(L)-mean values for the NRF2 oxidative stress response pathway, which was identified as an important pathway in furan's cancer MoA (discussed later), were 4.0 (2.7), 4.3 (2.9), and 5.0 (3.4) mg kg^{-1} per day for RNA-seq, microarray, and qPCR, respectively; and (3) the BMD(L)-mean values of the most sensitive pathways were 1.7 (1.0), 2.2 (1.5), 2.2 (1.6) mg kg^{-1} per day for RNA-seq, microarray, and qPCR, respectively. With respect to choosing a PoD, all of these approaches (mode, MoA-relevant pathway, and most sensitive pathway) produce doses that are in the appropriate range.

The work above has recently been supported by a second study conducted on male and female rats exposed to furan for 90 days.[57] Quantitative toxicogenomics analysis revealed a median BMD for pathway perturbations of 0.9 mg kg^{-1} per day in males, which is consistent with a BMD of 1.84 mg kg^{-1} per day for tumors in the same strain of rats.[58] In contrast, a BMD for female rats could not be derived due to lack of gene expression response in females, which is consistent with the significant increase in susceptibility of males over females for tumors.

Quantitative toxicogenomics is a relatively new approach and the concordance between cancer and transcriptional BMD values should be confirmed using additional chemicals, as should the robustness of approaches for choosing a PoD. However, the results presented here are very promising with respect to modeling the transcriptional dose response to identify a prognostic PoD dose.

12.2.4 Predictive Toxicogenomics

Predictive toxicogenomics begins to answer the question of what adverse health outcomes are produced by the compound and can be accomplished using open-source resources or paid software. For the former, the gene expression data of the test compound can be compared to any publically available dataset (*e.g.* any dataset in the National Center for Biotechnology Information's Gene Expression Omnibus (GEO)) using a hierarchical clustering package. This is an extremely powerful approach; however, it requires advanced bioinformatics and biostatistical skill sets. An alternative approach is to use commercially available software. For example, NextBio software provides a user-friendly interface that allows users to compare their gene expression datasets to all of the gene expression datasets in the NextBio knowledgebase, which houses a large number of disease- and chemical-specific expression profiles.

Using hierarchical clustering (Figure 12.2), we found that following a 21 day exposure to carcinogenic doses of furan, the changes in global gene expression that are produced by 8 mg kg^{-1} per day furan are most similar to those produced in spontaneous liver cancer and liver regeneration. The changes in global gene expression that are produced by the lower doses (1, 2, and 4 mg kg^{-1} per day) are most similar to those produced by nuclear factor-kappaB-mediated inflammation of the liver. Together, these two groups form a larger cluster with a group of hepatocarcinogens and a group of CYP2E1 ligands.[59] Therefore, by querying the transcriptomic changes produced in a large number of experimental conditions, it is possible to identify important, chemical-specific effects. We obtained similar results using NextBio software. The DiseaseAtlas indicated that furan exposure produced gene expression changes that are similar to injury of the liver (based on five studies), liver regeneration (based on five studies), breast cancer (based on 333 studies), and liver cancer (based on 111 studies). The ChemicalAtlas indicated that furan exposure produced gene expression changes that are similar to the carcinogens and CYP2E1 ligands: thioacetamide (based on 10 studies), 1,5-naphthalenediamine (based on three studies), and carbon tetrachloride (based on 14 studies); and the insecticides malathion (based on three studies) and dichlorobenzene (based on three studies).[59] These correlations are consistent with different aspects of furan's known MoA.

Also consistent with these findings, predictive toxicogenomics analysis of furan-exposed male rat livers using NextBio software revealed that the top disease profiles that were most correlated with the furan dataset were from rat liver cancer models, including transgenic myc-driven liver tumors and intrahepatic cholangiocarcinomas formed by inoculation of malignant rat cholangiocytes into liver.[57] In addition, liver regeneration, injury of the liver, inflammatory disease, and cirrhosis were highly positively correlated with

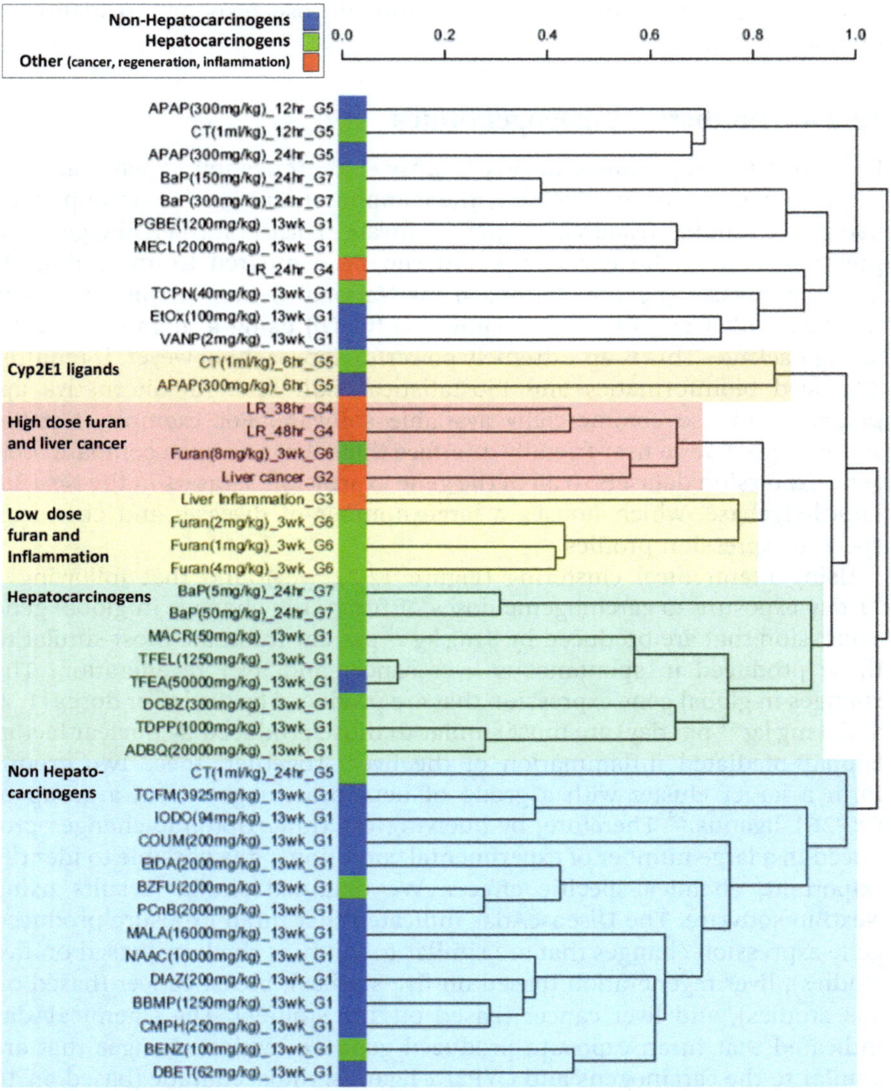

Figure 12.2 Cluster analysis comparing furan gene expression data with publically available datasets for liver cancer (hepatocellular carcinomas), liver regeneration (LR), Cyp2E1 ligand-, and hepatocarcinogen- and non-hepatocarcinogen-induced changes in gene expression in mouse liver. Datasets were downloaded from Gene Expression Ominbus (GEO), where G1 is GSE18858, G2 is GSE26538, G3 is GSE35934, G4 is GSE20427, G5 is GSE4874, G6 is GSE48644 (our furan study), and G7 is from another study conducted by our laboratory (Yauk, unpublished data).
This figure is reproduced from Jackson *et al.*[59]

furan treatment in male rats. Finally, chemicals inducing similar toxicity profiles to furan included thioacetamide, nafenopin, methapyrilene, mestranol, and chloroform, which are all non-genotoxic carcinogens. The non-genotoxic carcinogens carbon tetrachloride and bromobenzene also induce similar gene expression profiles to furan treatment in male rats. Cyp2E1 is involved in the metabolism of both thioacetamide and carbon tetrachloride.

Determining which diseases and chemicals produce changes in gene expression that are most similar to those produced by the test compound is especially useful for a read-across approach (when the effects of well-studied chemicals are used to infer the effects of a data-poor chemical). Like quantitative toxicogenomics, predictive toxicogenomics can be completed relatively rapidly and provides some insight into the potential health hazards of the compound.

12.2.5 Mechanistic Toxicogenomics

Mechanistic toxicogenomics is used to develop a deeper understanding of the molecular effects of the compound, but can be time consuming. Therefore, when time is limiting (*e.g.* in rapid screening applications), it may be more feasible to restrict analyses to the quantitative and predictive approaches described earlier. In fact, because of the strong correlation between apical and transcriptomic BMD values, it has been argued that mechanistic understanding of a compound is not always required for the use of toxicogenomics data in chemical risk assessment.[60] However, mechanistic data are important for establishing human health relevance.[5] Because a mechanistic approach is time-consuming, it is important to have clearly defined goals at the outset of toxicogenomic studies. Specifically, researchers and risk assessors should collaborate in problem formulation prior to the assessment to determine whether the assessment deadlines permit a mechanistic analysis and whether this information will significantly improve their ability to assess the compound.

A compound's MoA may be developed using functional and pathway analyses [using software such as Ingenuity pathway analysis (IPA) and the Database for Annotation, Visualization and Integrated Discovery (DAVID)] and intensive literature review. Furan's cancer MoA is cytotoxicity followed by inflammation and regenerative proliferation that, in extreme cases, lead to tumor development (discussed above). Gene expression analysis of mouse liver tissue following sub-chronic exposure to furan revealed enrichment of molecular processes that confirmed this MoA and provided additional mechanistic information (Figure 12.3A).

The predominant molecular pathway that was perturbed by furan was the Nrf2 oxidative stress response pathway. We believe that this is because the CYP 2E1 catalytic cycle is prone to uncoupling,[61] which leads to the production of ROS. Together with furan's primary metabolite (BDA), ROS are another important source of cytotoxicity (*via* lipid peroxidation, for example) as well as oxidative lesions to DNA (which are likely the main source of

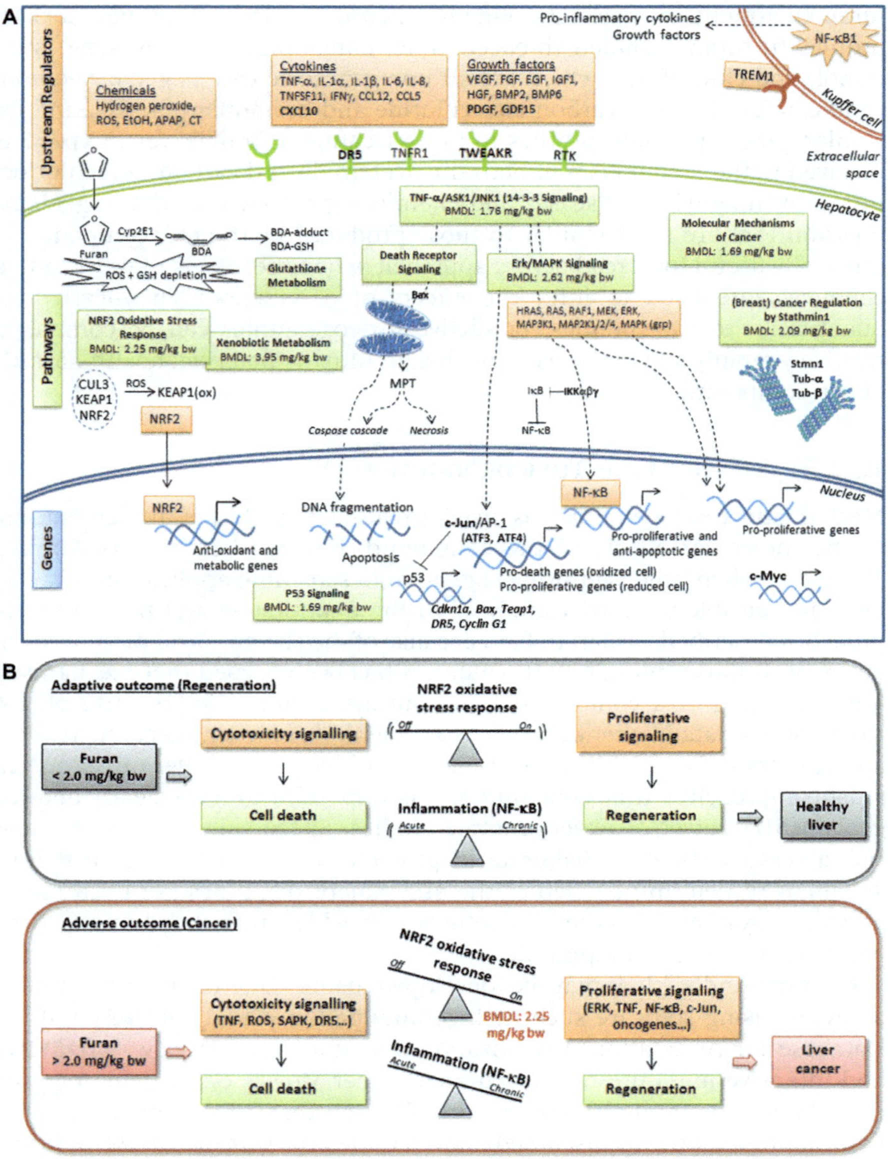

Figure 12.3 Furan mode of action (MoA). (A) Summary of predicted upstream regulators (orange), enriched molecular pathways (green) with lower confidence limits of benchmark dose (BMDL) values indicated, and selected differentially expressed genes (bold text) for the 8 mg kg^{-1} per day dose group. (B) Adaptive and adverse outcomes in response to non-carcinogenic and carcinogenic doses of furan. We propose that NRF2 activation and inflammation could represent important tipping points in the MoA and that, when chronically activated, drive the cancer outcome.
This figure is reproduced from Jackson *et al.*[59]

cancer-causing mutations). Redox-sensitive proteins are modified by oxidation and important changes in gene expression are carried out by the transcription factor Nrf2. Nrf2 increases the expression of cytoprotective genes, such as anti-oxidants and xenobiotic metabolism enzymes. At the same time, dying cells release pro-inflammatory signals and, together, these signals encourage regenerative proliferation of hepatocytes. However, when chronically activated, these molecular signals can produce dysregulated cellular proliferation during which normally cytoprotective cellular mechanisms are hi-jacked by pre-malignant and malignant cells (Figure 12.3B).[59]

Pathway analysis in rat livers exposed to furan was consistent with the above findings. The analysis supported that key events in furan-induced liver tumors in males involve gene expression changes related to oxidative stress, apoptosis and inflammatory response, while pathway changes in females are consistent with primarily adaptive responses.[57]

In addition to looking at mRNA profiles, we were also interested in changes in the profiles of long non-coding (lnc)RNA in response to furan treatment.[62] Recently, lncRNAs (RNA molecules that are >200 bases long) have become of broad interest because of their involvement in genome regulation and other key biological processes.[63] Like mRNA transcripts, the number of differentially expressed lncRNA transcripts increased in a dose-dependent manner in response to furan (with 83 lncRNAs being differentially expressed at the high dose). A number of "clusters of transcription" were identified in which lncRNAs were co-transcribed with neighboring genes. For example, lincRNA-p21 co-transcribed with *Cdnk1a*, and lncRNA chromosome (Chr.) 9: 78107225–78118850 co-transcribed with *Gsta1* and *Dppa5a*. While the exact function of these lncRNAs is still under investigation, it is interesting to note that *Cdkn1a* and *Gsta1* were two of the most highly elevated protein coding transcripts in response to furan. The identification of lncRNAs in a toxicogenomic study was novel and very much represents the potential of mechanistic studies to make novel discoveries regarding chemical-dependent effects.

Mechanistic analyses are limited by a number of parameters, including the expertise of the researcher, and the amount of ambiguity intrinsic to the dataset. Examples of the latter include increased expression of genes with conflicting functions; enrichment of pathways with conflicting (or unrelated) functions; functional redundancy across genes or pathways; the fact that gene expression data cannot (directly) capture post-translationally controlled processes (like phosphorylation during mitogen-activated protein kinase signaling cascades, or protein cleavage during caspase cascades) and therefore activation of these processes must be inferred through changes in downstream targets; and the fact that mRNA extracted from whole tissue reflects a heterogeneous cell population (*e.g.* liver is comprised of hepatocytes, BECs, Kupffer cells, *etc.*) that may react to the treatment in different ways. Ultimately, accurate interpretation of toxicogenomics data requires substantial understanding of molecular biology and cell physiology.

12.2.6 FFPE Toxicogenomics

Tissue repositories have immense untapped potential in toxicogenomic research. Applications of archival tissue resources range from biomarker discovery to large-scale studies of genome–disease interactions. Curated sample banks house diverse tissue types from different models or study populations, which are often linked with detailed clinical or pathological outcomes. Therefore, because these samples are very well characterized, they are an extremely valuable resource for retrospective, phenotypically anchored toxicogenomics studies. In many cases, tissue archives contain unique samples from animal bioassays, clinical trials, or epidemiological studies that may be impractical or impossible to repeat. However, even in cases where the animal study could be repeated, the use of archival samples is an experimental strategy that supports the 3Rs.

Despite their promise, the use of archival samples for transcriptomic profiling has been relatively limited. The majority of archival tissues are stored in FFPE blocks, which preserve tissue architecture for histopathological analysis and allow for tissue storage at room temperature. However, formalin treatment degrades RNA (through cross-linking, fragmentation, and adenine monomethylation), which significantly impairs molecular analyses.[64–66] These RNA effects can result in inconsistent genomic data and present important technical and analytical challenges when working with FFPE samples. Nevertheless, studies have demonstrated the value of using microarrays to analyze FFPE samples for understanding the molecular basis of a variety of cancers.[67–70] More recently, these approaches have been used for toxicogenomics studies.[71–73] Demonstration of the preservation of important, chemically dependent, disease-prognostic molecular signatures in FFPE samples will be fundamental to their use in predictive toxicogenomics. As technologies continue to improve, so too does the prospect of FFPE genomics.

During our furan experiment in mice, a sub-set of each liver sample was preserved in formalin for 18 hours or 3 weeks. Using these samples we assessed how FFPE-preserved samples correlated to samples that were flash-frozen on a variety of genomics platforms (one- and two-color DNA microarrays, poly-A RNA-seq and ribo-depletion RNA-seq). The analysis revealed that ribo-depletion RNA-seq is most amenable to degraded FFPE samples.[74] Importantly, ribo-depletion avoids the issue of the degraded poly-A tail (which is problematic in poly-A enrichment RNA-seq) and it does not rely on pre-defined probe sequences (which is the case for microarrays). These attributes increase the likelihood of capturing as much of the transcriptome as possible, despite its being degraded. However, we do recommend the addition of a technical replicate to the experiment in order to eliminate false-positive genes, which appeared to be an issue in our experiments, and expect that this approach would reduce the differences observed between paired samples. In our study, furan-dependent changes in gene expression were visible in FFPE samples (Figure 12.4).

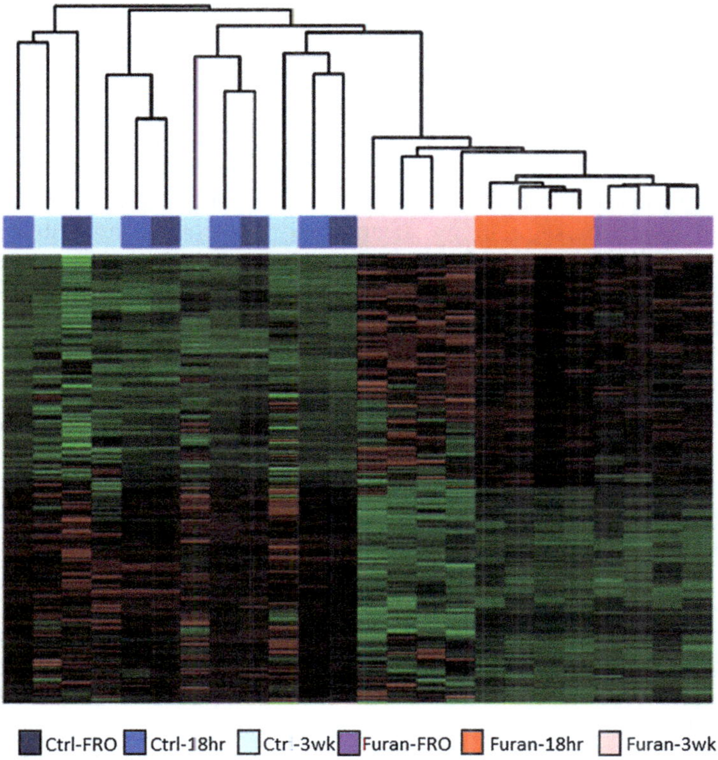

Figure 12.4 Hierarchical clustering of all differentially expressed genes from frozen (FRO) and formalin-fixed and paraffin-embedded (FFPE) livers analyzed using ribo-depletion RNA-seq (FDR $p < 0.05$, fold change $> \pm 2$). This figure is reproduced from Webster *et al.*[74]

We found that time in formalin greatly impacted the quality of data, with longer preservation leading to more degraded samples. However, the molecular signatures were still apparent even in samples stored in formalin for 3 weeks. We also assessed the impact of age of block, by evaluating archival control tissues that were 8, 19, and 26 years in blocks. Surprisingly, age in blocks did not impact the quality of the molecular signatures derived from these control samples. Overall, our work revealed important considerations for sample selection and experimental design for future work in archival samples. We expect that as next-generation sequencing technologies continue to improve, so too will their capacity to deal with degraded, archival samples.

12.3 Role for Toxicogenomics in Chemical Risk Assessment

Our furan case study used 50 mice (compared to ~800 mice and rats used by the NTP) that were exposed to four doses of furan plus control (compared to the three or two doses plus control in rats and mice, respectively), for 3 weeks (compared to 2 years). Approximately 1 year was required to complete the project, from laboratory work, to quantitative, predictive and mechanistic analyses, to manuscript publication.[59] The microarray experiment cost approximately US$10 000 (compared to US$2–4 million for a 2 year rodent bioassay). Alternatively, conducting this experiment using RNA-seq would cost approximately US$5000 (~US$200 per sample), although costs vary with depth of sequencing. Even with these improved efficiencies, it would not be practical to test all compounds using toxicogenomics; therefore, a strategy for prioritization of compounds for toxicogenomics testing is still required.

12.3.1 Guidelines for Using Toxicogenomics Data in Formal Risk Assessment

The implementation of toxicogenomics in standard toxicity testing will require: (1) increased communication between the scientists who generate the data and the risk assessors who apply it to decision-making; (2) a transition toward highly collaborative, multi-disciplinary teams that include biologists, bioinformaticians, and risk assessors; and (3) a clear and standardized approach for the production, analysis, and interpretation of toxicogenomics datasets. To this end, Bourdon-Lacombe *et al.*[75] recently published a comprehensive technical guide for evaluating and using gene expression data in human health risk assessment. The publication provides extensive practical guidance on interpreting toxicogenomic data. Despite the strong track record of technologies such as qPCR and microarray (with RNA-seq catching up quickly), they report that only a small amount of these data have been used in formal chemical risk assessments. The authors surveyed risk assessments that were completed between 2000 and 2013 by three groups: Health Canada's Existing Substances Risk Assessment Bureau and Guidelines for Canadian Drinking Water Quality program, and EPA's Integrated Risk Information System program; these groups used gene expression data in 2% (5/209), 0%, and 20% (12/59) of their assessments, respectively, with most of these within the past 5 years. The authors provide a strong rationale for the incorporation of genomics data into human health risk assessment, as well as the practical guidance to do so.

Bourdon-Lacombe *et al.*[75] provide guidance on how to evaluate the quality of genomics data, which is important for establishing inclusion criteria for using genomics data (from qPCR, microarray, and RNA-seq) from both *in vitro* and *in vivo* experiments in risk assessments. In particular, they stress the importance of sound experimental design, including the use of appropriate controls, adequate sample sizes, and controlling for confounding effects (for example, the use of a randomized block design in a microarray experiment to control for time of experiment and array location). Next, they list the criteria that are required for a good genomics study, including the use of high-quality RNA (RNA from FFPE samples is not considered in this manuscript), species appropriate microarrays or reference genomes, appropriate data normalization, statistical corrections for multiple comparisons, and setting appropriate thresholds for statistical comparisons. Most importantly, Bourdon-Lacombe *et al.*[75] provide practical examples (from the published literature) for how to include genomics data in a risk assessment. Many of these strategies are employed in this case study, for instance: (1) they describe how unsupervised hierarchical clustering and principal components analysis can be used to compare individual samples, treatment groups, or studies, in order to gain a better understanding of which data sets are similar and which are dissimilar; (2) they point out that molecular pathway, network, and upstream regulator enrichment (using software such as DAVID and IPA) are useful approaches for gaining mechanistic understanding of a toxicant; (3) they describe how meta-analyses using other disease- and toxicant-specific datasets can be used to narrow down MoA; and (4) they explain how the dose–response of genomics data can be used to identify a PoD. An important aspect for establishing human health relevance of a toxicant discussed by Bourdon-Lacombe *et al.*[75] (but not explored in the current case study) is the comparison of gene expression profiles produced in animal and human cells, which is particularly useful when there is a lack of epidemiological data for a given toxicant. Because the guidelines by Bourdon-Lacombe and colleagues are comprehensive and written in accessible language, they are expected to be very important for facilitating the uptake of toxicogenomics data into human health risk assessment.

12.3.2 FFPE Genomics in Risk Assessment

While additional case studies should be undertaken in order to confirm that transcriptional profiling of FFPE samples produces biologically accurate data, the aforementioned results suggest that these samples can be used to develop predictive signatures that could be used to help characterize data-poor chemicals (using a predictive toxicogenomics approach). The success of this work depends on storing the FFPE genomics datasets in a publically accessible format (such as a GEO Super Series) to ensure that all research groups conducting toxicogenomics experiments have access to the same set of standard chemical signatures (which, ideally, would include the >1500 cancer bioassay compounds). In addition, any corresponding apical data should also be made publically available. Importantly, the apical data need to be linked to the transcriptional data in a user-friendly way, to facilitate rapid analysis and decision-making. Altogether, archival samples from toxicity studies represent a massive opportunity to accelerate the use of toxicogenomics in chemical risk assessment. Ultimately, while the use of the tissues for genomics analyses might seem provocative, its huge potential should not be overlooked.

12.3.3 Adverse Outcome Pathways

It is expected that MoA development will be expedited by the development of the AOP wiki (aopkb.org/aopwiki). In 2012 the Organisation for Economic Cooperation and Development launched the AOP program, which is a truly collaborative effort that is crowd-sourcing experts from around the world to build, step-by-step, chemically agnostic pathways that link a molecular initiating event between a toxic substance and organism to an adverse outcome through a series of key events, which are sequentially linked by key event relationships (KERs). Importantly, an AOP is a simplified, linear pathway, not a detailed MoA. A key event must be measurable and essential to the pathway, but is not necessarily sufficient to cause the adverse outcome. A weight-of-evidence is built for each of these key events and KERs using modified Bradford-Hill criteria that include biological plausibility, empirical support, and essentiality.[76] Furthermore, key events that are shared between AOPs are also linked, thereby forming large meta-networks of possible adverse outcomes.[77]

AOPs are freely available online through a curated wiki. The AOP wiki currently lists 17 AOPs that are complete and open for commenting, and 64 AOPs that are under development. This web-based format, which is more flexible than standard publications, allows each AOP to exist as a living document that can be edited and expanded as new information becomes available. The interactive nature of the AOPs and AOP networks is expected to facilitate the assessment of new chemicals (for which there are limited data) by enabling researchers and risk assessors to fill-in-the-blanks by extrapolating from known chemical effects to possible AOs using the documented relationships found in existing AOPs. Published AOPs include a wide variety of etiologies.[78–82] Based in part on our experience using toxicogenomics to study furan, we are currently producing an AOP for CYP 2E1 activation, leading to liver cancer (Figure 12.5). The AOP program also represents a strategy toward reducing animal testing by directing hypotheses of chemical-induced toxicological effects, or preventing it altogether.[83,84] As the

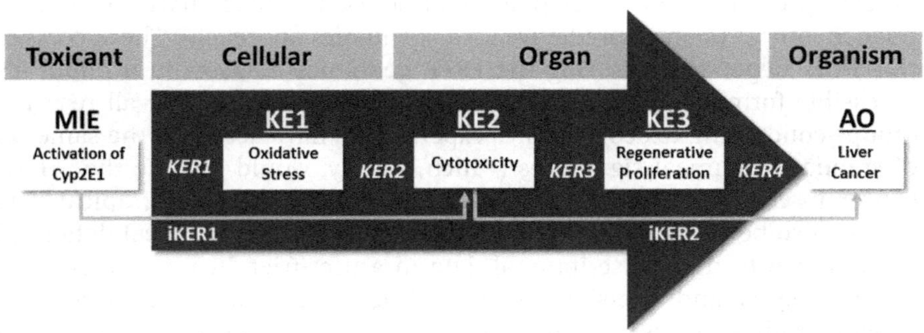

Figure 12.5 Adverse outcome pathway for CYP 2E1 activation leading to liver cancer. (MIE: molecular initiating event; KE: key event; KER: KE relationship; iKER: indirect KER; AO: adverse outcome).

AOP knowledge base grows, AOPs are expected to become an important resource in chemical risk assessment.

12.3.4 Toxicogenomics in Tiered Testing Strategies

Toxicity testing in the 21st century: a vision and a strategy is a "call to action" publication produced by the National Academy of Sciences (USA, 2007) that criticizes standard approaches for toxicity testing and proposes a provocative plan for the future of chemical testing that is heavily reliant on high-throughput, systems biology approaches that are analyzed using sophisticated bioinformatics. Furthermore, the vision proposes that toxicological assays should be primarily conducted *in vitro*, with a preference for human cell lines.[85] The report also emphasizes the identification and use of predictive toxicity pathways and dose–response modeling in the assessment of novel, putative toxicants. In the wake of this publication, the toxicology community remains in a state of upheaval as researchers collaborate (and compete) to establish a novel testing paradigm that can be universally applied (and agreed upon). A three-tiered paradigm for chemical testing and prioritization has been proposed, in which toxicogenomics appears in tier two.[6] This tiered strategy is heavily reliant on the calculation of MoE values, which are used to include/exclude compounds from next-tier testing based upon the quantification of perceived risk. An MoE is the ratio of the PoD dose to the estimated exposure; a high MoE indicates that the harmful dose is much larger than the expected exposure dose (which corresponds to low risk to human health). MoE cut-offs are typically set in a chemically specific manner. Well-characterized chemicals tend to have lower MoE cut-offs (100–1000) because there is less uncertainty associated with them, whereas poorly characterized chemicals have higher, more conservative MoE cut-offs (10 000) that correspond to a greater number of uncertainty factors. It is noteworthy that the tiered test strategy was developed for testing all types of toxicants; however, the focus of this discussion is its application to carcinogens.

Tier one is composed of five components that rely entirely on *in vitro* and *in silico* assays. (1) Chemicals are divided into "selective" (compounds with a clear MoA, *e.g.* receptor-mediated) and "non-selective" (compounds with an ambiguous MoA) using high-throughput *in vitro* assays that were developed for ToxCast (http://epa.gov/ncct/toxcast/) and which evaluate chemical selectivity against a large number of biological targets.[12,86,87] ToxCast comprises more than 700 assays that measure a compound's activity both in living cells and in non-cellular assays that use purified proteins, with the goal of understanding how the compound would behave in a more complex biological context. Non-selective compounds (*i.e.*, compounds that are active in a large number of the ToxCast assays) are expected to make up the majority of the test articles. (2) Next, chemicals are divided into "genotoxic" and "non-genotoxic" MoAs based on the results of *in vitro* genotoxicity assays (typically the Ames test and the mammalian cell micronucleus test).

(3) IVIVE pharmacokinetic modeling is used to derive an oral equivalent dose for a key event (selective MoA) or most sensitive *in vitro* assay (non-selective MoA), which can be used as the PoD (4). Together with a human exposure estimate (derived from biomonitoring data or probabilistic modeling), the PoD is used to calculate a chemical-specific MoE (5), which is used to determine whether the compound should continue into tier-two testing. Compounds that demonstrate a concerning MoA and relevant MoE move to tier-two testing.

Tier-two testing is comprised of short-term *in vivo* assays and, like tier one, has five components. (1) *In vivo* PoD values are derived for compounds with non-selective MoAs *via* BMD modeling of transcriptomic data from the eight tissues that are established as the primary sites of cancer in the 2 year rodent bioassay (*i.e.*, the liver, lung, kidney, stomach, mammary gland, urinary bladder, hematopoietic system, and vascular system, which account for majority of cancers in this assay); whereas (2) *in vivo* PoD values of compounds with selective MoAs are derived using dose–response data from *in vivo* assays that are relevant to the MoA (transcriptomic or otherwise). (3) *In vivo* and *in vitro* pharmacokinetic studies are used to identify the discrepancy between the applied and the internal dose, as well as to identify/predict potential primary metabolites. (4) Tier-one human exposure estimates are verified against any novel data. (5) MoE values are recalculated. Chemicals with MoEs below the cut-off continue to tier-three testing, whereas those with MoE values above the cut-off are not considered to pose sufficient risk to human health to warrant further testing. The authors estimate that additional testing could be avoided for 97% or 85% of chemicals by applying cut-offs of 100 or 1000, respectively.

Tier three represents standard, chronic *in vivo* testing (*e.g.* the 2 year rodent cancer bioassay). The authors of this framework do not propose any changes to the current long-term *in vivo* testing methodologies. Together, this framework promotes efficient chemical testing and reduction of animal use, and is consistent with the *Toxicity testing in the 21st century* vision.[85,88] It proposes an effective way to exclude chemicals from upper-tier testing, which makes doing these tests more realistic. Finally, it builds a weight of evidence that uses toxicogenomic data together with a large amount of data from short-term *in vitro* tests, such that informed, evidenced-based decisions can be confidently made regarding chemical toxicity (while, in most cases, avoiding the use of chronic *in vivo* assays).

12.4 Concluding Remarks

For implementation of the *Toxicity in the 21st century* vision, implementation of the 3Rs, and timely assessment of environmental hazards, changes are needed in the current chemical testing paradigm. Case studies such as the one presented here are important for providing evidence of the utility of toxicogenomics data in human health risk assessment. Moreover, the tiered testing strategy and technical guidance framework are important for the actual application of these data. To date, a number of toxicogenomics case studies have been completed[89–95] with promising results. While it is always possible to complete further validation exercises, there do not appear to be any clear advantages to delaying the inclusion of high-quality genomics data in present-day chemical risk assessments.

References

1. OECD, *OECD Guideline for the Testing of Chemicals: Carcinogenicity Studies (Test Guideline 451)*, Organisation for Economic Co-operation and Development, Paris, France, 2009.
2. M. D. Waters, M. Jackson and I. Lea, Characterizing and predicting carcinogenicity and mode of action using conventional and toxicogenomics methods, *Mutat. Res.,- Rev. Mutat. Res.*, 2010, **705**, 184–200.
3. S. S. Auerbach, R. R. Shah, D. Mav, C. S. Smith, N. J. Walker, M. K. Vallant, G. A. Boorman and R. D. Irwin, Predicting the hepatocarcinogenic potential of alkenylbenzene flavoring agents using toxicogenomics and machine learning, *Toxicol. Appl. Pharmacol.*, 2010, **243**, 300–314.
4. L. Yang, B. C. Allen and R. S. Thomas, BMDExpress: a software tool for the benchmark dose analyses of genomic data, *BMC Genomics*, 2007, **8**, 387.
5. M. P. Holsapple, H. C. Pitot, S. H. Cohen, A. R. Boobis, J. E. Klaunig, T. Pastoor, V. L. Dellarco and Y. P. Dragan, Mode of action in relevance of rodent liver tumors to human cancer risk, *Toxicol. Sci.*, 2006, **89**, 51–56.
6. R. S. Thomas, M. A. Philbert, S. S. Auerbach, B. A. Wetmore, M. J. Devito, I. Cote, J. C. Rowlands, M. P. Whelan, S. M. Hays, M. E. Andersen, M. E. Meek, L. W. Reiter, J. C. Lambert, H. J. Clewell 3rd, M. L. Stephens, Q. J. Zhao, S. C. Wesselkamper, L. Flowers, E. W. Carney, T. P. Pastoor, D. D. Petersen, C. L. Yauk and A. Nong, Incorporating new technologies into toxicity testing and risk assessment: moving from 21st century vision to a data-driven framework, *Toxicol. Sci.*, 2013, **136**, 4–18.
7. W. Russell and R. Burch, *The Principles of Human Experimental Technique*, Methuen, London, 1959.
8. N. Burden, F. Sewell and K. Chapman, Testing chemical safety: what is needed to ensure the widespread application of non-animal approaches?, *PLoS Biol.*, 2015, **13**(5), e1002156.
9. J. Liu, K. Mansouri, R. S. Judson, M. T. Martin, H. Hong, M. Chen, X. Xu, R. S. Thomas and I. Shah, Predicting hepatotoxicity using ToxCast *in vitro* bioactivity and chemical structure, *Chem. Res. Toxicol.*, 2015, **28**, 738–751.
10. N. C. Kleinstreuer, J. Yang, E. L. Berg, T. B. Knudsen, A. M. Richard, M. T. Martin, D. M. Reif, R. S. Judson, M. Polokoff, D. J. Dix, R. J. Kavlock and K. A. Houck, Phenotypic screening of the ToxCast chemical library

to classify toxic and therapeutic mechanisms, *Nat. Biotechnol.*, 2014, **32**, 583–591.

11. D. J. Dix, K. A. Houck, M. T. Martin, A. M. Richard, R. W. Setzer and R. J. Kavlock, The ToxCast program for prioritizing toxicity testing of environmental chemicals, *Toxicol. Sci.*, 2007, **95**, 5–12.
12. R. S. Judson, K. A. Houck, R. J. Kavlock, T. B. Knudsen, M. T. Martin, H. M. Mortensen, D. M. Reif, D. M. Rotroff, I. Shah, A. M. Richard and D. J. Dix, *In vitro* screening of environmental chemicals for targeted testing prioritization: the ToxCast project, *Environ. Health Perspect.*, 2010, **118**, 485–492.
13. S. Padilla, D. Corum, B. Padnos, D. L. Hunter, A. Beam, K. A. Houck, N. Sipes, N. Kleinstreuer, T. Knudsen, D. J. Dix and D. M. Reif, Zebrafish developmental screening of the ToxCast Phase I chemical library, *Reprod. Toxicol.*, 2012, **33**, 174–187.
14. M. Yoon, B. J. Blaauboer and H. J. Clewell, Quantitative in vitro to in vivo extrapolation (QIVIVE): An essential element for *in vitro*-based risk assessment, *Toxicology*, 2015, **332**, 1–3.
15. B. Z. Stanger, Cellular homeostasis and repair in the mammalian liver, *Annu. Rev. Physiol.*, 2015, **77**, 179–200.
16. K. Yanger and B. Z. Stanger, Facultative stem cells in liver and pancreas: fact and fancy, *Dev. Dyn.*, 2011, **240**, 521–529.
17. N. Oishi, T. Yamashita and S. Kaneko, Molecular biology of liver cancer stem cells, *Liver Cancer*, 2014, **3**, 71–84.
18. A. Jemal, F. Bray, M. M. Center, J. Ferlay, E. Ward and D. Forman, Global cancer statistics, *CA Cancer. J. Clin.*, 2011, **61**, 69–90.
19. T. W. Kensler, B. D. Roebuck, G. N. Wogan and J. D. Groopman, Aflatoxin: a 50-year odyssey of mechanistic and translational toxicology, *Toxicol. Sci.*, 2011, **120**(Suppl. 1), S28–S48.
20. M. C. Kew, Alatoxins as a cause of hepatocellular carcinoma, *J. Gastrointestin. Liver Dis.*, 2013, **22**, 305–310.
21. Z. G. Han, Functional genomic studies: insights into the pathogenesis of liver cancer, *Annu. Rev. Genomics Hum. Genet.*, 2012, **13**, 171–205.
22. D. Hanahan and R. A. Weinberg, Hallmarks of cancer: the next generation, *Cell*, 2011, **144**, 646–674.
23. D. Hanahan and R. A. Weinberg, The hallmarks of cancer, *Cell*, 2000, **100**, 57–70.
24. L. Min, Y. Ji, L. Bakiri, Z. Qiu, J. Cen, X. Chen, L. Chen, H. Scheuch, H. Zheng, L. Qin, K. Zatloukal, L. Hui and E. F. Wagner, Liver cancer initiation is controlled by AP-1 through SIRT6-dependent inhibition of survivin, *Nat. Cell Biol.*, 2012, **14**, 1203–1211.
25. Q. Ma and X. He, Molecular basis of electrophilic and oxidative defense: promises and perils of Nrf2, *Pharmacol. Rev.*, 2012, **64**, 1055–1081.
26. IARC, *Dry cleaning: Some chlorinated solvents and other industrial chemicals*, 1995.
27. J. A. Maga and I. Katz, Furans in foods, *Crit. Rev. Food Sci. Nutr.*, 1979, **11**, 355–400.

28. C. Pye and C. Crews, Furan in canned sardines and other fish, *Food Addit. Contam., Part B*, 2014, **7**, 43–45.
29. S. Moro, J. K. Chipman, J. W. Wegener, C. Hamberger, W. Dekant and A. Mally, Furan in heat-treated foods: Formation, exposure, toxicity, and aspects of risk assessment, *Mol. Nutr. Food Res.*, 2012, **56**, 1197–1211.
30. C. Crews, D. Roberts, S. Lauryssen and G. Kramer, Survey of furan in foods and coffees from five European Union countries, *Food Addit. Contam., Part B*, 2009, **2**, 95–98.
31. C. Crews, S. Hasnip, D. P. T. Roberts and L. Castle, Factors affecting the analysis of furan in heated foods, *Food Addit. Contam.*, 2007, **24**, 108–113.
32. S. Hasnip, C. Crews and L. Castle, Some factors affecting the formation of furan in heated foods, *Food Addit. Contam.*, 2006, **23**, 219–227.
33. F. Morisco, V. Lembo, G. Mazzone, S. Camera and N. Caporaso, Coffee and liver health, *J. Clin. Gastroenterol.*, 2014, **48**(Suppl. 1), S87–S90.
34. V. W. Setiawan, L. R. Wilkens, S. C. Lu, B. Y. Hernandez, L. Le Marchand and B. E. Henderson, Association of coffee intake with reduced incidence of liver cancer and death from chronic liver disease in the US multiethnic cohort, *Gastroenterology*, 2015, **148**, 118–125, quiz e15.
35. N. Khalaf, D. White, F. Kanwal, D. Ramsey, S. Mittal, S. Tavakoli-Tabasi, J. Kuzniarek and H. B. El-Serag, Coffee and Caffeine Are Associated With Decreased Risk of Advanced Hepatic Fibrosis Among Patients With Hepatitis C, *Clin. Gastroenterol. Hepatol.*, 2015, **13**, 1521.
36. P. Carthew, M. DiNovi and R. W. Setzer, Application of the margin of exposure (MoE) approach to substances in food that are genotoxic and carcinogenic: example: furan (CAS No. 110-00-9), *Food Chem. Toxicol.: an Int. J. Publ. Br. Ind. Biol. Res. Assoc.*, 2010, **48**, S69–S74.
37. NTP, *NTP Technical Report on the Toxicology and Carcinogenesis Studies of Furan in F344/N Rats and B6C3F1 Mice (Gavage Studies)*, 402, National Toxicology Program, NIH, Research Triangle Park, NC, 1993.
38. H. Nakagawa and S. Maeda, Molecular mechanisms of liver injury and hepatocarcinogenesis: focusing on the role of stress-activated MAPK, *Pathol. Res. Int.*, 2012, 172894.
39. M. K. Manibusan, M. Odin and D. A. Eastmond, Postulated carbon tetrachloride mode of action: A review, *J. Environ. Sci. Health, Part C: Environ. Carcinog. Ecotoxicol. Rev.*, 2007, **25**, 185–209.
40. M. V. Templin, K. C. Jamison, C. S. Sprankle, D. C. Wolf, B. A. Wong and B. E. Butterworth, Chloroform-induced cytotoxicity and regenerative cell proliferation in the kidneys and liver of BDF1 mice, *Cancer Lett.*, 1996, **108**, 225–231.
41. J. L. Larson, D. C. Wolf and B. E. Butterworth, Induced cytolethality and regenerative cell proliferation in the livers and kidneys of male B6C3F1 mice given chloroform by gavage, *Fundam. Appl. Toxicol.: Off. J. Soc. Toxicol.*, 1994, **23**, 537–543.

42. G. J. Moser, J. Foley, M. Burnett, T. L. Goldsworthy and R. Maronpot, Furan-induced dose–response relationships for liver cytotoxicity, cell proliferation, and tumorigenicity (furan-induced liver tumorigenicity), *Exp. Toxicol. Pathol.*, 2009, **61**, 101–111.
43. R. Fransson-Steen, T. L. Goldsworthy, G. L. Kedderis and R. R. Maronpot, Furan-induced liver cell proliferation and apoptosis in female B6C3F1 mice, *Toxicology*, 1997, **118**, 195–204.
44. L. T. Burka, K. D. Washburn and R. D. Irwin, Disposition of 14[C] furan in the male F344 rat, *J. Toxicol. Environ. Health*, 1991, **34**, 245–257.
45. M. I. Churchwell, R. C. Scheri, L. S. Von Tungeln, G. Gamboa da Costa, F. A. Beland and D. R. Doerge, Evaluation of Serum and Liver Toxicokinetics for Furan and Liver DNA Adduct Formation in Male Fischer 344 Rats, *Food Chem. Toxicol.*, 2015, **86**, 1–8.
46. G. L. Kedderis, M. A. Carfagna, S. D. Held, R. Batra, J. E. Murphy and M. L. Gargas, Kinetic analysis of furan biotransformation by F-344 rats *in vivo* and *in vitro*, *Toxicol. Appl. Pharmacol.*, 1993, **123**, 274–282.
47. F. J. Gonzalez, Role of cytochromes P450 in chemical toxicity and oxidative stress: studies with CYP2E1, *Mutat. Res., Fundam. Mol. Mech. Mutagen.*, 2005, **569**, 101–110.
48. M. Kellert, S. Wagner, U. Lutz and W. K. Lutz, Biomarkers of furan exposure by metabolic profiling of rat urine with liquid chromatography-tandem mass spectrometry and principal component analysis, *Chem. Res. Toxicol.*, 2008, **21**, 761–768.
49. L. A. Gates, D. Lu and L. A. Peterson, Trapping of cis-2-butene-1,4-dial to measure furan metabolism in human liver microsomes by cytochrome P450 enzymes, *Drug Metab. Dispos.: Biol. Fate Chem.*, 2012, **40**, 596–601.
50. L. A. Gates, M. B. Phillips, B. A. Matter and L. A. Peterson, Comparative metabolism of furan in rodent and human cryopreserved hepatocytes, *Drug Metab. Dispos.*, 2014, **42**, 1132–1136.
51. Y. Lu and A. I. Cederbaum, CYP2E1 and oxidative liver injury by alcohol, *Free Radical Biol. Med.*, 2008, **44**, 723–738.
52. A. Cederbaum, Nrf2 and antioxidant defense against CYP2E1 toxicity, *Expert Opin. Drug Metab. Toxicol.*, 2009, **5**, 1223–1244.
53. M. A. Carfagna, S. D. Held and G. L. Kedderis, Furan-induced cytolethality in isolated rat hepatocytes: correspondence with in vivo dosimetry, *Toxicol. Appl. Pharmacol.*, 1993, **123**, 265–273.
54. K. C. Hickling, J. M. Hitchcock, V. Oreffo, A. Mally, T. G. Hammond, J. G. Evans and J. K. Chipman, Evidence of oxidative stress and associated DNA damage, increased proliferative drive, and altered gene expression in rat liver produced by the cholangiocarcinogenic agent Furan, *Toxicol. Pathol.*, 2010, **38**, 230–243.
55. W. Ding, D. M. Petibone, J. R. Latendresse, M. G. Pearce, L. Muskhelishvili, G. A. White, C. Chang, R. A. Mittelstaedt, J. G. Shaddock, L. P. McDaniel, D. R. Doerge, S. M. Morris, M. E. Bishop, M. G. Manjanatha, A. Aidoo and R. H. Heflich, In vivo genotoxicity of

furan in F344 rats at cancer bioassay doses, *Toxicol. Appl. Pharmacol.*, 2012, **261**, 164–171.

56. A. F. Webster, N. Chepelev, R. Gagne, B. Kuo, L. Recio, A. Williams and C. L. Yauk, Impact of Genomics Platform and Statistical Filtering on Transcriptional Benchmark Doses (BMD) and Multiple Approaches for Selection of Chemical Point of Departure (PoD), *PLoS One*, 2015, **10**, e0136764.
57. H. Dong, S. Gill, I. H. Curran, A. Williams, B. Kuo, M. G. Wade and C. L. Yauk, Toxicogenomic assessment of liver responses following subchronic exposure to furan in Fischer F344 rats, *Arch. Toxicol.*, 2015, 1.
58. P. Carthew, M. DiNovi and R. Woodrow Setzer, Application of the Margin of Exposure (MOE) approach to substances in food that are genotoxic and carcinogenic: example: CAS No: 105650-23-5 PhIP (2-amino-1-methyl-6-phenylimidazo[4,5-b]pyridine), *Food Chem. Toxicol.*, 2010, **48**, S98–S105.
59. A. F. Jackson, A. Williams, L. Recio, M. D. Waters, I. B. Lambert and C. L. Yauk, Case study on the utility of hepatic global gene expression profiling in the risk assessment of the carcinogen furan, *Toxicol. Appl. Pharmacol.*, 2014, **274**, 63–77.
60. S. R. Thomas, S. C. Wesselkamper, N. C. Y. Wang, Q. J. Zhao, D. D. Petersen, J. C. Lambert, I. Cote, L. Yang, E. Healy, M. B. Black, H. J. Clewell III, B. C. Allen and M. E. Andersen, Temporal concordance between apical and transcriptional points of departure for chemical risk assessment, *Toxicol. Sci.*, 2013, **134**, 180–194.
61. L. D. Gorsky, D. R. Koop and M. J. Coon, On the stoichiometry of the oxidase and monooxygenase reactions catalyzed by liver microsomal cytochrome P-450. Products of oxygen reduction, *J. Biol. Chem.*, 1984, **259**, 6812–6817.
62. L. Recio, S. L. Phillips, T. Maynor, M. Waters, A. F. Jackson and C. L. Yauk, Differential Expression of Long Non-Coding RNAs in the Livers of Female B6C3F1 Mice Exposed to the Carcinogen Furan, *Toxicol. Sci.*, 2013, **135**, 369–379.
63. J. L. Rinn and H. Y. Chang, *Genome regulation by long noncoding RNAs*, 2012.
64. B. P. Bass, K. B. Engel, S. R. Greytak and H. M. Moore, A review of preanalytical factors affecting molecular, protein, and morphological analysis of formalin-fixed, paraffin-embedded (FFPE) tissue: how well do you know your FFPE specimen?, *Arch. Pathol. Lab. Med.*, 2014, **138**, 1520–1530.
65. R. Klopfleisch, A. T. Weiss and A. D. Gruber, Excavation of a buried treasure–DNA, mRNA, miRNA and protein analysis in formalin fixed, paraffin embedded tissues, *Histol. Histopathol.*, 2011, **26**, 797–810.
66. S. M. Farragher, A. Tanney, R. D. Kennedy and D. Paul Harkin, RNA expression analysis from formalin fixed paraffin embedded tissues, *Histochem. Cell Biol.*, 2008, **130**, 435–45.
67. A. M. Sadi, D. Y. Wang, B. J. Youngson, N. Miller, S. Boerner, S. J. Done and W. L. Leong, Clinical relevance of DNA microarray analyses using

archival formalin-fixed paraffin-embedded breast cancer specimens, *BMC Cancer*, 2011, **11**(253), 1–13.

68. T. A. Jacobson, J. Lundahl, H. Mellstedt and A. Moshfegh, Gene expression analysis using long-term preserved formalin-fixed and paraffin-embedded tissue of non-small cell lung cancer, *Int. J. Oncol.*, 2011, **38**, 1075–1081.
69. J. Budczies, W. Weichert, A. Noske, B. M. Muller, C. Weller, T. Wittenberger, H. P. Hofmann, M. Dietel, C. Denkert and V. Gekeler, Genome-wide gene expression profiling of formalin-fixed paraffin-embedded breast cancer core biopsies using microarrays, *J. Histochem. Cytochem.: Off. J. Histochem. Soc.*, 2011, **59**, 146–157.
70. K. Linton, C. Howarth, M. Wappett, G. Newton, C. Lachel, J. Iqbal, S. Pepper, R. Byers, W. J. Chan and J. Radford, Microarray gene expression analysis of fixed archival tissue permits molecular classification and identification of potential therapeutic targets in diffuse large B-cell lymphoma, *J. Mol. Diagn.*, 2012, **14**, 223–232.
71. B. A. Merrick, S. S. Auerbach, P. S. Stockton, J. F. Foley, D. E. Malarkey, R. C. Sills, R. D. Irwin and R. R. Tice, Testing an aflatoxin B1 gene signature in rat archival tissues, *Chem. Res. Toxicol.*, 2012, **25**, 1132–1144.
72. S. S. Auerbach, D. P. Phadke, D. Mav, S. Holmgren, Y. Gao, B. Xie, J. H. Shin, R. R. Shah, B. A. Merrick and R. R. Tice, RNA-Seq-based toxicogenomic assessment of fresh frozen and formalin-fixed tissues yields similar mechanistic insights, *J. Appl. Toxicol.*, 2014, **35**, 766–780.
73. A. F. Jackson, A. Williams, I. Moffat, S. L. Phillips, L. Recio, M. D. Waters, I. B. Lambert and C. L. Yauk, Preparation of archival formalin-fixed paraffin-embedded mouse liver samples for use with the Agilent gene expression microarray platform, *J. Pharmacol. Toxicol. Methods*, 2013, **68**(2), 260–268.
74. A. F. Webster, P. Zumbo, J. Fostel, J. Gandara, S. D. Hester, L. Recio, A. Williams, C. E. Wood, C. L. Yauk and C. E. Mason, Mining the archives: a cross-platform analysis of gene expression profiles in archival formalin-fixed paraffin-embedded (FFPE) tissue, *Toxicol. Sci.*, 2015, **148**, 460–472.
75. J. A. Bourdon-Lacombe, I. D. Moffat, M. Deveau, M. Husain, S. Auerbach, D. Krewski, R. S. Thomas, P. R. Bushel, A. Williams and C. L. Yauk, Technical guide for applications of gene expression profiling in human health risk assessment of environmental chemicals, *Regul. Toxicol. Pharmacol.*, 2015, **72**, 292–309.
76. R. A. Becker, G. T. Ankley, S. W. Edwards, S. W. Kennedy, I. Linkov, B. Meek, M. Sachana, H. Segner, B. Van Der Burg, D. L. Villeneuve, H. Watanabe and T. S. Barton-Maclaren, Increasing Scientific Confidence in Adverse Outcome Pathways: Application of Tailored Bradford-Hill Considerations for Evaluating Weight of Evidence, *Regul. Toxicol. Pharmacol.*, 2015, **72**, 514–537.
77. D. Knapen, L. Vergauwen, D. L. Villeneuve and G. T. Ankley, The potential of AOP networks for reproductive and developmental toxicity assay development, *Reprod. Toxicol.*, 2015, **56**, 52–55.

78. M. Vinken, B. Landesmann, M. Goumenou, S. Vinken, I. Shah, H. Jaeschke, C. Willett, M. Whelan and V. Rogiers, Development of an adverse outcome pathway from drug-mediated bile salt export pump inhibition to cholestatic liver injury, *Toxicol. Sci.*, 2013, **136**, 97–106.
79. E. C. Tonk, J. L. Pennings and A. H. Piersma, An adverse outcome pathway framework for neural tube and axial defects mediated by modulation of retinoic acid homeostasis, *Reprod. Toxicol.*, 2014, **55**, 104–113.
80. C. MacKay, M. Davies, V. Summerfield and G. Maxwell, From pathways to people: applying the adverse outcome pathway (AOP) for skin sensitization to risk assessment, *ALTEX*, 2013, **30**, 473–486.
81. C. L. Russom, C. A. LaLone, D. L. Villeneuve and G. T. Ankley, Development of an adverse outcome pathway for acetylcholinesterase inhibition leading to acute mortality, *Environ. Toxicol. Chem.*, 2014, **33**, 2157–2169.
82. C. L. Yauk, I. B. Lambert, M. E. Meek, G. R. Douglas and F. Marchetti, Development of the adverse outcome pathway "alkylation of DNA in male premeiotic germ cells leading to heritable mutations" using the OECD's users' handbook supplement, *Environ. Mol. Mutagen.*, 2015, **56**, 724–750.
83. N. Burden, F. Sewell, M. E. Andersen, A. Boobis, J. K. Chipman, M. T. Cronin, T. H. Hutchinson, I. Kimber and M. Whelan, Adverse Outcome Pathways can drive non-animal approaches for safety assessment, *J. Appl. Toxicol.*, 2015, **35**, 971–975.
84. B. Landesmann, M. Mennecozzi, E. Berggren and M. Whelan, Adverse outcome pathway-based screening strategies for an animal-free safety assessment of chemicals, *Altern. Lab. Anim.*, 2013, **41**, 461–471.
85. D. Krewski, D. Acosta Jr., M. Andersen, H. Anderson, J. C. Bailar 3rd, K. Boekelheide, R. Brent, G. Charnley, V. G. Cheung, S. Green Jr., K. T. Kelsey, N. I. Kerkvliet, A. A. Li, L. McCray, O. Meyer, R. D. Patterson, W. Pennie, R. A. Scala, G. M. Solomon, M. Stephens, J. Yager and L. Zeise, Toxicity testing in the 21st century: a vision and a strategy, *J. Toxicol. Environ. Health, Part B*, 2010, **13**, 51–138.
86. T. B. Knudsen, K. A. Houck, N. S. Sipes, A. V. Singh, R. S. Judson, M. T. Martin, A. Weissman, N. C. Kleinstreuer, H. M. Mortensen, D. M. Reif, J. R. Rabinowitz, R. W. Setzer, A. M. Richard, D. J. Dix and R. J. Kavlock, Activity profiles of 309 ToxCast chemicals evaluated across 292 biochemical targets, *Toxicology*, 2011, **282**, 1–15.
87. R. Judson, K. Houck, M. Martin, T. Knudsen, R. S. Thomas, N. Sipes, I. Shah, J. Wambaugh and K. Crofton, *In vitro* and modelling approaches to risk assessment from the U.S. Environmental Protection Agency ToxCast programme, *Basic Clin. Pharmacol. Toxicol.*, 2014, **115**, 69–76.
88. R. S. Settivari, N. Ball, L. Murphy, R. Rasoulpour, D. R. Boverhof and E. W. Carney, Predicting the future: opportunities and challenges for the chemical industry to apply 21st-century toxicity testing, *J. Am. Assoc. Lab. Anim. Sci.*, 2015, **54**, 214–223.
89. S. Nesnow, W. Ward, T. Moore, H. Ren and S. D. Hester, Discrimination of tumorigenic triazole conazoles from phenobarbital by transcriptional

analyses of mouse liver gene expression, *Toxicol. Sci.: Off. J. Soc. Toxicol.*, 2009, **110**, 68–83.
90. C. Q. Yao, S. D. Prokopec, J. D. Watson, R. Pang, C. P'ng, L. C. Chong, N. J. Harding, R. Pohjanvirta, A. B. Okey and P. C. Boutros, Inter-strain heterogeneity in rat hepatic transcriptomic responses to 2,3,7,8-tetrachlorodibenzo-p-dioxin (TCDD), *Toxicol. Appl. Pharmacol.*, 2012, **260**, 135–145.
91. R. M. Maertens, P. A. White, A. Williams and C. L. Yauk, A global toxicogenomic analysis investigating the mechanistic differences between tobacco and marijuana smoke condensates in vitro, *Toxicology*, 2013, **308**, 60–73.
92. H. Ellinger-Ziegelbauer, J. Aubrecht, J. C. Kleinjans and H. J. Ahr, Application of toxicogenomics to study mechanisms of genotoxicity and carcinogenicity, *Toxicol. Lett.*, 2009, **186**, 36–44.
93. J. A. Bourdon, A. Williams, B. Kuo, I. Moffat, P. A. White, S. Halappanavar, U. Vogel, H. Wallin and C. L. Yauk, Gene expression profiling to identify potentially relevant disease outcomes and support human health risk assessment for carbon black nanoparticle exposure, *Toxicology*, 2013, **303**, 83–93.
94. J. A. Bourdon, S. Halappanavar, A. T. Saber, N. R. Jacobsen, A. Williams, H. Wallin, U. Vogel and C. L. Yauk, Hepatic and pulmonary toxicogenomic profiles in mice intratracheally instilled with carbon black nanoparticles reveal pulmonary inflammation, acute phase response, and alterations in lipid homeostasis, *Toxicol. Sci.*, 2012, **127**, 474–484.
95. S. Labib, C. Yauk, A. Williams, V. M. Arlt, D. H. Phillips, P. A. White and S. Halappanavar, Subchronic oral exposure to benzo(a)pyrene leads to distinct transcriptomic changes in the lungs that are related to carcinogenesis, *Toxicol. Sci.*, 2012, **129**, 213–224.

CHAPTER 13

The Parallelogram Approach to Assess Human Relevance of Toxicogenomics-derived Toxicity Pathways in Human Health Risk Assessment

ANNE S. KIENHUIS,*[a] WIM C. MENNES,[b] MARJA DRIESSEN,[a] LEO T. M. VAN DER VEN,[a] JAN VAN BENTHEM[a] AND MIRJAM LUIJTEN[a]

[a] Centre for Health Protection, National Institute for Public Health and the Environment (RIVM), Bilthoven, The Netherlands; [b] Centre for Nutrition, Prevention and Health Services, National Institute for Public Health and the Environment (RIVM), Bilthoven, The Netherlands

*Email: Anne.Kienhuis@rivm.nl

13.1 Human Health Risk Assessment

In the European Union, policy on industrial-, non-food-chemical products and biocides, either newly introduced or already on the market, is regulated under REACH, the Regulation, Evaluation, Authorization and restriction of CHemicals. For chemical substances such as food additives, pesticides, food contact materials and medicinal products other legislation is in place, but also these substances have to be evaluated for safety before entering the market. To assess the safety of chemical substances, four aspects need

Issues in Toxicology No. 28
Toxicogenomics in Predictive Carcinogenicity
Edited by Michael D. Waters and Russell S. Thomas

Published by the Royal Society of Chemistry, www.rsc.org

to be covered. These are hazard identification, dose–response assessment, exposure assessment and risk characterization. This risk assessment process, which has been in place since the 1970s, is presented in Figure 13.1. In exposure assessment, the route (inhalation/dermal in non-food applications and in the work environment *versus* oral exposure for substances in food and drinking water) and level of exposure will be determined. For hazard identification, the nature of the toxic (adverse) effect(s) that a substance may elicit will be assessed. These hazards may vary depending on the exposure situation. Normally, acute toxicity data and local toxicity data have very little relevance for exposures related to food and drinking water, but they can be very relevant for work-related exposures or for exposures from non-food consumer products. In dose–response assessment, the lowest exposure level at which this effect is anticipated to occur is estimated, *i.e.* the lowest observed adverse effect level (LOAEL), or preferably the highest exposure level below which the effect will not occur, *i.e.* the no observed adverse effect level (NOAEL). A statistically more advanced and justified way to process results of toxicity testing is the benchmark dose (BMD) approach. The BMD may be considered as a dose level that is associated with a predefined level of effect. LOAEL, NOAEL or BMD can all serve as points of departure (PoD), which in turn are used, with the application of extrapolation factors, to derive reference dose (RfD) levels. The RfDs are used for risk characterization, to evaluate estimates of human exposure; exposures above the RfD are usually concluded to be of safety concern.[1]

The current system for safety or risk assessment largely relies on information obtained in studies using animal (mostly rodent) models. Acute toxicity and local toxicity studies are rather demanding on animal welfare.

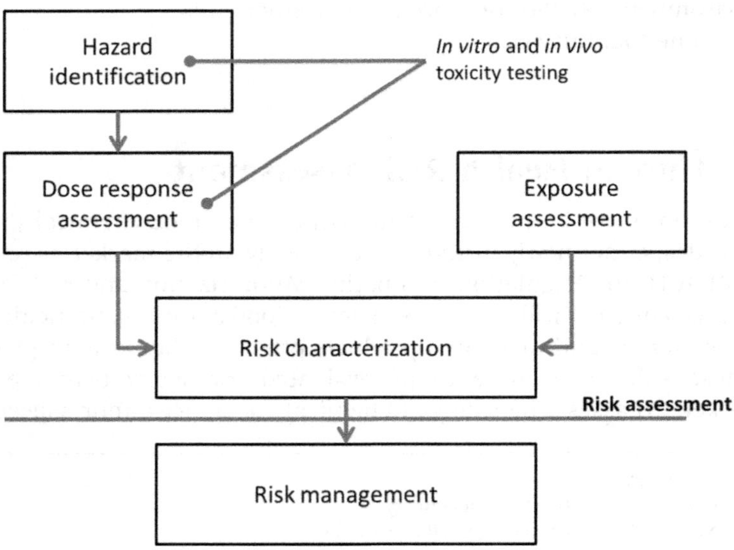

Figure 13.1 The process of chemical risk assessment.

Although these studies usually involve a relatively limited number of animals, they cause severe animal suffering. In contrast, (sub-)chronic and reproductive toxicity studies cause less suffering for the individual animals but will require a far larger number of animals. In terms of reducing use of experimental animals for the safety assessment of chemical substances, the major profit can be made when these large studies are replaced by alternative, animal-sparing approaches. In the past decade, significant advances have been made in the development of *in vitro* approaches that hold promise for more reliable prediction of human toxicity, at the same time avoiding animal testing. The applicability of an *in vitro* approach or technology in risk assessment is dictated by its ability to reliably identify (1) the putative adverse effect(s) of a substance; and (2) the level of exposure to that substance above which this adverse effect will occur. These two basic principles determine the criteria that any assay has to meet in order to replace the classical tests for toxicity in animals and to be included in risk assessment in a quantitative way. In this regard, toxicogenomics is proposed as a promising contribution to risk assessment due to its sensitivity and the ability to examine the molecular changes resulting from chemical exposure in more detail.[2] In this chapter, we propose approaches, the parallelogram approach and the concordance model, for (1) assessment of the relevance of molecular changes resulting from chemical exposure in animal studies *in vivo* and *in vitro* cell systems (of animal and human origin) for toxicity that would occur in humans *in vivo*; and (2) evaluation of (novel) *in vitro* assays for the assessment of toxicity. Studies are described that have been successful in using the parallelogram approach and concordance model for toxicogenomics-based hazard identification in risk assessment and for evaluation of *in vitro* test systems. The challenge that remains is application of these approaches in a quantitative manner for risk assessment purposes; this is also discussed in this chapter.

13.2 Toxicogenomics in Risk Assessment

Toxicogenomics is defined as the application of omics-based technologies in toxicology. In various legal frameworks for (industrial) chemicals, but particularly in REACH, the use of toxicogenomics in risk assessment is anticipated. Since changes in gene expression usually precede the manifestation of apical endpoints, *i.e.* traditional toxicity markers such as clinical chemistry and histopathology, toxicogenomic responses are recognized as a sensitive read-out. As such, toxicogenomics is expected to contribute to alternative testing strategies for the safety evaluation of chemical substances by reducing the number of animals required, shortening the study durations, and lowering the dose levels used in animal studies. To date, toxicogenomics is predominantly used by pharmaceutical companies for screening and prioritization of lead compounds to produce pharmaceuticals. Its limited use in regulatory frameworks of human health risk assessment is partly explained by significant concerns that unsubstantiated toxicogenomic data could be inappropriately extrapolated to toxicity that might occur in humans. Such an inadequate use of toxicogenomics data could evoke over-actions such as the unwarranted use of the precautionary principle.[3]

13.2.1 Toxicogenomics in Hazard Identification

From a scientific perspective, toxicogenomics has the potential to improve understanding of processes that underlie the modes of action (MoA) of chemical substances at the molecular level.[4] It has enhanced several areas in toxicology, including the hazard prediction of substances based on toxicogenomics biomarkers.[5–7] For example, toxicogenomics investigations have been applied for the hazard identification of substances with respect to inducing toxicogenomic endpoints indicative of carcinogenicity. These studies have been reviewed by Waters *et al.*[8] Study designs ranged from 1 day treatments to 14, 28, and 90 day treatments. Studies were performed *in vitro* (human and rodent cell lines) and *in vivo* (rodent). Cells and animals were treated with various genotoxic carcinogens, non-genotoxic carcinogens and/or non-carcinogens in concentrations/doses ranging from non-toxic to toxic up to the maximum tolerable dose. Results showed that the translation of gene expression changes into regulated pathways allowed for prediction of chemical carcinogenicity, at least in the respective model systems.[8]

13.2.2 Toxicogenomics in Dose–Response Modeling

In the past years, pragmatic approaches for the application of toxicogenomics in quantitative risk assessment, primarily in cancer risk assessment, have been suggested.[9–13] The word "pragmatic" here means that conventional methods were applied, with the exception that these studies use toxicogenomic data as an alternative to apical endpoints. These studies aimed to investigate the possible application of toxicogenomics in quantitative risk assessment by comparison of the lowest BMD derived from toxicogenomics data with PoDs derived from standard cancer and non-cancer apical endpoints. Based on the above-cited studies, Chepelev *et al.* have proposed an experimental design for quantitative risk assessment.[14] Suggestions for standardization of study designs for quantitative toxicogenomics do not substantially differ from those that apply to traditional toxicology study designs. For example, agreement is needed on the test models used. For *in vivo* studies, rodents will be the most obvious choice, and preferred species and strains need to be proposed. Furthermore, there needs to be agreement on dose selection and time points to be included as well as exposure duration. Once a unified study design is established, standardization of data processing tools and consistency in analysis of the results is required, including methods on how to derive BMD values from toxicogenomic datasets.[11,12] For more information on suggestions for and discussions on a unified study design, as well as methods to derive transcriptional BMD values, refer to the publications of Chepelev *et al.* (2015) and Thomas and colleagues.[11,12,14]

For implementation of quantitative toxicogenomics in a regulatory framework such as REACH, risk assessors and research scientists should collaborate and communicate, particularly to discuss and agree on interpretation of the data in the context of toxicity. This requires confidence in the human relevance of data as well as confidence in the biological significance of the data, *i.e.* selection of specific genes and pathways related to the MoA to be used to calculate BMDs. One of the biggest challenges is how to use the data to select PoDs that are based on dose–response assessment of critical effects and that are protective of adverse rather than adaptive responses.[14]

13.3 Assessing Biological Significance and Human Relevance of Toxicogenomics Data

For hazard identification of chemical substances, toxicogenomics studies using rodents *in vivo* have been conducted to search for early key events and toxicity pathways that constitute the MoA. Since toxicogenomics is regarded as a sensitive read-out, allowing for detection of toxicity before apical endpoints are observed, this type of toxicogenomics study is characterized by relatively short exposure durations in comparison to, for example, 2 year rodent carcinogenicity studies where tumors get sufficient time to develop. The use of well-characterized chemical substances (*i.e.* substances that are well known to induce apical endpoints as assessed in conventional *in vivo* studies) allows for phenotypic anchoring of early key events and toxicity pathways derived from short-term toxicogenomics studies.[8] The phenotypically anchored toxicity pathways of well-known chemicals could be collected in a database. By adding toxicogenomics data from several model systems (*in vitro* and *in vivo*) and species the likelihood that these substances will exhibit a similar MoA in humans can be established. This can be done by using the so-called parallelogram approach and the more complex concordance model, as described below. The database may also be used to compare toxicogenomics profiles of new, untested chemical substances, thereby linking unknown chemical substances to human relevant MoAs and apical endpoints.

13.3.1 The Parallelogram Approach

The parallelogram approach, introduced by Sobels in the late 1970s[15] and also presented for application in risk assessment by Sutter,[16] can be used to estimate robust, human-relevant adverse effects by extrapolating toxicogenomics results between *in vitro* and *in vivo* studies and across species. The parallelogram approach to predict the human relevance of toxicogenomics derived toxicity pathways is presented in Figure 13.2.

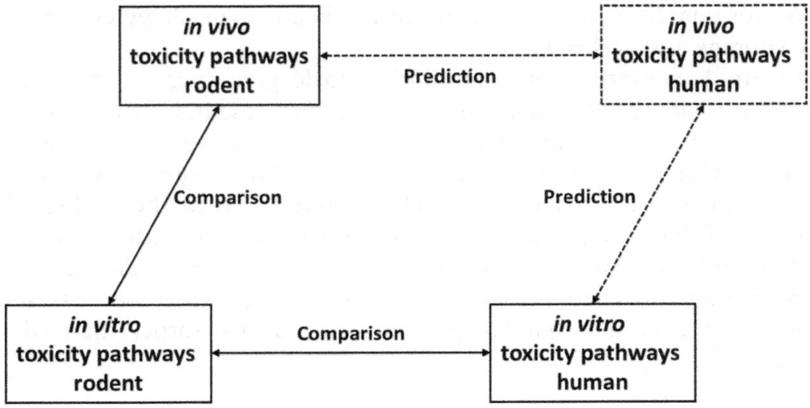

Figure 13.2 The parallelogram approach to predict relevance of toxicogenomics derived toxicity pathways.
Adapted from Sutter.[16]

The parallelogram approach was applied in a study to identify robust, human-relevant MoAs, using the hepatotoxicant acetaminophen (APAP) as a model compound.[17] Toxicogenomics data were generated from APAP-treated primary rat hepatocytes *in vitro*, primary human hepatocytes *in vitro* and rats *in vivo*. Rats *in vivo* were exposed to 1.5 $g\,kg^{-1}$ body weight APAP, an overtly toxic dose as confirmed using histopathology and clinical chemistry.[18] An APAP concentration of 5 mM did not result in loss of cell viability in rat and human hepatocytes *in vitro*, but 10 mM APAP resulted in 10% and 20% loss of cell viability in human and rat hepatocytes, respectively. All exposures resulted in deregulated genes and pathways that were compared between *in vitro* and *in vivo* and between species. *In vitro* to *in vivo* and rat to human extrapolation confirmed induction of oxidative stress, impairment of energy metabolism and mitochondrial function as the most relevant and robust toxicity pathways that are likely to drive APAP-induced liver injury in humans. In the APAP parallelogram study, no risk characterization, *i.e.* quantitative assessment of human hepatotoxic health risk, was performed. However, it was argued that the doses and concentrations used for the rat *in vivo* and rat and human *in vitro* exposures were relevant to humans despite the fact that these exposure values were one order of magnitude beyond the generally accepted hepatotoxic dose for an average adult within 24 hours after exposure (1.5 $g\,kg^{-1}$ body weight for rat *versus* 150 $mg\,kg^{-1}$ for humans). This conclusion was drawn since humans are regularly co-exposed to several factors, such as food constituents and alcohol consumption, that enhance APAP toxicity and decrease the hepatotoxic dose, even to a level below the maximum therapeutic dose.[17]

We have conducted several studies to elucidate toxicity pathways related to apical endpoints, specifically, cholestasis, steatosis and necrosis phenotypes in hepatotoxicity. Cholestasis is a chronic condition that is phenotypically characterized by bile accumulation as a result of changes in intra- or extracellular bile flow or bile composition.[19] Steatosis is a condition that may also occur chronically and is characterized as an increase in cellular lipid content due to an increase in *de novo* synthesis of fatty acids or reduced lipid secretion or oxidation.[20] Necrosis is an acute condition that is characterized by cell death, *e.g.* due to oxidative stress.[21] For phenotypic anchoring of the (sub-)chronic conditions cholestasis and steatosis, we employed a study design in which we selected a dose range at which the maximum dose caused phenotypic changes at 25 days after repeated exposure, as seen by changes in phenotype-specific clinical chemistry parameters and histopathology. This study design is presented in Figure 13.3. Snapshots of the onset and progression of hepatotoxicity were taken at 1, 4 and 11 days after repeated exposure by gene expression profiling. Toxicity pathways were deduced from these transcriptomic profiles using various data analysis methods, with the aim to identify gene expression changes that are truly related to development of hepatotoxic phenotypes.

In one of the methods applied, we performed phenotype-directed data analysis to isolate the cholestasis-inducing property of the model compound cyclosporin A, thereby disregarding other effects of the compound, such as its pharmaceutical immunosuppressive effect.[22] We therefore correlated changes in gene expression to changes in cholestasis-specific increases in a set of three clinical chemistry parameters: cholesterol, total bilirubin, and bile acids. These biliary constituents are measured clinically in cholestasis, as they are known to accumulate in hepatocytes during development of the phenotype.[23] They act as ligands for specific nuclear receptors that act as mediators of the cholestasis response, inducing all kinds of downstream pathways that as such are proposed to represent toxicity pathways of cholestasis. Analysis of the hepatic transcriptome of mice exposed to cyclosporin A revealed regulation of drug metabolism pathways, metabolism of bile acids and bile salts, lipid metabolism and the peroxisome proliferator-activated receptor signaling pathway. Key genes were identified that are involved in these pathways and regulated as downstream targets of the

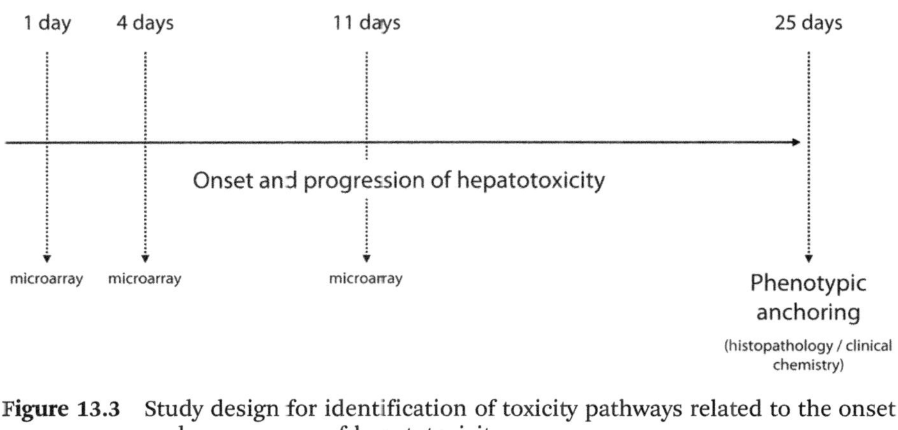

Figure 13.3 Study design for identification of toxicity pathways related to the onset and progression of hepatotoxicity.

nuclear receptors that are known to be induced by the increased biliary constituents.[22] A molecular concepts map, based on the results of the phenotype-directed data analysis, is depicted in Figure 13.4. Such a map facilitates the comparison of gene sets obtained from *in vivo* toxicogenomics studies with those from *in vitro* toxicogenomics studies, which is key to a parallelogram approach.

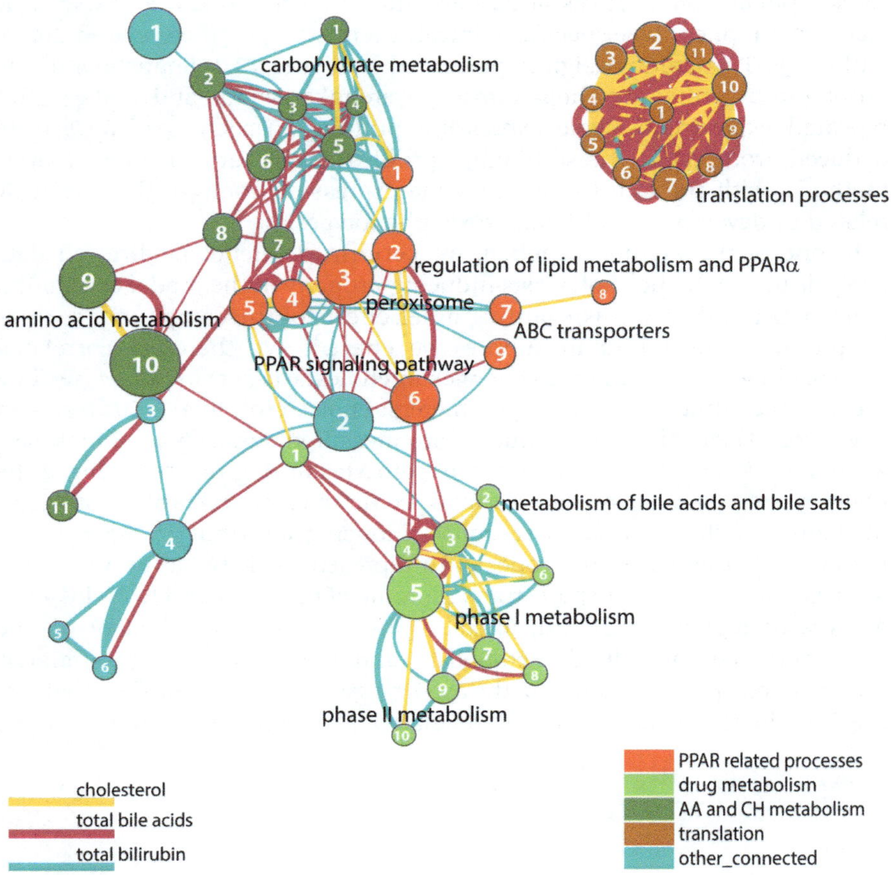

Figure 13.4 Molecular concepts map. Interconnection of cholestasis-related gene sets with respect to overlap in top correlated genes. Node sizes reflect number of genes per gene set, and edge sizes reflect number of genes overlapping between gene sets. Node colors refer to gene set categories, and edge color refers to serum parameters in which gene sets are correlated. Node numbers refer to specific gene sets within gene set categories, as listed in table 3 in Kienhuis *et al.*[22] PPAR: peroxisome proliferator-activated receptor.
Reprinted from Kienhuis *et al.*, Cyclosporine A treated *in vitro* models induce cholestasis response through comparison of phenotype-directed gene expression analysis of *in vivo* Cyclosporine A-induced cholestasis, *Toxicol. Lett.*, 2013, **221**, 225–236, with permission from Elsevier.

Similarly, phenotype-directed analysis was performed to identify toxicity pathways reflective of the development of steatosis in mice *in vivo*.[24] To substantiate the relevance of results for steatosis, mice were treated with three well-known steatosis-inducing compounds: amiodarone, valproic acid and tetracycline. In this study, only amiodarone treatment resulted in a convincing steatosis phenotype in mice, which allowed for the identification of toxicity pathways that correlated with phenotype development. Although no clinical phenotype was established in valproic acid treated mice, some of the phenotype-specific pathways found for amiodarone were enriched upon valproic acid treatment as well, confirming the sensitivity of disturbance in gene expression compared to changes in clinical chemistry and histology. The reduced severity of the steatosis response upon treatment with valproic acid was observed in the toxicogenomics response, since steatosis-related pathways were only perturbed from exposure duration of 11 days onwards, whereas most of these pathways were already perturbed after 4 days of amiodarone treatment. Pathways related to steatosis and commonly regulated upon amiodarone and valproic acid treatment, but not tetracycline treatment, were involved in fatty acid and steroid biosynthesis and metabolism (both hallmarks in steatosis as they lead to the increase in lipid deposition within the liver[25]), retinol biosynthesis and drug metabolism (Table 13.1).[24] Furthermore, enriched nuclear receptors that regulate the downstream signaling cascades involved in steatosis were identified. Eleven key genes were proposed to be steatosis markers as they were involved in the steatosis-specific pathways and are downstream targets of the enriched nuclear receptors.[24]

Both above-mentioned studies provide possible methods to retrieve toxicity pathways, nuclear receptors and gene expression markers reflective of hepatotoxic phenotypes in mice *in vivo*. The relevance of these pathways, receptors and genes to indicate these forms of hepatotoxicity in humans, however, remains unknown. To be predictive of hepatotoxicity in humans, these responses should be identified in *in vitro* studies with rat and/or mouse liver cells and similar responses should be observed in human hepatocytes *in vitro*, as can be assessed using the parallelogram approach.

13.3.2 The Concordance Model

The toxicogenomics-based parallelogram approach provides a valuable model for extrapolation between model systems and species to assess human relevance in hazard identification in risk assessment. However, some limitations should be addressed. The parallelogram approach does correct for rodent-specific effects that would not occur in humans and *in vitro* artifacts that would not occur *in vivo*. In addition, effects that can only be detected *in vivo* and human-specific effects will be missed.[26] To increase the confidence in toxicity pathways that are predictive of toxicity in humans, we have presented the concordance model for toxicogenomics-based extrapolation. Compared to the parallelogram approach, a larger number of species

Table 13.1 Pathways enriched during onset and progression of steatosis.[a]

KEGG pathway	AMD H1	AMD H4	AMD H11	AMD PD	VPA H1	VPA H4	VPA H11
Fatty acid metabolism	6.31	13.59	13.25	12.53		0.40	1.50
Biosynthesis of unsaturated fatty acids	4.07		9.60	6.05	3.74 ↓		0.57
PPAR signaling pathway		12.49				0.37	
Steroid biosynthesis		3.45			2.13		
Steroid hormone biosynthesis			0.20 ↓		1.87 ↓		1.45 ↓
Arachidonic acid metabolism	1.67	0.62	0.79	0.46	0.26 ↓	0.38	2.90
N-Glycan biosynthesis					1.34		
Valine, leucine and isoleucine degradation		3.50	2.94	3.53			0.33
Propanoate metabolism		2.56	4.96	1.91			1.08
beta-Alanine metabolism		1.31	2.39	1.21			
DNA replication		1.49					
ABC transporters		0.65	1.40	0.78	1.29 ↓		
Retinol metabolism	3.09	3.13	3.39	3.26	1.53 ↓	2.12	3.91
Starch and sucrose metabolism					1.52 ↓		
Metabolism of xenobiotics by cytochrome P450		2.07	2.36	3.10	3.80 ↓	0.31	6.76
Drug metabolism		1.85	1.45	1.94	2.91 ↓	0.34	7.17
Glutathione metabolism		3.19	1.92	0.72	1.56 ↓		3.90
Nitrogen metabolism					1.56 ↓		
Complement and coagulation cascades		4.93 ↓		2.41 ↓	0.19		2.28 ↓
Systemic lupus erythematosus		2.51 ↓		1.18 ↓			1.56 ↓
Prion diseases		0.94 ↓					1.54 ↓

[a]Shown are KEGG pathways with a Benjamini < 0.05 in a least one group from DAVID. Values are displayed as $-\log(\text{Benjamini})$, with a higher value meaning more enriched. Pathways are enriched with upregulated genes, except where ↓ signifies pathways enriched with downregulated genes. PD: phenotype directed; H1: high dose 1 day; H4: high dose 4 days; H11: high dose 11 days. Reprinted with kind permission from Springer Science + Business Media: A. P. Vitins *et al.*,[24] Mechanisms of amiodarone and valproic acid induced liver steatosis in mouse *in vivo* act as a template for other hepatotoxicity models, *Arch. Toxicol.*, 2014, **88**, 1573.

and (human) *in vitro* assays are included in this model.[26] The concordance model is presented in Figure 13.5. It can be used to predict a range of toxicity pathways decreasing in robustness and as such likeliness to be predictive for humans *in vivo*. At the same time, it provides additional evidence for effects that are only measurable *in vivo* (when found in multiple *in vivo* models) or in humans (when found in multiple human *in vitro* assays). We evaluated the concordance model using acetaminophen as a case study.[26] In this study, results obtained from other studies in rats and mice *in vivo* and rat and human cells *in vitro* substantiated the findings obtained with the rat–human *in vivo–in vitro* parallelogram approach. Furthermore, addition of data from mice *in vivo* suggested a species-specific response on apoptosis and immune

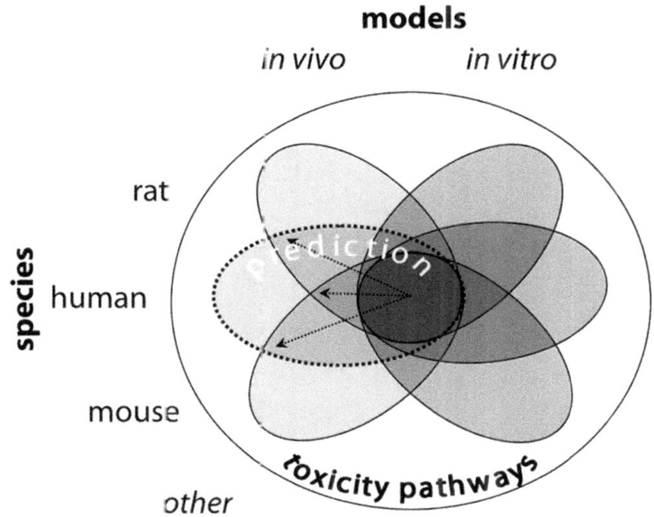

Figure 13.5 Concordance model for extrapolation of toxicogenomics-derived toxicity pathways. Arrows represent the decreasing conservation across species/models, and therefore the decreasing robustness of toxicity pathways.

response that was not found in rats *in vivo* and *in vitro* and in human cells *in vitro*.[26] Similarly, the predictability of the cyclosporin-induced cholestasis response[22] and steatosis response[24] was tested across multiple test models and species and was found to withstand the complexity of evolution and environment, underpinning the relevance of these toxicity pathways as indicators of hepatotoxicity in humans.

The concordance model was further explored in a study performed by Driessen *et al.*[27] In this study, rather than retrieving toxicity pathways for hepatotoxicity, the aim was to evaluate the zebrafish embryo as an alternative test model to study hepatotoxicity by using the concordance model.[27] Zebrafish embryos were exposed to three model compounds, each well-known inducers of specific hepatotoxic phenotypes: cyclosporin A to induce cholestasis, amiodarone to induce steatosis and acetaminophen to induce necrosis. Toxicogenomic studies were performed to compare effects of these compounds in the zebrafish embryo model with more conventional *in vivo* and *in vitro* assays, *in vivo* mouse and rat studies and primary mouse, rat and human hepatocytes *in vitro*. Zebrafish embryo and mouse *in vivo* studies were performed in our laboratories.[22,24,27] To compare with toxicogenomics data from others, optimal use was made of the studies deposited in the large publicly available transcriptomics databases such as the Gene Expression Omnibus database (www.ncbi.nlm.nih.gov/geo/) and ArrayExpress (www.ebi.ac.uk/arrayexpress/). The results obtained show that differential gene expression upon treatment with the hepatotoxic substances cyclosporin A, amiodarone and acetaminophen are primarily model specific, as visualized by a principal component analysis (Figure 13.6). For this analysis, only the top 500 ranked genes were used to ensure equal input for all the conditions and models evaluated. Closer evaluation of the data within

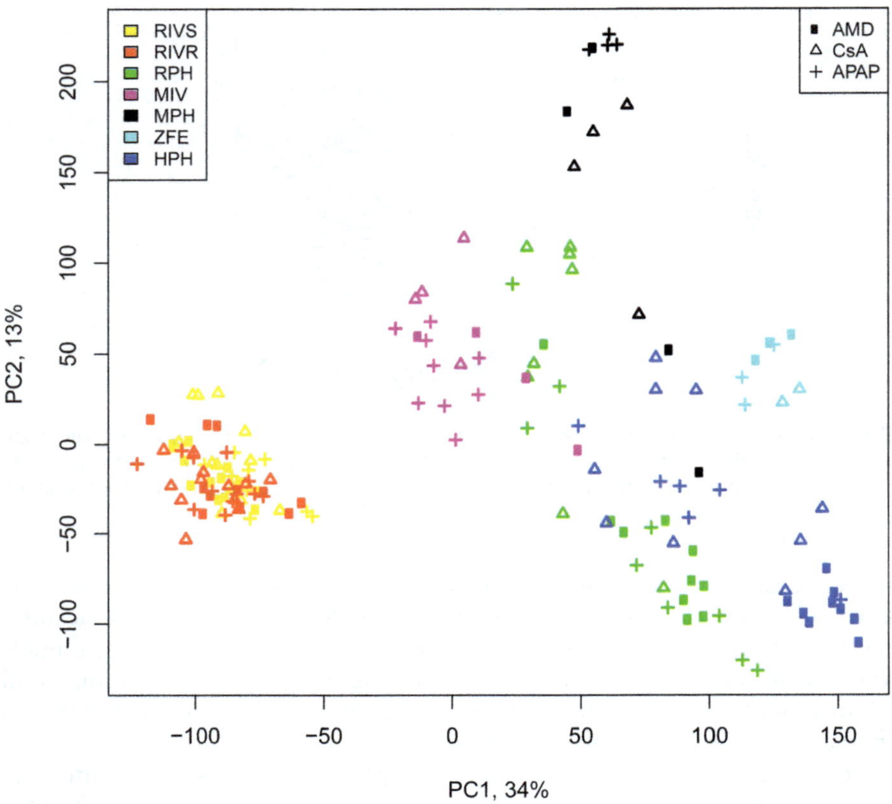

Figure 13.6 Principal component analysis based on a distance matrix using the overlap of the top 500 ranked genes between rat liver exposed *in vivo* (RIVS: single dose; RIVR: repeated dose), mouse liver *in vivo* (MIV), rat, mouse and human primary hepatocytes (RPH, MPH and HPH, respectively) and whole zebrafish embryo (ZFE), exposed to amiodarone (AMD), cyclosporine A (CsA) and acetaminophen (APAP). Percentages on the *x*- and *y*-axes indicate the explained variances for the two first principal axes. All *in vivo* models cluster on the left side in PC1, with a distinct position of both rat *in vivo* models. Subsequently, all *in vitro* models cluster on the right side, including ZFE. Herein, ZFE clusters most closely with HPH, whereas in PC2, ZFE aligns with the *in vivo* models. Overall, the human model, HPH clusters most closely with its rat counterpart RPH and with ZFE, and is most distinct from its mouse counterpart MPH and from the rat *in vivo* models.
Reprinted from Driessen *et al.*,[27] A transcriptomics-based hepatotoxicity comparison between the zebrafish embryo and established human and rodent *in vitro* and *in vivo* models using cyclosporine A, amiodarone and acetaminophen, *Toxicol. Lett.*, 2015, **232**, 403–412, with permission from Elsevier.

the models revealed clusters of regulated genes that allow for discrimination between compounds, dose and time treatment conditions.[27] Although the gene sets regulated upon exposure differed between models, common toxicity pathways could be identified. However, the study did not result in the identification of unique pathways per compound, conserved across species and reflective of a specific hepatotoxic phenotype. Rather, these pathways reflected a more general hepatotoxic response; the one pathway common across conditions (compound, dose and time) in all except one models was "bile acid regulation of glucose and lipid metabolism *via* farnesoid X receptor". A key gene of this pathway, CYP8B1, was regulated in all *in vivo* models but not *in vitro*. Interestingly, this particular pathway (and gene) was not found to be regulated in primary human hepatocytes upon exposure and as such, relevance for humans *in vivo* cannot be concluded from this study.[27] The study of Driessen *et al.* is an example of how to use the concordance model to evaluate an alternative model for its use in hazard identification. However, it did not aim to and succeed in retrieving toxicity pathways with relevance for humans *in vivo*. When applying the concordance model for that purpose, the focus should be on one disease phenotype and more human-based *in vitro* assays should be included.

13.4 Discussion

This chapter describes the application of the parallelogram approach and concordance model for hazard identification in risk assessment to identify toxicity pathways that are conserved across species and across models *in vivo* and *in vitro*. Increasing weight of evidence contributes to the identification of biologically significant and human relevant toxicity pathways. Furthermore, these approaches can be used to evaluate (novel) *in vitro* assays to replace conventional *in vivo* toxicity testing. The parallelogram approach is designed to compare one species and one type of *in vitro* assay with a human-based *in vitro* assay, thereby increasing confidence in the biological significance of toxicity pathways. The concordance model includes a larger number of species *in vivo* and a larger number of (human-based) *in vitro* assays. Common toxicity pathways identified with this model may be associated with higher levels of confidence in their biological significance. The added advantage of using human cells *in vitro* to substantiate the human relevance of toxicity pathways found in rodent *in vivo* studies, was also shown by others.[28] As such, the parallelogram approach and concordance model have shown to contribute to hazard identification in human health risk assessment. Recent advances have been made in the integration of toxicogenomics in quantitative human health risk assessment.[9–12,14] In this respect, the parallelogram approach and concordance model can contribute to selection of biolcgical significant and human relevant toxicity pathways that can be used to derive a lowest BMD in dose–response modeling. Although first studies on quantitative toxicogenomics show promising results, many challenges remain before quantitative toxicogenomics can be implemented in regulatory frameworks such as REACH. A discussion of the main challenges follows.

13.4.1 Pathway Analysis

A first problem to address is the interpretation of toxicogenomics data in terms of toxicity pathways. Many pathway analysis tools are available, and to some extent they all suffer from "annotation bias". For example, biological processes that have been investigated in more detail, such as apoptosis, metabolism and cancer, will have a relatively high number of associated genes compared to biological processes that are not as well studied. Consequently, these "more popular" processes will appear affected more frequently in the analysis. Furthermore, pathway analysis is sensitive to under- or over-interpretation, as the algorithm for the analysis used generally does not consider position and relevance of genes (or rather gene products) within pathways. Expression of a few strongly interacting key genes may biologically impact on a pathway, but not result in statistical overrepresentation, since most algorithms use number of genes and not relevance of genes within a pathway. In contrast, expression of several downstream genes may result in statistical overrepresentation, whereas biological impact of the pathway may be marginal. A limitation that is particularly important when pathway analysis would be used as the basis for risk assessment is the difficulty to discriminate between less or more relevant biological processes. Some genes have several roles and therefore may be involved in multiple biological pathways.[4]

Together, these considerations show the importance of research directed to improving pathway annotation, also with respect to roles of genes within pathways, *e.g.* providing information about different degrees of pathway regulation for one gene *versus* another.

13.4.2 Toxicity Pathway-derived BMDs

In recent studies, it was shown that BMD value based on the most sensitive toxicity pathways that were retrieved from short-term *in vivo* carcinogenicity studies in mice were in the same order of magnitude as BMD values that were based on apical endpoints assessed in cancer bioassays.[11,12] However, it was recognized that these most sensitive pathways are not necessarily of biological significance to the known or proposed MoA.[12,14] A more sophisticated pathway analysis tool that allows for selection of toxicity pathways that are relevant to the MoA can aid in a "MoA-guided derivation" of the lowest BMD and PoD from toxicogenomics data.[14] However, in this respect, care should be taken to avoid "annotation bias" and biological knowledge may become the rate-limiting step in developing MoAs.[14]

In the case of an unknown MoA and in the framework for toxicogenomic studies it may be recommendable to use the most sensitive pathway as the basis for the PoD as a precautionary measure.[9,11–13] It is indeed plausible to assume that toxicity endpoints are always preceded by changes in gene expression. However, the opposite is also true: transcriptional changes do not always result in the manifestation of toxicity endpoints. Gene expression changes may be reflecting an adaptive response towards a new state of homeostasis and therefore not inform on apical endpoints. As such, the use of BMDs based on the most sensitive toxicity pathways may result in a risk assessment that is too conservative. Increased confidence in the biological relevance of toxicity pathways in case of an unknown MoA may be achieved by using "negative controls" for which different MoAs and apical endpoints are expected. Overlapping toxicity pathways would then rather reflect more general toxicity that is not specific to the MoA and therefore could be neglected for deriving the lowest BMD.

13.4.3 Apical Endpoints

Apical endpoints from traditional toxicity tests such as the rodent carcinogenicity bioassay still serve as indispensable anchors to support findings in toxicogenomics studies. In the immediate future, toxicogenomics may be included in traditional assays to increase confidence in toxicity pathways by phenotypic anchoring to the traditional apical endpoints.[14] In the longer run, and when sufficiently validated, it may be anticipated that toxicogenomics replaces the need to determine apical endpoints *in vivo*. Compared to *in vivo* testing, *in vitro* assays can be performed much faster and at considerably lower cost. Furthermore, as also shown by using the parallelogram approach and concordance model, human-based assays are required to assess human relevance of results. However, at this point in time measurement of apical endpoints that correlate with toxicogenomic responses are still necessary to provide evidence that toxicity pathways actually reflect the MoA. Additionally, in particular for application of toxicogenomics in a quantitative manner in an *in vitro* setting, more knowledge on quantitative *in vitro* to *in vivo* extrapolation is needed, to examine the relationship between adverse effects observed in *in vitro* toxicity tests and the possible consequences of target tissue exposures in humans.[14]

13.4.4 Adversity

Discriminating adverse from adaptive effects when considering the relation of toxicity pathways to a certain MoA remains a challenge. However, the ability to discriminate between physiological adaptive response to retain homeostasis and adverse responses is pivotal for the implementation of toxicogenomics in risk assessment. Redefinition of adversity is recognized as a key step in the paradigm shift in human health risk assessment.[29] An adverse effect is traditionally defined as "a biochemical, morphological or physiological change (in response to a stimulus) that either singly or in combination adversely affects the performance of the whole organism or reduce the organism's ability to respond to an additional environmental challenge".[30] In contrast, non-adverse effects are defined as "those biological effects that do not cause biochemical, morphological, or physiological changes that affect the general well-being, growth, development or life span of an animal".[30] The definition of adversity can be used for human risk assessment, given that the observed change is reflective of the changes that may occur in humans upon chemical exposure.[29] The critique on this assumption is mainly based on the rodent species that are used as "golden standards" in regulatory frameworks for risk assessment that use a dose–response design in a fixed study duration, for example a 28 day or 90 day toxicity study, or a 2 year rodent carcinogenicity bioassay. It is argued that current testing strategies are based on consensus agreements, *e.g.* study duration, that do not necessarily allow evaluation of changes in apical endpoints that are of human relevance.[29] The redefinition of adversity in this context would include a complete shift from rodent *in vivo* studies to (human) *in vitro* assays, as advocated by the National Research Council[2] and also proposed recently in a more pragmatic form.[31]

13.5 Conclusion

In conclusion, human health risk assessment could certainly benefit from the incorporation of toxicogenomics in regulatory frameworks. The parallelogram approach and concordance model presented here have shown their applicability in hazard identification in risk assessment, substantiating biological significance and human relevance of toxicogenomics-retrieved toxicity pathways. Regarding the contribution of toxicogenomics to dose–response modeling in risk assessment, the parallelogram approach and concordance model may increase confidence in the specificity of toxicity pathways for a known or proposed MoA and thus aid in selecting the most sensitive biologically significant and human relevant pathway to derive a BMD. Furthermore, these approaches can be used to evaluate *in vitro* assays for assessment of toxicity. Although quantitative toxicogenomics studies have shown promising results, particularly in cancer risk assessment, and have led to advances towards unified study designs and protocols for data analysis and interpretation, many challenges remain for implementation of toxicogenomics in regulatory frameworks such as REACH. Key to the acceptance and use of toxicogenomics in regulatory frameworks is collaborative work and dialogue between risk assessors and research scientists.

References

1. IPCS, Principles of methods for the risk assessment of chemicals in food, 2009, World Health Organization, Environmental Health Criteria 240.
2. NRC, *Toxicity Testing in the 21st Century: A Vision and a Strategy*, 2007, National Research Council, Committee on Toxicity Testing and Assessment of Environmental Agents.

3. J. Tuomisto, Is the precautionary principle used to cover up ignorance?, *Basic Clin. Pharmacol. Toxicol.*, 2004, **95**, 49–52.
4. A. K. Goetz, B. P. Singh, M. Battalora, J. M. Breier, J. P. Bailey, A. C. Chukwudebe and E. R. Janus, Current and future use of genomics data in toxicology: opportunities and challenges for regulatory applications, *Regul. Toxicol. Pharmacol.*, 2011, **61**, 141–153.
5. H. Ellinger-Ziegelbauer, H. Gmuender, A. Bandenburg and H. J. Ahr, Prediction of a carcinogenic potential of rat hepatocarcinogens using toxicogenomics analysis of short-term *in vivo* studies, *Mutat. Res.*, 2008, **637**, 23–39.
6. W. Gao, Y. Mizukawa, N. Nakatsu, Y. Minowa, H. Yamada, Y. Ohno and T. Urushidani, Mechanism-based biomarker gene sets for glutathione depletion-related hepatotoxicity in rats, *Toxicol. Appl. Pharmacol.*, 2010, **247**, 211–221.
7. M. J. Jonker, O. Bruning, M. van Iterson, M. M. Schaap, T. V. van der Hoeven, H. Vrieling, R. B. Beems, A. de Vries, H. van Steeg, T. M. Breit and M. Luijten, Finding transcriptomics biomarkers for *in vivo* identification of (non-)genotoxic carcinogens using wild-type and Xpa/p53 mutant mouse models, *Carcinogenesis*, 2009, **30**, 1805–1812.
8. M. D. Waters, M. Jackson and I. Lea, Characterizing and predicting carcinogenicity and mode of action using conventional and toxicogenomics methods, *Mutat. Res.*, 2010, **705**, 184–200.
9. J. P. Bercu, R. A. Jolly, K. M. Flagella, T. K. Baker, P. Romero and J. L. Stevens, Toxicogenomics and cancer risk assessment: a framework for key event analysis and dose-response assessment for nongenotoxic carcinogens, *Regul. Toxicol. Pharmacol.*, 2010, **58**, 369–381.
10. A. F. Jackson, A. Williams, L. Recio, M. D. Waters, I. B. Lambert and C. L. Yauk, Case study on the utility of hepatic global gene expression profiling in the risk assessment of the carcinogen furan, *Toxicol. Appl. Pharmacol.*, 2014, **274**, 63–77.
11. R. S. Thomas, H. J. Clewell, 3rd, B. C. Allen, S. C. Wesselkamper, N. C. Wang, J. C. Lambert, J. K. Hess-Wilson, Q. J. Zhao and M. E. Andersen, Application of transcriptional benchmark dose values in quantitative cancer and noncancer risk assessment, *Toxicol. Sci.*, 2011, **120**, 194–205.
12. R. S. Thomas, H. J. Clewell, 3rd, B. C. Allen, L. Yang, E. Healy and M. E. Andersen, Integrating pathway-based transcriptomic data into quantitative chemical risk assessment: a five chemical case study, *Mutat. Res.*, 2012, **746**, 135–143.
13. R. S. Thomas, S. C. Wesselkamper, N. C. Wang, Q. J. Zhao, D. D. Petersen, J. C. Lambert, I. Cote, L. Yang, E. Healy, M. B. Black, H. J. Clewell, 3rd, B. C. Allen and M. E. Andersen, Temporal concordance between apical and transcriptional points of departure for chemical risk assessment, *Toxicol. Sci.*, 2013, **134**, 180–194.
14. N. L. Chepelev, I. D. Moffat, S. Labib, J. Bourdon-Lacombe, B. Kuo, J. K. Buick, F. Lemieux, A. I. Malik, S. Halappanavar, A. Williams and

C. L. Yauk, Integrating toxicogenomics into human health risk assessment: lessons learned from the benzo[a]pyrene case study, *Crit. Rev. Toxicol.*, 2015, **45**, 44–52.
15. F. H. Sobels, Some problems associated with the testing for environmental mutagens and a perspective for studies in "comparative mutagenesis", *Mutat. Res.*, 1977, **46**, 245–260.
16. T. R. Sutter, Molecular and cellular approaches to extrapolation for risk assessment, *Environ. Health Perspect.*, 1995, **103**, 386–389.
17. A. S. Kienhuis, M. C. van de Poll, H. Wortelboer, M. van Herwijnen, R. Gottschalk, C. H. Dejong, A. Boorsma, R. S. Paules, J. C. Kleinjans, R. H. Stierum and J. H. van Delft, Parallelogram approach using rat-human *in vitro* and rat *in vivo* toxicogenomics predicts acetaminophen-induced hepatotoxicity in humans, *Toxicol. Sci.*, 2009, **107**, 544–552.
18. A. N. Heinloth, R. D. Irwin, G. A. Boorman, P. Nettesheim, R. D. Fannin, S. O. Sieber, M. L. Snell, C. J. Tucker, L. Li, G. S. Travlos, G. Vansant, P. E. Blackshear, R. W. Tennant, M. L. Cunningham and R. S. Paules, Gene expression profiling of rat livers reveals indicators of potential adverse effects, *Toxicol. Sci.*, 2004, **80**, 193–202.
19. M. Wagner, G. Zollner and M. Trauner, New molecular insights into the mechanisms of cholestasis, *J. Hepatol.*, 2009, **51**, 565–580.
20. N. Anderson and J. Borlak, Molecular mechanisms and therapeutic targets in steatosis and steatohepatitis, *Pharmacol. Rev.*, 2008, **60**, 311–357.
21. H. Jaeschke, M. R. McGill and A. Ramachandran, Oxidant stress, mitochondria, and cell death mechanisms in drug-induced liver injury: lessons learned from acetaminophen hepatotoxicity, *Drug Metab. Rev.*, 2012, **44**, 88–106.
22. A. S. Kienhuis, A. P. Vitins, J. L. Pennings, T. E. Pronk, E. N. Speksnijder, M. Roodbergen, J. H. van Delft, M. Luijten and L. T. van der Ven, Cyclosporine A treated *in vitro* models induce cholestasis response through comparison of phenotype-directed gene expression analysis of in vivo Cyclosporine A-induced cholestasis, *Toxicol. Lett.*, 2013, **221**, 225–236.
23. G. Zollner and M. Trauner, Mechanisms of cholestasis, *Clin. Liver Dis.*, 2008, **12**, 1–26, vii.
24. A. P. Vitins, A. S. Kienhuis, E. N. Speksnijder, M. Roodbergen, M. Luijten and L. T. van der Ven, Mechanisms of amiodarone and valproic acid induced liver steatosis in mouse in vivo act as a template for other hepatotoxicity models, *Arch. Toxicol.*, 2014, **88**, 1573–1588.
25. D. E. Amacher, The mechanistic basis for the induction of hepatic steatosis by xenobiotics, *Expert Opin. Drug Metab. Toxicol.*, 2011, **7**, 949–965.
26. A. S. Kienhuis, J. G. Bessems, J. L. Pennings, M. Driessen, M. Luijten, J. H. van Delft, A. A. Peijnenburg and L. T. van der Ven, Application of toxicogenomics in hepatic systems toxicology for risk assessment: acetaminophen as a case study, *Toxicol. Appl. Pharmacol.*, 2011, **250**, 96–107.

27. M. Driessen, A. P. Vitins, J. L. Pennings, A. S. Kienhuis, B. Water and L. T. van der Ven, A transcriptomics-based hepatotoxicity comparison between the zebrafish embryo and established human and rodent *in vitro* and *in vivo* models using cyclosporine A, amiodarone and acetaminophen, *Toxicol. Lett.*, 2015, **232**, 403–412.
28. H. H. Li, D. R. Hyduke, R. Chen, P. Heard, C. L. Yauk, J. Aubrecht and A. J. Fornace, Jr., Development of a toxicogenomics signature for genotoxicity using a dose-optimization and informatics strategy in human cells, *Environ. Mol. Mutagen.*, 2015, **27**, 243–252.
29. K. Boekelheide and M. E. Andersen, A mechanistic redefinition of adverse effects – a key step in the toxicity testing paradigm shift, *ALTEX*, 2010, **27**, 243–252.
30. R. W. Lewis, R. Billington, E. Debryune, A. Gamer, B. Lang and F. Carpanini, Recognition of adverse and nonadverse effects in toxicity studies, *Toxicol. Pathol.*, 2002, **30**, 66–74.
31. R. S. Thomas, M. A. Philbert, S. S. Auerbach, B. A. Wetmore, M. J. Devito, I. Cote, J. C. Rowlands, M. P. Whelan, S. M. Hays, M. E. Andersen, M. E. Meek, L. W. Reiter, J. C. Lambert, H. J. Clewell, 3rd, M. L. Stephens, Q. J. Zhao, S. C. Wesselkamper, L. Flowers, E. W. Carney, T. P. Pastoor, D. D. Petersen, C. L. Yauk and A. Nong, Incorporating new technologies into toxicity testing and risk assessment: moving from 21st century vision to a data-driven framework, *Toxicol. Sci.*, 2013, **136**, 4–18.

CHAPTER 14

Bioinformatics of Genomics in the Assessment of Cancer

PIERRE R. BUSHEL

Biostatistics and Computational Biology Branch, National Institute of Environmental Health Sciences, P.O. Box 12233, RTP, North Carolina 27709, USA
Email: bushel@niehs.nih.gov

14.1 Introduction

In the early days of the advent of genomics, it became apparent that in order to make sense of the volumes of data and share it among the research community, there was a pressing need to engage experts with specialized and broad-based knowledge in biology, statistics and information technology to interface with the data generators, data analysts and data providers. Bioinformatics can be loosely defined in one way as the scientific field dedicated to the collection, storage, analysis and provision of large-scale biological data.

As early as the 1970s, investigators began crafting computational algorithms to assist with the analysis of biological data.[1,2] For example, the Needleman–Wunsch dynamic programming algorithm was developed to find global alignments between two nucleic acid sequences. Figure 14.1 illustrates the way two DNA sequences can be compared given matches or mismatches of bases and the best alignment of the two identified by tracing back high-scoring pairs within the matrix. Without such an essential bioinformatics methodology, generators of DNA and RNA sequences would never know how similar their sequences might be.

Issues in Toxicology No. 28
Toxicogenomics in Predictive Carcinogenicity
Edited by Michael D. Waters and Russell S. Thomas

Published by the Royal Society of Chemistry, www.rsc.org

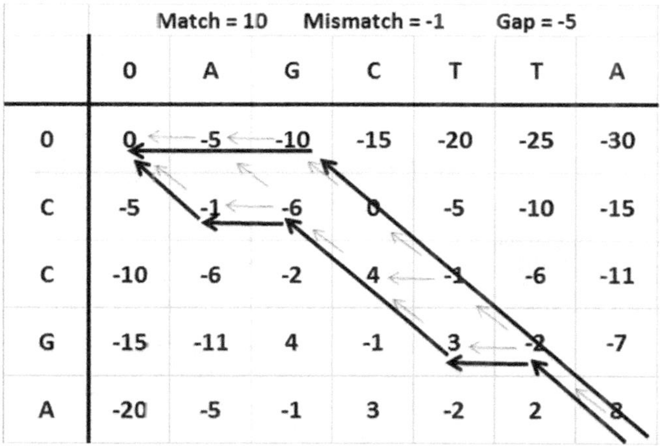

Figure 14.1 Needleman–Wunsch algorithm trace back for sequence alignment. X-axis is DNA sequence #1, Y-axis is DNA sequence #2.

The explosion of data generated from improved DNA and protein sequencing technologies bolstered the need and demand for bioinformatics and ways to compare, analyze and manage the respective sequences. If we fast track to the mid-1990s when the first microarrays were invented, we see another explosion of genomic data. Microarrays are interrogation platforms to measure the expression of genes.[3] Early on, the gene content on microarrays was a few thousand. Today, microarrays are genome-wide, covering the entire transcriptome of an organism. Next-generation sequencing of messenger (m)RNA (RNA-Seq) offers a wide variety of advantages over microarrays in terms of measuring gene expression. However, once the transcripts are annotated, given a gene model, properly normalized and quantified, RNA-Seq gene expression data can be readily interpreted in a similar fashion to microarray gene expression data. Thus, for all intents and purposes, and in most cases, RNA-Seq data presented in a similar space (*i.e.* ratio values) as microarray data can be analyzed with similar bioinformatics, analytical algorithms and computational tools developed for microarray.

Of interest to investigators was the ability to discover patterns in the gene expression data across biological conditions, time or dose points. Clustering gene expression measurements using a two-dimensional matrix of the data with the genes in the rows and samples in the columns (Figure 14.2) became a bioinformatics indispensable tool for investigators to reveal patterns associated with the experimental design.

Cancer is a disease that impacts a wide swath of the population. As shown in Figure 14.3, in the US from 2007 to 2011, prostate cancer in males and breast cancer in females had the highest incidence rates. Obviously, these top cancers by gender are sex-related. However, the mechanisms underlying various types of cancers are poorly understood. In fact, some have claimed cancer risk is simply due to bad luck associated with random mutations and can be largely explained by the number of stem cell divisions within the

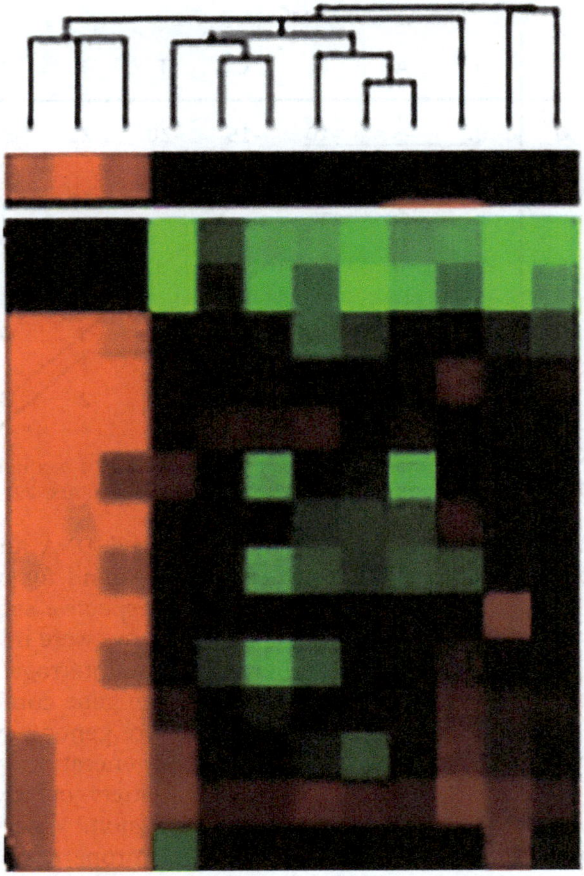

Figure 14.2 Hierarchical clustering of gene expression data with genes in the rows and samples in the columns. Red signifies up-regulation, green down-regulation and black unchanged.

tissue type.[4] Armed with the genomics technologies to assay DNA and RNA on a large scale and bioinformatics strategies to manage and analyze the data, investigators capitalized on these tools to wage the war against cancer. However, rate-limiting steps in predictive carcinogenicity are managing the data, analyzing and integrating high dimensional disparate data and interpreting the results.[5] Bioinformatics brings together the expertise to store, disseminate, amalgamate and analyze biological and genomics data in highly informative ways for better biological understanding, leading to improved therapeutic targeting.

This chapter is divided into three sections: (1) classification and prediction; (2) over-represented pathways, enriched gene sets and gene regulatory networks; and (3) bioinformatics and computational biology for integrative genomics. The material included is not intended to be exhaustive of the

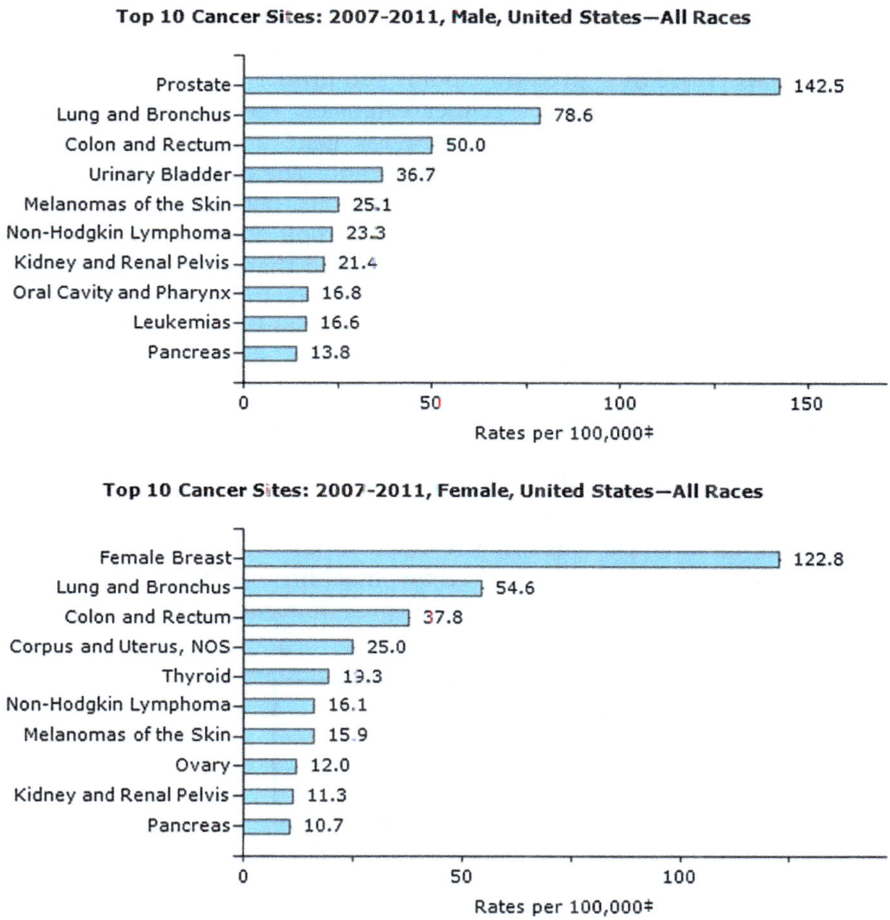

Figure 14.3 Incidence rates of the top 10 cancers in males and females separately in the US in 2007–2011.
Data obtained from the National Program of Cancer Registries at the US Centers for Disease Control and Prevention. NOS: not otherwise specified.

bioinformatics of genomics used to study cancer, rather a review of an assortment of a few major contributions in bioinformatics and computational biology designed specifically to enhance the ability to discern cancer using genomic data.

14.1.1 Classification and Prediction

In the early days of molecular biology, investigators needed to clone a gene, derive of a primer and perform a Northern blot to assess the expression of a gene. The determination of RNA produced by the gene was certainly qualitative, but often times quantitative by approximating the amount of hybridized probe detected as compared to a reference. Other forms of ascertaining gene expression included nuclear run-ons to identify the genes that are transcribed at a given time point and primer extension to map the 5′ end of RNA. For example, primer extension was used to identify a bidirectional promoter within the c-Ha-ras gene and revealed that the oncogene is regulated by proximal *cis* elements.[6] The common limitation to these assays was that they were low-throughput, permitting only tens of genes to be monitored at a time. Given the complexity of cancer and polygenic root to the disease, investigators surmised that a higher-throughput gene expression assay was needed to better understand cancer biology.

Following the advent of the microarray to monitor the expression of thousands of genes,[3] investigators used an array platform to profile the expression of 16 000 genes in more than 200 tumor tissue samples.[7] With over 3 million data points, it became very obvious that bioinformatics strategies were desperately needed to distill the data to a manageable form and to make predictions about the 14 tumor types. A self-organizing map (SOM)[8] was used to cluster the tumors into partitions, albeit some nodes intermixed, hierarchical clustering revealed unknown similarities between tumors and supervised support vector machines (SVM) predicted the multiple classes of tumors with a high degree of accuracy (Figure 14.4).

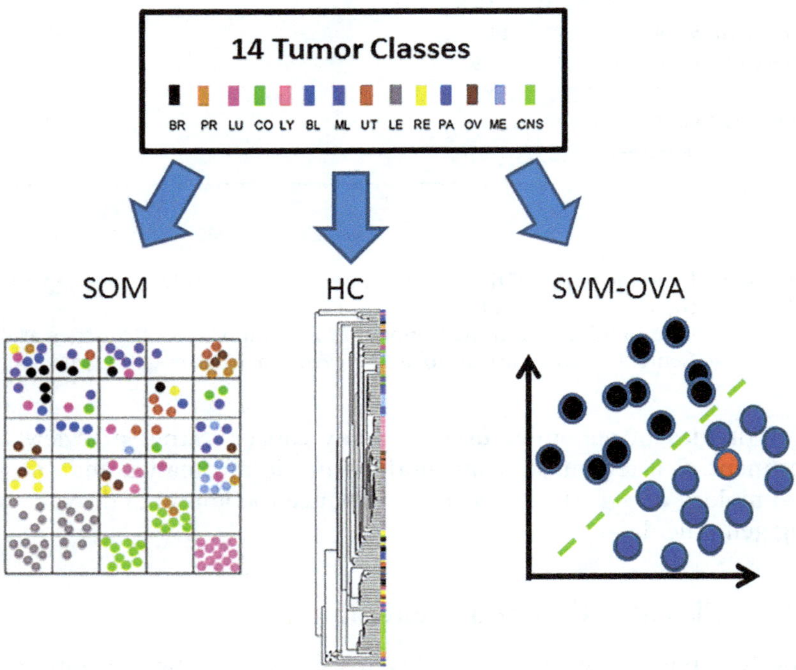

Figure 14.4 Multiclass cancer tumor types diagnosis using self-organizing maps (SOM), hierarchical clustering (HC) and a one-*vs.*-all (OVA) support vector machine (SVM).

An SOM is an unsupervised learning neural network consisting of nodes (or neurons) each with weights assigned to the samples being clustered. Each SOM has a topology which defines the number of clusters into which to partition the data. Thus, mapping high-dimensional data into a more compacted form in which the clusters represent ordered sets of highly similar features. SVMs are supervised learning models whereby, based on the classifications of training data samples by groups, the best linear (or non-linear) separation between two groups is constructed in a classifier with the features closest to the separation plane being the support vectors. When the classifier is applied to a test data set of unknown classification, the samples can be predicted by determination of the group they are closest to. From these bioinformatics strategies, gene expression data may be phenotypically anchored to known cancer markers, thereby providing a better understanding of the possible transcriptional components tied to cancer endpoints. A comparable investigation was pursued by the National Institutes of Health (NIH) National Cancer Institute (NCI) in partnership with Stanford University.[9] A panel of 60 human cancer cell lines was profiled for gene expression using microarrays and then the data correlated with the cells' responses to drugs in order to discern the drug sensitivity, resistance and pharmacologic responsiveness of the cancer cells.[10]

Caution is needed when classification and prediction algorithms are used, especially when the number of features far exceeds the number of samples in the data. The classic problem in transcriptomics is that there are typically tens of thousands of genes but only a few tens or hundreds of samples. This ordinarily will lead to overfitting of classifiers, resulting in a prediction model that is too restrictive to the training data rather than learning the nuance differences between the groups. The outcome is a prediction where the accuracy from the training is artificially high, causing the accuracy of the testing to be compromised and hence subject to "the curse of dimensionality".[11]

The MicroArray Quality Control (MAQC) consortium led by the US Food and Drug Administration set out to provide guidelines for the use of transcriptomics in classifier prediction and recommendations for use by regulatory agencies (MAQC-II).[12] What is certain is that to derive of an unbiased predictor, a full leave-one-out cross-validation strategy with inner (nested) and outer loops for gene selection and classifier testing/optimization (*i.e.* n-fold validation) is highly recommended (Figure 14.5).

Despite good intentions to adhere to proper protocol in terms of building classifiers for prediction, the end-result or accuracy may be dependent on the data set at hand (*i.e.* the endpoint that is being predicted). For example, as shown in Figure 14.6, most classifiers predicted neuroblastoma-positive, rat liver necrosis and multiple myeloma (MM)-positive relatively well, according to the Matthews correlation coefficient (MCC) as a measure of prediction accuracy (median MCC > 0.8). However, the prediction accuracy of neuroblastoma-negative, mouse lung tumor and MM-negative was less than a median MCC 0.1. Minimization of the feature size has been shown to be advantageous in identifying putative biomarkers related to cancer.[13]

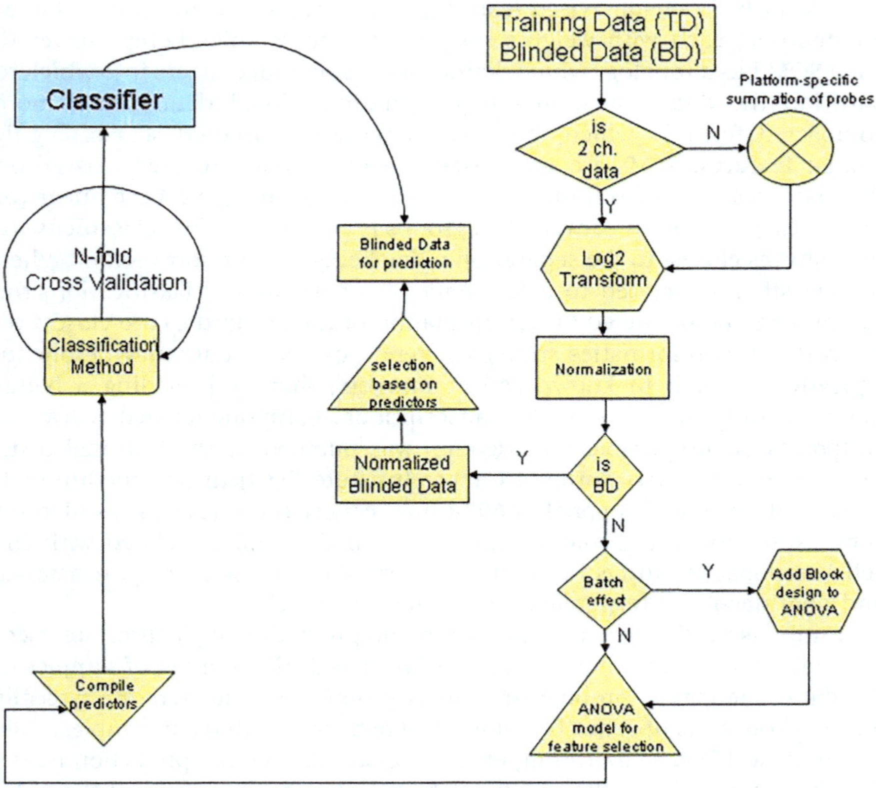

Figure 14.5 Classifier optimization for prediction.

The ability to correlate gene expression profiles from cancer cells and associate them with phenotypic markers does shed some light on the similarities and differences between tumor types. However, mechanistic details are needed to sort out the biological processes underlying tumorigenesis. The correlation of gene expression profiles into patterns is akin to "guilt by association". In other words, if a set of genes are correlated in their gene expression across various tumor types, then it is likely that they are controlled by a similar regulatory factor or complex of transcription regulators. As depicted in Figure 14.7, a regulatory cascade believed to be involved in prostate cancer is the translocation of the androgen receptor to the nucleus following activation by dihydrotestosterone or *via* Ras/MEK/ERK1/2 signaling and then binding of the androgen receptor to its recognition element.[14]

Determining the co-expression across samples of varied conditions or treatment groups corroborates shared functional mechanisms with a set of genes with similar expression profiles. As mentioned in the introduction, clustering gene expression data reveals groups of genes that are regulated across samples of similar types. A common feature of clustering is the

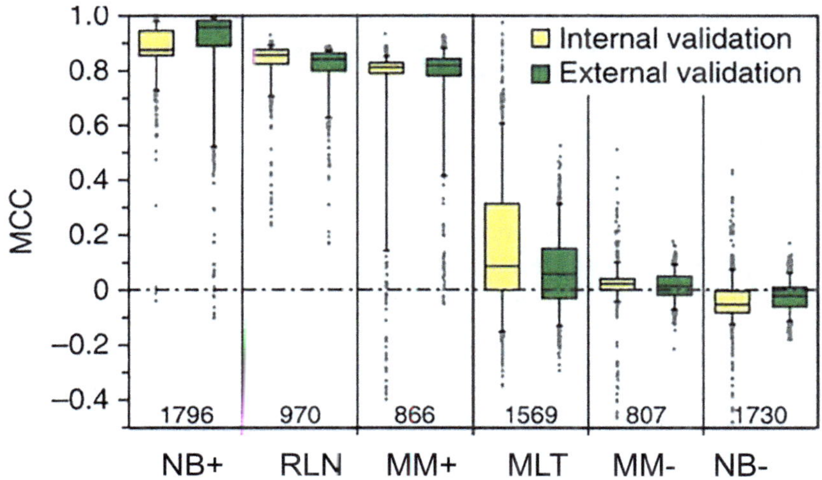

Figure 14.6 Prediction of samples by endpoint. MCC: Matthews correlation coefficient; NB: neuroblastoma; RLN: rat liver necrosis; MM: multiple myeloma; MLT: mouse lung tumor.
Modified from Shi *et al.*, 2010.[12]

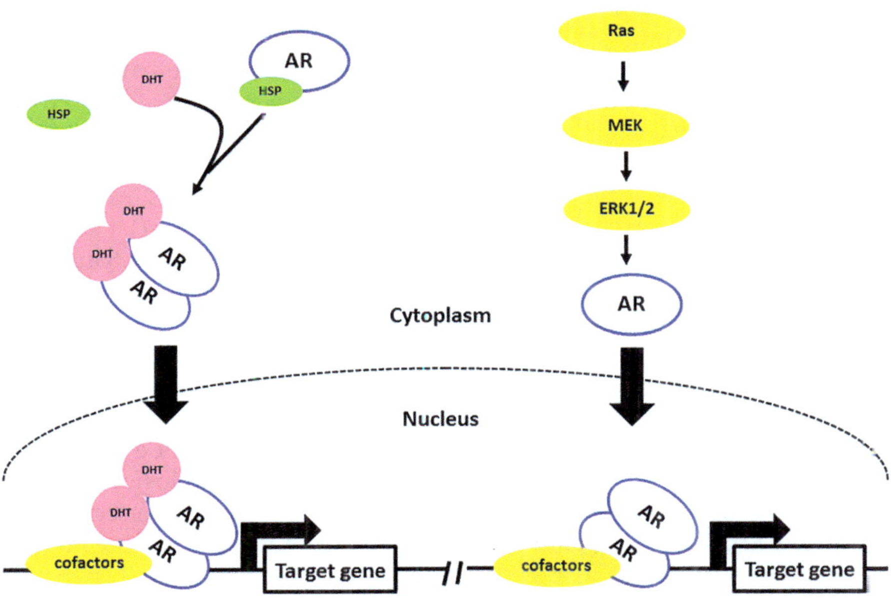

Figure 14.7 Androgen receptor (AR) cascade. DHT: dihydrotestosterone; HSP: heat shock protein; MEK: mitogen-activated protein kinase extracellular kinase; ERK: extracellular signal-regulated kinase.
Depicted from Lonergan and Tindall, *J. Carcinog.*, 2011.[14] PMID 21886458.

utilization of some form of metric to measure how dissimilar genes are in terms of their expression across the samples. Euclidean distance, 1-Pearson correlation (r) and Spearman rank dissimilarity are a few common clustering metrics. Often the sets of differentially expressed genes (DEGs) defined by a fold change threshold and/or *p*-value cut-off, are grouped together and clustered according to their gene expression. An advantage of this strategy is that all the significant genes that are perturbed within a given condition or treatment are surveyed for their co-expression. The drawback is that it can miss key genes which did not make the DEG list because the fold change or *p*-value for the gene did not meet the cut-off. For instance, since the expression of a transcription factor typically only has to change slightly to impact downstream target (DST) genes, transcription factors are often not detected as differentially expressed. However, the transcription factor is likely to have salient co-expression with its target genes and other transcription factors. Therefore, to detect the totality of co-expressed genes, it is more prudent to cluster the gene expression data based on all the genes in the transcriptome. There are several challenges to this strategy: (1) how do we know how many clusters there are in the data; (2) what additional metrics other than correlation are used to cluster the genes; and (3) how is significance assigned to the co-expression clustering?

Chou *et al.* (2007) developed a method called extracting patterns and identifying co-expressed genes (EPIG).[15] Using Pearson correlation, a signal-to-noise ratio (SNR) and the magnitude of fold change of the gene, EPIG first finds all the possible patterns across conditions/treatments in the data set and then categorizes (clusters) the gene profiles to those patterns. The patterns of co-expressed genes are determined to be significant based on the probability that a set of genes within a pattern will have a particular correlation and overall SNR by chance. Figure 14.8 shows the EPIG analysis of microarray gene expression data from the livers of rats exposed to chemical agents according to mode of action (MoA) (Table 14.1).

The parameters used were Pearson correlation $r = 0.8$, $\mathrm{SNR} = 2.5$ and fold change $= 0.5$. Although both the aryl hydrocarbon receptor (AhR) and cytotoxic MoA chemicals can cause cancer in rats, EPIG patterns #1 and #2 contain co-expressed genes that are relatively unchanged by AhR MoA-treated chemicals, but increased or decreased, respectively, in expression by cytotoxic chemicals. Thus, potentially there exists a common regulatory mechanism perturbed by the chloroform (CHL), thioacetamide (THI)- and carbon tetrachloride (CAR)-responsive genes that are exclusive to their shared MoA. Some of the genes (Jun, p21, Myc and NRG1) in pattern #1 are within the ErbB signaling pathway containing four receptor tyrosine kinases known to be structurally related to the epidermal growth factor receptor and associated with the development of tumors. Conversely, some of the genes (Cyp1a2, FMO1 and Ugt2b1) in pattern #2 function to metabolize drugs.

In cases where the correlation of the gene expression is not consistent across all the samples, there may be a cluster of genes which are co-expressed only in a subset of the conditions/treatments. In that case,

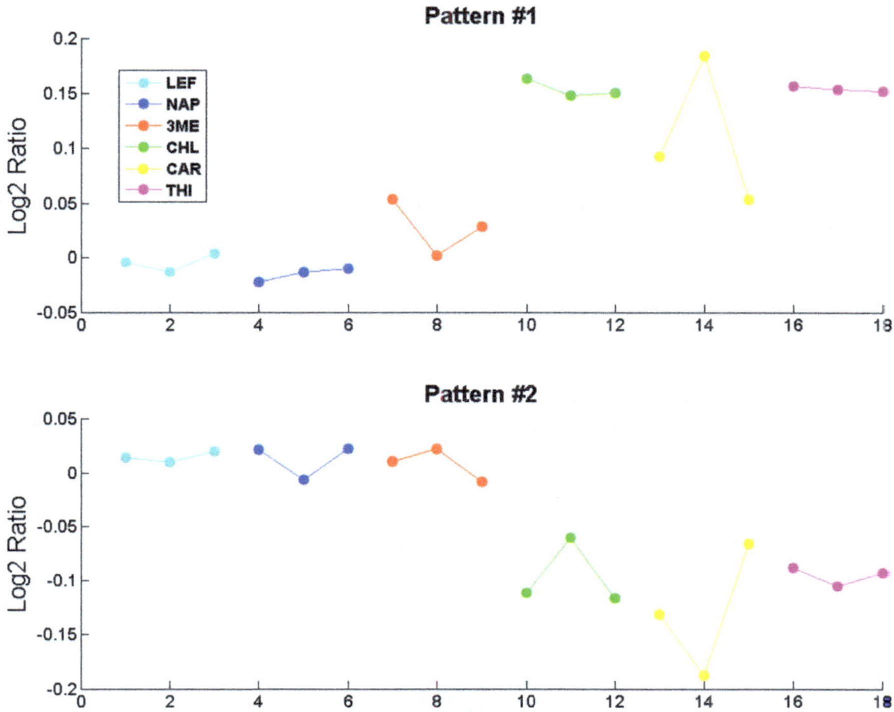

Figure 14.8 Co-expressed EPIG (extracting patterns and identifying co-expressed genes) patterns from rat liver mode of action microarray gene expression data. LEF: leflunomide; NAP: β-naphthoflavone; 3ME: 3-methylchloranthrene; CHL: chloroform; CAR: carbon tetrachloride; THI: thioacetamide.

informatically, one would want what is called a "bicluster" of the two-dimensional gene expression matrix.[16–19] Biclustering algorithms identify sub-matrices [subsets of rows (genes) and subsets of columns (samples) from an original data matrix] of gene expression that are coherent and possess homogeneity of co-expression. Hence, the biclusters contain co-expressed genes that represent distinct biological responses related to mechanistic changes in samples within the subset of conditions/treatments. As depicted in Figure 14.9, the two-dimensional matrix of data contains groups of features (rows) that are highly similar within a group of samples (columns).

Chou and Bushel (2009) developed coherent co-expression biclustering (cc-biclustering) as a bioinformatics tool to discern mechanisms of hepatotoxicity *via* biological processes that are over-represented by co-expressed genes.[20] In an unsupervised fashion or supervised by anchoring the co-expressed genes to serum levels of alanine aminotransferase (ALT), cc-biclustering of a microarray gene expression data set containing 318 liver samples from rats exposed to hepatotoxicants discerned 1,4 dichlorobenzene (a non-toxic isomer at low and mid-doses) from the other toxic

Table 14.1 Exposures of rats to chemicals according to mode of action.

Mode of action	Agent	Dose ($mg\,kg^{-1}$ body weight)	Duration (days)	Agent type
Aryl hydrocarbon receptor	3-Methylcholanthrene	300	5	Carcinogen
	Leflunomide	60	5	Antirheumatic drug
	beta-Naphthoflavone	1500	5	Putative chemopreventive agent
	Chloroform	600	5	Organic compound
Cytotoxic	Thioacetamide	200	5	Carcinogen
	Carbon tetrachloride	1175	7	Solvent for cleaning products, refrigerant

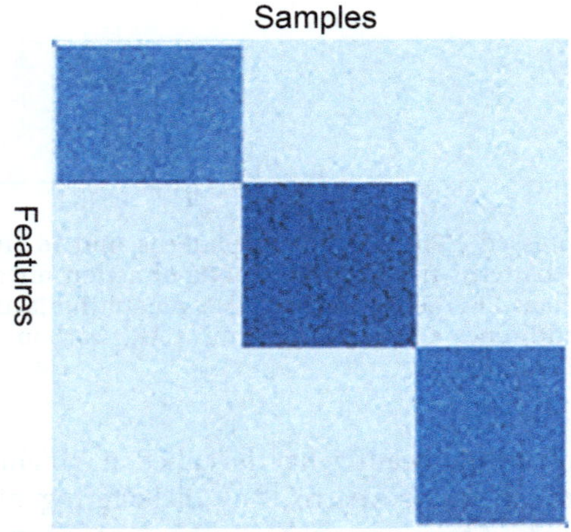

Figure 14.9 Biclusters.

chemicals. It also identified pathways and biological processes that underlie liver injury and extracted a bicluster with genes up-regulated in an early response to toxic exposure. Iterative binary biclustering of gene sets transforms the gene expression profiles into binary collections to permit the analysis of data from different array platforms and has been applied to breast cancer data sets to uncover novel gene set–phenotype associations that predicted tumor metastases within tumor subtypes.[21]

When there is an offset in gene expression time series data, phase-shift analysis methods are useful for detecting overlapping co-regulated genes which have a lag.[22,23] One approach, phase-shift analysis of gene expression (PAGE) supports data from both dose and time-course studies simultaneously.[24] Another supports the analysis of periodic data by using the

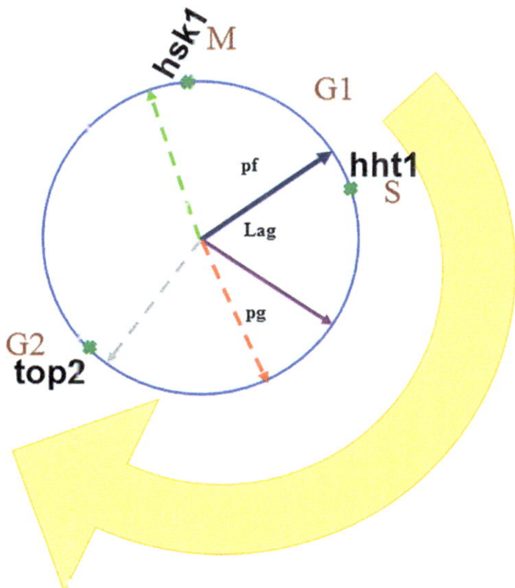

Figure 14.10 Transcription factors gene expression phases mapped on a circle.

phase angles of the gene expression and rather than statistical testing the data in linear space, it uses circle regression for inference. For example, in many cases, cancer drivers are transcription factors that govern the cell cycle by regulating proliferation and maintaining homeostasis. Interpolating the expression of transcription factors as points on a circle representative of the cell cycle phases is a natural consideration and is intuitive (Figure 14.10). If order of the phase angles is preserved across a pair of transcription factors, then one can align the two by simply rotating one of the circles and then moving the points within the circle until the two circles are as close to each other as possible. This can be accomplished using circle–circle regression.[25]

14.1.2 Over-represented Pathways, Enriched Gene Sets and Gene Regulatory Networks

Understanding the signal transduction pathways involved in the promotion of cancer is paramount in predictive carcinogenicity as a single transcription factor can regulate several hundred DSTs. The Biobase Transfac[26,27] and Transpath[28] databases are bioinformatics knowledgebases which detail experimentally proven binding of transcription regulators [transcription factors, cofactors, micro (mi)RNAs, *etc.*] to gene targets and gene regulatory networks/signal transduction respectively. As depicted in Figure 14.11, based on the 2011.1 version of Transfac, there are several human transcription regulators that regulate more than one DST gene. In fact, three transcription regulators, all stimulating proteins, regulate 80 or more DSTs.

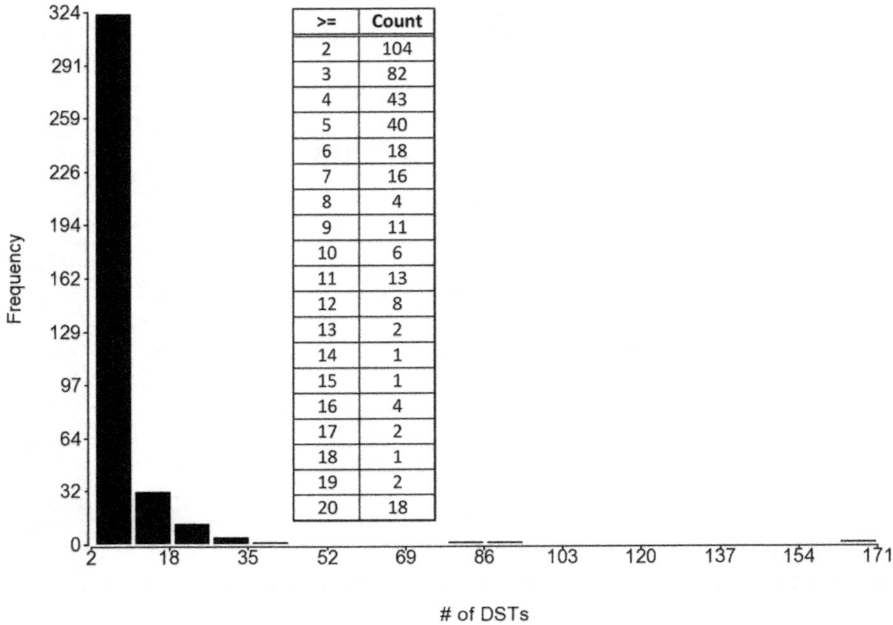

>=	Count
2	104
3	82
4	43
5	40
6	18
7	16
8	4
9	11
10	6
11	13
12	8
13	2
14	1
15	1
16	4
17	2
18	1
19	2
20	18

Figure 14.11 Distribution of the number of genes as downstream targets (DSTs) of human transcript regulators.

Compiling and utilizing this mapping information is advantageous for cancer informatics. For example, signal transduction pathway profiling of a gene expression data set derived from breast tissue samples from 117 patients (of which 46 developed metastases) revealed that the estrogen receptor tumor necrosis factor-α induced genes, and other transcription regulators are significantly dysregulated, as are various regulatory components in their signaling pathways.[29] By assuming that the set of DST genes for a transcription regulator are co-regulated, their pairwise gene expression profile correlation across the breast cancer samples should be significantly greater that the correlation of randomly paired genes. Furthermore, by annotating the array probes and then collapsing the genes to a particular gene model, the DSTs of a transcription regulator can be identified through Transfac. Transcription regulators can then be mapped to signaling pathways *via* Transpath. Subsequently, for each gene expression data set, significant transcription regulators can be identified through the strength of its group correlation score:

$$GCS = \sum_{i \neq j} r_{i,j}^2$$

defined as the sum of the squares of the Pearson correlations (r) among all pairs of genes i and j determined to be DSTs of the transcription regulator.

The significant pathways made up of transcription regulators are based on an exclusive group correlation score:

$$EGCS = \sum_{TR(i) \cap TR(j) = 0} r_{i,j}^2$$

defined as the sum of the squares of r over all pairs of genes i and j in a pathway that do not share any transcription regulator. This eliminates the contribution of the co-expression of DSTs that share transcription regulators. The p-value for a score is determined from a non-parametric distribution of correlation scores obtained from random cases (for example, $B = 10\ 000$ reshuffles of the genes) and the number of times (n) one of these permuted scores is greater than the observed correlation score. Thus,

$$p - \text{value} = n/B.$$

For both correlation scores, the null hypothesis keeps the structure and overlap of all pathways fixed, but changes the identity of the genes. Table 14.2 lists the top three transcription regulators from a gene expression study in four human cell lines treated with a cancer therapeutic and compared to controls. The transcription regulators include p53, a p53 isoform and the E2F-DP complex—each of which are components of signaling transduction pathways associated with the etiology of cancer. These transcription regulators, and the signaling pathways they are a part of, have been targets for the development of cancer-related drugs.[30,31]

Chemical agents and cancer profiles have been represented in terms of the genes that are differentially expressed. As indicated earlier, perturbed genes are components of signaling pathways. In addition, these DEGs carry out specific molecular function and have well-defined roles in biological processes. Enriching pathways, molecular reactions and biological processes according to the DEGs helps to facilitate the interpretation of the biological categories impacted by adverse conditions or toxic exposures.

To grasp the use of enrichment analysis in toxicogenomics, one has to have an appreciation of the efforts put forward to catalog and annotate genes to biological categories. Gene ontology is an example of such a categorization. The same principle can be applied to most any collection of biological annotation, *i.e.* the Kyoto Encyclopedia of Genes and Genomes

Table 14.2 Top three group correlation scores for transcription regulators using data from the gene expression from four human cell lines treated with a cancer therapeutic.[a]

TRID	TR	# of DSTs	GCS	p-value
T00221	E2F:DP	9	23.99	9E-05
T08321	p53-isoform1	31	134.23	9E-05
T00671	p53	17	44.8	2E-04

[a]TRID: transcription regulator ID from the Transfac database; DST: downstream target; GCS: group correlation score.

knowledgebase,[32,33] Ingenuity® pathway analysis ontology and canonical pathways[34] or the Panther pathway classification system.[34,35]

The Gene Ontology Consortium (GOC) was established in 1998 to develop a shared, structured vocabulary (an ontology) for the annotation of molecular characteristics across different organisms. It is a collaborative effort to address the need for consistent descriptions of gene products in different databases. The original members of the GOC were participants from the *Saccharomyces* genome database (SGD), the *Drosophila* genome database (FlyBase) and the Mouse Genome Informatics databases (MGD/GXD). Currently there are more than 30 organizations contributing to Gene Ontology.

The two primary purposes of an ontology are to facilitate communication between people and organizations and to improve upon the interoperability between systems. For example, a good ontology would create a common understanding that "translation of mRNA" is equivalent in terminology to "amino acid/protein synthesis". The goals of the Gene Ontology initiative are (1) to compile a comprehensive structured vocabulary of terms describing different elements of molecular biology that are shared among life forms; (2) to describe biological objects (in the model organism database of each contributing member) using these terms; (3) to provide tools for querying and manipulating these vocabularies; and (4) to provide tools enabling curators to assign Gene Ontology terms to biological objects. However, Gene Ontology is not a way to unify biological databases, nor a dictated standard derived from the self-interest of users to mandate nomenclature across databases, and it does not serve to define homologies between gene products from different organisms. The aforementioned are outside the scope of Gene Ontology; however, a shared vocabulary is indeed a step towards unification of systems.

The three tasks of the GOC are (1) writing and maintaining the ontologies; (2) creating cross-links between the ontologies and the gene and gene products; and (3) providing tools for creation, maintenance and use of the ontologies. The ontologies are structured vocabularies in the form of directed acyclic graphs that represent a network (not a tree) in which each term may be a child of one or more than one parent. The relationships of child to parent can be of the "is a" type or the "part of" type (Figure 14.12). For instance, the term transmembrane receptor protein-tyrosine kinase is a subclass both of the parent transmembrane receptor and of the parent

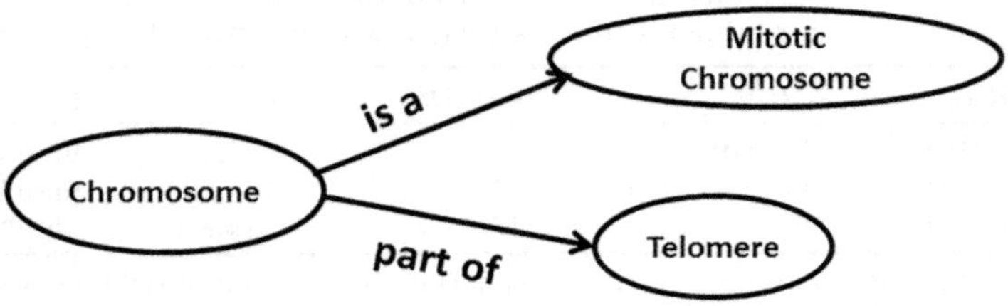

Figure 14.12 Child-to-parent relationship in a directed acyclic graph.

protein tyrosine kinase. It is critical for Gene Ontology that each term has its own identifier (accession) that uniquely distinguishes it.

The GOC developed and maintains three ontologies: (1) *cellular component,* referring to the place in the cell (*i.e.* the location) where a gene product is found; (2) *molecular function,* describing activities, such as catalytic or binding activities, at the molecular level (*i.e.* enzyme, transporter, ligand, adenylate cyclase or Toll receptor ligand); and (3) *biological process* (GOBP), referring to a biological objective to which the gene product contributes, such as cell growth and maintenance, signal transduction, pyrimidine nucleotide metabolism or α-glucoside transport (although it is clear that a biological process is not a pathway which would have a chain of processes and/or substrates/ dependencies—Gene Ontology does not capture this).

As a standard practice, all annotations of gene products to Gene Ontology terms must incorporate controlled statements of the type of evidence that supports the relationship, as well as appropriate citations. In other words, the annotations of gene products to the Gene Ontology vocabularies are attributed to a source, which may be a literature reference, another database or a computational analysis. Furthermore, all attributes of Gene Ontology must be accompanied by appropriate indications of evidence. See Table 14.3 for the codes (acronyms) denoting the type of evidence or citation.

As an example of the use of the Gene Ontology with gene expression data, a portion of the MoA study was used.[36] Using gene expression data from the livers of rats exposed to chemical agents according to MoA (Table 14.1), enriched GOBPs are identified using a hypergeometric distribution to model the occurrence of DEGs within the ontology categories

$$p(X<k)=1-\sum_{i=0}^{k-1}\frac{\left(\dfrac{M}{i}\right)\left(\dfrac{N-M}{n-i}\right)}{\dfrac{N}{n}}.$$

This models the probability of observing k genes from a set of n DEGs by chance within a biological process category containing M members from a total genome (reference or array) size of N genes. The closer the probability (p-value) is to 0, the more unlikely the chance is that the majority of the DEGs have the same biological function.

Comparison of the various enriched GOBPs is simply done by taking the union of the p-values from all the enriched GOPBs from all the samples, imputing the missing p-values and then clustering the data. As shown in Figure 14.13, the enrichment of the GOPB for AhR signaling was predominant in the 3-methylcholanthrene (3ME) treatment, whereas p38 mitogen-activated protein kinase signaling, p53 signaling and NRF2-mediated stress response were highly enriched exclusively by the THI treatment. Both these chemicals cause cancer in rats in the doses shown in Table 14.1; however, they have different MoAs. 3ME exerts its effect through AhR whereas THI is non-receptor-mediated and causes cytotoxicity. In fact,

Table 14.3 Gene ontology evidence codes.

Code	Evidence
	Experimental
EXP	Inferred from experiment
IDA	Inferred from direct assay
IPI	Inferred from physical interaction
IMP	Inferred from mutant phenotype
IGI	Inferred from genetic interaction
IEP	Inferred from expression pattern
	Computational
ISS	Inferred from sequence or structural similarity
ISO	Inferred from sequence orthology
ISA	Inferred from sequence alignment
ISM	Inferred from sequence model
IGC	Inferred from genomic content
IBA	Inferred from biological aspect of ancestor
IBD	Inferred from biological aspect of descendant
IKR	Inferred from key residues
IRD	Inferred from rapid divergence
RCA	Inferred from reviewed computational analysis
	Author statement
TAS	Traceable author statement
NAS	Non-traceable author statement
	Curatorial
IC	Inferred by curator
ND	No biological data available
	Automatically assigned
IEA	Inferred from electronic annotation

although three agents share a MoA, the clustering of enriched GOBPs reveals that there are unique biological processes impacted by treatment with each agent. This type of mining of the data is extremely useful for finding novel effects on biological systems.

For more data-driven enrichment, the gene set enrichment analysis (GSEA) approach was developed[37,38] whereby, instead of enriching for pathway based on a set of DEGs, all genes are considered based on their ranking and expression response within a particular pathway (Figure 14.14). The workflow begins by selecting an ontology to categorize genes. Then, based on the gene expression data, a test statistic or particular data type (*i.e.* ratio of treated to control) or metric is computed for ranking the genes. The next step entails calculating the enrichment score for each category in the gene set based on a non-parametric, weighted Kolmogorov–Smirnov-like statistic and storing the maximum enrichment scores (MES) as the observed enrichment value. Then, through repeated ranking and computing of enrichment scores from permuted data, a non-parametric distribution of MES scores is generated, from which a p-value for each category in the gene set is obtained.

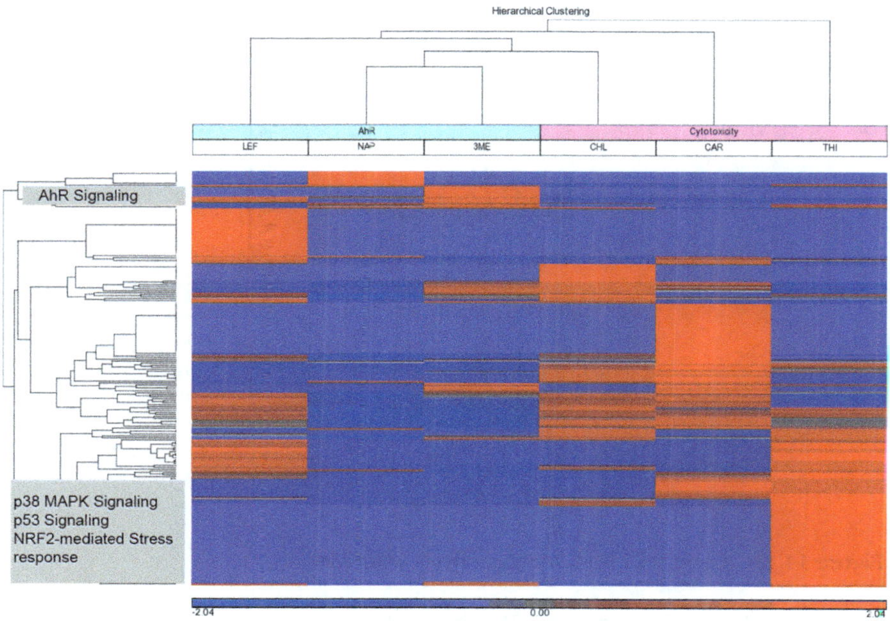

Figure 14.13 Clustering and heat map of enriched Gene Ontology biological processes (GOBPs) based on the −log base 10 (p-values) of the 212 GOBPs for each chemical treatment. The GOBPs are on the y-axis, the treatments are on the x-axis. The enriched GOBPs were obtained from limma analysis differentially expressed genes (absolute fold change >1.5 and p-value <0.05). Significance of enrichment was set to $p<0.05$. Missing values were imputed as 0 (−log base 10(1), *i.e.* $p=1$). Clustering of the pathways were performed using data in the rows standardized to mean of 0 and standard deviation of 1, Euclidean distance and average linkage grouping. The color in the legend denotes the significance of enrichment. The more red the heat map color, the more significant the enrichment. AhR: aryl hydrocarbon receptor; LEF: leflunomide; NAP: β-naphthoflavone; 3ME: 3-methylchloranthrene; CHL: chloroform; CAR: carbon tetrachloride; THI: thioacetamide; MAPK: mitogen-activated protein kinase.

The Broad Institute maintains the Molecular Signatures Database;[38] currently a collection of eight major gene sets (Table 14.4). One is specific for DEGs from cancer studies. Incidentally, ONCOMINE is a repository of cancer-related gene expression studies and offers a user-friendly interface to query the database for genes linked to clinical pathology.[39]

One output from GSEA is an enrichment plot that displays the distribution of the enrichment scores for the genes within a particular category of a gene set. Figure 14.15 shows the enrichment of the peroxisome proliferator-activated receptor (PPAR) signaling pathway with genes mostly responding to PPAR-γ activation by rosiglitazone.

A mammalian organism is made up of more than 200 types of specialized cells. Each cell type carries out a specific task integral to maintaining the

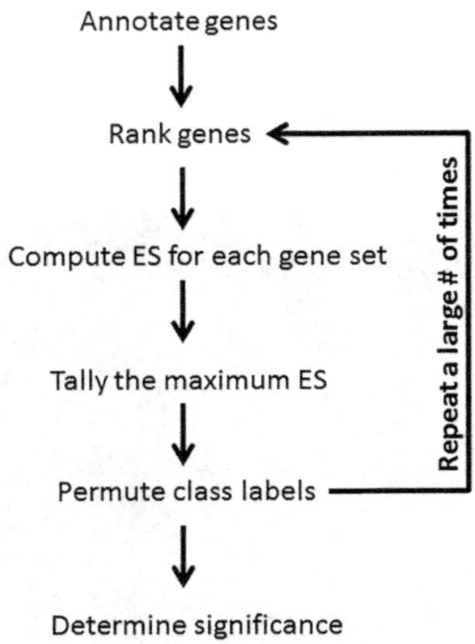

Figure 14.14 Gene set enrichment analysis workflow.

Table 14.4 Molecular Signatures Database (MSigDB) gene set collections.

Collection	Gene set	Description
H	Hallmark	Gene are coherently expressed signatures derived by aggregating many MSigDB gene sets to represent well-defined biological states or processes
C1	Positional	One for each human chromosome and cytogenetic band
C2	Curated	Obtained from online pathway databases, publications in PubMed, and knowledge of domain experts
C3	Motif	Based on conserved *cis*-regulatory motifs from a comparative analysis of the human, mouse, rat and dog genomes
C4	Computational	Defined by mining large collections of cancer-oriented microarray data
C5	GO	Gene ontologies
C6	Oncogenic	Defined directly from microarray gene expression data from cancer gene perturbations
C7	Immunologic	Defined directly from microarray gene expression data from immunological studies

homeostasis of the organism. Cell types can vary by morphology, structure, lifespan, functional ability and much more. Different cell or tissue types acquire their diversity by driving differentially coordinated expression patterns through interacting gene networks. The attractor hypothesis proposed by Kauffman (2004) describes how cell fate transitions between cell types occur through coordinated changes in genome-wide gene expression.[40] Colleagues at the Dana–Farber Cancer Institute and the Harvard School of Public Health utilized the attractor process (Figure 14.16) to elucidate core network pathways that define cellular phenotypes.[41] The GSEA analysis of

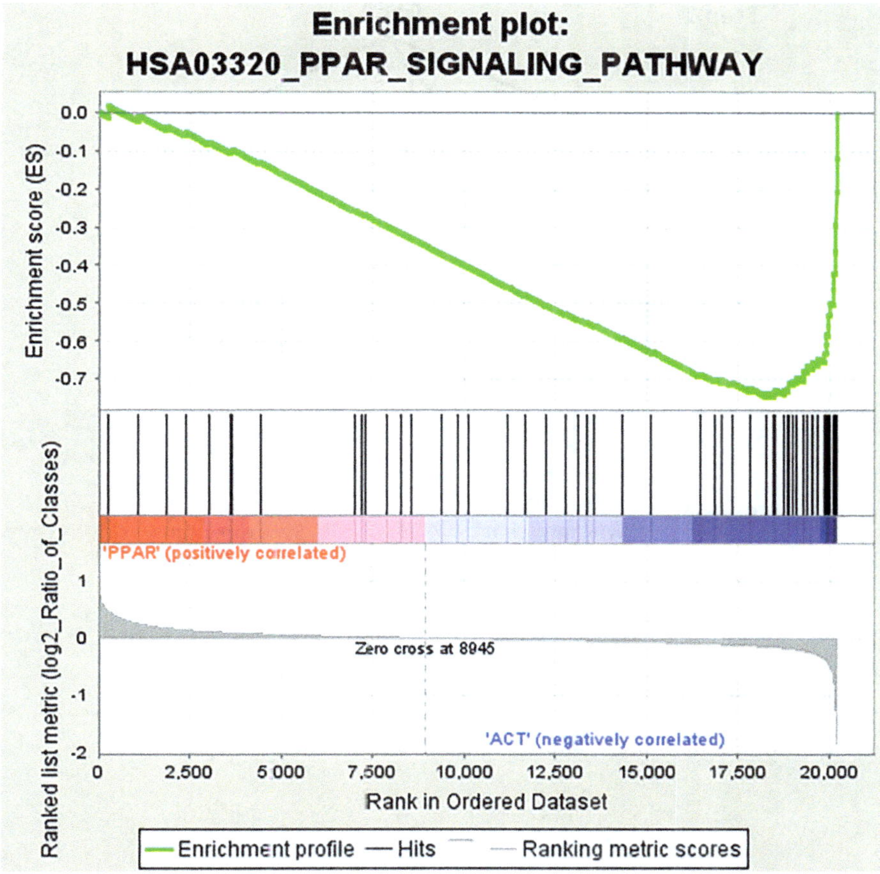

Figure 14.15 Gene set enrichment analysis enrichment of the peroxisome proliferator-activated receptor (PPAR) signaling pathway.

variance (ANOVA)-based strategy models each gene expression profile by a factor representing the phenotypic classes of the samples. In step one, GSEA is performed to enrich biological categories (attractors). Secondly, significant "synexpression" groups[42] (coordinately expressed genes) are detected based on ANOVA modeling of the genes in the attractor modules. Step three entails building gene sets that are highly correlated with the synexpression patterns. Note that steps two and three are similar to the two steps of EPIG. Finally, cellular networks are derived from the correlated gene set.

Adverse outcome pathways (AOPs) are biological mechanisms linking molecular initiating events, intermediate steps and key event relationships to adverse outcomes relevant to risk assessment.[43,44] Ascertaining AOPs enables investigators to be more informed of the cause and effect of toxic/environmental exposures and to better assess risk.[45,46] The Organisation for Economic Co-operation and Development (OECD) spearheaded a collaborative effort to create a knowledgebase of AOPs whereby the scientific

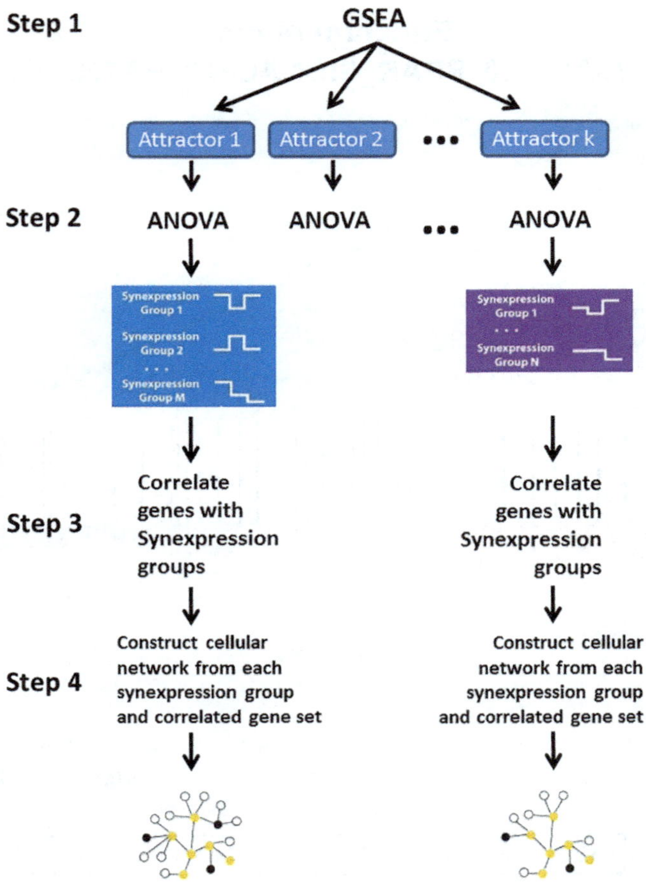

Figure 14.16 Gene set enrichment analysis (GSEA) attractor workflow. ANOVA: analysis of variance.

community could enter, share and discuss AOP-related details at one central location. The site is currently at https://aopkb.org/aopwiki. The OECD provides instructions, guidance and policies for contributing to and sharing AOPs.[47] In addition, to aid in the development and assessment of AOPs, a handbook is available which contains a template to help to portray an AOP and provides focused and practical instructions for developers and assessors intended to assist in identifying, organizing and evaluating critical information on key events and linkages within the AOP.[48] Furthermore, the material provides insight on how to assess the weight of evidence supporting the overall AOP. An example of an AOP is shown in Figure 14.17, depicting estrogen receptor antagonism leading to reproductive dysfunction. Using bioinformatics analysis of genomic data such as gene expression measurements integrated with toxicologic data will certainly help to facilitate the identification of AOPs from a purely computational perspective (see Section 14.1.3 for more details).

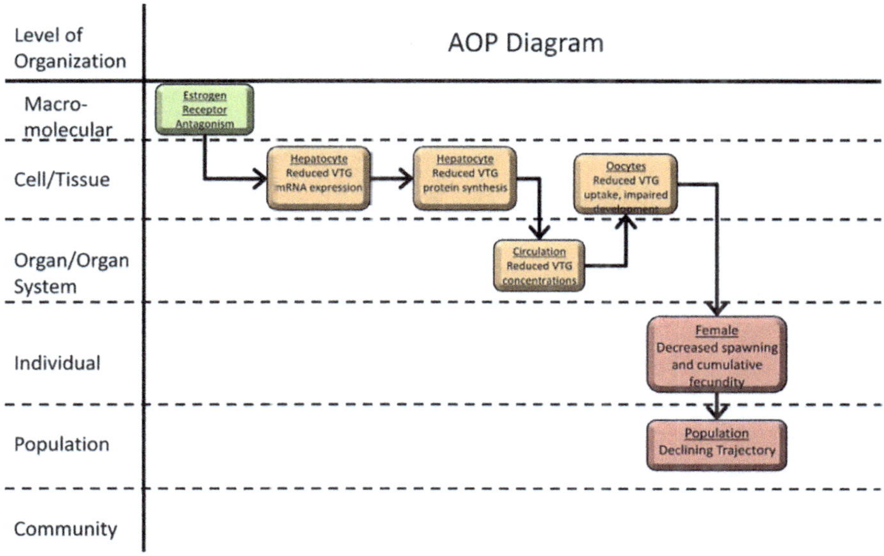

Figure 14.17 Adverse outcome pathway (AOP) for estrogen receptor antagonism leading to reproductive dysfunction. VTG: vitellogenin; mRNA: messenger RNA.

As we have seen thus far, biological processes, pathways, signal transduction cascades and regulatory mechanism are all informative and useful as bioinformatics tools to explore the underlying biology of cancer. However, often, regulatory connections within the system at hand are of interest to be inferred from the data. Gene regulatory networks (GRNs) are creative ways of identifying altered molecular interactions within perturbed systems. The network is composed of nodes defining a genes or molecules and connections denoting interactions (Figure 14.18).

The network can be derived as a static one where genes are assumed to be expressed in a steady state, or dynamic one where the altering patterns of the genes and their influence over time, varying conditions or other type of series of perturbations are resolved. The latter type affords the ability to ascertain directionality with respect to a molecule inhibiting or activing another or itself. Table 14.5 lists a few of the common ways of inferring GRNs;[49] most methods are derivations of these methods or hybrid strategies.

The bottom line is that these approaches simply derive the causal relationships between molecules and are essentially inferred and not actual networks. A dynamic GRN was constructed from gene expression data to infer the gene interactions during hepatocellular carcinoma progression.[50] Recently, a GRN for breast cancer was inferred from gene expression data obtained from patient samples and then evaluated biologically using Gene Ontology analysis of its structural components.[51] Consensus GRNs can be estimated by identifying the most consistent network structure from multiple time-course gene expression data sets.[52,53]

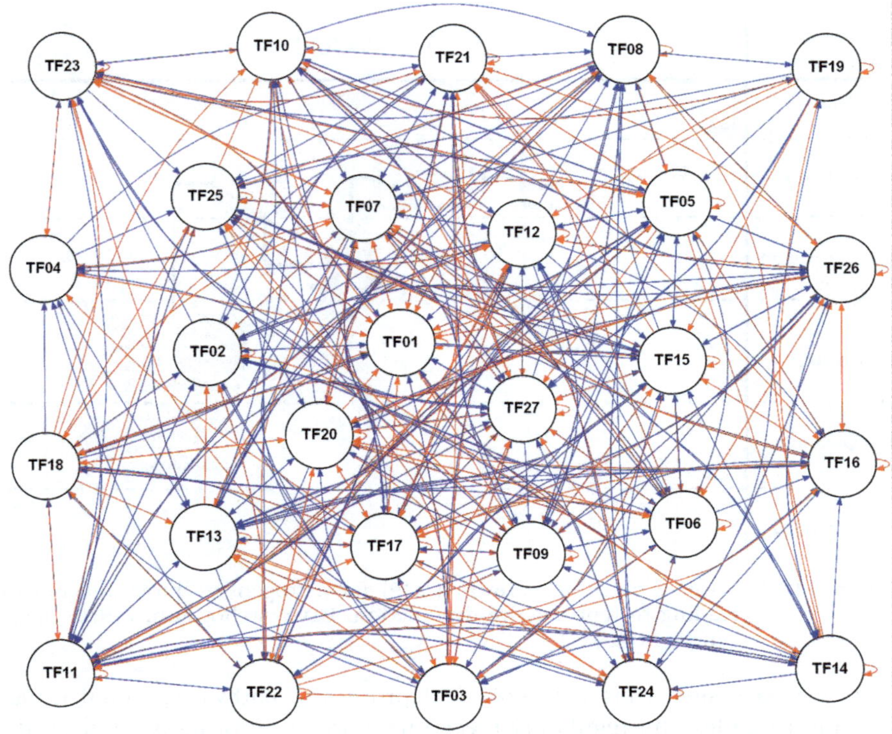

Figure 14.18 Gene regulatory network. Circles represent genes, arrows denote interactions, red is activation and blue is inhibition.

Table 14.5 Some basic ways of deriving gene regulatory networks (GRNs).

GRN type	Methodology
Boolean	Molecules have two states based on their data: off and on
Distance based	Molecules are associated as interacting based on the distance or correlation of their data
Ordinary differential equations	A series of differential equations and derivatives are used to mathematically link molecules
Bayesian	Conditional dependencies between molecules are derived from the data using associations to conditions, diseases or outcomes
Graph theory	Molecules and their interactions are depicted in graph G defined as a pair (V, E) where V is a set of vertices representing the nodes and E is a set of edges representing the connections between the nodes. Highly connected sub-graphs are inferred
Information theory	Molecules are associated conditionally as interacting based on mutual information and entropy of the data

In toxicology the dose–response relationship between the amount of a chemical administered and the degree of the response of the biological system characterizes the exposure to an individual in a graded/continuous-scaled fashion, or to a population of individuals defined by the quantal, all-or-none condition where a distinction is made between an individual being a responder or a non-responder to the given dose of the chemical agent. The dose–response relationship assumes that the response is due to the chemical treatment, the magnitude of the response is related to the dose and there exists a precise, quantifiable method of measuring and expressing the level of toxicity elicited.

In general terms, the minimally effective dose of a chemical to evoke a yes or no response is referred to as the threshold dose.[54] In other words, there potentially is a dose of a chemical below which the probability of responding to the exposure is zero. The essential facts for interpreting the existence of a threshold in the dose–response relationship are common to all types of effects and are as follows[55]: (1) the number and identity of the targets; (2) the mode of the interaction of the acting agent with the target; and (3) the type and sensitivity of the endpoint used to quantify the effect. In the case of individual dose–response relationships, thresholds do exist for most toxic agents with an acute biological response. However, it is challenging to threshold the no observable adverse effect level of a chemical given the individual variation in response to the exposure. In addition, there is wide disagreement (differences of opinion and interpretation) and conflicting evidence regarding the existence of thresholds for chronic responses, especially in the case of chemical carcinogenesis. For example, it has been suggested that there will be no threshold for chemical carcinogenesis if the mechanism that produces tumors in control animals is the same as the mechanism that gives rise to cancer in treated animals. However, if the mechanism producing the cancer in the control samples is different than those in the treated, then it is plausible to argue the possibility of thresholds in chemical carcinogenesis.[56] Moreover, it is reasoned that, since carcinogenesis is a stochastic event, by definition the process should not reveal a threshold of toxicity.[57,58] The debate between proponents and opponents of putative thresholds in the dose–response relationship for chemical carcinogenesis presumably stems from differences in the definition of a threshold and results from a different understanding of the etiology of chemical carcinogenesis.[59]

Complex genetic alterations and irreversible changes in gene expression are hallmark consequences of the "progression" stage of carcinogenesis. Theoretically, chemical carcinogenic agents capable of damaging DNA in a "single hit, single target" paradigm, are not expected to exhibit a threshold in a dose–response relationship. Genotoxic effects can lead to spontaneous cancer incidences in a linear fashion at the low dose–response range (see chapter 6 for low-dose threshold responses in toxicogenomics). Conversely, epigenetic modes of action of chemical carcinogens do not alter DNA or induce mutations yet elicit cancer after long-term (prolonged) administration by affecting, for example, cell differentiation, toxicity, receptor-mediated pathways and cell turnover. Non-genotoxic carcinogens are postulated to have a threshold in a dose-responsive manner.[60]

Not until recently has convincing scientific proof for the existence of thresholds in chemical carcinogenesis been brought forward. Waddell[61] presents a reanalysis of the ED01 Study data which measured the effect of 2-acetylaminoflurene (2-AAF) tumor incidences in the liver and the bladder of a large sample of female mice.[62,63] The original conclusion of the study suggested that there was a threshold for 2-AAF-induced bladder neoplasmas, but not for liver tumor incidence, since the latter showed a linear response over the experimental dose range of the treatment. The reanalysis of the ED01 Study data used a procedure that plotted the percent of tumor response linearly against the logarithmic of the dose amount to demonstrate a threshold for chemical carcinogenicity of 2-AAF in both liver and bladder tissue types. This reanalysis and recent finding dispels the notion that a threshold level for a carcinogen cannot be determined without any degree of accuracy in favor of the strong support of a threshold for chemical carcinogenesis below which no tumor will be produced. However, extrapolation of these results to larger populations is baseless, and there are fundamental differences in the threshold of response to chemical carcinogens as well as in the variety of tissues targeted, cells types perturbed and molecular and/or cellular systems/mechanisms altered.[64]

The advent of toxicogenomics heralded the bridging of genomics, expression analysis and toxicology for modern-day assessment of chemical effects on biological systems.[65] Judiciously, Lobenhofer *et al.* (2004) utilized microarray gene expression to characterize a no observed transcriptional effect level (NOTEL) in a study of low doses of estrogen in the MCF-7 breast cancer cell line.[66] NOTEL, as shown in Figure 14.19, is the dose at which there is no significant change in gene expression baseline level from exposure to a chemical.

Whole-genome expression arrays, statistical analyses and bioinformatics approaches are currently available to build a compendium of data and information regarding the biological consequences of and responses to, chemical and environmental exposures.[67] Total coverage gene expression analysis of a variety of tissue samples exposed to chemical agents in both a dose–response fashion and a time-series manner can be a powerful tool (database of expression data) to construct a network of interacts between gene targets perturbed in the course of chemical carcinogenesis, *e.g.* the Chemical Effects in Biological Systems database.[68] In addition, capturing genome-wide expression data during the course of the pathogenesis of neoplasia can presumably permit the "phenotypic anchoring"[69] of gene expression profiles to the endpoints observed from long-term toxic exposure resulting in chemical carcinogenesis. Ultimately, the cellular and molecular mechanisms of the neoplastic development will be better understood and utilized to pinpoint the signature profiles, within chemically exposed cells, that provide value-added information (are more informative) for extrapolating chemical carcinogenesis thresholds of toxic agents from typical dose–response relationships. Thomas *et al.* (2013) incorporated benchmark dose (BMD) estimation with genomic data to try to bracket the point of

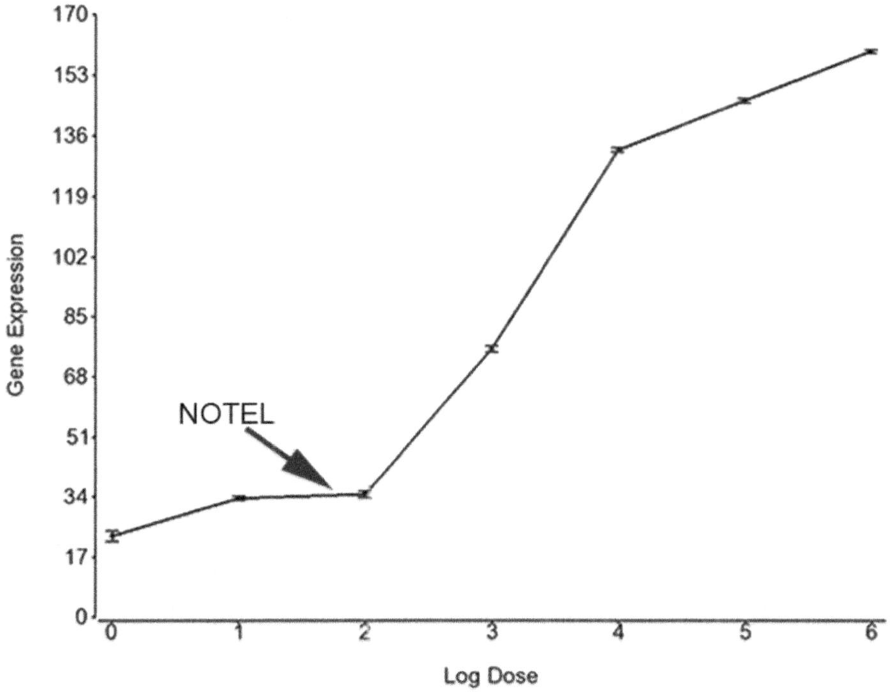

Figure 14.19 Dose–response curve denoting no observed transcriptional effect level (NOTEL).

departure based on gene expression responses to chemical treatments.[70] BMDExpress is a tool to perform BMD analysis of genomic data.[71] As alluded to in chapter 5, in practice, BMD extrapolation with gene expression data extends to sets of genes which share biological function, map to the same mechanistic pathway or are DSTs of a particular transcription regulator. Limitations to this approach are the vast number of genes with unknown functions in several model organisms of interest. Even though a gene or genomic region of the chromosome may be associated with maintaining a threshold for a given chemical carcinogenesis, the biological function of the gene or genomic region needs to be revealed for the threshold assessment to be biologically meaningful. In addition, differentiation of splice variants across genome arrays is currently elusive. This limitation has consequences to determining thresholds in chemical carcinogenesis since it is known that different forms of the same transcript product can have different biological functions. Finally, although expression analysis may reveal a potential marker related to maintaining/pinpointing a threshold in chemical carcinogenesis, the true mechanism may involve protein modification and not transcription. These limitations in utilizing and interpreting gene expression analysis for determining thresholds in chemical carcinogenesis warrant the simultaneous utilization of genomic and protein expression data to supplement and support the interpretation of transcriptomic data.

14.1.3 Bioinformatics and Computational Biology for Integrative Genomics

Aristotle, one of the greatest philosophers and authors in history, stated that "the whole is greater than the sum of its parts". His quote is applicable in many aspects of life, including biology whereby organisms sustain life through a system of inner-working parts. Systems biology is the encompassing of various molecules, cells/tissue types, biological components, systems, and mechanisms of an organisms or entire species for exploratory investigation.[72,73] A key to advancing systems biology is the ability to effectively and efficiently integrate a wealth of data representative of the entire entity being investigated.

Integration of data takes at least three forms[74,75] (Figure 14.20). (1) Early: combining separate data into a single matrix and analyzing it using a particular statistical function; (2) intermediate: keeping the data separate, applying a specific/optimized statistical function for each data set and then combining the functions; and (3) late: analyzing each data set with an appropriate statistical function and then merging the results. Each strategy has its own advantages and disadvantages. Early integration is simple to perform, provided that sufficient computational means are available to merge the data. However, combining the data in such a way loses some if not all of the unique properties of each data set and often times the data types

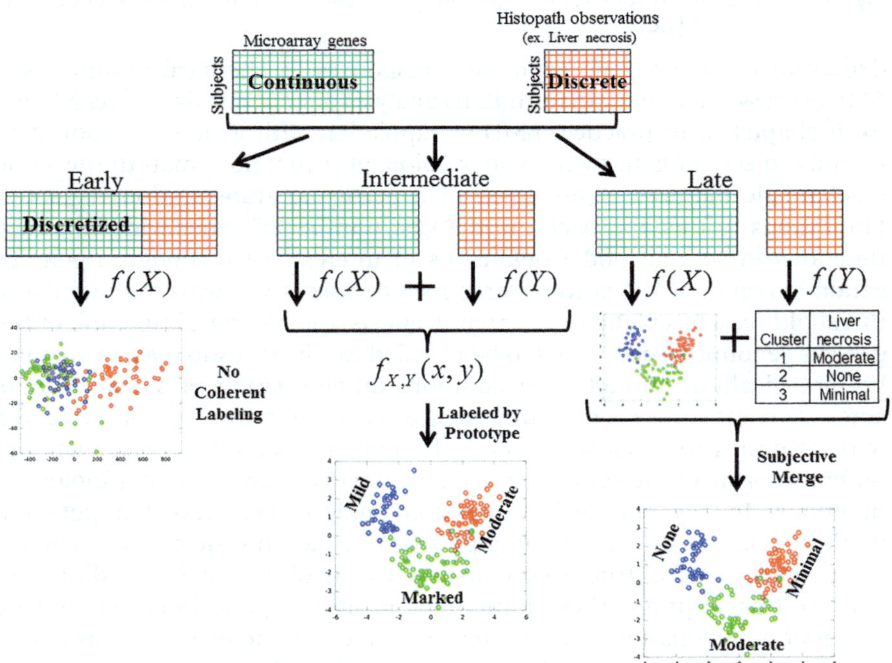

Figure 14.20 Ways to integrate data: early, intermediate and late integration.

and scales in each data domain (*i.e.* gene expression ratios and histopathology observations) do not permit the use of a single analytical function. Intermediate integration benefits from applying a particular analytical function to each data domain optimized for the data type and scale. Little to no transformation of the data is required and relationships between the data sets are preserved. The disadvantage is that construction of the various statistical functions can be complex, there may not be a specific analytical function for a particular data domain and ways of merging the functions may not be immediately intuitive to many or tractable. Late integration permits data-specific analytical functions to be applied to each respective data set. However, the end results require *ad hoc* and subjective overlays to interpret the data, and key relationships like correlations, interactions and dependencies among the data sets are ignored or extremely difficult to comprehend. In many ways, intermediate integration overcomes the limitations of the other two approaches and therefore is be the focus of this section, as it applies to leveraging multiple genomic, clinical and/or toxicological data for analysis and interpretation.

To begin with, let's look at the finer details of intermediate integration of data in the toxicogenomics realm. One of the simplest and intuitive ways of performing a comprehensive analysis of toxicogenomics data is by clustering. As mentioned in Section 14.1, clustering of data uses a dissimilarity metric and merging algorithm to group the data. If for instance, from the same samples, gene expression data, clinical chemistry data and histopathology observations are obtained from the livers of rats exposed to acetaminophen, a separate dissimilarity function for each data domain can be applied and then a common grouping algorithm employed to merge the groups. Bushel *et al.* (2007) devised the ModKPrototypes algorithm[76] which uses the K-means distance metric for microarray and clinical chemistry continuous data, but the K-modes for the histopathology categorical data. Each domain of data was weighted dynamically to adjust the contribution of each during the merging according to how much impact each had for optimizing the clustering based on necrosis of the liver as an external indicator of each sample. As depicted in Figure 14.21, after 100 iterations of the algorithm, more than 75% of the clustering results resolved highly consistent clusters (adjusted Rand index 0.85). The projection of the high-dose 2000 mg kg^{-1} for 24 h samples were set apart from the other samples (Figure 14.22).

Furthermore, the integrated analysis labeled that cluster of samples with the histopathology observations describing moderate degeneration of the hepatocytes and the clinical chemistry ALT at approximately 5000 (U/I). To improve on obtaining the best clustering solution, a simulated annealing optimization routine version of the algorithm (SAModKPrototypes) is available, but takes considerably longer to run.[77] Also, a multi-step decision tree (MSDT) method for integrating gene expression, demographics information and clinical data has been developed.[78] When applied to the Mechanistic Indicators of Childhood Asthma study,[79] the MSDT excelled at separating

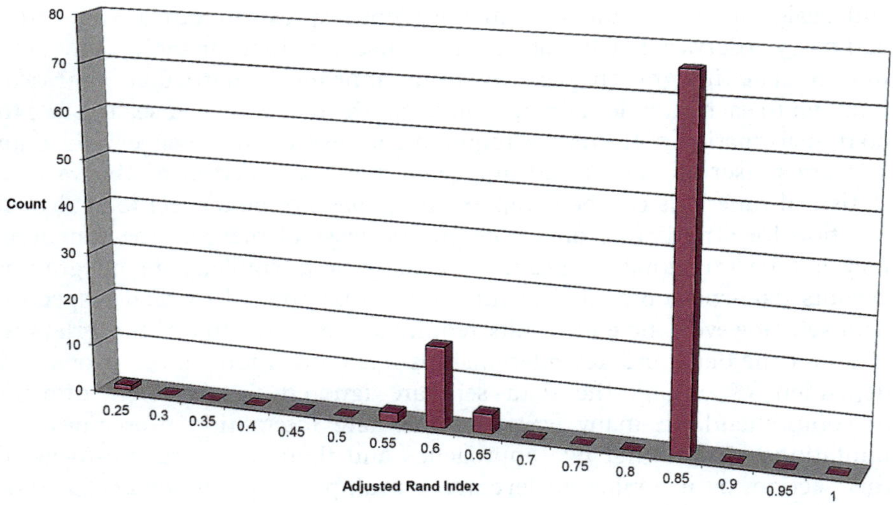

Figure 14.21 ModKPrototype clustering consistency of acetaminophen mixed data.

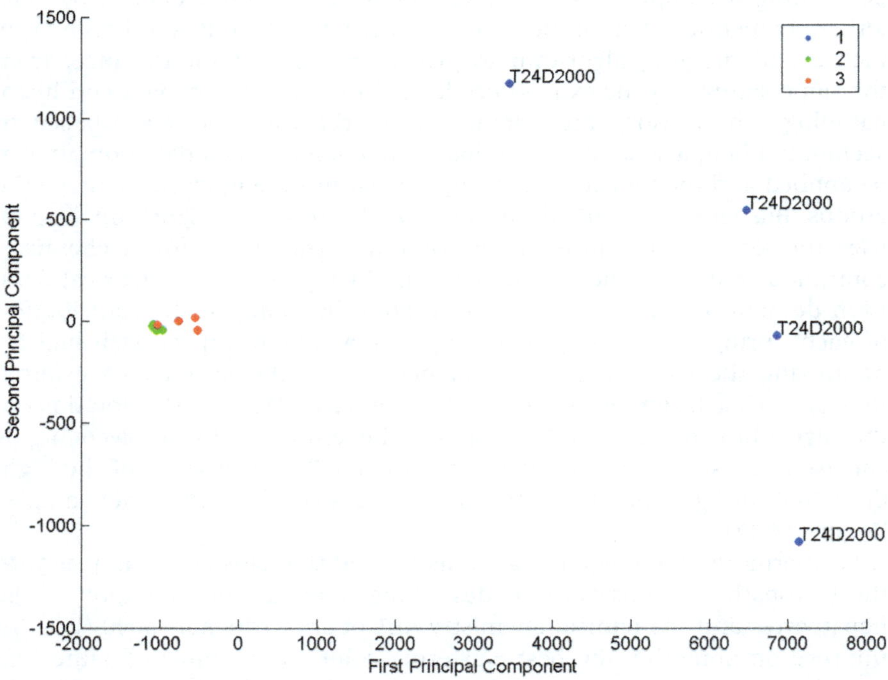

Figure 14.22 Principal component analysis of the acetaminophen mixed data samples grouped by ModKPrototype clustering. Clusters 1, 2 and 3, colored blue, green and red, respectively, denote the groups of rats with the prototypes revealed as having moderate, no and minimal centrilobular necrosis of the liver, respectively.

asthmatics from non-asthmatics and also extracted descriptive endotypes for the complex disease based on the clinical covariates. In addition, a hybrid GSEA-like, multi-task learning algorithm was created to integrate genotype data with microarray gene expression data in order to identify single nucleotide polymorphism (SNP) loci and gene sets that are predictive of estrogen receptor-positive and -negative breast cancer samples.[80] Additionally, transcript regulator expression quantitative trait loci (TReQTLs) were identified from the integration of microarray gene expression data and genotype data *via* a multivariate linear regression to regress a SNP with genes that are DSTs of a given transcription regulator.[81] As shown in Figure 14.23, the DSTs of a transcription regulator are co-regulated and the co-expression may be associated with a *cis*-acting SNP and/or *trans*-acting ones. Testing one SNP-gene pair at a time or simply averaging co-expressed genes ignores the expression relationship between the genes. Incorporating the covariance of the gene expression for co-regulated genes in the TReQTLs bolsters inference for detecting expression quantitative trait loci[82,83] under the control of a master regulator. Ultimately, pleiotropic effects, adverse mechanistic pathways and dysregulated signaling cascades of complex diseases are revealed from the structure of the biological process subtree and gene interaction network within the TReQTL.

Clearly, the amalgamation of data for analysis provides a more comprehensive interpretation of biological circuitry. A prime example is the evolution of the breast cancer molecular fingerprint and its use for clinical treatment.[84] Using hierarchical clustering of expression profiles of 534 "intrinsic" genes from human breast cancer tissues, tumors could be classified into five subtypes representative of their pervasiveness, prognosis and survival: luminal A, luminal B, triple negative/basal-like, HER2 type and rare normal-like.[85,86] Subsequently, additional genomic data were generated on

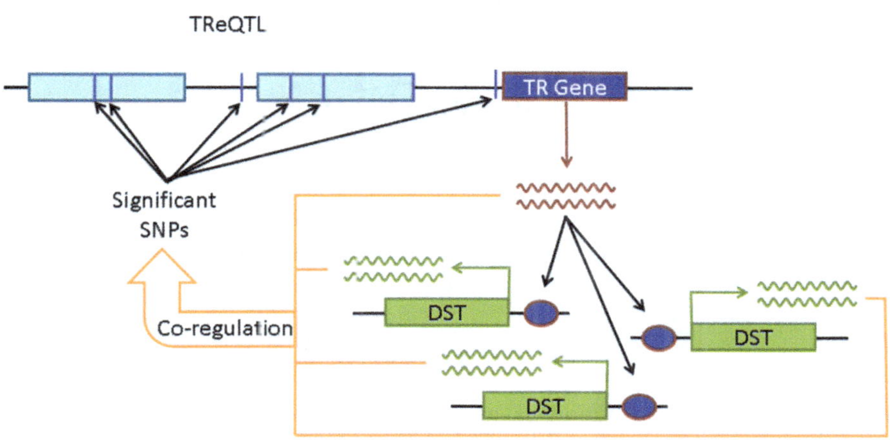

Figure 14.23 Transcript regulator expression quantitative trait loci (TReQTL) schematic illustrating *cis*- and *trans*-regulation of a transcription regulator which affects the expression of its downstream targets.

these and other tumor samples. The Cancer Genome Atlas (TCGA) is a data portal containing DNA copy number arrays, DNA methylation, exome sequencing, mRNA arrays, miRNA sequencing and reverse-phase protein arrays data from tumor and normal samples, many of which are matched to the same individual.[87] Using the "late" form of integration, investigators found a few somatic mutations in common across all breast cancer tumor types, numerous subtype-associated and novel gene mutations, and specific signaling pathways within a subtype. Early integration and analysis of TCGA miRNA and mRNA gene expression data and association with DNA methylation data (Figure 14.24) revealed 30 prognostic genes associated with overall survival across different clinical and molecular subclasses of a cohort of invasive ductal carcinomas from TCGA.[88]

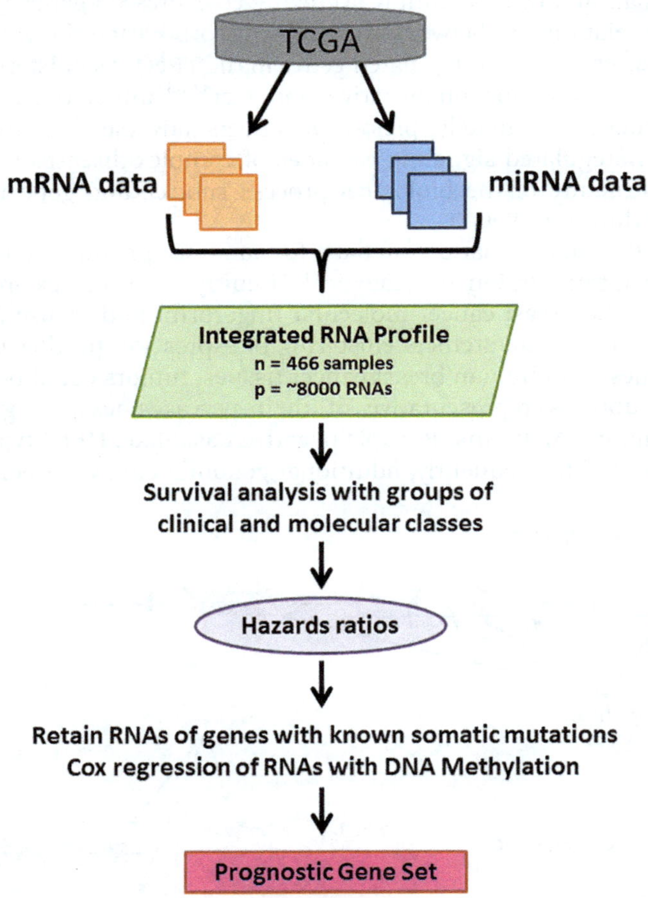

Figure 14.24 Early integration of micro (mi)RNA and messenger (m)RNA The Cancer Genome Atlas (TGCA) breast cancer data to identify molecular prognosticators.
Adapted from Volinia and Croce, *Proc. Natl. Acad. Sci. U. S. A.*, 2013.[88] PMID: 23589849.

"Intermediate" integration and subsequent analysis of TCGA mRNA, miRNA and methylation data results in better separation of the four subtypes of glioblastoma multiforme samples (11 classical, 16 proneural, 17 mesenchymal and 7 neural) than if mRNA gene expression data alone is used (Figure 14.25) and reveals that two clusters with predominantly proneural subtype samples differ according to their mean days of survival and percentage of necrosis (Table 14.6). Inspection of the clusters of mixed genomic data reveals that several miRNAs (hsa-let-7) discern the groups of samples. A previous study has shown that let-7 has an anti-tumorigenic effect by inhibiting proliferation of glioblastoma cells.[89]

Access to, and analysis of, integrated genomics and clinical data empowers investigators to more effectively identify target genes, pathways and biological systems dysregulated in cancer. Undoubtedly, taking a holistic approach to aggregate multi-systems data that are intrinsic to cancer improves the chances that effective therapeutics will ultimately be developed and translated to individualized patient treatment, precision medicine and tailored care. Bioinformatics, computational biology and meta-informatics play crucial roles in making sense of the data and triaging it in ways which are comprehensive and sensible, informative and reliable, synergistic and impactful. Efforts are currently on the way to assemble data commons to federate cancer genomic data. The University of Chicago along with the NIH-NCI and the NIH National Heart, Lung, and Blood Institute have announced such efforts.

14.2 The Future can be Now

Envision silos containing petabytes of genomic, clinical and other biological data and information. Means to traverse through the data and make sensible correlations are computationally challenging. These systems will be programmed to translate data and information into knowledge about a patient's individual cancer outlook and also serve to make overall assessments about a given population or cohort. Just the mere vastness of the data will certainly give rise to false discoveries and associations by chance. Experts with sound statistical cognizance will be critical in the data mining and knowledge extraction stages. Computer scientists with capabilities to outfit the data commons with failsafe technology and warp speed, user-friendly search engines will need to be implemented in the design and development processes. To harmonize such massive amounts of biological data, teams of bioinformaticians, biostatisticians, biologists, genomicists and clinicians will undoubtedly need to collaborate in order to strategize how best to integrate the data in sensible ways and leverage crowdsourcing expertise to promote open science, *i.e.* Sage Bionetworks.[90] Cutting-edge science such as nanoinformatics[91] to bridge nanotechnology, epigenomics/genetics, cancer biology and bioinformatics will flourish out of the need to further the war against cancer. Finally, public policy makers, ethics practitioners and publicists/public relations professionals will play key roles in order to gain the

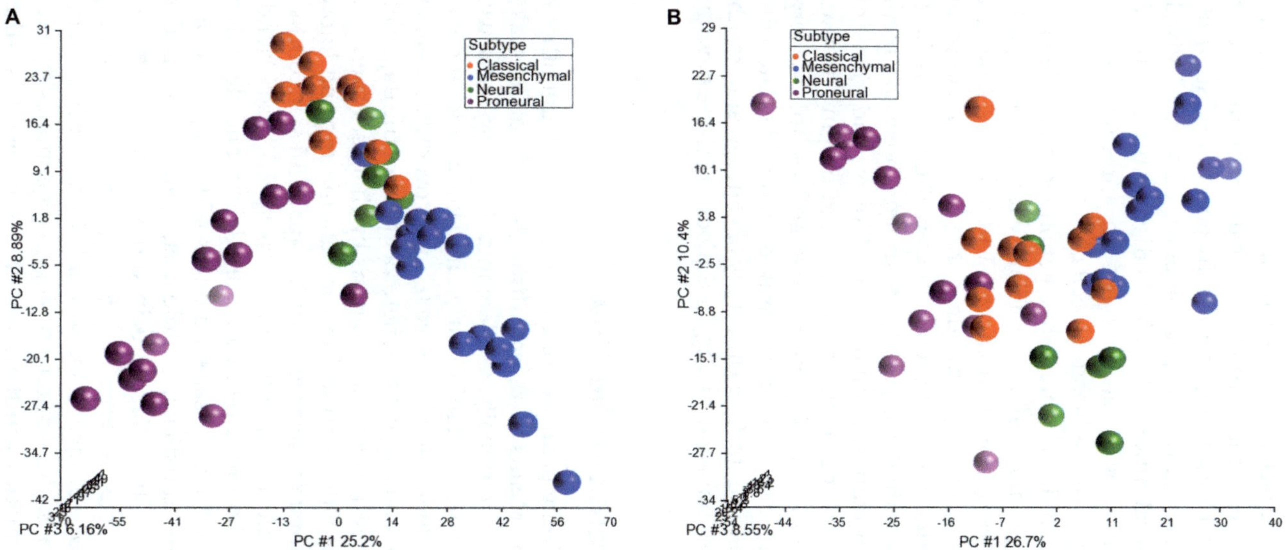

Figure 14.25 Principal component analysis to project 51 glioblastoma multiformeGBM samples in 3D space. (A) Using 1740 microarray gene expression profiles, 534 micro (mi)RNA expression profiles and methylation indications (+1, 0 or −1) from 1412 CpG sites from the 51 samples with matched data. (B) Using only the data from the 1740 microarray gene expression profiles from the 51 samples.

Table 14.6 Integrated clustering of glioblastoma multiforme samples.

Feature	Cluster 1	Cluster 2	Cluster 3
Prototype	Proneural	Proneural	Mesenchymal
Samples n	5	18	28
Silhouette width separation score	0.31	0.18	0.19
Mean age at first diagnosis	39.4	48.33	57.11
Mean survival days	1669.8	299.11	377.21
Tumor nuclei %	99	98.47	97.05
Necrosis (cell death)%	3	10.61	10.18

trust of the stakeholders and to ensure that the data is utilized in accordance with guidelines. In one sense, such undertakings can appear to be a huge venture in terms of resources, management and sustainability. Conversely, given the aforementioned implied burdens of cancer on the public well-being and human health, endeavoring on such a monumental mission is quite understandable, certainly defensible and obviously essential to galvanize genomic data for the next generation of predictive carcinogenicity.

Acknowledgements

This research was supported (in part) by the Intramural Research Program of the National Institute of Environmental Health Sciences, National Institutes of Health.

References

1. S. B. Needleman and C. D. Wunsch, A general method applicable to the search for similarities in the amino acid sequence of two proteins, *J. Mol. Biol.*, 1970, **48**, 443–453.
2. R. Staden, Sequence data handling by computer, *Nucleic Acids Res.*, 1977, **4**, 4037–4051.
3. M. Schena, D. Shalon, R. W. Davis and P. O. Brown, Quantitative monitoring of gene expression patterns with a complementary DNA microarray, *Science*, 1995, **270**, 467–470.
4. C. Tomasetti and B. Vogelstein, Cancer etiology. Variation in cancer risk among tissues can be explained by the number of stem cell divisions, *Science*, 2015, **347**, 78–81.
5. R. Simon, Bioinformatics in cancer therapeutics—hype or hope?, *Nat. Clin. Pract. Oncol.*, 2005, **2**, 223.
6. N. F. Lowndes, J. Paul, J. Wu and M. Allan, c-Ha-ras gene bidirectional promoter expressed in vitro: location and regulation, *Mol. Cell Biol.*, 1989, **9**, 3758–3770.
7. S. Ramaswamy, P. Tamayo, R. Rifkin, S. Mukherjee, C. H. Yeang, M. Angelo, C. Ladd, M. Reich, E. Latulippe, J. P. Mesirov, T. Poggio, W. Gerald, M. Loda, E. S. Lander and T. R. Golub, Multiclass cancer

diagnosis using tumor gene expression signatures, *Proc. Natl. Acad. Sci. U. S. A.*, 2001, **98**, 15149–15154.
8. P. Tamayo, D. Slonim, J. Mesirov, Q. Zhu, S. Kitareewan, E. Dmitrovsky, E. S. Lander and T. R. Golub, Interpreting patterns of gene expression with self-organizing maps: methods and application to hematopoietic differentiation, *Proc. Natl. Acad. Sci. U. S. A.*, 1999, **96**, 2907–2912.
9. D. T. Ross, U. Scherf, M. B. Eisen, C. M. Perou, C. Rees, P. Spellman, V. Iyer, S. S. Jeffrey, M. Van de Rijn, M. Waltham, A. Pergamenschikov, J. C. Lee, D. Lashkari, D. Shalon, T. G. Myers, J. N. Weinstein, D. Botstein and P. O. Brown, Systematic variation in gene expression patterns in human cancer cell lines, *Nat. Genet.*, 2000, **24**, 227–235.
10. U. Scherf, D. T. Ross, M. Waltham, L. H. Smith, J. K. Lee, L. Tanabe, K. W. Kohn, W. C. Reinhold, T. G. Myers, D. T. Andrews, D. A. Scudiero, M. B. Eisen, E. A. Sausville, Y. Pommier, D. Botstein, P. O. Brown and J. N. Weinstein, A gene expression database for the molecular pharmacology of cancer, *Nat. Genet.*, 2000, **24**, 236–244.
11. R. Bellman, Rand Corporation and John Ptak Computer Science Book Collection (North Carolina State University), *Dynamic Programming*, Princeton University Press, Princeton, 1957.
12. L. Shi, G. Campbell, W. D. Jones, F. Campagne, Z. Wen, S. J. Walker, Z. Su, T. M. Chu, F. M. Goodsaid, L. Pusztai, J. D. Shaughnessy Jr., A. Oberthuer, R. S. Thomas, R. S. Paules, M. Fielden, B. Barlogie, W. Chen, P. Du, M. Fischer, C. Furlanello, B. D. Gallas, X. Ge, D. B. Megherbi, W. F. Symmans, M. D. Wang, J. Zhang, H. Bitter, B. Brors, P. R. Bushel, M. Bylesjo, M. Chen, J. Cheng, J. Cheng, J. Chou, T. S. Davison, M. Delorenzi, Y. Deng, V. Devanarayan, D. J. Dix, J. Dopazo, K. C. Dorff, F. Elloumi, J. Fan, S. Fan, X. Fan, H. Fang, N. Gonzaludo, K. R. Hess, H. Hong, J. Huan, R. A. Irizarry, R. Judson, D. Juraeva, S. Lababidi, C. G. Lambert, L. Li, Y. Li, Z. Li, S. M. Lin, G. Liu, E. K. Lobenhofer, J. Luo, W. Luo, M. N. McCall, Y. Nikolsky, G. A. Pennello, R. G. Perkins, R. Philip, V. Popovici, N. D. Price, F. Qian, A. Scherer, T. Shi, W. Shi, J. Sung, D. Thierry-Mieg, J. Thierry-Mieg, V. Thodima, J. Trygg, L. Vishnuvajjala, S. J. Wang, J. Wu, Y. Wu, Q. Xie, W. A. Yousef, L. Zhang, X. Zhang, S. Zhong, Y. Zhou, S. Zhu, D. Arasappan, W. Bao, A. B. Lucas, F. Berthold, R. J. Brennan, A. Buness, J. G. Catalano, C. Chang, R. Chen, Y. Cheng, J. Cui, W. Czika, F. Demichelis, X. Deng, D. Dosymbekov, R. Eils, Y. Feng, J. Fostel, S. Fulmer-Smentek, J. C. Fuscoe, L. Gatto, W. Ge, D. R. Goldstein, L. Guo, D. N. Halbert, J. Han, S. C. Harris, C. Hatzis, D. Herman, J. Huang, R. V. Jensen, R. Jiang, C. D. Johnson, G. Jurman, Y. Kahlert, S. A. Khuder, M. Kohl, J. Li, L. Li, M. Li, Q. Z. Li, S. Li, Z. Li, J. Liu, Y. Liu, Z. Liu, L. Meng, M. Madera, F. Martinez-Murillo, I. Medina, J. Meehan, K. Miclaus, R. A. Moffitt, D. Montaner, P. Mukherjee, G. J. Mulligan, P. Neville, T. Nikolskaya, B. Ning, G. P. Page, J. Parker, R. M. Parry, X. Peng, R. L. Peterson, J. H. Phan, B. Quanz, Y. Ren, S. Riccadonna, A. H. Roter, F. W. Samuelson, M. M. Schumacher, J. D. Shambaugh,

Q. Shi, R. Shippy, S. Si, A. Smalter, C. Sotiriou, M. Soukup, F. Staedtler, G. Steiner, T. H. Stokes, Q. Sun, P. Y. Tan, R. Tang, Z. Tezak, B. Thorn, M. Tsyganova, Y. Turpaz, S. C. Vega, R. Visintainer, J. von Frese, C. Wang, E. Wang, J. Wang, W. Wang, F. Westermann, J. C. Willey, M. Woods, S. Wu, N. Xiao, J. Xu, L. Xu, L. Yang, X. Zeng, J. Zhang, L. Zhang, M. Zhang, C. Zhao, R. K. Puri, U. Scherf, W. Tong, R. D. Wolfinger and M. Consortium, The MicroArray Quality Control (MAQC)-II study of common practices for the development and validation of microarray-based predictive models, *Nat. Biotechnol.*, 2010, **28**, 827–838.

13. C. Chang, J. Wang, C. Zhao, J. Fostel, W. Tong, P. R. Bushel, Y. Deng, L. Pusztai, W. F. Symmans and T. Shi, Maximizing biomarker discovery by minimizing gene signatures, *BMC Genomics*, 2011, **12**(Suppl. 5), S6.
14. P. E. Lonergan and D. J. Tindall, Androgen receptor signaling in prostate cancer development and progression, *J Carcinog.*, 2011, **10**, 20.
15. J. W. Chou, T. Zhou, W. K. Kaufmann, R. S. Paules and P. R. Bushel, Extracting gene expression patterns and identifying co-expressed genes from microarray data reveals biologically responsive processes, *BMC Bioinf.*, 2007, **8**, 427.
16. K. O. Cheng, N. F. Law, W. C. Siu and A. W. Liew, Identification of coherent patterns in gene expression data using an efficient biclustering algorithm and parallel coordinate visualization, *BMC Bioinf.*, 2008, **9**, 210.
17. A. L. Shaffer, A. Rosenwald, E. M. Hurt, J. M. Giltnane, L. T. Lam, O. K. Pickeral and L. M. Staudt, Signatures of the immune response, *Immunity*, 2001, **15**, 375–385.
18. S. Yoon, C. Nardini, L. Benini and G. De Micheli, Discovering coherent biclusters from gene expression data using zero-suppressed binary decision diagrams, *IEEE/ACM Trans. Comput. Biol. Bioinform.*, 2005, **2**, 339–354.
19. S. C. Madeira and A. L. Oliveira, Biclustering algorithms for biological data analysis: a survey, *IEEE/ACM Trans. Comput. Biol. Bioinform.*, 2004, **1**, 24–45.
20. J. W. Chou and P. R. Bushel, Discernment of possible mechanisms of hepatotoxicity *via* biological processes over-represented by co-expressed genes, *BMC Genomics*, 2009, **10**, 272.
21. D. Gusenleitner, E. A. Howe, S. Bentink, J. Quackenbush and A. C. Culhane, iBBiG: iterative binary bi-clustering of gene sets, *Bioinformatics*, 2012, **28**, 2484–2492.
22. L. Ji and K. L. Tan, Identifying time-lagged gene clusters using gene expression data, *Bioinformatics*, 2005, **21**, 509–516.
23. C. J. Langmead, A. K. Yan, C. R. McClung and B. R. Donald, Phase-independent rhythmic analysis of genome-wide expression patterns, *J. Comput. Biol.*, 2003, **10**, 521–536.
24. E. Leung and P. R. Bushel, PAGE: phase-shifted analysis of gene expression, *Bioinformatics*, 2006, **22**, 367–368.

25. D. Liu, C. R. Weinberg and S. D. Peddada, A geometric approach to determine association and coherence of the activation times of cell-cycling genes under differing experimental conditions, *Bioinformatics*, 2004, **20**, 2521–2528.
26. E. Wingender, X. Chen, E. Fricke, R. Geffers, R. Hehl, I. Liebich, M. Krull, V. Matys, H. Michael, R. Ohnhauser, M. Pruss, F. Schacherer, S. Thiele and S. Urbach, The TRANSFAC system on gene expression regulation, *Nucleic Acids Res.*, 2001, **29**, 281–283.
27. E. Wingender, X. Chen, R. Hehl, H. Karas, I. Liebich, V. Matys, T. Meinhardt, M. Pruss, I. Reuter and F. Schacherer, TRANSFAC: an integrated system for gene expression regulation, *Nucleic Acids Res.*, 2000, **28**, 316–319.
28. M. Krull, N. Voss, C. Choi, S. Pistor, A. Potapov and E. Wingender, TRANSPATH: an integrated database on signal transduction and a tool for array analysis, *Nucleic Acids Res.*, 2003, **31**, 97–100.
29. T. Breslin, M. Krogh, C. Peterson and C. Troein, Signal transduction pathway profiling of individual tumor samples, *BMC Bioinform.*, 2005, **6**, 163.
30. R. C. Bargou, C. Wagener, K. Bommert, W. Arnold, P. T. Daniel, M. Y. Mapara, E. Grinstein, H. D. Royer and B. Dorken, Blocking the transcription factor E2F/DP by dominant-negative mutants in a normal breast epithelial cell line efficiently inhibits apoptosis and induces tumor growth in SCID mice, *J. Exp. Med.*, 1996, **183**, 1205–1213.
31. Z. Wang and Y. Sun, Targeting p53 for Novel Anticancer Therapy, *Transl. Oncol.*, 2010, **3**, 1–12.
32. M. Kanehisa, S. Goto, Y. Sato, M. Kawashima, M. Furumichi and M. Tanabe, Data, information, knowledge and principle: back to metabolism in KEGG, *Nucleic Acids Res.*, 2014, **42**, D199–205.
33. M. Kanehisa and S. Goto, KEGG: kyoto encyclopedia of genes and genomes, *Nucleic Acids Res.*, 2000, **28**, 27–30.
34. H. Mi, A. Muruganujan, J. T. Casagrande and P. D. Thomas, Large-scale gene function analysis with the PANTHER classification system, *Nat. Protoc.*, 2013, **8**, 1551–1566.
35. H. Mi, A. Muruganujan and P. D. Thomas, PANTHER in 2013: modeling the evolution of gene function, and other gene attributes, in the context of phylogenetic trees, *Nucleic Acids Res.*, 2013, **41**, D377–386.
36. C. Wang, B. Gong, P. R. Bushel, J. Thierry-Mieg, D. Thierry-Mieg, J. Xu, H. Fang, H. Hong, J. Shen, Z. Su, J. Meehan, X. Li, L. Yang, H. Li, P. P. Labaj, D. P. Kreil, D. Megherbi, S. Gaj, F. Caiment, J. van Delft, J. Kleinjans, A. Scherer, V. Devanarayan, J. Wang, Y. Yang, H. R. Qian, L. J. Lancashire, M. Bessarabova, Y. Nikolsky, C. Furlanello, M. Chierici, D. Albanese, G. Jurman, S. Riccadonna, M. Filosi, R. Visintainer, K. K. Zhang, J. Li, J. H. Hsieh, D. L. Svoboda, J. C. Fuscoe, Y. Deng, L. Shi, R. S. Paules, S. S. Auerbach and W. Tong, The concordance between RNA-seq and microarray data depends on chemical treatment and transcript abundance, *Nat. Biotechnol.*, 2014, **32**, 926–932.

37. V. K. Mootha, C. M. Lindgren, K. F. Eriksson, A. Subramanian, S. Sihag, J. Lehar, P. Puigserver, E. Carlsson, M. Ridderstrale, E. Laurila, N. Houstis, M. J. Daly, N. Patterson, J. P. Mesirov, T. R. Golub, P. Tamayo, B. Spiegelman, E. S. Lander, J. N. Hirschhorn, D. Altshuler and L. C. Groop, PGC-1alpha-responsive genes involved in oxidative phosphorylation are coordinately downregulated in human diabetes, *Nat. Genet.*, 2003, **34**, 267–273.
38. A. Subramanian, P. Tamayo, V. K. Mootha, S. Mukherjee, B. L. Ebert, M. A. Gillette, A. Paulovich, S. L. Pomeroy, T. R. Golub, E. S. Lander and J. P. Mesirov, Gene set enrichment analysis: a knowledge-based approach for interpreting genome-wide expression profiles, *Proc. Natl. Acad. Sci. U. S. A.*, 2005, **102**, 15545–15550.
39. D. R. Rhodes, J. Yu, K. Shanker, N. Deshpande, R. Varambally, D. Ghosh, T. Barrette, A. Pandey and A. M. Chinnaiyan, ONCOMINE: a cancer microarray database and integrated data-mining platform, *Neoplasia*, 2004, **6**, 1–6.
40. S. Kauffman, A proposal for using the ensemble approach to understand genetic regulatory networks, *J. Theor. Biol.*, 2004, **230**, 581–590.
41. J. C. Mar, N. A. Matigian, J. Quackenbush and C. A. Wells, Attract: A method for identifying core pathways that define cellular phenotypes, *PLoS One*, 2011, **6**, e25445.
42. C. Niehrs and N. Pollet, Synexpression groups in eukaryotes, *Nature*, 1999, **402**, 483–487.
43. G. T. Ankley, R. S. Bennett, R. J. Erickson, D. J. Hoff, M. W. Hornung, R. D. Johnson, D. R. Mount, J. W. Nichols, C. L. Russom, P. K. Schmieder, J. A. Serrrano, J. E. Tietge and D. L. Villeneuve, Adverse outcome pathways: a conceptual framework to support ecotoxicology research and risk assessment, *Environ. Toxicol. Chem.*, 2010, **29**, 730–741.
44. M. Vinken, The adverse outcome pathway concept: a pragmatic tool in toxicology, *Toxicology*, 2013, **312**, 158–165.
45. K. J. Groh, R. N. Carvalho, J. K. Chipman, N. D. Denslow, M. Halder, C. A. Murphy, D. Roelofs, A. Rolaki, K. Schirmer and K. H. Watanabe, Development and application of the adverse outcome pathway framework for understanding and predicting chronic toxicity: II. A focus on growth impairment in fish, *Chemosphere*, 2015, **120**, 778–792.
46. M. Vinken, Adverse Outcome Pathways and Drug-Induced Liver Injury Testing, *Chem. Res. Toxicol.*, 2015, **28**, 1391–1397.
47. D. L. Villeneuve, D. Crump, N. Garcia-Reyero, M. Hecker, T. H. Hutchinson, C. A. LaLone, B. Landesmann, T. Lettieri, S. Munn, M. Nepelska, M. A. Ottinger, L. Vergauwen and M. Whelan, Adverse outcome pathway (AOP) development I: strategies and principles, *Toxicol. Sci.*, 2014, **142**, 312–320.
48. D. L. Villeneuve, D. Crump, N. Garcia-Reyero, M. Hecker, T. H. Hutchinson, C. A. LaLone, B. Landesmann, T. Lettieri, S. Munn, M. Nepelska, M. A. Ottinger, L. Vergauwen and M. Whelan, Adverse

outcome pathway development II: best practices, *Toxicol. Sci.*, 2014, **142**, 321–330.
49. S. Aluru, *Handbook of Computational Molecular Biology*, Chapman & Hall/CRC, Boca Raton, FL, 2006.
50. H. Zhu, R. S. Rao, T. Zeng and L. Chen, Reconstructing dynamic gene regulatory networks from sample-based transcriptional data, *Nucleic Acids Res.*, 2012, **40**, 10657–10667.
51. F. Emmert-Streib, R. de Matos Simoes, P. Mullan, B. Haibe-Kains and M. Dehmer, The gene regulatory network for breast cancer: integrated regulatory landscape of cancer hallmarks, *Front. Genet.*, 2014, **5**, 15.
52. Y. Wang, T. Joshi, X. S. Zhang, D. Xu and L. Chen, Inferring gene regulatory networks from multiple microarray datasets, *Bioinformatics*, 2006, **22**, 2413–2420.
53. L. Z. Liu, F. X. Wu and W. J. Zhang, A group LASSO-based method for robustly inferring gene regulatory networks from multiple time-course datasets, *BMC Syst. Biol.*, 2014, **8**(Suppl. 3), S1.
54. L. J. Casarett, J. Doull and C. D. Klaassen, *Casarett and Doull's toxicology: the basic science of poisons*, McGraw-Hill Medical Pub. Division, New York, 6th edn, 2001.
55. M. Kirsch-Volders, M. Aardema and A. Elhajouji, Concepts of threshold in mutagenesis and carcinogenesis, *Mutat. Res.*, 2000, **464**, 3–11.
56. I. F. Purchase and T. R. Auton, Thresholds in chemical carcinogenesis, *Regul. Toxicol. Pharmacol.*, 1995, **22**, 199–205.
57. R. J. Fry, The question of thresholds for carcinogenesis, *Cancer Invest*, 1989, **7**, 299–300.
58. A. C. Upton, M. L. Randolph, J. W. Conklin, M. A. Kastenbaum, M. Slater, G. S. Melville Jr., F. P. Conte and J. A. Sproul Jr., Late effects of fast neutrons and gamma-rays in mice as influenced by the dose rate of irradiation: induction of neoplasia, *Radiat. Res.*, 1970, **41**, 467–491.
59. W. K. Lutz, A true threshold dose in chemical carcinogenesis cannot be defined for a population, irrespective of the mode of action, *Hum. Exp. Toxicol.*, 2000, **19**, 566–568, discussion 571–562.
60. W. N. Aldridge, The biological basis and measurement of thresholds, *Annu. Rev. Pharmacol. Toxicol.*, 1986, **26**, 39–58.
61. W. J. Waddell, Thresholds of carcinogenicity in the ED01 study, *Toxicol. Sci.*, 2003, **72**, 158–163.
62. N. A. Littlefield, J. H. Farmer, D. W. Gaylor and W. G. Sheldon, Effects of dose and time in a long-term, low-dose carcinogenic study, *J. Environ. Pathol. Toxicol.*, 1980, **3**, 17–34.
63. J. H. Farmer, R. L. Kodell, D. L. Greenman and G. W. Shaw, Dose and time responses models for the incidence of bladder and liver neoplasms in mice fed 2-acetylaminofluorene continuously, *J. Environ. Pathol. Toxicol.*, 1980, **3**, 55–68.
64. P. Kotin, Carcinogenesis. Dose-response relationship and threshold concepts, *Ann. N. Y. Acad. Sci.*, 1976, **271**, 22–28.

65. E. F. Nuwaysir, M. Bittner, J. Trent, J. C. Barrett and C. A. Afshari, Microarrays and toxicology: the advent of toxicogenomics, *Mol. Carcinog.*, 1999, **24**, 153–159.
66. E. K. Lobenhofer, X. Cui, L. Bennett, P. L. Cable, B. A. Merrick, G. A. Churchill and C. A. Afshari, Exploration of low-dose estrogen effects: identification of No Observed Transcriptional Effect Level (NOTEL), *Toxicol. Pathol.*, 2004, **32**, 482–492.
67. T. R. Hughes, M. J. Marton, A. R. Jones, C. J. Roberts, R. Stoughton, C. D. Armour, H. A. Bennett, E. Coffey, H. Dai, Y. D. He, M. J. Kidd, A. M. King, M. R. Meyer, D. Slade, P. Y. Lum, S. B. Stepaniants, D. D. Shoemaker, D. Gachotte, K. Chakraburtty, J. Simon, M. Bard and S. H. Friend, Functional discovery via a compendium of expression profiles, *Cell*, 2000, **102**, 109–126.
68. M. Waters, S. Stasiewicz, B. A. Merrick, K. Tomer, P. Bushel, R. Paules, N. Stegman, G. Nehls, K. J. Yost, C. H. Johnson, S. F. Gustafson, S. Xirasagar, N. Xiao, C. C. Huang, P. Boyer, D. D. Chan, Q. Pan, H. Gong, J. Taylor, D. Choi, A. Rashid, A. Ahmed, R. Howle, J. Selkirk, R. Tennant and J. Fostel, CEBS–Chemical Effects in Biological Systems: a public data repository integrating study design and toxicity data with microarray and proteomics data, *Nucleic Acids Res.*, 2008, **36**, D892–900.
69. R. Paules, Phenotypic anchoring: linking cause and effect, *Environ. Health Perspect.*, 2003, **111**, A338–339.
70. R. S. Thomas, S. C. Wesselkamper, N. C. Wang, Q. J. Zhao, D. D. Petersen, J. C. Lambert, I. Cote, L. Yang, E. Healy, M. B. Black, H. J. Clewell 3rd, B. C. Allen and M. E. Andersen, Temporal concordance between apical and transcriptional points of departure for chemical risk assessment, *Toxicol. Sci.*, 2013, **134**, 180–194.
71. L. Yang, B. C. Allen and R. S. Thomas, BMDExpress: a software tool for the benchmark dose analyses of genomic data, *BMC Genomics*, 2007, **8**, 387.
72. M. D. Mesarović and Case Institute of Technology, Systems Research Center., Systems theory and biology, *Proceedings of the 3rd Systems Symposium at Case Institute of Technology*, Springer, Berlin, New York etc., 1968.
73. L. Alberghina and H. V. Westerhoff, Systems biology: definitions and perspectives, Springer, Berlin, New York, 2005.
74. T. Lengauer and ACM Special Interest Group for Algorithms and Computation Theory, ACM, New York, Editon edn., 2001, p. 1 online resource.
75. M. D. Ritchie, E. R. Holzinger, R. Li, S. A. Pendergrass and D. Kim, Methods of integrating data to uncover genotype-phenotype interactions, *Nat. Rev. Genet.*, 2015, **16**, 85–97.
76. P. R. Bushel, R. D. Wolfinger and G. Gibson, Simultaneous clustering of gene expression data with clinical chemistry and pathological evaluations reveals phenotypic prototypes, *BMC Syst. Biol.*, 2007, **1**, 15.

77. P. R. Bushel, Clustering of gene expression data and end-point measurements by simulated annealing, *J. Bioinform. Comput. Biol.*, 2009, **7**, 193–215.
78. C. R. Williams-DeVane, D. M. Reif, E. C. Hubal, P. R. Bushel, E. E. Hudgens, J. E. Gallagher and S. W. Edwards, Decision tree-based method for integrating gene expression, demographic, and clinical data to determine disease endotypes, *BMC Syst. Biol.*, 2013, **7**, 119.
79. J. Gallagher, E. Hudgens, A. Williams, J. Inmon, S. Rhoney, G. Andrews, D. Reif, B. Heidenfelder, L. Neas, R. Williams, M. Johnson, H. Ozkaynak, S. Edwards and E. C. Hubal, Mechanistic indicators of childhood asthma (MICA) study: piloting an integrative design for evaluating environmental health, *BMC Public Health*, 2011, **11**, 344.
80. B. D. Bennett, Q. Xiong, S. Mukherjee and T. S. Furey, A predictive framework for integrating disparate genomic data types using sample-specific gene set enrichment analysis and multi-task learning, *PLoS One*, 2012, **7**, e44635.
81. P. R. Bushel, R. McGovern, L. Liu, O. Hofmann, A. Huda, J. Lu, W. Hide and X. Lin, Population differences in transcript-regulator expression quantitative trait loci, *PLoS One*, 2012, **7**, e34286.
82. M. Morley, C. M. Molony, T. M. Weber, J. L. Devlin, K. G. Ewens, R. S. Spielman and V. G. Cheung, Genetic analysis of genome-wide variation in human gene expression, *Nature*, 2004, **430**, 743–747.
83. E. E. Schadt, S. A. Monks, T. A. Drake, A. J. Lusis, N. Che, V. Colinayo, T. G. Ruff, S. B. Milligan, J. R. Lamb, G. Cavet, P. S. Linsley, M. Mao, R. B. Stoughton and S. H. Friend, Genetics of gene expression surveyed in maize, mouse and man, *Nature*, 2003, **422**, 297–302.
84. C. M. Perou, T. Sorlie, M. B. Eisen, M. van de Rijn, S. S. Jeffrey, C. A. Rees, J. R. Pollack, D. T. Ross, H. Johnsen, L. A. Akslen, O. Fluge, A. Pergamenschikov, C. Williams, S. X. Zhu, P. E. Lonning, A. L. Borresen-Dale, P. O. Brown and D. Botstein, Molecular portraits of human breast tumours, *Nature*, 2000, **406**, 747–752.
85. T. Sorlie, C. M. Perou, R. Tibshirani, T. Aas, S. Geisler, H. Johnsen, T. Hastie, M. B. Eisen, M. van de Rijn, S. S. Jeffrey, T. Thorsen, H. Quist, J. C. Matese, P. O. Brown, D. Botstein, P. E. Lonning and A. L. Borresen-Dale, Gene expression patterns of breast carcinomas distinguish tumor subclasses with clinical implications, *Proc. Natl. Acad. Sci. U. S. A.*, 2001, **98**, 10869–10874.
86. T. Sorlie, R. Tibshirani, J. Parker, T. Hastie, J. S. Marron, A. Nobel, S. Deng, H. Johnsen, R. Pesich, S. Geisler, J. Demeter, C. M. Perou, P. E. Lonning, P. O. Brown, A. L. Borresen-Dale and D. Botstein, Repeated observation of breast tumor subtypes in independent gene expression data sets, *Proc. Natl. Acad. Sci. U. S. A.*, 2003, **100**, 8418–8423.
87. N. Cancer Genome Atlas, Comprehensive molecular portraits of human breast tumours, *Nature*, 2012, **490**, 61–70.
88. S. Volinia and C. M. Croce, Prognostic microRNA/mRNA signature from the integrated analysis of patients with invasive breast cancer, *Proc. Natl. Acad. Sci. U. S. A.*, 2013, **110**, 7413–7417.

89. S. T. Lee, K. Chu, H. J. Oh, W. S. Im, J. Y. Lim, S. K. Kim, C. K. Park, K. H. Jung, S. K. Lee, M. Kim and J. K. Roh, Let-7 microRNA inhibits the proliferation of human glioblastoma cells, *J. Neurooncol.*, 2011, **102**, 19–24.
90. J. Kaiser, Profile: Stephen Friend. The visionary, *Science*, 2012, **335**, 651–653.
91. V. Maojo, M. Fritts, D. de la Iglesia, R. E. Cachau, M. Garcia-Remesal, J. A. Mitchell and C. Kulikowski, Nanoinformatics: a new area of research in nanomedicine, *Int. J. Nanomed.*, 2012, **7**, 3867–3890.

Subject Index

Abcb1a 226, 227
Acaa1b 227
Acadl 227
acetaminophen 5, 428, 432, 433, 469, 470
acetyl-CoA acyltransferase gene 227
2-acetylaminofluorene 132, 446
2-acetylaminoflurene (2-AAF) 466
N-acetyl-S-(1,2-dichlorovinyl)-L-cysteine (NAcDCVC) 253, 255
N-acetyltransferase 262, 263, 264
activating transcription factor 3 gene (*ATF3*) 49, 51, 57, 62, 118, 133, 136, 139, 141, 142, 143, 144
acute lymphocytic leukemia (ALL) 352, 363
acute myeloid leukemia (AML) 352, 354, 355, 359, 360, 362, 363, 363–4, 368, 369
acyl-CoA dehydrogenase gene 227
ADAM8 117
adaptive effects 437
    adverse effects and, discriminating 437
    benchmark doses and 437
    benzene 356
    furan 406
adverse responses (adverse health effects/outcome pathways) 6, 18, 437–8, 461–2
    furan 396, 399, 403, 406, 412–13
    lowest observed adverse effect level (LOAEL) 155, 156, 185, 198, 424
    mode of action *see* mode of action
    no observed adverse effect level (NOAEL) 155, 156, 424
    noncancer, with trichloroethylene 255–7
    suitability of gene expression studies for 201–2
AEN 124, 142, 144
aflatoxin B1 131, 397–8
agricultural azoles, *C. albicans* inhibition 288
AHSP 127, 129, 141
Akr1b7 307, 321
AKT1 122
ALAS2 127, 129, 141
albendazole 134
aldehyde dehydrogenase-1 family (ALDH-1), subfamily A7 321
aldehyde dehydrogenase-2 (ALDH-2) 262, 263, 264
aldo-keto reductase family 1 (*Akr1b7*) 307, 321
*O*6-alkylguanine DNA alkyltransferase gene (*O*-methylguanine DNA methyltransferase gene; *MGMT*) 116, 120–1, 131, 200
all-*trans*-retinoic acid (atRA) and propiconazole 320, 322, 326, 328
alpha globin gene cluster (HBA1) 127, 129
Ames (*Salmonella*) test 22, 59, 66, 92, 114, 174
amiodarone 431, 433

amitrole 281, 303
AML12 cells, conazole studies 318, 322, 323
analysis of variance (ANOVA) 461
transcriptomic dose–response data 156
androgen receptor 448
androstane receptor, constitutive (CAR) 293–4, 298, 300, 307, 316, 320, 327, 330, 333, 334
angiogenesis, pathogenic 213–14
animal toxicity studies 5–6, 14, 15, 18
developmental toxicity 237
extrapolation to human health risks 18
reducing use and finding alternatives 393–6
risk assessment 18
ANOVA, transcriptomic dose–response data 156
antifungals, azole *see* conazoles
antimitotic agents 50, 59–61
apical endpoints 169, 426, 427, 437
propiconazole 325
apoptosis, increased resistance to 213
apoptosis-enhancing nuclease gene (*AEN*) 124, 142, 144
aromatic hydrocarbons
halogenated 167
polycyclic 259, 263
aryl hydrocarbon receptor (ARH) 167, 168, 174, 175, 452
Aspergillus fumigatus and conazoles 286, 287
ataxia telangiectasia mutated (ATM) 116, 188
ATF3 and 49, 51, 57, 62, 118, 133, 136, 139, 141, 142, 143, 144
ATM (ataxia telangiectasia mutated) 116, 188
ATP6V1D 123, 131
attractor hypothesis 460–1
azole antifungals *see* conazoles

BALB/c 3T3 cells (mouse), predictive *in vitro* toxicogenomic studies 86
BAX 116, 118, 141, 144
BCL2-associated X protein gene (*BAX*) 118
benchmark dose (BMD)
method 156–73, 436–7, 466–7
benzene 360–1
furan 402
integrating dose–response 197–9
point of departure values *see* point of departure values
benzene (and occupational exposure) 352–89
application of toxicogenomics 352
transcriptomic studies 353, 354, 356–68
benzo[*a*]pyrene (BaP) 173, 174
betacellulin gene 228
biclustering methods 63, 451–2
Big Blue mouse 114
conazole studies 319, 326
bile (inc. bile acids)
stasis *see* cholestasis
synthesis 321, 323, 396–7
Biobase Transfac and Transpath 453, 454
biochemical assays in high-throughput screening 12
bioinformatics (and computational biology) 11, 442–83
benzene-related gene fusions 363
definition 442
for integrative genomics 468–73
biological process in gene ontology 457
biological significance, assessing 427
biologically-based dose-response (BBDR) models 16
biomarkers (signatures; fingerprints) 10–11, 391–2
benzene toxicity 354
considerations in identification of 47–8
definition 10–11

biomarkers (*continued*)
gene expression *see* transcriptomics
genomic/genotoxic 39–75, 78, 80–97, 113–53, 191–5
comparing signatures for chemicals causing different types of DNA Damage 191–5
"general" 116–47
in vitro 80–97
in vivo 113–53
molecular epidemiology and 259–60
non-genotoxic chemicals 113–53
pathway analysis *see* pathways
biomonitoring equivalents 16–18
birth defects and environmental exposures or pharmaceuticals 237
bitertanol 287, 288, 304
bladder (mouse), 2-acetylaminoflurene effects 466
bleomycin 42, 49, 144
blood (and blood cells) 140–3
benzene and 353, 354, 355, 365–8
mononuclear cells *see* peripheral blood mononuclear cells
bone marrow 140–3
suppression 127–9
signature (BSS) 129, 133, 136, 137, 138, 139, 142, 143, 147
breast cancer 322, 452, 454, 463, 466, 471, 472, *see also* mammary tissue
Broad Institute's Molecular Signatures Database 459, 460
bromobenzene 405
Btc 228
BTG2 116, 121, 131, 144

cis-2-butene-1, 4-dial (BDA) 401, 405
caffeine (tri-methylxanthine) 57, 59, 66
Cancer Genome Atlas, The (TCGA) 472–3
Candida and conazoles 286, 287, 288
carbon tetrachloride 131, 210, 399, 403, 405, 450, 452
carboxylesterase 1 gene 7
carboxylesterase 6 gene 227
carcinogenicity, predictive (basics) 21–6
cardiac genotoxicity signatures 137–40
CBFA2T3–GLIS2 gene fusion 363
Ccl5 228
CCNG1 116, 118, 141, 142, 144
CDK *see* cyclin-dependent kinase; cyclin-dependent kinase inhibitor 1A
cell(s)
apoptosis, increased resistance to 213
in gene ontology, cellular component 457
mammalian cell types 459–60
proliferation *see* proliferation
senescence 214
signaling *see* signaling
cell-based systems/assays (incl. cell cultures) 258
genomic biomarkers 39–75
in high-throughput screening 12
human embryonic stem cells 238–9
in vitro predictive gene expression signatures for differentiating genotoxic mechanisms in 81, 82–91
cell cycle 43, 133, 162, 196, 221–4
transcription factors regulating 453
cell lines (and their choice) 96
propiconazole mode of carcinogenic action studied in 315, 323, 328
TGx-28.65 studies 48–9
Ces1 7

Ces6 7
Chemical Abstracts Service (CAS) 39
chemical classes 20–1
ChemicalAtlas 403
chemokines 228
chloral hydrate 257
chlorinated solvents 251–7
β-chloroprene in rodent lung 162–3
cholestasis 428–9
cyclosporin-induced 218, 226, 429, 431, 433
FK506-induced 218
cholesterol and conazoles 282, 300–2, 316, 321–4, 327, 328, 333–5
chromosome aberrations (structural and numerical) and their detection 259
benzene 354, 363
conazoles 302, 303, 314
in vitro 40, 57, 59, 65, 66, 302
chronic lymphocytic leukemia (CLL) 355, 368
CIBERSORT 367
CIITA–DEXI gene fusion 363–4
Cip1 (*CDKN1A*) 43, 57, 116, 119
circular RNA (cRNA) 356, 367
cis-regulation of a transcription regulator 471
classification
chemicals 20–1
tumors, and classifiers 445–53
climbazole 295
clotrimazole 281, 286, 288, 290, 291, 293, 294, 295, 301, 328
clusters/clustering of genes 53–4, 63–5, 97
biclustering methods 63, 451–2
bioinformatics and 443, 446–7, 448–52, 469, 472, 473
hierarchical clustering 52, 53, 193, 403, 411, 446, 471
CMAP (Connectivity Map) 65, 365
co-expressed genes 448–50, 451, 455, 471
coffee, furan in 399
coherent co-expression biclustering (cc-biclustering) 451
colon (mouse), signatures of genotoxicity 144
comparison methods for non-genotoxic carcinogen identification 215–28
case study 217–28
complex mixtures of agents 14
computational biology *see* bioinformatics
conazoles/azole antifungals (incl. imidazoles and triazoles) 280–351
genotoxicity 302
historical development 280–1
P450 enzymes and *see* cytochrome P450 enzymes
tumorigenic effects 302–6
concentration
in comparison approach with cyclosporin and FK506
range 219–20
selection 218
highest appropriate for *in vitro* studies 96
concordance model for toxicogenomics-based extrapolation 431–5
congenital (birth) defects and environmental exposures or pharmaceuticals 237
Connectivity Map (CMAP) 65, 365
costs (financial) of environmental regulations 175
coupled two-way clustering (CTWC) 53, 65
cross-species extrapolation/comparisons/differences *see* species
Cryptococcus neoformans and conazoles 286, 297
Csad 227
curcumin 188, 196, 197, 198
Cxcl2 228
Cxcl10 228
Cxcl16 357
cyclin-dependent kinase(s) 118

cyclin-dependent kinase inhibitor 1A (*CDKN1A*; *Cip1*; *p21*; *Waf1*) 43, 57, 116, 119
    furan and 407
cyclin G1 gene (*CCNG1*) 116, 118, 141, 142, 144
cyclosporin A 216–28, 229, 230, 429, 433
cyproconazole 286, 287, 288, 295, 298, 300, 301, 303, 304, 308–9, 330
cysteine sulfinic acid decarboxylase gene 227
cytochrome P450 enzymes (CYP) 281–300
    conazoles and 281–300, 317, 318
        induction of CYPs 293–4, 316–17, 328
        inhibition of CYP51 281–93, 320
    furan and 397
        CYP2E1 401, 403, 405, 412
    non-genotoxic carcinogens and 211–12, 226, 227
    polymorphisms 262, 263, 264
    trichloroethylene and 251–3

2,4-D (2,4-dichlorophenoxyacetic acid) 267
daunorubicin 138, 144, 145
DDB2 119–20
DDT 267
debrisoquine 261–2
demethylation (and demethylase), lanosterol 282, 284, 287, 306, 321
developmental and reproductive toxicity 14, 237–8, 239, 240, 425
    conazoles 334
DEXI–CIITA gene fusion 363–4
DF (transcription factor) 455
DHH–RHEBL1 gene fusion 363
diazinon 267
diazoxon 267
dichloroacetate (DCA) 251, 253, 256, 257
dichlorobenzene 168, 403, 451
dichlorodiphenyltrichloroethane (DDT) 267
2,4-dichlorophenoxyacetic acid (2,4D) 267
S-(1,2-dichlorovinyl) glutathione (DCVG) 253, 255
di(2-ethylhexyl)phthalate (DEHP) 264, 267
difenoconazole 303, 304, 306
differentially expressed genes (DEGs) 218, 221, 450, 455, 457, 458, 459
DiseaseAtlas 403
DNA 7
    adducts 40, 41
        conazoles 319–20, 326, 327
        furan 398, 401
    agents interacting with 40–1
    damage 23, 40–2, 44, 186–202, 465, *see also* genotoxicity
        agents causing 40–2, 80, 92, 191–5
        biomarkers *see* biomarkers
        dose–response 186–202
    methylation 472, 473
        benzene and 355, 369
    repair (of damage) 186, 199, 200–1
        centers 200–1
        mechanisms 188
    transcription *see* transcription
dose (and dosimetry) 6, 16–18
    maximum tolerated dose *see* maximum tolerated dose
    signatures of genotoxicity and 145–6
    thresholds *see* thresholds
    transcriptomics 185–208
        benchmark dose method *see* benchmark dose
        TGx-28.65 transcriptomic biomarker 49
dose–response relationships (and their assessment) 14, 424, 426–7, 464–5

biologically-based dose-response (BBDR) models 16
transcriptomic analysis 154–84, 391–2
example chemicals 160–6, 360–1
in mode of action assessment 159–66
traditional statistical methods 156
downstream target genes (DSTs) 450, 453–5, 467, 471
doxorubicin 135, 137, 138, 139, 141, 144
Dppa5a and furan 407
duration *see* timescale

E2F 308, 309, 455
early growth response gene (*EGR1*) 120
early integrative genomics 468–9, 472
econazole 281, 286
economic costs of environmental regulations 175
ectodermal-neural cortex 1 gene (*ENC1*) 126, 131, 138, 143
ectodysplasin A2 receptor gene (*EDA2R*) 123, 131
ED01 study 466
EDA2R 123, 131
EGF receptor family 228
EGR1 120
embryonic stem cells 236–46
human (hES) 236–46
HES-Hep in predictive *in vitro* toxicogenomic studies 83
xenobiotic impact on genome and epigenome of 236–46
mouse, non-genotoxic carcinogen identification 216–17
ENC1 126, 131, 138, 143
endoplasmic reticulum (ER)
ingenuity pathway analysis 64
stressors 54, 65
non-genotoxic carcinogens 217, 224, 227
TGx-28.65 48, 54, 64
enriched gene sets 453–67
environment
complex mixtures in (exposure to) 14
diseases relating to 236–7
gene interactions with 14
regulations *see* regulations
Environmental Protection Agency (EPA) ToxCast 2, 3, 14–15, 98, 267, 294–5, 413
epidermal growth factor (EGF) receptor family 228
epigenetics 14, 19, 237, 241, 465
epigenomes 8, 236–46
human embryonic stem cells, xenobiotic impact 236–46
epirubicin 134, 138
epoxiconazole 281, 286, 287, 303, 304, 306, 308–9
Epstein–Barr virus (EBV) infection and benzene exposure 364
ErbB signaling pathway 450
Erk 323, 324
estrogen receptor (Esr)
antagonism 462
Esr1 (estrogen receptor 1) 308
etaconazole 303, 304, 306, 308
etoposide 138, 188, 196, 197, 198, 369
exclusive group correlation score (EGCS) 455
exposomics 369
exposure assessment 13, 424
extracting patterns and identifying co-expressed genes (EPIG) method 450, 461

farnesoid X receptor (FXR)
conazoles and 316, 333
non-genotoxic carcinogens and 221, 226, 227
farnesylation of Ras 323, 326

FAS 117
Fbp2 227
FDXR 120
female cancer incidence by site 445
fenbuconazole 295, 298, 300, 303, 304, 306
fenofibrate, receptor-mediated non-genotoxic carcinogenesis 165–6
ferredoxin reductase gene (*FDXR*) 120
financial costs of environmental regulations 175
fingerprints *see* biomarkers
FK506 (tacrolimus) 216–28, 229, 230
fluconazole 281, 284, 285, 286, 287, 288, 289, 290, 291, 292, 294, 298, 302, 303, 330
flusilazole 281, 288, 303, 304
food, furan in 398–9
formaldehyde exposure, nasal epithelium (rat) 160–2
formalin-fixed and paraffin-embedded (FFPE) toxicogenomics 391
    furan and 408–9, 411
FOSB 120
fructose biphosphatase 2 gene 227
fungicides, conazole *see* conazole
furan 94, 174, 396–415
FusionMap 364
Fyn 227

Gadd45a 49, 57, 63, 125
    conazoles and 307
Gadd45b and conazoles 307
gap junctional intercellular communication and non-genotoxic carcinogens 211
GAS5 123–4
GDF15 121, 131, 144
gel-based proteomics 10
gene(s)
    clusters *see* clusters
    differentially expressed (DEGs) 218, 221, 450, 455, 457, 458, 459
    downstream target (DST) 450, 453–5, 467, 471
    enriched sets of 453–67
    environmental interaction with 14
    expression 443–73, *see also* transcription
        benzene exposure and 357–60
        biomarkers (=genomic biomarkers) 39–75, 78
        co-expression 448–50, 451, 455, 471
        correlating 448–50, 454–5
        profiling *see* transcriptomics
        variability/variation in 14, 19, 228
    fusions, benzene exposure and 363–4
    in "general" signature of genotoxicity 117–26
    in non-genotoxic carcinogenesis 211–30
    polymorphisms *see* polymorphisms
    regulatory networks (GRNs) 463, 464
Gene Ontology (GO) 146, 158, 160–1, 169, 171, 193, 456–7
gene set enrichment algorithm/analysis (GSEA) 65, 215, 220, 221, 224, 226, 227, 359, 458, 460–1
GeneGo MetaCore 220, 222–5, 226, 308, 331
genetic instability 214
genetic variation/variability 19, 261
genome
    biomarkers *see* biomarkers
    human embryonic stem cells, xenobiotic impact 236–46
genomics (genomic technology and methods) 7–12, 114–15
    bioinformatics and computational biology for integrative genomics 468–73

conazoles 306–9
mouse liver 306–9, 328–34
rat liver 313
rat thyroid 313–14, 408, 410–11
furan 401–2, 411
genotoxicity (genotoxic chemicals) 22–3, 25, 26, 40–4, 48–67, 79–87, 113–53, 185–208, 465, *see also* DNA, damage
adverse responses *see* adverse responses
biomarker/signatures of *see* biomarkers
dose–response, and use of transcriptomics to evaluate thresholds 185–208
low-dose, evaluation methods 189–90
mechanisms of action 40–2, 80–93
stress responses *see* stress
toxicogenomic predictors 80–97
rationale for the need 70
traditional tests 39–40, 79, 115
glioblastoma 473, 474, 475
GLIS2–CBFA2T3 gene fusion 363
global transcriptional analysis 51–3
globin gene loci/clusters
alpha (*HBA1*) 127, 129
beta (*HBB1*/*HBA2*) 127, 129
glutathione (GSH) 199
formaldehyde and 161, 163
furan and 401
naphthalene and 165
trichloroethylene and 253, 255, 256
glutathione reductase genes 226
glutathione transferase (GST) genes 226, 253–4
conazoles and 318
furan and 407
polymorphisms 262
glyphosate 267
gold standard 19, 21, 390
groove-binding agents 41, 42
group correlation score 455–6
growth arrest-specific-5 gene 123–4
growth differentiation factor 15 gene (*GDF15*) 121, 131, 144
Gsr 226
Gst/*GST* *see* glutathione transferase

halogenated aromatic hydrocarbons 167
hazard
assessment *see* risk assessment
identification 326, 424
screening 13
HBA1 and *HBA2* 127, 129
HBB 127, 129
health (human)
adverse responses *see* adverse responses
risk assessment *see* risk assessment
Health and Environmental Sciences Institute (HESI) 25, 26, 79, 93, 94, 97
heart, signatures of genotoxicity 137–40
heatmap, transcriptome 52, 53, 54, 55, 59
hematotoxicity, benzene 353, 367
hepatic... *see* liver *and entries below*
hepatocellular tumors (adenomas and carcinomas) 463
conazoles and 334–5, 337
furan and 397
hepatocyte(s)
human primary, dose–response analysis 167
human primary, predictive *in vitro* toxicogenomic studies 95
surrogates (HepaRG/HepG2 etc.) 81, 82, 83, 84, 85, 88, 90, 91, 92, 93, 95
mouse
conazoles effects 300, 308–9, 315, 321–4, 326, 327, 329, 329–30

hepatocyte(s) (*continued*)
non-genotoxic carcinogen identification 216–17, 218, 219, 220, 224, 228–30
predictive *in vitro* toxicogenomic studies 82, 86, 87
rat *see* rat
hepatocyte nuclear factor 4-a gene (*Hnf4a*) 308, 309
hES-Hep, predictive *in vitro* toxicogenomic studies 83, 85
hexaconazole 281, 287, 288, 303, 304
hierarchical clustering 52, 53, 193, 403, 411, 446, 471
high-throughput screening (HTS) and data analysis 2, 11–12, 265–7
in vitro 13, 14, 97, 176
relative role 12–15
Hill model 158
histone modification inhibitors 50
deacetylase (HDAC) inhibitors 52, 64, 65
Hnf4a 308, 309
homeostasis 18, 199–201
hormone receptors, nuclear *see* nuclear hormone receptors
hsalet-7 microRNAs 473
human(s)
CYP51 compared with fungal CYP51 287–8
embryonic stem cells *see* embryonic stem cells
health risk assessment *see* risk assessment
hepatocytes *see* hepatocytes
propiconazole mode of carcinogenic action and its relevance 327–34
relevance of toxicogenomic data 427–35
signatures of genotoxicity
bone marrow and blood 141–3
heart 139–40
kidney 137
liver 135
skeletal muscle 140
spleen 143
other tissues 144
transcriptome, complexity and analysis 356–7
variability (inter-individual) in 14, 19, 261–4
Human Genome Project 7
3-hydroxy-3-methylglutaryl-CoA (HMG) reductase and conazoles 282, 301, 323, 328
hypertrophy, hepatic 294, 298, 299, 312, 327, 329, 330

IARC *see* International Agency for Research on Cancer
Igf1 gene 227
imatinib 133, 134, 136
imazalil 294, 295, 296, 298, 304
imidazole antifungals *see* conazoles
immunosuppression 213
drugs inducing (immunosuppressants), as non-genotoxic carcinogens 216–28
in silico assessment methods 413
genotoxic chemical 114
non-genotoxic chemicals 215
in vitro toxicity assays 76–112, 258–9, 394–5, 413–14
conventional 5
limitations 395
toxicogenomic 3, 4, 15, 76–112
chromosome aberration assays 40, 57, 59, 65, 66, 302
high-throughput screening (HTS) 13, 14, 97, 176
in vivo toxicity assays 3, 4, 113–53, 395–6, 414
conventional 5, 6
genotoxic chemicals 113–53
micronucleus assay 66, 141

non-genotoxic chemicals 113–53
organs studies 24, 129–44
incidence rate by site of cancer 443–4
infection and benzene exposure 364–5
ingenuity pathway analysis (IPA) 53, 405
conazoles 307–8, 331
mode of action and 405, 411
TGx-28.65 and 63, 64
insulin-like growth factor 1 gene 227
integrative genomics, bioinformatics and computational biology for 468–73
intercalating agents, DNA 41, 42
intermediate integrative genomics 468, 469
International Agency for Research on Cancer (IARC) 22
Monographs Programme 247–79
non-genotoxic carcinogens 210, 215
intestine (colon or jejunum) of mouse, signatures of genotoxicity 144
isoniazid 261–2
isopropyl methanesulfonate (iPMS) 57, 59

jejunum (mouse), signatures of genotoxicity 144
Jun 308, 309, 314, 357, 398, 450

KEGG pathways 128, 158, 166, 432
ketoconazole, structure 281
kidney, signatures of genotoxicity 135–7
Kyoto Encyclopedia of Genes and Genomes (KEGG) pathways 128, 158, 166, 432

L1000 approach 357, 358, 365–6
L15178Y cells (mouse), predictive *in vitro* toxicogenomic studies 88, 89, 90, 91
lanosterol 281, 284, 301, 306, 321
late intermediate integrative genomics 468, 469, 472
let-7 microRNA 473
leukemia (and benzene) 354, 363, 364, 368, 369
acute lymphocytic 352, 363
acute myeloid 352, 354, 355, 359, 360, 362, 363, 363–4, 368, 369
chronic lymphocytic 352, 363
Library of Integrated Network-based Cellular Signatures (LINCS) program 365–6
lindane 267
lipids and conazoles 300–2
literature trees with mechanistic data 264
liver 129–35
acetaminophen effects 428, 433, 434
2-acetylaminoflurene effects 466
conazole effects 294–300, 303–34
furan effects 396–414
hypertrophy (hepatic/hepatocellular) 294, 298, 299, 312, 327, 329, 330
mouse *see* mouse
necrosis *see* necrosis
peroxisome proliferation mechanistic event in 266
physiology and carcinogenesis 396–8
rat *see* rat
regeneration 397, 398, 401, 403, 405, 406, 407
signatures of genotoxicity 129–35
steatosis 429, 431, 432, 433
Virtual Liver project 14
liver X receptor (LXR) and conazoles 316, 333
lobules of liver 396
location (site) of tumor, incidence rates by 443–4

long non-coding RNA 9, 241–3
benzene exposure and 356, 360
furan exposure and 407
stem cells and 239, 241, 242
long-timescale *in vitro* assays (chronic testing) 4, 414
low-dose effects
genotoxicity, evaluation methods 189–90
risk assessment and the challenge of 354–5
lowest observed adverse effect level (LOAEL) 155, 156, 185, 198, 424
lung, rodent (mouse and/or rat), *see also* respiratory system
β-chloroprene 162–3
signatures of genotoxicity 144
tumors 163, 164, 447
Lutz (and Lutiz) model 189–90, 191, 196
Ly6d 226
lymphocyte antigen 6 complex, locus D 226
lymphocytic leukemia
acute (ALL) 352, 363
chronic (CLL) 355, 368
lymphoma and benzene 352, 364, 368

malaoxon 267
malathion 267, 403
male cancer incidence by site 445
mammals (other than humans)
cell culture systems *see* cell-based systems/assays
cell types 459–60
CYP inhibition by conazoles 288–93
mammary tissue, mouse, signatures of genotoxicity 144
MAPK (mitogen-activated protein kinase)
ciclosporin and FK506 and *Mapk4* 217, 226
conazoles and 308, 323, 326
MAQC (MicroArray Quality Control) consortium 97, 358, 447
margin of exposure (MoE) and furan 399, 402, 413, 414
mass spectrometry (MS) in metabolomics 10
Matthews correlation coefficient (MCC) 447
maximum enrichment scores (MESs) 458
maximum tolerated dose
benchmark dose method in transcriptomics 169
signatures of genotoxicity and 145
MDM2 116, 120, 131, 138, 142, 144, 145, 186
melphalan 365
men (males), cancer incidence by site 445
messenger RNA *see* RNA
metabolic activation 93–5
metabolism (in general)
inter-individual variation in 261
liver 396–7
metabolomics 7, 10–11
conazole-treated mouse liver samples 310–12
MetaCore (from GeneGo) 220, 222–5, 226, 308, 331
metastasis 214
methapyriline, receptor-mediated non-genotoxic carcinogenesis 165–6
methyl methanesulfonate (MMS) 46, 53, 198
methylation of DNA *see* DNA
3-methylcholanthrene 452, 457
methylene chloride 168, 169
*O*6-methylguanine DNA methyltransferase gene (*O*-alkylguanine DNA alkyltransferase gene; *MGMT*) 116, 120–1, 131, 200
mevalonic acid and conazoles 323–4, 326, 328
MGMT 116, 120–1, 131, 200

mice *see* mouse
miconazole 280, 281, 286, 288, 289, 290, 291, 292
microarray(s) 8–9, 154–5
benzene exposure 356–7, 357–61, 369
RNA-seq's advantage over 361
biomarkers of cancer risk in molecular epidemiology 259–60
cyclosporin and FK506 218–19
furan exposure 410
overcoming limitations 155
MicroArray Quality Control (MAQC) consortium 97, 358, 447
micronucleus assay 187, 188, 190–1
conazoles 302, 303, 314
dose–response 190–1
in vivo 66, 141
microRNA *see* RNA
microtubules 121
inhibitors 49, 64
minimally effective dose *see* thresholds
mitogen-activated protein kinase *see* MAPK
mitosis inhibitors (antimitotic agents) 50, 59–61
mixtures of agents 14
mode of action (mechanism of action; MoA; mechanistic toxicogenomics) 13, 78, 392, 413–14
benzene 354
biological parameters in prediction of 42
conazoles 314–34
CYP51 inhibition 284–5, 320–1
human relevance 327–34
furan 399–402, 405–7
gene ontology and 457
genotoxicity 40–2, 80–93
information 13
non-genotoxic chemicals *see* non-genotoxic chemicals
novel and multiple data streams 247–79
critical information provided by 247–57
future directions 274–7
transcriptomic analysis 159–66
triage 13
unknown 436–7
ModKPrototypes 469, 470
molecular concepts map of cholestasis 430
molecular epidemiology, biomarkers of cancer risk 259–60
molecular function in gene ontology 457
Molecular Signatures Database (Broad Institute) 459, 460
Monographs Programme, IARC 247–79
mouse
2-acetylaminoflurene and the liver and bladder of 466
Big Blue *see* Big Blue mouse
β-chloroprene in lung of 162–3
conazoles and the liver 300, 303–34
furan and the liver 399–401, 402
non-genotoxic carcinogen identification 216, 217–18
phenobarbital and the liver 328–34
predictive *in vitro* toxicogenomic studies
BALB/c 3T3 cells 86
hepatocytes 82, 86, 87
L15178Y cells 88, 89, 90, 91
signatures of genotoxicity
bone marrow and blood 141
heart 138–9
kidney 137
liver 133
skeletal muscle 140
spleen 143
other tissues 144
t,t-muconic acid (*t,t*-MA) 359–60

multiple data streams, integrating across 247–57
multiple myeloma 352, 355, 447, 449
muscle (skeletal), signatures of genotoxicity 140
mutagenicity assessment 247–79
Myc 43, 308, 309, 311, 324, 403, 450
myclobutanil 286, 287, 288, 294, 296, 298, 300, 301, 303, 304, 306, 307, 309, 312, 313, 319, 326
myelodysplastic syndrome (MDS) and benzene 352, 354, 362, 369
myeloid leukemia, acute (AML) 352, 354, 355, 359, 360, 362, 363, 363–4, 368, 369
myeloma, multiple 352, 355, 447, 449

Nanostring's nCounter 357, 358, 365, 366, 370
naphthalene 168
    nasal epithelium (rat) 163–5
nasal epithelium (rat)
    formaldehyde exposure 160–2
    naphthalene exposure 163–5
National Research Council (US), *Toxicity Testing in the 21st Century: A Vision and a Strategy* 5, 189, 230, 413, 414
National Toxicology Program (NTP), US 2, 4, 20, 21, 24, 25, 391
    furan 399, 409
    olfactory and respiratory epithelium 165
nCounter (Nanostring's) 357, 358, 365, 366, 370
necrosis, hepatic 429, 473
    conazoles and 329, 399
Needleman–Wunsch dynamic programming algorithm 442
neoangiogenesis, pathogenic 213–14
neocarzinostatin 188, 191, 195, 198, 200
neuregulin-1 gene 228
neuroblastomas 164, 165, 447
next generation sequencing 409–10
    whole-transcriptome *see* RNA-seq
NextBio 116, 403
    furan 403
    furan exposure 403
    human blood 142
    human kidney 137
    human liver 135
    mouse blood 141
    mouse heart 139
    mouse kidney 137
    rat bone marrow 141
    rat heart 138
    rat kidney 135
    rat liver 403
NH32 cells 45, 61, 87
NHEJ1 124
3-nitropropionic acid (3-NP) 57, 59
no observed adverse effect level (NOAEL) 155, 156, 424
no observed transcriptional effect level (NOTEL) 466
non-coding RNA *see* RNA
non-genotoxic chemicals 22–4
    defining 210
    identification methods 215–28
    in vivo signatures 113–53
    modes of action 209–35
        receptor-mediated *see* receptor-mediated non-genotoxic carcinogenesis
    regulations 209–10
nonhomologous end-joining factor-1 124
NRF2 401, 402, 405, 407, 457
Nrg1 228
nuclear hormone receptors and conazoles 293–4, 306, 309, 313, 333
    receptor activation 316, 326
nuclear magnetic resonance (NMR) spectroscopy in metabolomics 10
Nupra 227–8

occupational (workplace) exposure 249, 352–89
    benzene *see* benzene
    trichloroethylene 251

OECD *see* Organisation for Economic Co-operation and Development
olfactory epithelium, naphthalene exposure 164–5
-omics technologies 1–12
ONCOMINE 459
one-*vs.*-all (OVA) support vector machine (SVM) 445
ontology, gene (and Gene Ontology/GO) 146, 158, 160–1, 169, 171, 193, 456–7
oral mucosa, human, signatures of genotoxicity 144
Organisation for Economic Co-operation and Development (OECD) guidelines 20
  adverse outcome pathways 461
  genotoxicity 80, 96
organs and tissues
  incidence rates for cancer involving various 443–4
  target *see* target organs and tissues
oxidative stress 196, 212, 398, *see also* reactive oxygen species
  conazoles and 317–18
  furan and 401, 402, 405, 407
oxon metabolites 267

P-glycoprotein gene (*Abcb1a*) 226
p21 (*CDKN1A*) 43, 57, 116, 119
p53 (TP53 gene) 43, 122, 128, 186, 187, 455
  dose–response 191, 193, 196, 198
  TGx-28.65 and 49, 53, 57, 60, 61, 63, 64
P450 enzymes *see* cytochrome P450 enzymes
paracetamol (acetaminophen) 5, 428, 432, 433, 469, 470
parallelogram approach 427–31
paraoxon 267
paraoxonase-1 gene polymorphisms (and paraoxon) 262
parathion 264, 267
pathways (detection and analysis) 46–7, 63–4, 173–4, 258–9, 436–7
  adverse outcomes *see* adverse responses
  benzene 357–60
  conazoles 306–35
  concordance model and 431–5
  cyclosporin 220–4
  FK506 220–4
  furan 407
  ingenuity *see* ingenuity pathway analysis
  over-represented pathways 444, 453–67
  parallelogram approach and 428–31
  steatosis 432
  transcriptional benchmark dose method in 169–73
Pdia3/4/6 A3/A4/A6 227
penconazole 281, 287, 288, 305
perfluorooctanoic acid (PFOA) 265–7
peripheral blood mononuclear cells
  benzene and 357, 359, 361–2, 364, 366, 367, 369
  predictive *in vitro* toxicogenomic studies 84
peroxisome proliferation mechanistic event in liver 266
peroxisome proliferator-activated receptors (PPARs) 459
  PPARα 166
    conazoles and 316
  PPARγ 459
    conazoles and 314
pesticide genotoxicity 302
PF4 357
pharmaceutical azoles, *C. albicans* inhibition 288
phase-shift analysis of gene expression 452
phenobarbital, mouse liver responses 328–34
phenotypic anchoring 18–19
phenotypic-directed analysis in steatosis 431

PHLDA3 122
phospholamban gene (*PLN*) 121, 131
physiologically based pharmacokinetic (PBPK) models 16, 160, 163
tetrachloroethylene 256, 261
plasminogen activator gene (*Plat*) 228
PLN 121, 131
point of departure (PoD) values 155, 156, 158, 164, 169, 171, 172, 173, 174, 197, 424, 426, 427, 436
benzene 360, 361, 368, 369
furan 402, 403, 411, 413, 414
polo-like kinase 2 gene (*PLK2*) 122, 144
polyacrylamide gel electrophoresis (PAGE)-based proteomics 10
polyamine biosynthesis and conazoles 311
polycyclic aromatic hydrocarbons 259, 263
polymerase chain reaction 8
polymorphisms 262–4
single-nucleotide 471
PON1 polymorphisms 262, 263
postgenomic technology 7–12
post-translational processes 323
DNA damage and 189, 199–201
PPAR *see* peroxisome proliferator-activated receptor
PRDM16–SKI gene fusion 364
predictive toxicology/toxicogenomics (basic references only) 15, 78–9, 392, 445–53
additional considerations for experimental design 95–7
carcinogenicity 21–6
classifiers in 445–53
furan 403–5
genotoxicity *see* genotoxicity
pregnane X receptor (PXR) 293–4, 307, 313, 316, 327, 328, 333
prenylation of proteins 323
principal component analysis 313, 332, 411, 434, 470, 474
microarray analyses 218–19
transcriptomic and micronucleus endpoints 196
prioritization exercises, modeling aiding 260–1
prochloraz 281, 287, 288, 305, 306
progression stage of carcinogenesis 465
proliferation (cell)
hepatocyte
conazoles and 300, 308–9, 321–4, 326, 327, 329–30
furan and 397, 398, 399, 401, 407
with non-genotoxic carcinogens 211
propiconazole 298, 306–34
cell proliferation and 300
cholesterol and triglycerides and 301
genotoxicity 302
mode of carcinogenic action 314–34
human relevance 327–34
structure 281
toxicogenomic mouse studies 306–27
tumorigenicity 299, 303, 305, 306
other (minor) mentions 286, 287, 288, 294, 296, 297, 299
propylene glycol mono-*t*-butyl ether 168
prostate cancer 443, 448
conazoles and 300, 311, 328
protein
prenylation 323
synthesis, basic process 7
protein disulfide isomerases A3/A4/A6 genes 227
proteomics 7, 9–10
conazole-treated mouse liver samples 309
prototypic compounds 20–1
pseudouridine levels (hepatic) and conazoles 311–12
psoralens 42

quantitative structure–activity response (QSAR) models 215, 260
quantitative toxicogenomics 392, 426, 427, 435–6, 438
  furan 402–3
quercetin 188, 191, 198

Ras 313, 323, 324, 326
rat
  bone marrow and blood, signatures of genotoxicity 141
  heart, signatures of genotoxicity 138
  hepatocytes
    dose–response analysis 167
    predictive *in vitro* toxicogenomic studies 84, 91, 93
  kidney, signatures of genotoxicity 135–7
  liver 23, 130–3
    conazole effects 286, 306, 313, 317, 324
    furan effects 401, 402, 403, 407
    S9 preparations, predictive *in vitro* toxicogenomic studies 81, 82, 84, 85, 94
    signatures of genotoxicity 130–3
  nasal epithelium *see* nasal epithelium
  respiratory system (incl. lung)
    β-chloroprene 162–3
    signatures of genotoxicity 144
  skeletal muscle, signatures of genotoxicity 140
  thyroid, conazole effects 312, 313–14
REACH 391, 423, 425, 427, 436, 438
reactive oxygen species (ROS), *see also* oxidative stress
  conazoles and 307, 317–18, 319, 326, 328
  furan and 401, 405
receptor-mediated non-genotoxic carcinogenesis 211
  fenofibrate and methapyriline and 165–6
reduction (in 3Rs animal toxicity testing) 393, 396
refinement (in 3Rs animal toxicity testing) 393, 396
regulations (environmental)
  financial costs 175
  high-throughput screening and 5
  non-genotoxic chemicals 209–10
replacement (in 3Rs animal toxicity testing) 393, 396
reproductive toxicity *see* developmental and reproductive toxicity
respiratory nasal epithelium, naphthalene exposure 164–5
respiratory system, signatures of genotoxicity 144, *see also* lung
retinoid X receptor (RXR) 293, 309, 316)
retinol/retinoic acid metabolism and propiconazole 320
RHEBL1–DHH gene fusion 363
risk assessment (hazard/safety assessment) 13–14, 76, 247–79, 423–41
  animal studies 18
  biomarkers in *see* biomarkers
  challenges with potential genotoxicants 185–7
  four aspects (processes) of 423–4
  furan 409–14
  non-genotoxic carcinogens 209–10
  novel data streams (for mutagenicity and carcinogenicity) 247–79

risk assessment (*continued*)
transcriptomic analysis 159, 166–75
benzene and other occupational exposures 353, 354–5, 368
compared with traditional methods 173–4
cross-species extrapolation 166–8
significance of integrating data from 175–6
TGx-28.65 biomarker 57–67
risk characterization 424
RNA
circular (cRNA) 356, 367
messenger, *see entry below*
microRNA (miRNA) 241–3, 473
benzene exposure and 367
stem cells and 9
xenobiotics and 241–3
non-coding (ncRNA) 241–2, 356
long *see* long non-coding RNA
small 356
non-human, benzene exposure and 364–5
mRNA (messenger RNA/mRNA) 356
next generation sequencing *see* next generation sequencing
synthesis 7, *see also* transcription
synthesis 7, benzene and alternative splice variants 356, 361, 362–4
translation 456
non-coding RNAs and 242
RNA CoMPASS 364
RNA-seq (whole-transcriptome next generation sequencing) 2, 9, 115, 153, 443
benzene exposure 357, 358, 361–4, 366–7
furan exposure 402, 408, 410
rodent cancer bioassay (2-year) 6, 15, 21, 25, 390, 391, *see also* mouse; rat
furan 399
RUNX1–USP42 gene fusion 364

S100 calcium binding protein G gene 227
S5000 approach 357, 358, 365–6, 370
Saccharomyces cerevisiae, predictive *in vitro* toxicogenomic studies 82, 86, 87
safety assessment *see* risk assessment
Salmonella (Ames) test 22, 59, 66, 92, 114, 174
SAModKPrototypes 469, 470
sarcosine biosynthesis and conazoles 311
SDS-PAGE gel electrophoresis-based proteomics 10
self-organizing maps (SOMs) 446–7
senescence, cellular 214
Sequencing Quality Control (SEQC) project 95, 358
Serp1 227
SESN2 (sestrin) 126
SEURAT-1 396
short-timescale (short-term) assays 22
genotoxicity 393
signatures of 146
in vitro 4
in vivo 414
in tiered testing strategies 414
shotgun proteomics 10
shrunken centroids method 56–7
signaling (cell)
genotoxic stress and 42–3
non-genotoxic carcinogens and 213

signatures *see* biomarkers
simulated annealing optimization routine version of ModKPrototypes 469
single-nucleotide polymorphisms 471
site (of tumor), incidence rates by 443–4
skeletal muscle, signatures of genotoxicity 140
SKI–PRDM16 gene fusion 364
Slco1a4 227
small non-coding RNA (sncRNA) 356
species, extrapolation across or comparisons and differences between 13–14, 160, 166–8, 431–5
  concordance model and 431–5
  propiconazole mode of action 327–8
spleen, liver, signatures of genotoxicity 143
statistical methods in data analysis
  comparison methods for non-genotoxic carcinogen identification 215–16
  transcriptomic dose–response data 156
steatosis 429, 431, 432, 433
sterol metabolism and conazoles 282–3, 301
stress
  cell responses to 43, 48
    endoplasmic reticulum *see* endoplasmic reticulum
  genotoxic 42–3, 44, 45, 53, 79, 81, 191
    epigenome and 243
  oxidative *see* oxidative stress
stress-associated endoplasmic reticulum protein 1 gene 227
structure–activity response (SAR) models 114
  quantitative 215, 260
subclusters, gene 53–4
sulconazole 288, 289, 290, 291, 292
superparamagnetic clustering (SPC) 53, 65
support vector machines (SVMs) 446–7
systems toxicology 16

T-statistics with comparison methods for non-genotoxic carcinogen identification 215
tacrolimus (FK506) 216–28, 229, 230
target organs and tissues 21, 23–4, 129–44
  breadth for surveying cancer and noncancer effects 173–4
  selection 146–7
TCDD (2,3,7,8-tetrachlorodibenzo-*p*-dioxin) 167, 168, 175, 250
TCGA (The Cancer Genome Atlas) 472–3
tebuconazole 281, 286, 287, 288, 298, 303, 305, 306, 330
telomerase 214
TempO-Seq 366
2,3,7,8-tetrachlorodibenzo-*p*-dioxin (TCDD) 167, 168, 175, 250
1,1,1,2-tetrachloroethane 257
1,1,2,2-tetrachloroethane 257
tetrachloroethylene 251, 256–7
tetraconazole 288, 305, 306
tetracycline 431
TGF *see* transforming growth factor
TGx-28.65 48–67, 92, 94, 95, 98, 99
The Cancer Genome Atlas (TCGA) 472–3
thioacetamide 131, 132, 133, 403, 405, 450, 452
thioacetamide (THI) 450, 451, 457, 459
thresholds (threshold dose; minimally effective dose) 40, 465–7
  transcriptomics in evaluation of 185–208
thymidylate synthetase gene (*TYMS*) 126, 131
thyroid (rat), conazole effects 312, 313–14

tiered testing strategies 98, 174
furan 413–14
timescale/duration/term of studies/ treatments 146–7, 414
cellular stress responses and 48
in vitro assays 4, 82–91
short-term *see* short-timescale assays
signatures of genotoxicity and 145
TGx-28.65 transcriptomic biomarker 49
tioconazole 288, 289, 290, 291, 292
tissues and organs, target *see* target organs and tissues
TK6 cells 25, 258
TGx-28.65 and 48, 49, 59, 61, 63, 65, 92
various predictive in vitro toxicogenomics studies 80, 82, 83, 86, 87, 90, 96
topoisomerase inhibitors 41, 48, 50, 187, 188
Tox21 2, 79, 239, 260, 265, 267, 366
ToxCast 2, 3, 14–15, 98, 267, 394–5, 413
toxicity testing, conventional/ traditional 4, 5–6, 391, 391–6
animal use reduction 393–6
genotoxicity 39–40, 79, 115
relative role 12–15
Toxicity Testing in the 21st Century: A Vision and a Strategy 5, 189, 230, 413, 414
toxicogenomics (basic aspects)
definition 7–8
relative role 12–15
TP53I3 122, 142
TP53INP1 124–5, 131
training compounds in predictive *in vitro* toxicogenomic studies 81, 82–91, 92
trans-regulation of a transcription regulator 471
transcript regulator expression quantitative trait loci (TReQTLs) 471
transcription 7
dose–response analysis 154–84
genotoxic assays based on responses in 43–4
genotoxic stress affecting 42, 79
no observed transcriptional effect level (NOTEL) 466
regulation and regulators 199–201, 453–5, 467, 471
-omics technologies in studies of 26
transcription factors 450
cell cycle-related 453
conazoles and phenobarbital and 332
nuclear receptors as 293
transcriptomics (gene expression profile/signature determination) 1–2, 8–9, 43–8, 76–112, 154–208
benzene (and other occupational exposures) 353, 354, 356–68
biomarkers/signatures 76–112
in vitro 76–112
TGx-28.65 48–67, 79–87, 95, 98, 99
dose–response relationships *see* dose–response relationships
global transcriptomic response 51–3
heatmap 52, 53, 54, 55, 59
risk assessment *see* risk assessment
Transfac 453, 454
transforming growth factor (TGF)-α 308–9
transforming growth factor (TGF)-β 162, 217–18, 221, 222, 224, 227–8
translation *see* post-translational processes; mRNA *(look under RNA)*

Transpath 453, 454
triadimefon 281, 287, 288, 294, 297, 298, 300, 301, 303, 305, 306, 307, 308, 309, 312, 313, 314, 319, 330, 332, 333
triadimenol 3–5, 281, 286, 287, 303
triazole antifungals *see* conazoles
tricarboxylic acid (TCA) cycle and conazoles 310, 326
trichloroacetate (TCA) 252, 253, 256, 257
trichlorobenzene 171
trichloroethyelene (TCE) 251–7, 261, 263
tri-methylxanthine (caffeine) 57, 59, 66
TUBB2A (tubulin, beta 2A class IIa gene) 121
tumor protein p43 inducible protein 3 (TP53I3) 122, 142
tumor protein p53 inducible nuclear protein (TP53INP1) 124–5, 131
TYMS 126, 131

uniconazole 302, 303, 305, 306
urea cycle and conazoles 311
USP42–RUNX1 gene fusion 364

vacuolation, hepatocyte 298, 329
validation 19–20
    in vitro toxicogenomic predictors of genotoxic MoAs 80–93
    TGx-28.65 biomarker in human health risk assessment 57–67
valproic acid 132, 431
vanin gene 227
variance analysis *see* analysis of variance
vinblastine 61, 134
viral infection and benzene exposure 364–5
Virtual Liver project 14
vitamin D receptor gene 226
Vnn 227
voriconazole 281, 285, 286, 287, 290, 292

WA09 cell line 240
Waf1 (*CDKN1A*) 43, 57, 116, 119
warfarin 264
whole-transcriptome next generation sequencing *see* RNA-seq
Wnt 162, 211, 307
women (females), cancer incidence by site 445

X receptors (nuclear) 293, 316
xenobiotic impact on genome and epigenome of human embryonic stem cells 236–46

zebrafish (embryo/developmental) testing 395
    hepatotoxicity 433
ZNF331 357